CRC Handbook of Eicosanoids: Prostaglandins and Related Lipids

Volume I
Chemical and Biochemical Aspects
Part A

Editor

Anthony L. Willis, Ph.D.
Principal Scientist
Atherosclerosis and Thrombosis Section
Institute of Experimental Pharmacology
Syntex Research
Palo Alto, California

 **CRC Press**
Taylor & Francis Group
Boca Raton London New York

CRC Press is an imprint of the
Taylor & Francis Group, an **Informa** business

First published 1987 by CRC Press
Taylor & Francis Group
6000 Broken Sound Parkway NW, Suite 300
Boca Raton, FL 33487-2742

Reissued 2018 by CRC Press

© 1987 by Taylor & Francis
CRC Press is an imprint of Taylor & Francis Group, an Informa business

No claim to original U.S. Government works

A Library of Congress record exists under LC control number: 86018779

Publisher's Note
The publisher has gone to great lengths to ensure the quality of this reprint but points out that some imperfections in the original copies may be apparent.

Disclaimer
The publisher has made every effort to trace copyright holders and welcomes correspondence from those they have been unable to contact.

ISBN 13: 978-1-138-10572-0 (hbk)
ISBN 13: 978-1-138-55947-9 (pbk)
ISBN 13: 978-0-203-71266-5 (ebk)

Visit the Taylor & Francis Web site at http://www.taylorandfrancis.com and the
CRC Press Web site at http://www.crcpress.com

PREFACE

I am greatly honored that Professors Ulf S. von Euler and George O. Burr, originators of the eicosanoid area, have written preface contributions to this handbook. This *CRC Handbook of Eicosanoids* is dedicated to these two great men and to the memory of Prof. Ulf von Euler, Nobel Laureate, who died in 1983.

To adequately cover the literature in this enormous and rapidly expanding field has proved to be a monumental task. Thus, I gratefully acknowledge the dedication and fortitude of the various authors who have contributed to this handbook. I also thank the publishers and our coordinating editor, Ms. Amy Skallerup, for their great patience.

Finally, I thank the members of our editorial board and many others for reviewing the manuscripts and Ms. Joan M. Fisher and Ms. Becky Nagel who have given so freely of their time in acting as editorial assistants.

Anthony L. Willis

EICOSANOIDS

U. S. von Euler*

When the high biological significance of certain unsaturated fatty acids was revealed through the discovery of the essential fatty acids in the 1930s by Prof. George Burr this opened the door for studies on a new class of physiological important compounds.

Polyunsaturated fats in the diet later moved into the center of interest of nutritionists, based on their alleged role in preventing the development of atherosclerosis and related conditions.

One of the essential fatty acids of the early studies, arachidonic acid, was later to become the cornerstone in a large chemical complex, comprising numerous pharmacologically active eicosanoids. In 1935, however, when the strong pharmacological actions of extracts of the sheep vesicular gland and of human seminal fluid was found to be due to fatty acids, apparently unsaturated, virtually nothing was known about the pharmacology of such compounds. In retrospect it seems that this finding should have prompted a contact with Dr. Burr which might well have contributed to the identification of the new class of pharmacologically active substances at an earlier date. Very likely it would have revealed the biosynthesis of the active compounds in the vesicular gland, using arachidonic acid as precursor.

At any rate the findings in due time added to the significance of the eicosanoids which in an unexpected way have proven to comprise compounds possessing biological actions of the most varying kind. Who could have foreseen 25 years ago that the main function of our most widely used remedy was to block synthesis of prostaglandins and thromboxanes? Or who could then have believed that nanogram doses of such eicosanoids could provoke asthma or send forth inflammatory signs or the local defense forces of the organism?

Through intensive chemical, physiological, and pharmacological work it has become clear that the chemical backbone for a large series of bioregulators is an eicosanoid. A question which is almost unavoidable in this context is why this class of substances have been chosen by nature to perform such a variety of functions. When more knowledge has accumulated, in the relatively new field involving the laws determining the interaction of agonists and receptors at the molecular level, we may find the answer to this question and may understand better why eicosanoids have been selected for so many important tasks.

Ulf S. von Euler
August 23, 1982

---

* Prof. von Euler died in 1983.

ESSENTIAL FATTY ACIDS

While Prof. von Euler and I were chatting at the Golden Jubilee International Congress in 1980, we found that our paths had crossed at the laboratory of Prof. Thunberg in Lund, Sweden, in 1933. But we did not meet. If we had met it is likely that we would have discussed our then current work on essential fatty acids and prostaglandins without noting any significant relationship between the two. Now, 50 years later, these compounds are being treated as a unit for the solution of their biological functions.

The presentations at the Golden Jubilee Congress came as a revelation to me who viewed them with the perspective gained by an absence of more than 35 years from the center of activity.

> "Perspective is a pleasant thing!
> It keeps the windows back of sills
> And puts the sky behind the hills."
>
> David McCord

It also pans the sand and gravel from the gold.

If the 165 papers presented at the 1980 Golden Jubilee Congress (on essential fatty acids and prostaglandins) are a harbinger of what is to come in this complex field of study, there is clearly a need for a handbook to collate the mass of data.

George O. Burr
March 27, 1982

THE EDITOR

Anthony L. Willis, Ph.D., is a Principal Scientist and Head of the Atherosclerosis and Thrombosis Section, Institute of Experimental Pharmacology, Syntex Research, Palo Alto, California.

Born in Penzance, Cornwall, England, Dr. Willis obtained degrees in pharmacology at Chelsea College, London, and at the Royal College of Surgeons of England, London. Dr. Willis is a member of several learned societies, including the British Pharmacology Society (of which he is a Sandoz Prizewinner), and a member of the Council on Atherosclerosis of the American Heart Association.

With the exception of brief sojourns at Stanford University, California, and Leeds University, England, Dr. Willis has spent his entire research career in the pharmaceutical industry: at Lilly Research (England), Hoffman-La Roche (U.S. and England) and now Syntex (U.S.)

Dr. Willis has made many fundamental and applied contributions of the area of prostaglandins and related substances now collectively termed the *eicosanoids*. He was among the first to delineate the role of prostaglandin as mediators of inflammation, including the first description of their presence in inflammatory exudate. Later, in work done alone and in collaboration, he shared in the discovery that platelets of human individuals synthesize and release prostaglandins and labile endoperoxides that induce platelet aggregation and that the mode of action of aspirin in inflammation, fever, and platelet aggregation was via inhibition of prostaglandin synthesis. His work included establishing isolation procedures of labile PG endoperoxides and description of deficiency prostaglandins and endoperoxide responsiveness in hemostatic disorders.

Later, Dr. Willis pursued the now very topical idea that thrombosis and other disorders may be preventable by redirecting eicosanoid biosynthesis addition to the diet of pure biochemical precursors of certain prostaglandins. This work led to the conclusion that there was considerable species variation in the enzymatic desaturation of unsaturated essential fatty acids and that metabolic pools of eicosanoid precursors may be of importance in basal production of prostaglandins by most tissues.

Most recently, Dr. Willis has developed several novel methods of thrombotic and atherosclerotic processes that allow rapid and predictive evaluation of test compounds.

This work has led to the development of potentially antiatherosclerotic prostacyclin analogs and a new description of this potential activity via inhibition of mitogen release and cholesterol metabolism.

Dr. Willis is author or co-author of almost 100 scientific articles and review articles, including a previous compendium of the properties of prostaglandins, which served as the starting point for this handbook.

ADVISORY BOARD

CONTRIBUTORS

Y. S. Bakhle, D.Phil.
Department of Pharmacology
Royal College of Surgeons of England
London, England

Laszlo Z. Bito, Ph.D.
Professor of Ocular Physiology
Department of Ophthalmology
College of Physicians and Surgeons
Columbia University
New York, New York

Pierre Borgeat, Ph.D.
Inflammation and Immunology-
  Rheumatology Research Unit
Centre Hospitalier de l'Universite Laval
Quebec, Canada

Rodolfo R. Brenner
Director
Instituto de Investigaciones Bioquímicas
  de la Plata
CONICET — University of La Plata
La Plata, Argentina

Holm Holmsen, Ph.D.
Professor
Department of Biochemistry
University of Bergen
Bergen, Norway

Richard J. Kulmacz
Department of Biological Chemistry
University of Illinois at Chicago
Chicago, Illinois

James E. Mead, Ph.D.
Professor Emeritus
Biological Chemistry and Nutrition
Laboratory of Biomedical and
  Environmental Science
University of California, Los Angeles
Los Angeles, California

Cecil Pace-Asciak, Ph.D.
Professor
Research Institute
Department of Pediatrics
Hospital for Sick Children and
  Department of Pharmacology
University of Toronto
Toronto, Ontario, Canada

L. Jackson Roberts, II, M.D.
Associate Professor of Medicine and
  Pharmacology
Department of Pharmacology
Vanderbilt University
Nashville, Tennessee

Donald L. Smith, Ph.D.
Staff Researcher
Atherosclerosis and Thrombosis Section
Institute of Experimental Pharmacology
Syntex Research
Palo Alto, California

William L. Smith, Ph.D.
Professor
Department of Biochemistry
Michigan State University
East Lansing, Michigan

K. John Stone, Ph.D.
Group Head
Preclinical Development
Roche Products Ltd.
Hertsfordshire, England

Carmen Vigo, Ph.D.
Principal Scientist
Liposome Technology
Menlo Park, California

Anthony L. Willis, Ph.D.
Principal Scientist
Atherosclerosis and Thrombosis Section
Institute of Experimental Pharmacology
Syntex Research
Palo Alto, California

TABLE OF CONTENTS

Volume I

Volume II

Introduction to the Field

THE EICOSANOIDS: AN INTRODUCTION AND AN OVERVIEW

A. L. Willis

INTRODUCTION

This chapter is intended to serve both as an introduction to newcomers and as a framework of reference for the various specialized chapters in this handbook. It was based originally upon a compendium on the eicosanoids prepared by Willis and Stone in 1976.[1]

The field of prostaglandins and related eicosanoids* is one of the most complicated areas of biological research that has ever existed. There is a bewildering array of chemical structures and an equally bewildering (and often widely differing) spectrum of biological activity. Often markedly different or opposing biological effects can be produced by closely related prostaglandins; the nature of these effects may further depend upon the species examined. With the recent discovery of the thromboxanes and prostacyclin, as well as the leukotrienes and other lipoxygenase products, the situation is even more complex. Thus, lipoxygenase products may modulate thromboxane production,[2] and a prostacyclin analog can modify leukotriene formation,[3] and so on.

The voluminous and exponentially expanding literature of the eicosanoids (Figure 1) has made production of a state-of-the-art handbook very difficult, but we have attempted to include vital data and have allowed all of the individual authors an opportunity to provide an "eleventh hour" update of their contributions.

THE ESSENTIAL FATTY ACIDS (EFA)

Commencing in 1929, the classic papers of Burr and Burr[4,5] introduced the concept of the EFAs, which are necessary for normal physiological function in animals and man. The "essentiality" of the EFAs lies not only in their physiological importance, but in the fact that, like many vitamins, they cannot by synthesized *de novo*. They are derived either directly or in partially elaborated precursor form from the diet.

Nomenclature

The so-called "derived" EFAs that can be oxidatively converted into eicosanoids are generally of a longer chain (usually C-20) and are more unsaturated than their common dietary precursors (Figure 2). Dietary sources, interconversion, and other more detailed aspects of the EFAs are dealt with in the chapters by Brenner and by Mead and Willis. All of the EFAs are generally of 18 to 20 carbons in chain length and are monocarboxylic polyunsaturated fatty acids. They all have the so-called "skipped", "polyallenic", "methylene interrupted", or "divinyl-methane" pattern of unsaturation. The EFAs have the general formula:

$$CH_3[CH_2]_k[\overset{\text{cis}}{\overline{CH=CH\cdot CH_2}}]_m[CH_2]_n \cdot COOH$$

For linoleic acid, the indirect precursor for the PG_1 and PG_2 series of prostaglandins, $k = 4$, $m = 2$, and $n = 6$. For α-linolenic acid, indirect precursor for the PG_3 series of prostaglandins, $k = 1$, $m = 3$, and $n = 6$. The eicosanoid precursor role of the various derived EFA is outlined in Figure 3.

---

* Eicosanoids are oxygenated metabolites of arachidonic acid and other polyunsaturated fatty acids of about 20 carbons in length ("Eicosa" indicating "20"). Prostaglandins are only one family of the eicosanoids.

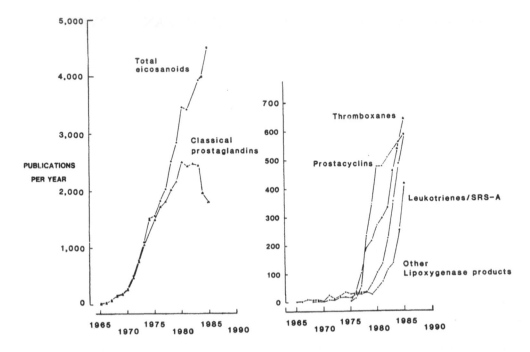

FIGURE 1.   Explosion of literature in the eicosanoid field. At least 10 publications now appear every day of the year in the eicosanoid field. Although interest in the primary prostaglandins has lately abated, this has been more than made up for in the newer areas of thromboxanes, prostacyclins, and leukotrienes. This data is from a Medline® search carried out by Ms. V. Garlow of the Syntex Library, Palo Alto, Calif.

The systematic nomenclature of fatty acids defines both the position and geometry of each double bond, counting from the terminal carboxylic acid function (alpha end of the molecule). Thus, arachidonic acid is *cis*-5,*cis*-8,*cis*-11,*cis*-14-eicosatetraenoic acid and is sometimes abbreviated to all-*cis*-$\Delta^{5,8,11,14}$-eicosatetraenoic acid. The Δ (delta) is an abbreviation which denotes position of a double bond numbered from the terminal carboxyl group; usually it is omitted as being obvious.

Throughout this chapter (and also used throughout the rest of this book), the shorthand nomenclature recommended by Holman[6] has been used. This system designates the chain length, number of double bonds, and the position of the double bond nearest to the terminal methyl group. This terminal carbon atom is referred to as the omega (ω) carbon atom, as it occurs at the opposite end of the molecule to the alpha (α) carbon atom which bears the terminal carboxyl function. A methylene-interrupted sequence in which the double bonds all have *cis* geometry is assumed. This shorthand system is most useful when considering the biochemical interrelationships between EFAs. Chain elongation and insertion of double bonds (desaturation) occurs only at the carboxyl end of the molecule. Thus, all ω6 EFAs are biochemically interrelated, and all members of the ω3 series are also related to each other.* No interconversion can take place between the ω6 and ω3 series of fatty acids in animals, although such interconversions can take place in plants.

The Greek letters used in trivial names for isomers of linolenic acid should not be confused with those used to define position of carbon atoms, and thus the shorthand nomenclature.

On the basis of chronological discovery, α-linolenic acid (18:3ω3, sometimes just called linolenic acid) is found in several vegetable oils, most notably, linseed oil from which its name derives.[7] However, its occurrence was apparently** accompanied by smaller amounts

---

*   Sometimes the term n-6, n-3, etc. is used instead of ω6 or ω3 to describe positions of the double bond nearest the terminal methyl group.

** "β-Linolenic acid" may have been an artifact of the early isolation procedures used.[284]

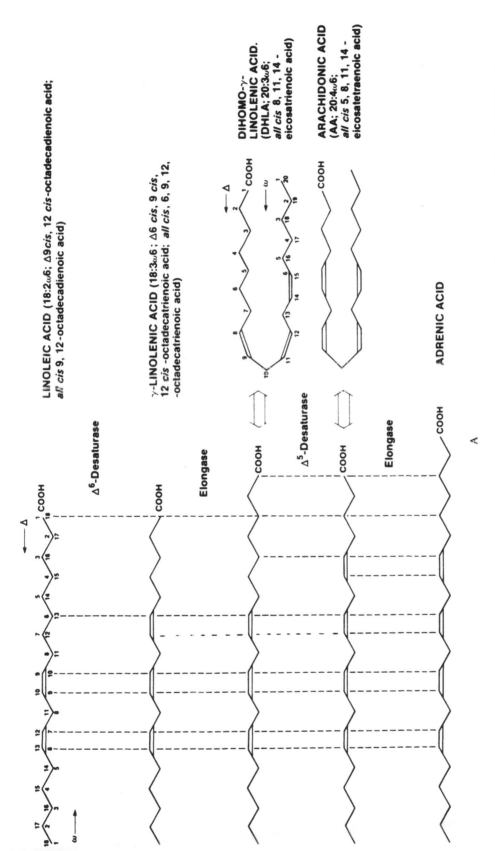

FIGURE 2.   Nonoxidative metabolism of eicosanoid precursors. (A) Derivation of ω6 long-chain polyunsaturated eicosanoid precursors from linoleic acid. (B) Shared pathways of desaturation/elongation and the concept of substrate competition.

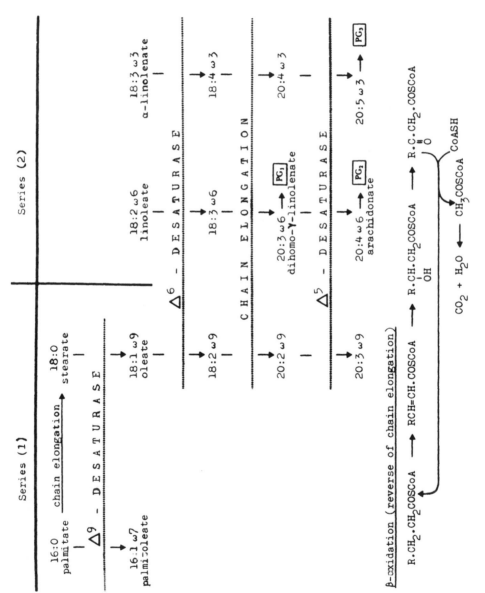

FIGURE 2B

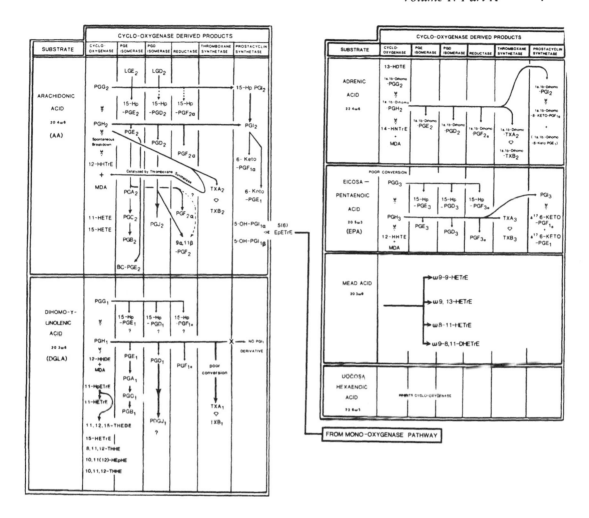

A

FIGURE 3.   Current spectrum of various eicosanoids derived from arachidonic acid and other precursors (to late 1986). The contribution by D. L. Smith in this volume of the handbook contains structures, systematic nomenclature, and also explanation of our new standarized shorthand nomenclature for oxygenated fatty acids ("HETEs" etc.) in addition to the most recent references. (A) *Products of cyclooxygenase.* With the exception of the keto- metabolites of prostacylin (PGI$_2$) and the recently described 5-hydroxy PGIs, all of the cyclized cyclo-oxygenation products of arachidonic acid are of the "2-series" (e.g. PGE$_2$) with regards to unsaturation in the upper and lower side chains. Similarly, "1-series" PGs, TXs, etc. are generally derived from dihomo-γ-linolenic acid and "3-series" compounds from eicosapentaenoic acid. Compounds formed from adrenic acid are similar to those from arachidonic acid, except for having an upper chain that is two carbons longer. Recently several noncyclized products have been shown to be produced by the action of cyclooxygenase. In such cases, co-oxygenation processes might be involved. These compounds may have important biological activity (see addendum to the chapter by Mead and Willis). (B) *Products of lipoxygenases and monooxygenases.* It can now be seen that arachidonic acid and similar fatty acids can be acted upon sequentially by more than one lipoxygenase, sometimes involving more than one cell type (e.g. interaction between the 12 and 5-lipoxygenases). In general, leukotrienes are formed via the 5-lipoxygenase pathway and contain the same number of double bonds as their precursor fatty acid (e.g. Leukotriene A$_4$-F$_4$ from arachidonic acid). However, there are now exceptions to these generalizations. Thus, a type of 3-series leukotrienes can be formed from dihomo-γ-linolenic acid via lipoxygenation at the 8-position, while 4-series leukotrienes of a different type can be formed from arachidonic acid via action of the 12-lipoxygenase. The lipoxins and lipoxenes are derived via the 15-lipoxygenase pathway. Not shown on the figure is the ability of the products of one pathway to interfere with another, e.g. the inhibition 5-lipoxygenase by 15-hydroxy derivatives. The monoxygenases are interesting in that their products may be further converted to PG-like compounds (e.g. 5-hydroxy PGI-derivatives) by the action of cyclo-oxygenase, and its related enzymes.

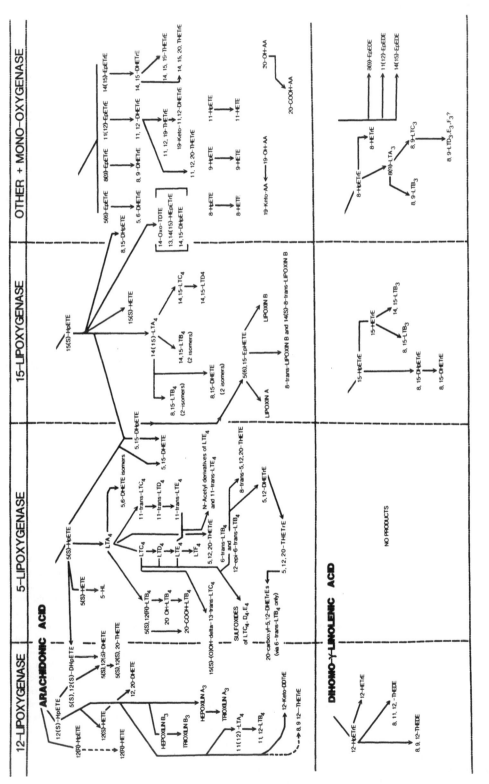

FIGURE 3B

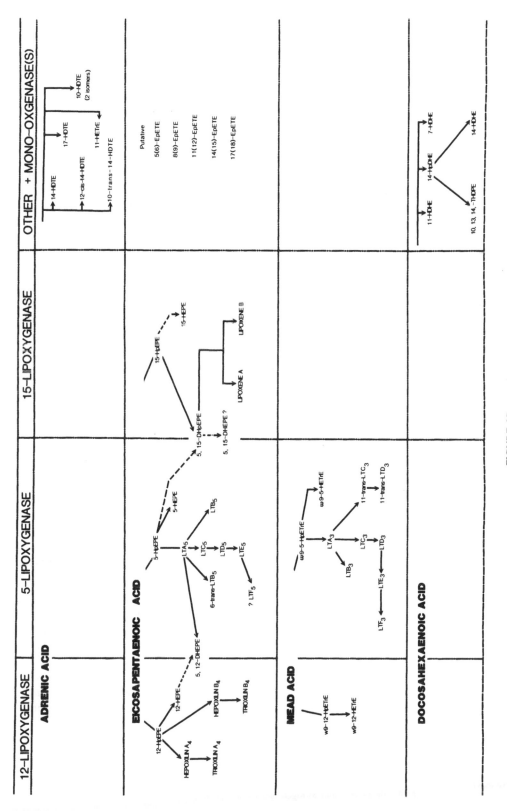

FIGURE 3B (continuation)

of an $18:3\omega3$ isomer designated β-linolenic acid.[8] The subsequent identification of $18:3\omega6$, which occurs in the seeds of the evening primrose,[9] led to its designation as γ-linolenic acid. The addition of two extra methylene groups to the carboxyl end (α-end) of γ-linolenic acid has resulted in the trivial name of dihomo-γ-linolenic acid being given to this precursor of PGE_1 biosynthesis. In some early literature, this compound was inaccurately termed homo-γ-linolenic acid, suggesting incorrectly that only one extra methylene group had been added.

Eicosanoid Precursor Role

The link between the EFAs and prostaglandins was forged in the mid 1960s. The groups of Van Dorp[10] and Bergstrom[11] simultaneously reported that prostaglandins (E_2 and $F_{2\alpha}$) could be formed biosynthetically from arachidonic acid ($20:4\omega6$). Eicosanoids derived from arachidonic acid are generally the most commonly occurring in mammalian tissues. This is not surprising, given that cell membrane phospholipid is rich in $20:4\omega6$ that is readily liberated by phospholipases (see Figures 6 to 7).[1,12,13]

There are, however, some exceptions. For instance, sheep vesicular gland phospholipids are rich in dihomo-γ-linoleic acid ($20:3\omega6$),[12] explaining why in this tissue and in semen, PGE_1 (and 19-hydroxy PGE_1) predominates.[13]

Reasons for the specific accumulation of arachidonic acid or some other eicosanoid precursor in different organs has never been adequately investigated.[13] Even more intriguing is the role of other lipid pools in eicosanoid biosynthesis. For example, in the thyroid,[14] triglyceride can serve as a source of arachidonate for prostaglandin biosynthesis. In the adrenal cortex, cholesteryl arachidonate is present.[15,16] This is the precursor for PGE_2 biosynthesis stimulated during the release of steroids in response to adrenocorticotrophic hormone (ACTH).[17]

However, amounts of other eicosanoid precursors in cholesteryl esters of the adrenal may approach or exceed those of arachidonate.[15,16] These include[13,15,16] dihomo-γ-linolenic acid ($20:3\omega6$), precursor for PGE_1, docosatrienoic acid ($22:3\omega6$), precursor for a dihomo-PGE_1, and adrenic acid ($22:4\omega6$), which has been shown to be a precursor for dihomo PGE_2, and dihomo-thromboxane B_2 (see Figure 3).[18-21]

Both arachidonic acid and adrenic acid are also found in the interstitial cell lipid droplets of the renal medulla. These fatty acids are present mainly incorporated into triglycerides but also in free acid form.[22] Interestingly, administration of indomethacin to rabbits (thus blocking renal prostaglandin biosynthesis) causes an increase in both physical size and prostaglandin precursor composition of the lipid droplets[23] indicating that the free acid $20:4\omega6$ and $22:4\omega6$ may function in the normal endogenous turnover of renal prostaglandins. "Metabolic pools" of free acid PG precursor also seem to be involved in basal turnover of PGs in brain and other tissues.[24-26]

Now, the stimulation of eicosanoid biosynthesis occurring in response to physiological or pathophysiological stimuli is triggered by the release of arachidonic acid (or other precursor) by acyl hydrolases from its incorporated form in phospholipids (or other esterified form). However, in most mammalian tissues there is also a constant basal turnover of PGs that is largely reflected in the "tissue content" of unstimulated* tissue. An important source for this basal turnover of PGs may be the small amounts of free $20:3\omega6$ and $20:4\omega6$ spilling over from "metabolic pools" of the fatty acids present in the cascade of desaturation/elongation involved in their biosynthesis from dietary linoleate.

Evidence for this hypothesis was obtained in experiments in which rabbits were placed

---

* In determining "tissue content" precautions have to be taken to avoid stimulating eicosanoid biosynthesis. One way is to snap freeze the tissue in liquid nitrogen and extract prostaglandins from the frozen material into chemicals that inactivate prostaglandin production.[24,25]

on a fat-free (EFA-deficient) diet for 2 weeks and then the phospholipid PG precursor content and "tissue content" of prostaglandins were examined.[24,25] During this short time period, the animals had managed to conserve the 20:4ω6 and 20:3ω6 content of their phospholipids although content of linoleate (18:2ω6) was diminished. Surprisingly, tissue content of prostaglandins was also markedly reduced. This seemed to link basal prostaglandin content with availability of 18:2ω6 and, by inference, size of metabolic pools of 20:3ω6 and 20:4ω6.

There are other examples where basal PG content is dissociated from eicosanoid precursor content of phospholipids, in brain[25,27] and in platelets.[26,28] The best documented example is in human platelet suspensions. Under basal (unstimulated) conditions, the platelets synthesize more PGE_1 than PGE_2, and this can be shown to derive from ^{14}C-linoleate.[26] This ratio of PGE_1 to PGE_2 is also reflected in the amounts of their respective precursors, 20:3ω6 and 20:4ω6, present in free acid pools but *not* in platelet phospholipids where the ratio of 20:4ω6 to 20:3ω6 is greater than tenfold.[13,26,28] Upon stimulation of the platelets to aggregate by addition of thrombin, there is a massive production of arachidonate (not 20:3ω6)-derived metabolites including thromboxane B_2 and PGE_2; this now reflects the availability and enymatic release of arachidonate from platelet phospholipids.[26,28]

Animal studies with EFA-deficient diets may also, of course, be used to define the role of membrane phospholipids in prostaglandin production. Thus, in rats that are sufficiently EFA deficient to have diminished phospholipid content of arachidonate, platelet production of prostaglandins in response to stimulation is markedly diminished.[13,26]

A similar association between diminished renal biosynthesis of prostaglandins and phospholipid content of arachidonate was described by Van Dorp.[18] Finally, in rabbits maintained on EFA-deficient diets for a short time (and in which vascular phospholipid 20:4ω6 acid was not yet diminished), the liberation of PGI_2 in response to cutting the aorta into rings was not changed from control.[24]

From the above discussion, it is apparent that the control of both basal and stimulated production of eicosanoids is an extremely complex process. Ultimately, the amount and type of eicosanoid produced will depend upon the dietary intake of both ω6 and ω3 EFAs. Other important factors include competition between different fatty acids at sites of both desaturation and uptake into lipid pools: there is competition between the liberated precursors or oxidative metabolism. There is also considerable selectivity of substrates for cyclooxygenase and lipoxygenase pathways. In addition, cyclooxygenase products may interfere with lipoxygenase pathways and vice versa. The impact of drugs that modify any of the above processes would result in even more complexity. For these reasons, it is inevitable that successful therapeutic application of eicosanoid research should involve a closer understanding of the links between EFA and eicosanoid metabolism.

DISCOVERY OF THE PROSTAGLANDINS AND EVOLUTION OF THEIR NOMENCLATURE

"Prostaglandin" was the name given by von Euler[29] to the agent responsible for depression of blood pressure, contraction of intestinal smooth muscle, and relaxation of uterine smooth muscle. This activity had previously been found in human semen[29-31] or in sheep prostate or vesicular gland.[29] Its name (actually a misnomer) arose from the belief that it originated in the prostate. The story of the early discoveries in this area are charmingly described by von Euler[32] in the proceedings of the 1980 Golden Jubilee Congress on Essential Fatty Acids and Prostaglandins. Von Euler recognized the unsaturated acidic lipid nature of "prostaglandin", but it was not until the 1950s that chemical isolation and thus subdivisions of nomenclature occurred. Prostaglandins E and F were isolated from extracts of sheep vesicular gland[33,34] and were identified by the first functioning gas chromatography/mass spectrometry (GC/MS) instrument developed by Rhyhage.[35]

Over the years, many more prostaglandins and their metabolites have been isolated, and although seemingly complicated, their nomenclature is rather simple to remember when placed into historical context.

THE MAIN PROSTAGLANDIN SERIES (CLASSIFIED BY RING STRUCTURE)

The principal "families", "series", or "types" of prostaglandin are denoted by the letter E, F, etc. These now extend through the alphabet to the letter J. Initially by design, and sometimes by coincidence, this suffix has reflected the means by which the prostaglandins were first discovered. Thus, the E- and F-type prostaglandins were initially named because on partition between *E*ther and phosphate buffer (*F*osfat in Swedish), the E types tended to remain extracted into the ether and F types in the phosphate buffer.[33,34,36] Chemically, PGE is characterized by the presence of a keto group at position 9 on the cyclopentane ring, while PGF has this keto function replaced by an hydroxyl group. A-type prostaglandins are formed from members of the E series by treatment with *A*cid, while B type prostaglandins are formed by treatment with *B*ase (Figure 4A). These PGE dehydration products are both chromophores and exhibit characteristic UV absorption bands at 217 nm (for PGA) and 278 nm (for PGB). Thus, in some early literature they were denoted by the subscripts -217 and -278; hence PGE_{1-278} ($\equiv PGB_1$).[37] Prostaglandin A_2 isolated from extracts* of kidney medulla[38,39] and prior to its identification by MS was termed "medullin".[38] Chemical identification showed it to be identical[38,39] to the PGE_{2-217} (PGA_2) present in extracts of human seminal plasma.[40] In the early 1970s Jones[41-43] recognized that PGA can be enzymatically rearranged to PGB through an intermediate (PGC) with a potent ability to lower blood pressure. This isomerization of PGA to PGC is catalyzed by plasma of the cat but not by that of the human.[43]

During the biosynthesis of E- and F-type prostaglandins the endoperoxide (PGH) is isomerized enzymatically or nonenzymatically,[44,45] not only to PGE, but also to an isomer in which the 9-keto and 11-hydroxy groups are transposed (Figure 4B). These 11-dehydro-PGF compounds (once referred to by this nomenclature) have now been designated D-type prostaglandins.

In the mid-1960s the groups of Van Dorp and Bergström simultaneously reported that prostaglandins were biosynthesized from arachidonic acid.[10,11] Studies with isotopic oxygen indicated that a labile endoperoxide intermediate was involved in this cyclooxygenase ("prostaglandin-synthetase") reaction.[46] In the early 1970s, three groups[44,45,47,48] isolated these endoperoxides and studied their biological activities, principally induction of platelet aggregation and contraction of isolated vascular tissue. Eventually, endoperoxides with a hydroperoxy group at position 15 were designated as PGG, while those with a 15-OH group were designated PGH (Figure 7).[47] In very early papers in this area, PGH was referred to as PGR,[45] since it contracted the isolated rabbit aorta. Alternatively it was referred to as labile aggregation stimulating substance (LASS), since it was completely responsible for appearance of this biological activity in incubates of arachidonic acid with phenol-activated sheep vesicular gland.[48]

In the mid 1970s, Moncada et al.[49] described a labile prostaglandin (temporarily designated PGX) formed from the dienoic endoperoxides by a vascular microsomal enzyme. It was shown that this substance was an intermediate in the formation of 6-keto $PGF_{1\alpha}$ via a pathway described several years earlier by Pace-Asciak and Wolfe.[277,278] Upon determination of its unique bicyclic structure,[50] it was termed "prostacyclin", later shortened to PGI_2. Thus, I-type prostaglandins are all prostacyclins of some kind. Prostacyclin now has the approved

---

*   A-type prostaglandins may be formed artefactually by the acidic conditions used in prostaglandin extraction procedures (see text following).

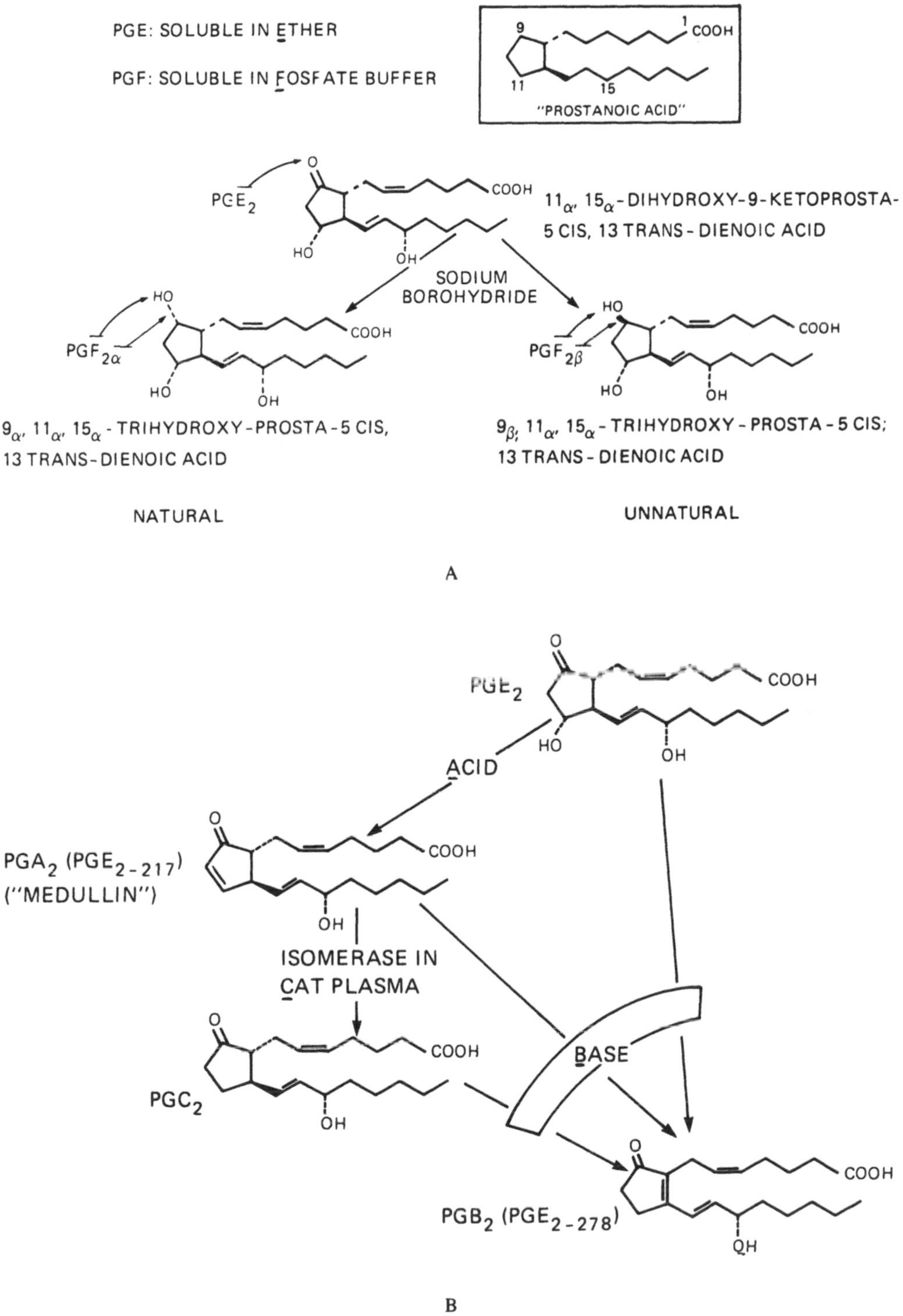
PGE: SOLUBLE IN ETHER
PGF: SOLUBLE IN FOSFATE BUFFER
9
1
COOH
11
15
"PROSTANOIC ACID"
PGE2
O
COOH
HO
OH
11α, 15α-DIHYDROXY-9-KETOPROSTA-5 CIS, 13 TRANS-DIENOIC ACID
SODIUM BOROHYDRIDE
PGF2α
HO
COOH
HO
OH
PGF2β
HO
COOH
HO
OH
9α, 11α, 15α-TRIHYDROXY-PROSTA-5 CIS, 13 TRANS-DIENOIC ACID
9β, 11α, 15α-TRIHYDROXY-PROSTA-5 CIS; 13 TRANS-DIENOIC ACID
NATURAL
UNNATURAL
A
O
PGE2
COOH
HO
OH
ACID
O
PGA2 (PGE2-217)
("MEDULLIN")
COOH
OH
ISOMERASE IN CAT PLASMA
O
COOH
PGC2
OH
BASE
O
COOH
PGB2 (PGE2-278)
OH
B

FIGURE 4. Interrelationships between classic ("primary") prostaglandins and derivation of nomenclature.

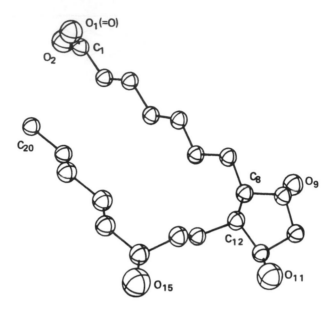

FIGURE 5.   Stereochemical configuration of $PGF_{1\beta}$ (reprinted from Rabinowitz et al., 1971).

name (for clinical use) of epoprostenol. Research on prostacyclin has recently been reviewed extensively;[51] biochemistry of the area is described in the contribution of Pace-Asciak to this volume.

There has recently been a resurgence of interest in the so-called classic (primary) prostaglandins. Thus, PGD_2 can be dehydrated to a "reverse A_2" compound, just as PGE was previously shown to be converted to PGA. This compound has been named PGJ_2 by Fukushima[52] and has been now well characterized both chemically and biologically.[53,54] While corresponding formation of a reverse B is chemically unlikely, it could be formed by total synthesis just as reverse PGC compounds could theoretically also be synthesized. If so, the PGK and PGL nomenclatures await them.

Classification According to Degree of Unsaturation

The usual naturally occurring prostaglandins have a $\Delta^{13\text{-}14}$ *trans* double bond. In prostaglandins of the 1 series, (PGE_1, A_1, etc.), this is the only double bond in the side chains. Prostaglandins of the 2 series have an additional $\Delta^{5\text{-}6}$ *(cis)* double bond, and member of the 3 series have a further $\Delta^{17\text{-}18}$ *(cis)* double bond.

Classification According to Stereoconfiguration

In the early days of prostaglandin research it was shown that incubation of PGE_1 with sodium borohydride[35] reduced it to PGF, but that two isomers were formed, with different stereochemistry of the newly formed 9-hydroxy group. These are referred to as $PGF_{1\alpha}$, which is also naturally occurring, and $PGF_{1\beta}$, which is not. In Figure 4B this transformation is illustrated by the identical conversion of PGE_2 to $PGF_{2\alpha}$ and $PGF_{2\beta}$. In order to define stereoconfiguration of the prostaglandins, the plane of reference is taken as the average plane of the five carbon atoms in the ring. Substituents projecting below the plane of the ring are designated alpha (α) while those projecting above the reference plane are designated beta (β). This nomenclature is most commonly used in reference to the geometry of hydroxyl groups, particularly at position 9.

The structure shown in Figure 5 will be recognized from the cover of the journal *Prostaglandins*. It is a computer drawing by Rabinowitz et al.[55] from X-ray diffraction data of $PGF_{1\beta}$; the stereoconfiguration is clearly apparent.

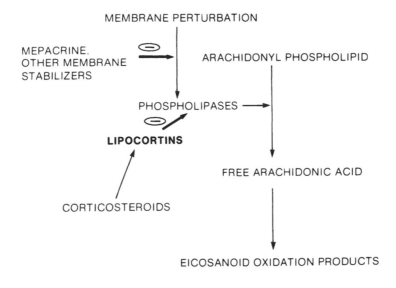

A

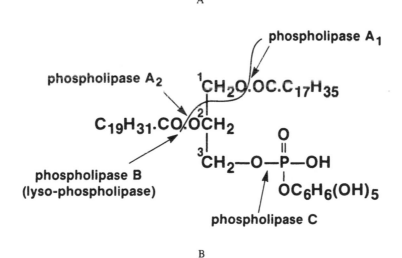

B

FIGURE 6. Phospholipase involvement in liberation of eicosanoid precursor fatty acids. (A) Membrane stabilizer drugs inhibit activation of phospholipases. Corticosteroids induce synthesis of peptides (lipocortins) that inhibit phospholipase activity. Free arachidonic acid is necessary for eicosanoid production. Similar mechanisms may be involved in release of other eicosanoid precursors. (B) Points of attacks by phospholipases. Shown is the structure of phosphatidyl inositol. Arachidonate and other eicosanoid precursors are predominantly located at the "2" position with saturated fatty acids (in this case, stearate) in position 1. In general, phospholipase A_2 is involved in cleavage of arachidonate. In platelets, however, sequential activation of phospholipase C and diglyceride lipase seem involved. There is also the possibility of sequential action of phospholipase A_1 and phospholipase B. There is such a phospholipase A_1 in platelet lysosomes (see Reference 134).

An alternative nomenclature has been used by some authors, particularly when referring to the stereochemistry at carbon C-15. With this system, a hydroxyl group in the α-position at C-15 confers the *S* configuration to this asymmetric center. This absolute system of Cahn, Ingold, and Prelog[56] is preferable to conventional systems that utilize the letters D and L to describe configurations which relate to D- and L-glyceraldehyde (e.g., 12-HETE was described[57]

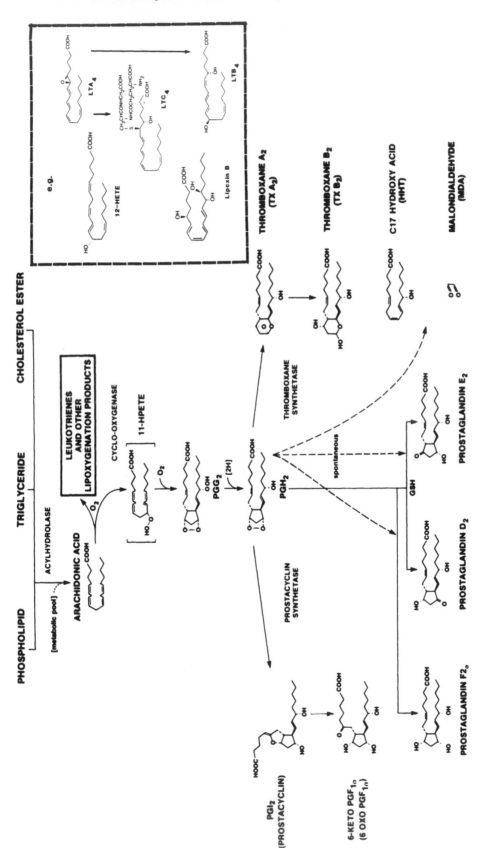

FIGURE 7.   Principal products of cyclooxygenase. (PGs, thromboxanes, and prostacyclin). It has recently been reported that 5(6) epoxides of arachidonate acid (formed by cytochrome P-450) can be converted by cyclooxygenase to 5-hydroxy PGI₁ (see Figure 3).

as 12-L-hydroxy-5,8,10,14-eicosatetraenoic acid). The prefixes *erythro* and *threo* refer to geometry of substituents on adjacent carbon atoms. This system is derived from carbohydrate nomenclature and relates to the configuration of erythrose and threose, respectively.[58]

When drawing structures of prostaglandins or other eicosanoids, substituents in the α position are conventionally shown attached to the carbon backbone by a broken line, while substituents in the β position are shown attached by heavy lines with a reverse taper. This system is universal. Where configuration of the substituent is not known, a straight line* is usually used, but where configuration of a substituent is unknown, the symbol Xi (ξ) is sometimes used (e.g., the PGE$_1$ analog ONO 747 is 17 (ξ)-ethyl-11,15-dihydroxy-9-keto-prosta-2-*trans*,13-*trans*-dienoic acid).[59] Some naturally occurring prostaglandins have a β-hydroxyl (15-ROH) at the 15 position and are termed 15-*epi*-PGE$_1$, 15-*epi*-PGF$_{2\alpha}$, etc. Epimers may also occur with altered geometry about C-8, where the carboxylic acid side chain joins the ring. When the chains are both in the β position in relation to the ring, the resulting compounds are often referred to as 8-*iso*-PGs, e.g., 8-*iso*-PGE$_1$, 8-*iso*-PGE$_2$, etc. (see Figure 5 and Table 6 of Reference 1).

It is also possible that antipodes of the prostaglandins might exist. An antipode is the mirror image of the related primary compound. The above considerations have been discussed further by Andersen.[60]

Systematic Nomenclature of Prostaglandins

The usual prostaglandin nomenclature is traditionally based upon the backbone of a theoretical "prostanoic acid" (Figure 4A) so that, for instance, PGF$_{2\alpha}$ becomes 9α,11α,15α-trihydroxyprosta-5-*cis*,13-*trans*-dienoic acid. This system was devised to avoid the confusion of an absolute systematic nomenclature system, where the carboxylic acid side chain would be regarded as the parent compound. With this system, PGF$_{2\alpha}$ would be regarded as 7-{3α,5α-dihydroxy-2β-[(C3S)-3-hydroxy-*trans*-1-octenyl]-1α-cyclopent}-*cis*-5-heptenoic acid.

Usually, the prostanoic acid-based nomenclature system, or trivial nomenclature (PGF$_{2\alpha}$, PGF$_{2\beta}$, etc.) is used for natural PGs, their metabolites, and many analogs. Moreover, in describing synthetic analogs (see chapters by Willis and Stone and by Muchowski), the alternative "prostane" backbone system of Andersen[60] is often used.

In studies on the metabolism of PGs and other eicosanoids, stereochemistry of each metabolite has usually been assumed to be identical with that of the prostaglandin substrate, unless proven otherwise. Similarly, the geometry of hydroxyl groups introduced onto the ω2 (or ψ carbon atom, carbon adjacent to the ω-carbon atom) has been assumed to be the same as that for the 19-hydroxyl group of the hydroxy-PGB isolated from human seminal fluid.[61]

Nomenclature Regarding Chain Length

Unless otherwise stated, chain length of all eicosanoids is assumed to be of 20 carbons, as derived from arachidonic acid, or a similar polyunsaturated fatty acid precursor. Because of the interposing of a ring structure in prostaglandins and thromboxanes, the molecule can be considered to have an upper and lower chain.

The number 1 (terminal carboxyl) carbon can also be defined as the α carbon, the second, the β carbon, etc. At the other end of the molecule, the carbon of the terminal methyl group is called the ω carbon by an analogous alphabetic terminology. Thus, metabolic removal of 2 carbon units (by attacking the number 2 or β carbon atom) is referred to as β-oxidation, while removal of single carbons at the other end of the molecule is referred to as ω-oxidation.

When a prostaglandin molecule has been shortened at the carboxyl end of the molecule,

---

* When a racemic mixture of isomers is known to be present, configuration of a prostaglandin substituent may be indicated by a wavy line.

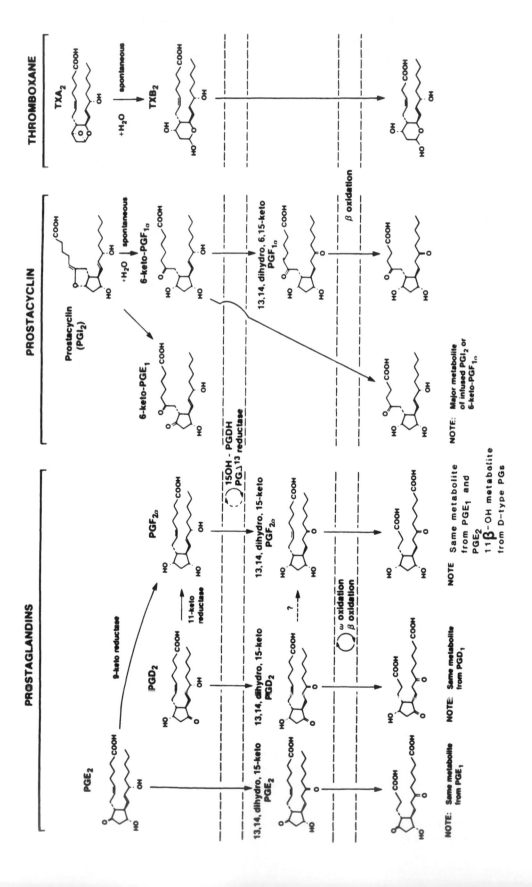

FIGURE 8. Main pathways in the metabolism and urinary excretion of primary prostaglandins, prostacyclin, and thromboxanes. *Primary Prostaglandins:* These compounds can (after uptake into cells) be rapidly metabolized. The principal pathway is dehydrogenation of the 15-OH group, followed by reduction of the 13-14 double bond (although these events can occur to some extent, in the opposite order). Oxidation of upper (β-oxidation) and lower (ω-oxidation side chain then occurs). An estimate of whole body PGE turnover can be obtained by measurement of the major (PGEM) tetranor metabolite. Likewise for the tetranor metabolite of PGD compounds. The largest amounts of tetranor metabolite have the PGF ring structure (PGFM). However, occurrence of this compound reflects not only turnover of PGF compounds but also PGE and PGD compounds that can be enzymatically converted to F-type prostaglandins. Indeed, in vivo metabolic studies in the monkey have shown that most polar metabolites of PGD$_2$ found in urine have a PGF ring structure. *Prostacyclin:* Prostacyclin (PGI$_2$) spontaneously breaks down to 6-keto-PGF$_{1\alpha}$. Both PGI$_2$ and 6-keto-PGF$_{1\alpha}$ can be acted upon by PG-15-dehydrogenase and Δ^{13} reductase, although PGI$_2$ seems to be a much better substrate for the 15-OH dehydrogenase. However, dinor-6-keto PGF$_{1\alpha}$ seems to be the main urinary metabolite of both prostacyclin and 6-keto-PGF$_{1\alpha}$ in man, accounting for ~20% of the material infused intravenously. Another important exception for PGI metabolism may be its localized conversion to 6-keto-PGE, which has similar biological effects on platelets. *Thromboxanes:* The hydrolytic conversion of TXA$_2$ into TXB$_2$ is nonenzymatic but very rapid. In both monkey and man, the major urinary metabolite of TXB$_2$ is a dinor product formed by a single step of β-oxidation. The thromboxane produced seemed to be derived mainly from platelets. Thus, in normal donors, urinary thromboxane B$_2$ metabolites were markedly reduced by doses of aspirin that were (presumably) only sufficient to affect platelets. See References 191, 198, 199, and chapters by Jackson-Roberts and by D. L. Smith et al. for more detail.

the prefix "nor" is used; thus, removal of two carbons by β-oxidation results in a "dinor" prostaglandin, further removal of 2 carbons results in a "tetranor" prostaglandin, etc. Analogous ω dinor compounds, etc. also exist. The "PGEM" metabolite shown in Figure 8 is a tetranor, ω carboxy prostaglandin.

If the chain length of a prostaglandin is abnormally long (e.g., if biosynthesized from adrenic acid, 22:4ω6; Figure 3)[19] or produced as a synthetic analog, then the prefix "homo" is used which implies extension by one carbon atom. Extension by two carbons is termed dihomo (or less commonly, bishomo). If this extension is at the ω end of the molecule, it would be ω-dihomo, etc. Use of this Greek nomenclature of prostaglandins should not be confused with trivial nomenclature of EFAs [see previous section, Essential Fatty Acids (EFAs)] where, for instance, dihomo-γ-linolenic acid merely means extension by two methylene units at the Δ^1, carbon, or α end of the molecule of γ-linolenic acid, 18:3ω6 (all cis-5,8,11-octadeca trienoic acid).

The Prostacyclins[51]

In 1976, Moncada et al.[49] discovered the bicyclic prostaglandin now known as prostacyclin (PGI$_2$). Before its structure was assigned,[50] it had tentatively been designated PGX.[49] As for the thromboxanes (see next section, "The Thromboxanes") and the classical prostaglandins, PGI$_2$ is derived enzymatically from prostaglandin endoperoxides of the 2 series (Figure 7). Formation of PGI$_2$ is most abundant in vascular and gastric tissue. In these locations it might have physiological roles in reducing tendency to thrombosis and gastric ulceration, respectively.[51] An analogous monoenoic prostacyclin (PGI$_1$) has been chemically synthesized and used to produce antibodies[62,63] with which to explore the physiological role of endogenous PGI$_2$ as a circulating antithrombotic hormone.[64] However, PGI$_1$ cannot be produced biosynthetically, since monoenoic endoperoxides lack the necessary Δ^5 unsaturation. A trienoic prostacyclin (PGI$_3$) has also been synthesized either chemically[65] or biosynthetically from PGH$_3$ formed by incubation of eicosapentaenoic acid (20:5ω3) with sheep vasicular gland cyclooxygenase.[66]

The characteristic ether bridge between carbons 6 and 9 of the prostanoic acid chain is indicative of the PGI compounds in general. The hydroxyl group at position 15 is assumed to be in the α position. Nomenclature of PGI compounds is adapted from that of classic prostaglandins and has been discussed in detail by Johnson et al.[67]

THE THROMBOXANES

In 1975, workers at the Karolinska Institute in Stockholm discovered a new family of hydroxylated, cyclized, polyallelic fatty acids that lacked the prostanoic acid backbone of prostaglandins. These were designated thromboxanes (TX)[68] because of their formation and probable role in aggregating platelets and because of the characteristic oxane ring structure. TXA$_2$ and other A-type thromboxanes are bicyclic and are derived enzymatically from prostaglandin endoperoxides by thromboxane synthetase, an enzyme that has been solubilized and partially purified.[69] A-type thromboxanes are highly unstable in aqueous (especially acidic) conditions. They may be partially stabilized by plasma constituents[70] or by dissolving in cold dry ether at neutral pH.[71] Their structure was established by chemical trapping — experiments that result in characteristic methano or azido derivatives when incubates containing TXA$_2$ were mixed with methanol or sodium azide, respectively.[68,70] Although the structure of thromboxane A$_2$ had previously been indirectly assumed, it has recently been chemically synthesized, stabilized, and shown to have the expected biological actions.[283] B-type thromboxanes are stable products derived spontaneously by hydroxylation of A-type thromboxanes.

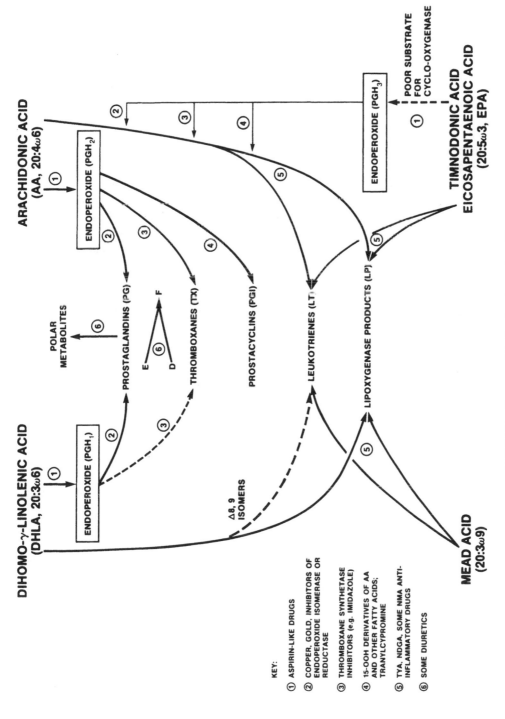

FIGURE 9. Some possible diet-drug interactions.[249]

Substrate Specificity

Interestingly, dihomo-γ-linolenic acid (20:3ω6) is not converted in significant amounts to monoenoic thromboxanes (TXA$_1$ and TXB$_1$)[72-74] since PGH$_1$ is a poor substrate for the thromboxane synthetase enzyme.[72,75]

Significant amounts of TXA$_3$ (and thus TXB$_3$) can be formed from purified preparations of the trienoic endoperoxide PGH$_3$.[72] However, the latter endoperoxide may itself be formed in only small amounts from endogenous 20:5ω3 since this is a poor substrate for cyclooxygenase.[76] The thromboxanes also differ in their biological properties: TXA$_2$ is a potent vasoconstrictor and platelet aggregator.[68,72] In contrast, TXA$_3$, although it does have vasoconstrictor properties, seems lacking in platelet aggregating activity.[72] Any small amount of TXA$_1$ that may be formed seems devoid of both of these biological effects.[48,72,77]

Nomenclature of the Thromboxanes

Although a "thrombanoic acid" type of chemical nomenclature has ever been proposed for the thromboxanes and the strict systematic nomenclature describing TXB$_2$ refers to a heptaenoic acid with an attached oxane ring structure between carbons 8 and 9.

The general classification of thromboxanes into 1 or 2 series, etc., depending on the number of double bonds is still applicable, and the A or B suffix describes the ring structure, which in the case of TXA$_2$ is 2,7-dioxabicyclo [3.1.1] heptan-4α-yl, and in the case of TXB$_2$ is (3α-hydroxy-1-*trans*-octenyl)-3α-tetrahydropyranyl. The hydroxyl group at position 15 is assumed to be in the α position. Before the existence of TXA$_2$ was recognized, TXB$_2$ was designated PHD, presumably an abbreviation for polar hydroxylated derivative.

BY-PRODUCTS OF PROSTAGLANDIN AND THROMBOXANE BIOSYNTHESIS

Malondialdehyde (MDA)

MDA is probably the best-known by-product of prostaglandin biosynthesis. It can be formed by cleavage from the ring structure of cyclic prostaglandin endoperoxide (Figure 7) when this breaks down nonenzymatically to the C-17 hydroxy fatty acid, 12-HHTrE ("HHT"). There is also evidence that 12-HHTrE (and thus MDA) formation can be catalyzed by (although not overtly produced by) thromboxane synthetase of platelets.[78] This would explain why addition of arachidonic acid (20:4ω6) to human platelet-rich plasma results in the formation of much larger amounts of MDA than when equivalent amounts of dihomo-γ-linolenic acid (20:3ω6) are added,[79] even though both are equally good substrates for cyclooxygenase. This is because in platelets 20:4ω6 is more readily converted to TXB$_2$ than to PGE$_2$, whereas 20:3ω6 is converted more readily to PGE$_1$ and PGD$_1$ than to TXB$_1$.[73-75]

MDA is also formed by autooxidation of unsaturated fatty acids[80] including linoleic acid (18:2ω6), which is not a direct precursor for prostaglandin production. However, PGE$_1$ and PGE$_2$, respectively, can be nonenzymatically formed during autooxidation of 20:3ω6 and 20:4ω6;[48,81] this suggests that MDA formation during autooxidation might arise from decay of prostaglandin endoperoxides that are formed nonenzymatically. MDA also arises during free radical-induced damage to DNA,[82] a process that is not obviously linked to prostaglandin biosynthesis.[82]

C-17 Hydroxy Fatty Acids

As described earlier in the chapter, the dienoic endoperoxides (PGG$_2$ and PGH$_2$) derived by cyclooxygenation of arachidonic acid can spontaneously decompose to MDA and HHT 12(S)-hydroxy-5-*cis*,-8-*cis*,10-*trans*-heptadecatrienoic acid (12-HHTrE, "HHT"), and this process seems catalyzed by activity of thromboxane synthetase.[78]

The monoenoic endoperoxides (PGG$_1$ and PGH$_1$) derived from dihomo-γ-linolenic acids

(20:3ω6) can similarly also break down to MDA and HHD 12(*S*)-hydroxy-8-*cis*,10-*trans*-heptadecadienoic acid (12-HHDE, "HHD"). Since much HHD is formed upon incubation of ^{14}C-20:3ω6 with human platelets,[83] but little TXB$_1$ or MDA is formed from 20:3ω6,[73-75,79] it is possible that this reaction is *not* catalyzed by thromboxane synthetase.*

The trienoic endoperoxides (PGG$_3$ and PGH$_3$) derived from eicosapentaenoic acid (20:5ω3) can also break down to a C-17 hydroxy fatty acid analogous to HHT and HHD.[72] This material, 12(*S*)-hydroxy-5-*cis*,8-*cis*,10-*trans*,17-*cis*-heptadecatetraenoic acid(12-HHTE, "HHTet") is likely not formed in vivo in large quantities, since 20:5ω3 is generally a rather poor substrate for cyclooxygenases.[66,76] Furthermore, levels of 12-HHTE were not markedly elevated in rats maintained on a fish oil diet rich in 20:5ω3.[84]

THE LEUKOTRIENES

Slow-reacting substance of anaphylaxis (SRS-A) was partially purified and studied for more than 15 years by Brocklehurst,[85] who also obtained evidence for its probable role in human allergic disease. The characteristic biological nature of SRS-A and other SRS compounds is discussed further in the addendum to this chapter. The peptido-lipid leukotrienes have now been found to be responsible for the biological activity of SRS-A (see chapter by Borgeat).

Leukotrienes are named for their occurrence and possible role in leukocytes and for their characteristic conjugated triene structure. As with the prostaglandins, the series or type of leukotriene denotes degree of unsaturation. The 4-series leukotrienes thus have four double bonds (three of them conjugated), the 3-series, three double bonds, etc. Unlike the case with the prostaglandins and thromboxanes, the number of double bonds in the leukotriene molecule is the same (not two less) than that present in the precursor fatty acid, although there are exceptions (Figure 3).[86-88]

The lettered suffixes A, B, C, D, E, and F of the leukotrienes refer both to the order in which they are biosynthesized and also to the character of the substituents. Thus, LTA$_4$ (a Δ^{5-6} epoxide) is the first of the leukotrienes formed from arachidonic acid via a 5-hydroperoxy intermediate. The next product formed from LTA$_4$ is LTB$_4$, which is a 5,12-dihydroxy leukotriene. Compounds from C through F are all peptido lipids, with a peptide linkage at position 5. LTC compounds have glutathione at this position, and γ-glutamyl transpeptidase then removes glutamic acid from LTC compounds to form LTDs which have cysteinyl glycine as the peptide moiety.

Leukotrienes can also be formed from fatty acids other than arachidonic acid. Conventional 5-hydroxy,6-peptido lipid leukotrienes are formed from 5,8,11-eicosatrienoic acid (Mead acid, 20:3ω9), and 5,8,11,14,17-eicosapentaenoic acid (EPA, 20:5ω3). These are, respectively, the 3- and 5-series leukotrienes.[87,88]

It is now known, however, that some (8,9)-LTC$_3$ can be formed from 20:3ω6. The positions of the double bonds of 20:3ω6 preclude oxygenation at C-5. The product formed is thus a positional isomer of LTC$_3$ with the hydroxy group at C-8, glutathione at C-9, and the triene at Δ.[10,12,14] Analogous leukotrienes formed via 11,12-LTA$_4$ can be formed from arachidonic acid.[88] In addition, Jubiz and Nolan[89] have reported the formation of $\Delta^{8,15}$-LTB$_3$ and $\Delta^{14,15}$-LTB$_3$ from incubations of 20:3ω6 with human polymorphonuclear (PMN) leukocytes.

---

* According to Needleman and colleagues,[83] 12-HHDE is virtually the sole product formed in platelets from 20:3ω6. However, this assumption was based solely upon results of experiments with radiolabeled 20:3ω6 and conflicts with a previous definitive report by Falardeau et al.,[74] who used GC/MS analysis of products derived from unlabeled 20:3ω6. Such discrepancies are not uncommon and probably involve displacement and compartmentalization of eicosanoid precursor fatty acids.

LIPOXYGENASE PRODUCTS OTHER THAN LEUKOTRIENES

Although lipoxygenases of plant origin have been known for many years, it was not until 1975 that Hamberg and Samuelsson[57] and, independently, Nugteren[73,90] described an animal lipoxygenase. This was found in the cytosol of blood platelets and converted arachidonic acid first to a hydroperoxy compound [12(S)-hydroperoxy-5-*cis*,8-*cis*,10-*trans*,14-*cis*-eicosatetraenoic acid], and then to the corresponding 12-(S)-hydroxy compound. For the sake of convenience, initials of the (somewhat trivialized) chemical nomenclature have been used as shorthand nomenclature. Thus, the two compounds mentioned above can be referred to as 12-HpETE and 12-HETE, respectively. Nugteren[73] showed that several *cis* polyunsaturated fatty acids of 18 to 22 carbons could serve as a substrate for the platelet lipoxygenase, but that *cis* unsaturation at both ω9 and ω12 positions is obligatory. Since then, several other arachidonic acid lipoxygenase products have been discovered. In rabbit or human PMN leukocytes, 5-HpETE is not only precursor for leukotriene synthesis but can also be converted into 5-HETE.[91-93] Again, in PMN leukocytes,[91] lymphocytes,[94] and reticulocytes,[95] 15-hydroperoxy and hence 15-hydroxy derivatives of arachidonic acid can be formed. The chapter by D. L. Smith and Figure 3 gives more detail.

There is an increasing awareness of possible cross-regulation both between lipoxygenase pathways and between lipoxygenase and cyclooxygenase dependent pathways. Thus, 15-HETE can *activate* a cryptic 5-lipoxygenase in basophils[96] and can *inhibit* not only the 12-lipoxygenase of platelets[97] but also the 5-lipoxygenase responsible for leukotriene formation in rat PMN leukocytes.[98] The latter effect is also seen with 15-HETrE that is derived from dihomo-γ-linolenic acid.[99] 12-HETE has also been reported to inhibit 5-lipoxygenase.[100]

By contrast, 12-HpETE (formed via the platelet lipoxygenase) appears to inhibit platelet thromboxane synthetase.[2] There is also at least one report[101] that the lipoxygenase products 11-HETE and 15-HETE may be produced via the cyclooxygenase pathway: their production in rat vascular smooth muscle cells was blocked by aspirin or indomethacin (see Figure 3).[101]

Arachidonic acid (and dihomo-γ-linolenic acid) can also be converted by lipoxygenases into dihydroxy and trihydroxy compounds, presumably via one or more hydroperoxy intermediates.[102,103] Bild et al.[104] showed that 8,15-dihydroperoxy derivatives of both arachidonic acid and dihomo-γ-linolenic acid potently inhibit aggregation of human platelets. These hydroperoxy derivatives are readily formed by plant lipoxygenases.[105] However, addition of dihomo-γ-linolenic acid (20:3ω6) to human platelet-rich plasma results in the formation of a similar 8,15-dihydroxy derivative, strongly suggesting that the corresponding hydroperoxy intermediate was formed. Formation of this compound might thus be involved in the well-known in vitro antiaggregatory effects of 20:3ω6.

Also in platelets, 12-HpETE is converted (probably via an 11(12)-epoxide) to a 10-hydroxy-11(12)-epoxy compound once called EPHETA[102] or 10,11,12-HEPA[103] (but which might be identical to Hepoxilin B₃). In turn, this is converted to Trioxilin B₃.* Also formed (by similar mechanisms?) is 8,9,12-THETrE (see Figure 3 and chapter by D. L. Smith).

The hepoxilins and trioxilins, produced in pancreatic islet cells may have a role in controlling secretion of insulin.[279-281]

Preceding the discovery of the above arachidonate metabolites, Falardeau et al.[14] had described similar compounds formed in human platelets from dihomo-γ-linolenic acid (20:3ω6).

Possible factors controlling platelet lipoxygenase pathways are levels of glucose and

---

* The shorthand nomenclature used by the groups of both Walker et al.[102] and Bryant[103,106,107] is based upon a trivialized nomenclature as follows: (1) EPHETA is a shorthand form of 11,12-epoxy,10-hydroxy-eicosatrienoic acid. This compound has also been designated as 10,11,12-HEPA which is an abbreviated form of the 10-hydroxy,11,12-epoxy acid.

selenium.[106,107] Thus, conversion of arachidonic acid to the 10-hydroxy-11(12)-epoxy compound and its derivatives is inhibited by glucose, but not by galactose or lactate.[106] In addition, selenium deficiency in rats results in diversion of 12-HpETE metabolism away from 12-HETE and toward Hepoxalin and Trioxilin-type compounds. This action is attributable to inhibition of glutathione peroxidase activity.[107]

NEW EICOSANOIDS DERIVED VIA CYTOCHROME P-450 MONOOXYGENASES AND THE LIPOXINS

Cytochrome P-450s are ubiquitous but occur largely in liver, renal cortex, lungs, and gut.[108] Arachidonic acid can be metabolized by cytochrome P-450 (and PB-B$_2$) to four different *cis* unsaturated triene epoxides in which the oxide bridge may be present at each of the four original double bonds of arachidonic acid.[109,110]

One of these epoxides, *cis*,5(6)-oxido-20:3 seems involved in mediating the release of luteinizing hormone (LH) by luteinizing hormone releasing hormone (LHRH).[111] Such epoxides may produce this and other biological actions by inducing alterations in intracellular Ca^{2+} transport. For instance, these epoxides induce release of $^{45}Ca^{2+}$ from canine aortic smooth muscle microsomes.[112]

In addition, Oliw[113] showed that the 5(6) epoxide of arachidonic acid is converted by cyclooxygenase into 5-hydroxy PGIs. Future studies on such conversions of the epoxides obviously may reveal a whole new series of biologically active eicosanoids.

Another new family of eicosanoids have just been described, named the lipoxins by Samuelsson's group.[114] These are formed in PMN leukocytes from 15-HpETE which is converted to two novel trihydroxy compounds with four conjugated double bonds in contrast to leukotrienes which have three double conjugated bonds. The 5(R)-14,15-hydroxylated compound has been designated "lipoxin B" and the 5,6,15(S)-hydroxylated compound has been designated lipoxin A (see Figure 3 and chapter by D. L. Smith).

CELLULAR INTERACTIONS IN EICOSANOID PRODUCTION

As already described in part (in the section "Lipoxygenase Products Other Than Leukotrienes"), different eicosanoid products may modulate each others' biosynthesis. In some cases this may involve interactions between different cells. For instance, the 15-HETE formed in leukocytes may inhibit the formation of 12-HETE in platelets.[97] For another example, leukocyte-derived 15-HpETE (formed as an intermediate in 15-HETE biosynthesis) is known to inhibit prostacyclin synthetase[51] (see the section "Inhibitors of Prostacyclin Synthase").

There are also more definitively documented examples. In early studies on the biosynthesis of PGI$_2$, Dusting et al.[51] postulated the "endoperoxide steal" hypothesis. This suggested that platelets aggregating adjacent to the vascular wall could liberate prostaglandin endoperoxide (PGH$_2$) that was then converted into PGI$_2$ by prostacyclin synthetase of the vascular wall, thus limiting growth of the thrombus. This hypothesis was based upon results of platelet aggregation admixture experiments in which indomethacin was added to platelet-rich plasma and/or arterial segments.

Subsequently this hypothesis became the subject of controversy until Marcus et al.[115] did the definitive radiolabel experiments. It was demonstrated that after incorporation of [^{3}H]-arachidonate into platelets, admixtures of platelets and cultured vascular endothelial cells could form radiolabeled PGI$_2$. This clearly implicated "steal" of platelet-derived PGH$_2$; however, an optimal ratio of platelets to endothelial cells and absence of albumin were critical in revealing this effect, casting some doubt upon in vivo relevance.

More recently, Aiken et al.,[116] using an in vivo model of thrombotic occlusion[117] did obtain in vivo evidence for the endoperoxide steal hypothesis. Thus, topical application of a prostacyclin synthetase inhibitor (15-hydroperoxy arachidonic acid, 15-HpETE) blocked the antithrombotic effects of a systemically administered thromboxane synthetase inhibitor, although topical application of a cyclooxygenase inhibitor (indomethacin) did not. However, (as expected) systemic administration of indomethacin did block the thromboxane synthesis inhibitor effects. Such results are analogous to early in vitro results showing that the addition of indomethacin to platelets and blood vessel admixtures blocked release of PGI_2-like material, while addition of the indomethacin only to the vessel segment did not.[51]

Cellular interactions between platelets and neutrophils have also been described.[118,119] Human platelets with [^{3}H]-arachidonate incorporated into their phospholipids were mixed with unlabeled human neutrophils and stimulated by the calcium ionophore A23187. Several [^{3}H]-lipoxygenase products were formed, derived from both cells acting in concert.[118]

The platelet-derived arachidonate* served as a precursor for the neutrophil-derived eicosanoids LTB_4, 5-HETE, and a DHETE; the latter could also be formed via platelet-derived 12-HETE.[118,119] The DHETE material (12(S)-20-dihydroxyeicosatetraenoic acid) had a previously unrecognized structure and seemed possibly involved in coagglutination reactions between the platelets and neutrophils.[119]

The principal implication of the above considerations are that "no cell is an island". Obviously then, study of eicosanoid production and actions in single cell types may not always produce clear evidence for biological role for this eicosanoid. However, in vitro studies of cell mixtures usually involve highly artificial conditions. In vivo experiments with specific antibodies to selectively remove one or more cell types remain the more definitive way of studying such cellular interactions.

DRUGS ACTING VIA EICOSANOID SYSTEMS

As research on the biochemical formation, metabolism, and actions of eicosanoids has progressed, it has become clear that many commonly used drugs have actions that impinge on one or another of the enzyme steps involved (Figure 9). In the case of aspirin-type drugs, their mode of action can be ascribed to inhibition of cyclooxygenase.[120-122]

Inhibitors of Eicosanoid Precursor Release

Mepacrine, an antimalarial drug, was one of the first recognized inhibitors of phospholipase-mediated release of 20:4ω6 in platelets or spleen slices.[123,124] Other work showed that anti-inflammatory steroids could inhibit production of prostaglandins by homogenates of human skin,[125] cultures of mouse fibroblasts,[126] or perfused lungs of guinea pig.[127,128] This effect of the steroids is now attributed to inhibition of arachidonic acid release from phospholipids[126] by inducing synthesis of one or more peptide inhibitors of phospholipase activation, most notably macrocortin[129] or lipomodulin.[130] RNA and protein synthesis is thus necessary for the anti-inflammatory corticosteroids to exert their effect.[131] Such proteins have now been isolated from several cell types, including monocytes, neutrophils, and renal medullary cells. The predominant active form has a molecular mass of 40,000 Daltons. This material has recently been sequenced and cloned.[285]

Inhibitors of Cyclooxygenase

In 1971, Vane et al.[132,133] and Smith and Willis[134] independently and simultaneously reported that aspirin and indomethacin could inhibit prostaglandin biosynthesis. Vane[132] further showed that this effect was exerted directly on the enzymatic conversion of arachidonic acid to prostaglandins. This finding eventually led to an adequate explanation for the mode

---

* The interaction between platelets and neutrophils is not unidirectional. Thus, addition of [^{3}H]-5-HETE (a neutrophil product) to unlabeled platelets also resulted in formation of a DHETE product.[118]

of action of such drugs in reducing inflammation, fever, hyperalgesia (tenderness to pain), and platelet aggregation.[120-122] Side effects of aspirin-type drugs, including water retention (antidiuretic action) and gastric ulceration also have been ascribed to their ability to inhibit prostaglandin biosynthesis;[120] prostaglandins have diuretic actions in the kidney and antiulcer activity in the stomach.[1,135]

Sodium salicylate, which is devoid of the acetyl group, does not markedly inhibit prostaglandin biosynthesis in vitro[121,132-134] or ex vivo,[26,121,136] but generally does produce such an effect in vivo.[121,137] Indeed, the only marked pharmacological difference between sodium salicylate and aspirin is the ability of aspirin to inhibit human platelet aggregation[138,139] in man. Since the inhibitory effects of aspirin on platelet aggregation are irreversible, the effect remains until fresh platelets are produced in the bone marrow.[26,136,139] This irreversible action of aspirin is due to acetylation of a platelet protein found to be identical to the platelet cyclooxygenase,[140] whereas sodium salicylate obviously cannot produce such acetylation and consequently does not markedly inhibit platelet aggregation or prostaglandin and thromboxane production.[26,136,138,139]

An additional mechanism has recently been reported[141] for salicylic acid, as opposed to aspirin which does not act similarly; this is inhibition of hydroperoxide conversion to a monohydroxy derivative of arachidonic acid of unknown structure. This substance (LOX) has potentially important antiplatelet adhesive properties.

Acetaminophen (4'-hydroxyacetanilide) is marketed under the brand name Tylenol® (in the U.S.) or Paracetamol (in the U.K.). It has antipyretic and analgesic actions similar to those of aspirin but does not interfere with platelet function or inhibit inflammation.[120,121,138] Correspondingly, the drug does not inhibit prostaglandin production in human platelets in vitro or ex vivo,[26,134] neither does it interfere with splenic prostaglandin biosynthesis,[142] or the appearance of PGE_2 in rat inflammatory exudate.[121] Its mode of action does, however, appear to be exerted via inhibition of prostaglandin biosynthesis, but with actions restricted to neural tissue. Thus, it strongly inhibits cyclooxygenase of mouse, gerbil, or rabbit brain.[121,143] It also inhibits the appearance of PGE in cerebral ventricular fluid during pyrogen-induced fever in cats.[144]

Factors Influencing the Ratio of Prostaglandin Products

Stone et al.[145] examined the effects of several drugs upon the relative amounts of PGE and PGF derived via cyclooxygenation of arachidonic acid. The common cyclooxygenase inhibitors (aspirin and indomethacin) suppressed biosynthesis of both E- and F-type prostaglandins. At low concentrations only, however, phenylbutazone selectively inhibited PGE synthesis while that of PGF was increased. These findings may be relevant to acute localized anti-inflammatory effects of $PGF_{2\alpha}$,[146] but have not yet been confirmed in vivo. The same authors also showed that the antirheumatic gold salt, sodium aurothiomalate, markedly and selectively stimulated PGE biosynthesis.[145] Daily injection of PGE_1 inhibited experimentally induced arthritis in rats.[147] Correspondingly, the fibroproliferative changes of carrageenin inflammation are exacerbated in rats that are EFA deficient (and thus lacking prostaglandins), even though inflammatory exudation is reduced.[148] Finally, a diet rich in evening primrose oil that would be expected to increase endogenous PGE synthesis (and possibly inhibit leukotriene synthesis)[13,99,149] reportedly does reduce adjuvant arthritis.[150] This is a finding that we have confirmed in subcutaneous injection experiments.[286]

It is thus tempting to speculate that biological stimulation of PGE biosynthesis by gold salts is related to its therapeutic benefit in arthritis. If so, an adequate intake (and any required Δ^6 desaturation) of ω6 EFAs would also be necessary. An examination of possible interrelationships between aspirin (which reduces prostaglandin production) and gold salts (which stimulates PGE production) may be in order.

PGD_2 has potential antithrombotic inhibitory effects on aggregation of platelets.[151] On the other hand, low concentrations of PGE_2 (which has the 9α-OH and 11-keto positions of PGD_2 reversed) can potentiate aggregation.[152]

As illustrated in Figures 3 and 7, both PGE_2 and PGD_2 are derived enzymatically or nonenzymatically from their common endoperoxide intermediate (PGH_2). Thus, agents controlling this ratio could be of physiological importance. Hamberg and Fredholm[153] reported that the nonenzymatic conversion of PGH_2 to PGD_2 is catalyzed by serum albumin. It would be interesting to examine whether fatty acids or drugs bound to the albumin alter its ability to catalyze this conversion.

Inhibitors of Thromboxane Synthesis

The platelet thromboxane synthetase that converts dienoic endoperoxide (PGG_2/PGH_2) to the prothrombotic substance, TXA_2, is a potential point of attack for putative antithrombotic agents.[26,135]

Thromboxane synthesis from prostaglandin endoperoxide is inhibited by imidazole and its analogs, certain prostaglandin endoperoxide analogs, and several other classes of compounds.[135] These compounds have proved useful as pharmacological tools for in vitro studies. One such thromboxane synthetase inhibitor, Dazoxiben (UK 37248), can produce desirable diversion of endoperoxide metabolism toward PGI_2 in animals[154] or man.[155] Dazoxiben reportedly has significant antithrombotic properties in the rabbit[154] but not in the baboon.[156]

Inhibitors of Prostacyclin Synthetase

Of several compounds originally tested by the Wellcome group, 15-HpETE was notable for its ability to inhibit the enzymatic conversion of PGH_2 to PGI_2.[157] Since then, hydroperoxy derivatives of other unsaturated fatty acids have been found to inhibit vascular prostacyclin synthetase.[158] Prostacyclin synthetase is also suppressed by tranylcypromine, a clinically used antidepressant and brain monoamine oxidase inhibitor.[51,159] This compound has been used as an experimental tool in attempting to delineate endogenous prostacyclin involvement in, for instance, control of intraocular pressure and pupil size.[160]

Drugs that Modulate Lipoxygenase Pathways

TYA (or ETYA, ETA) is the abbreviation for 5,8,11,14-eicosatetraynoic acid, an acetylenic analog of arachidonic acid. It was originally prepared as an intermediate in the total synthesis of arachidonic acid.[1,161] Downing and colleagues[162,163] discovered that TYA inhibited both plant lipoxygenase and the synthesis of prostaglandins from arachidonic acid; interestingly, these findings predated the discovery that aspirin-like drugs inhibited prostaglandin synthesis.[132-134] TYA has long been useful as an experimental tool in examining the role of arachidonate metabolites in platelets[164] and other tissues,[165] although it seems limited in bioavailability in vivo. With the discovery of the 12-lipoxygenase of platelets[57,73,90] interest arose in inhibition of such animal lipoxygenases. This interest received further impetus with the discovery of the leukotrienes derived via the 5-lipoxygenase pathway.

Wellcome researchers discovered that the common photographic reagent phenidone was a dual inhibitor of both the lipoxygenase and cyclooxygenase pathways,[166] an action shared more potently by its analog, BW755C.[167] The latter compound (as discussed later in this chapter) has been useful in probing the in vivo role of lipoxygenase products.

Nordihydroguaiaretic acid (NDGA) is an inhibitor (at least in vitro) of plant and animal lipoxygenases that does not inhibit the cyclooxygenase pathways.[168-170] Other fatty-acid related inhibitors of lipoxygenase include 5,8,11-eicosatriynoic acid (acetylenic analog of Mead acid, 20:3ω9).* This compound was reported to inhibit the 12-lipoxygenase of platelets but not the cyclooxygenase.[171] Since then it has also been shown to inhibit the 5-lipoxygenase

---

* Confusingly, this has also been termed ETYA on some occasions.

pathways responsible for leukotriene production[172] and also formation of the newly identified lipoxygenase product 13-HODE by vascular endothelium.[141] For another example, 5,6-methano LTA$_4$ is also a potent and specific inhibitor of 5-lipoxygenase.[173] Other agents reported to inhibit leukotriene biosynthesis include various retinoids,[174] the flavonoid biacalein,[175] and diphenyl disulfide.[176]

Also reported to inhibit the 5-lipoxygenase pathway is the clinically active* anti-inflammatory drug benoxaprofen.[177] Ibuprofen is also a clinically active anti-inflammatory drug, and was previously thought to exert its effects solely via inhibition of cyclooxygenase. It is therefore of considerable interest that ibuprofen has recently been shown to also potently inhibit the 5-lipoxygenase pathway, while simultaneously stimulating activity of the 15-lipoxygenase pathway.[178]

The same group had previously shown that 15-HETE was a potent inhibitor of the 5-lipoxygenase responsible for leukotriene formation; it is possible, therefore, that the ibuprofen-induced inhibition of the leukotriene pathway was secondary to the stimulated formation of 15-HETE.

Finally, it should be realized that inhibition of leukotriene production can also be produced by PGE$_1$ and PGE$_2$ that may act via elevation of cyclic AMP (cAMP) to inhibit arachidonic release and so production of all derived eicosanoids.[179]

Interestingly, production of leukotrienes has also been reported to be inhibited by U60257, a synthetic analog of PGI$_1$, even though PGI$_2$ and PGI$_1$ themselves were without effect.[3,180] In Willis' view, such an approach via prostaglandin actions on leukotriene production is likely to be the most promising therapeutic approach to this area. It might also be produced indirectly via prostaglandin precursor diets and/or use of thromboxane synthesis inhibitors that increase primary prostaglandin production.

Inhibitors of Cellular Uptake of Eicosanoids and their Precursors

Bito and collaborators'[181,182] (see chapter in this volume) first introduced the concept that prostaglandins are actively transported across cell membranes. There is thus competition for uptake between different prostaglandins and the uptake process is also inhibited by synthetic prostaglandin analogs and inhibitors of organic anion uptake, probenecid, and bromocresol green.[182-185] Use of the latter two compounds in perfused organ preparations showed that active uptake of prostaglandins was necessary for metabolism of E- or F-type prostaglandins in lung or kidney;[183,185] in the latter case urinary excretion of the prostaglandins was also inhibited by probenecid, showing that an active tubular secretion process was involved.[185]

Spagnuold et al.[186] obtained evidence for the in vivo selectivity in blockade of prostaglandin transport. Thus, following injections of pentamethylene tetrazole to rats, the elevations of brain cortex PGF$_{2\alpha}$ levels produced were both elevated and prolonged, whereas those of TXB$_2$ were not. Such active transport also seems to be involved in the gastrointestinal absorption of arachidonic acid, which is similarly inhibited by probenecid.[187] Active transport of leukotrienes into rat hepatocytes has also been reported.[188]

What are the implications of these findings? It is unlikely that probenecid will ever be used to prolong the actions of synthetic prostaglandin analogs as it was once used to inhibit renal tubular excretion of penicillin. However, it should be borne in mind that some synthetic prostaglandin analogs might owe part of their pharmacological effects (and/or side effects) to enhancing biological importance of endogenous prostaglandins, whose metabolism may be inhibited by uptake blockade.**

---

* Now withdrawn from clinical use.

** It is also possible that in a racemic mixture of different isomers of a synthetic prostaglandin analog, important interactions could occur at sites of cellular uptake. Thus, a biologically inactive isomer might inhibit vascular uptake, and hence delay metabolism, of an active isomer. Conversely, an isomer with poor oral absorption characteristics could block oral absorbtion of another isomer that might otherwise be very efficacious via the oral route. Thus a single isomer may not necessarily behave in vivo like the parent mixture of isomers.

Finally, when intraveneously administered dyes are used in animal experimentation, possible effects on inhibition of prostaglandin metabolism should be guarded against. However, adequate binding of the dye to serum albumin does seem to minimize ability to interfere with prostaglandin uptake.[189]

Inhibitors of Prostaglandin Metabolism

There are many naturally occurring eicosanoids and enzymes that sequentially metabolize them. However, prostaglandin 15-dehydrogenase (PGDH) is probably the most important single metabolic step in metabolism of E- and F-type prostaglandins (see chapters by Jackson Roberts and by D. L. Smith et al.).

Ferreira and Vane[190] showed that E- and F-type prostaglandins lost >95% of their smooth muscle-contracting activity in one passage through the pulmonary circulation. In lungs this "inactivation" of the prostaglandins is attributable to their uptake and metabolism.[183,184,191] The PGDH seems to act first to oxidize the 15-hydroxy group of the prostaglandin to yield a 15-keto derivative. Then, Δ^{13}-reductase reduces the double bond between the 13-14 carbon atoms. The 15-keto-13,14-dihydro prostaglandin metabolites persist in the circulation with a half-life of about 8 min.[193,194] Following ω- and β-oxidation in the liver, a variety of mono- and dicarboxylic metabolites are detected in the urine, the main one being the tetranor derivative* shown in Figure 8.

Metabolism and excretion of prostacyclins and thromboxanes seems less complex[191,198] with increased importance of oxidative chain shortening (Figure 8). However, biological actions of prostacyclin might, in some instances, be more prolonged in vivo than expected from its short circulating half-life in blood. This is because of its conversion to 6-keto-PGE$_1$,[199,200] which has a similar ability to inhibit platelet aggregation but (opposite to PGI$_2$) has a pronounced spasmogenic effect on the coronary artery.[201,202] Platelets may spontaneously generate small amounts of 6-keto-PGE$_1$.[203] However, if such conversion does occur, it is likely to be localized, since infusion of PGI$_2$ into human volunteers[287] did not produce detectable increases in 6-keto-PGE$_1$.

Less is known concerning the metabolic fate and principal urinary metabolites of the leukotrienes and other lipoxygenase products. However, an early stage appears to be a process analogous to omega oxidation of prostaglandins[88] (see Borgeat chapter); peptide cleavage can also occur.[204] Up-to-date summary of this area is given in the chapter by D. L. Smith.

Pathways also exist for the metabolism of A- and J-type prostaglandins although these substances may not truly occur endogenously in mammalian systems.**

Paulsrud and Miller[209] showed that PGDH could be inhibited in vitro by low (approximating to therapeutic) concentrations of some nonthiazide diuretic drugs, including furosemide and ethacrynic acid. Other agents that have been reported to inhibit PGDH in vitro are polyphloretin phosphate, sulfhydryl blocking agents, indomethacin, and sulphasalazine.[208,210,287]

---

*   This same tetranor metabolite is formed from both PGE$_1$ and PGE$_2$ and so measurement of this metabolite can only indicate the bodily turnover of E-type prostaglandins as a class. This metabolite is therefore sometimes called PGEM (M = major metabolite). From measurements of this metabolite, secretion rate of PGE$_1$ + PGE$_2$ was estimated at between approximately 20 to 200 μg/24 hr.[191,194] This excretion rate was markedly enhanced in subjects who had orally ingested ethyl arachidonate.[197]

**   Although A-type prostaglandins might be formed artificially from PGE during their extraction,[205,206] specific enzymatic pathways exist for PGA metabolism. A plasma isomerase (absent in humans) can convert PGA to PGC.[41-43] PGAs can also be inactivated by an enzyme in mammalian erythrocytes,[207] and a PGA-specific 15-dehydrogenase exists in rabbit kidney.[208] Recently, PGD$_2$ has been shown to spontaneously decompose to a "reverse PGA" compound (PGJ$_2$)[52-54] and hence to a Δ^{12} decomposition product.[53,54] Since the biological activity of PGJ$_2$ activity (inhibition of platelet aggregation) immediately disappears when it is injected intravenously into the guinea pig,[54] it is in doubt whether PGJ$_2$ has a significant endogenous role. Thus, it can be considered as an unstable metabolic intermediate of PGD$_2$.

Some inhibitors of prostaglandin metabolism may also selectively increase accumulation of the E-type prostaglandins. Furosemide and some other diuretics can inhibit the 9-keto reductase enzyme that converts PGE to PGF.[211] This enzyme has been detected in blood, heart, liver, kidney, brain, skin, and blood vessels.[1,212-216] Furosemide does appear to exert its diuretic actions via increased renal prostaglandin accumulation in vivo. However, it is not clear whether these effects are indeed due to the above-described inhibition of PGE metabolism.

Thus, in human subjects, urinary prostaglandins (including the tetranor PGEM metabolite) are elevated upon administration of furosemide.[217-220] Furthermore, this elevation and the diuretic effects of furosemide are inhibited by indomethacin;[220] however, an action of furosemide via inhibition of prostaglandin metabolism is not completely supported by other experimental data. Thus, it can elevate PGEM, while not inhibiting intrauterine prostaglandin metabolism.[220,221] An alternative explanation is that the drug increases availability of free arachidonic acid as substrate for cyclooxygenase.[218,288]

Agents that Modify Biological Effects of the Eicosanoids

It is probably true that most or all actions of the various eicosanoids (including lipoxygenase products) are exerted through activation of specific receptors.[1,222] Recently, the role of phosphatidylinositol diphosphate and protein kinase C has been postulated in some of these actions.[223,224,300] Numerous groups have attempted to develop specific receptor antagonists for eicosanoids. Some of the receptor antagonists have found use as experimental tools in developing receptor classification,[294] but so far the eventual goal of employing the agents to treat prostaglandin-mediated disease has not materialized.

Agents shown to possess prostaglandin-antagonist effects (vs. PGE and/or PGF in vitro on smooth muscle) are polyphloretin phosphate, SC 19220, and 7-oxaprostynoic acid.[1,225-227] In general, these agents have not been shown to have significant in vivo effects. Although polyphloretin phosphate and SC 19220 inhibit localized inflammatory processes that are thought to be mediated by localized overproduction of prostaglandins.[225-227]

SRS-A may now be classified as an eicosanoid (one or more of the amino acid-containing leukotrienes).[86-88,228] Interestingly, an SRS-A antagonist (FPL 55712) discovered solely by bioassay studies on impure SRS-A[229] was successfully used as an aid in identifying the leukotriene nature of SRS-A.[86-88,228] Recently, other leukotriene antagonists (SKF 55046 and KC 404), with possible therapeutic potential have been described.[230,231] KC 404 is particularly interesting, since it is orally active.[231] This area has recently been reviewed by Musser.[289]

Another prototype receptor antagonist that shows most promise in the direction of new therapeutic agents is pinane TXA_2. Intravenous infusions of this compound protects against artificially induced myocardial infarction[232] or traumatic shock[233] in experimental animals. It appears to act in this way by blocking the vasoconstrictor and platelet aggregatory actions of endogenously produced TXA_2, although it also inhibits endogenous thromboxane generation, and inhibits the release of lysosomal enzymes.[232,233]

In many cases, the inhibitory actions of prostaglandins (suppression of platelet aggregation and cellular secretion, relaxation of vascular or tracheal smooth muscle) are all associated with elevation of intracellular cAMP.[222,234,235] Such prostaglandin-induced actions are produced via activation of specific receptors leading to activation of adenylate cyclase. This process is attenuated by enzymatic removal of cAMP by phosphodiesterases[235] which may also sometimes be sequentially activated by prostaglandins[236] (see Alvarez chapter in Volume II).

Phosphodiesterase inhibitors (for example, theophylline, isobutylmethylxanthine, or dipyridamole) thus potentiate the cAMP-mediated prostaglandin effects. In platelets, for instance, PGE_1 (and analogs thereof), PGD_2, and PGI_2 all act via the cyclase.[1,135,234-236]

The concept[237,238] that prostacyclin normally exists in the circulation in significant concentrations has not been confirmed.[64,239-241] Nevertheless, the finding that therapeutic blood concentrations of dipyridamole act to potentiate the antithrombotic effects of prostacyclin[238] may still be valid. Localized vascular damage would not only result in thrombus formation but also production of prostacyclin by the injured vessel. The presence of a phosphodiesterase inhibitor would enhance the effects of even small amounts of prostacyclin produced locally and hence tend to reverse thrombus formation.

AVAILABILITY AND NATURE OF EICOSANOID PRECURSORS IN DETERMINING DRUG ACTIONS

The above considerations introduce the concept of "diet-drug interactions". Clearly one might expect actions of a phosphodiesterase inhibitor to be partially dependent upon an adequate supply of dietary prostaglandin precursors and appropriate in vivo availability for conversion into eicosanoids. One might also speculate that an ω6 EFA-rich diet might lead to increased clinical benefit of phosphodiesterase inhibitors as well as certain diuretics[209,211] and hypotensive agents[242,290] that may act in part via prostaglandin-dependent mechanisms.

Differences in dietary intake of EFA and/or their enzymatic elaboration into long-chain eicosanoid precursors is an often overlooked factor in the design of animal and human studies. This is especially important when studying endogenous eicosanoid production or effects of drugs (e.g., aspirin-type drugs) that modulate production or actions of the eicosanoids.[13] For instance, it is not generally recognized that the cat[13,243] or animals (or man?) that are diabetic[244] may necessarily have to obtain their eicosanoid precursors in pre-elaborated form from meat components of the diet. This is because they lack normal Δ^6 desaturation of 18:2ω6. Studies in the Greenland Eskimo provide the best documented example of such a potentially confusing drug/diet interaction. Aspirin has been reported to *shorten* bleeding time in Eskimos as opposed to its usual effect of prolonging bleeding time in people subsisting on a normal land-based Western diet.[245]

Aspirin normally inhibits hemostasis by blocking cyclooxygenase conversion of arachidonate (20:4ω6) to PGH_2 and TXA_2 that are prohemostatic.[26] It cannot do this in Eskimos, since platelet content of 20:4ω6 has been largely replaced by eicosapentaenoic acid (20:5ω3)* from the predominantly marine diet.[246] Instead, it might block conversion of 20:5ω3 to antiplatelet aggregatory metabolites such as PGI_3 or PGD_3.[13,245,247,248] Alternatively, it may act in concert with 20:5ω3 itself (a poor substrate but partial inhibitor of cyclooxygenase) to inhibit conversion of remaining vascular 20:4ω6 into PGI_2.[249]

CONCLUDING REMARKS: THERAPEUTIC PROSPECTS OF EICOSANOID RESEARCH

Unfortunately, the full therapeutic promise derived from basic eicosanoid research has still not yet been realized. However, we do know how aspirin-like drugs act and have a rational basis for understanding their limitations.

Furthermore, orally active thromboxane synthetase inhibitors are now being clinically investigated and the search is on for drugs that may beneficially interfere with leukotriene pathways and the formation of biologically active epoxides.

Also, for the first time, we are beginning to understand the possible mechanisms involved in the beneficial effects of both ω6 and ω3 series polyunsaturated fat diets in reducing blood pressure[252,253] and thrombotic tendency.[254-257] Although the cholesterol and triglyceride-lowering properties of such diets[253,255-257] cannot yet be seen to involve eicosanoid mecha-

---

* Such hypotheses are further complicated by the presence in marine diets of docosahexaenoic acid (22:6ω3). This is an eicosanoid precursor in fish[250] and yet inhibits mammalian cyclooxygenase, not lipoxygenase pathways.[251]

nisms,[257] a synthetic PGE analog (enprostil) has been reported to reduce plasma cholesterol levels in man.[292]

At the moment, most clinical promise lies with the prostaglandins themselves as therapeutic agents. Thus prostaglandins (E_2 or $F_{2\alpha}$) or their analogs have been successfully used for the induction of labor and as abortifacients.[258-260] Synthetic PGF analogs have been used for some time to synchronize estrus or to induce abortion in cattle.[261,262] In neonatal medicine, infusions of PGE_1 have been employed to keep the ductus arteriosus patent.[263]

Infusions of prostacyclin have been used to reduce thrombotic deposits in extracorporeal shunts (hemodialysis) and it has also been successfully used in the treatment of peripheral vascular disease.[51]

Indeed, infusions of either PGE_1 or prostacyclin produce long-term remission from the limb ischemia of both atherosclerosis obliterans[264,265] or of Raynaud's disease.[266,267] This is remarkable, given that the acute vasodilator and antiplatelet effects of the prostaglandins disappears upon cessation of infusion (see Addendum).

Since natural prostaglandins serve endogenously as local hormones, they are not ideally suited as drugs. Hence, they are rapidly metabolized, thus having to be infused, and have a multitude of biological actions that usually result in distressing side effects.

The answer to this dilemma would lie with development of synthetic prostaglandin analogs that are long lived, free of side effects, and are bioavailable via the oral or other convenient route. Sustained delivery systems may also be useful.

Although the pharmaceutical industry has devoted over 15 years to the synthesis and testing of prostaglandin analogs, it is only now that therapeutically useful compounds are emerging.

In the field of gastroenterology, there are now potent antiulcer PGE analogs.[268] Enprostil (Syntex) is remarkable for its long duration and cytoprotective effects in the wall of the stomach and duodenum.[269]

Lederle Laboratories[270] has synthesized a long-lived potent vasodilator and hypotensive 16-vinyl PGE analog (Viprostol) that is active either orally or when applied topically to the skin; this compound is without effect on platelet function. It is now being clinically investigated for the treatment of baldness.[303]

The ONO company has introduced both a prostacyclin analog (ONO 41483) and PGE_1 analog (ONO 1206) that are orally active in inhibiting platelet aggregation.[271,272] The latter compound reportedly has an extremely long duration of action[272] and has antianginal properties.[273] Both compounds are reported to be orally active in man in the microgram per kilogram dose range.[271,294]

At least one PGI_2 analog (ZK 36374, Iloprost, Schering) has been reported to exert a cardioprotective effect when infused during experimentally induced myocardial infarction.[275] Such an effect had previously been reported for intravenous infusions of prostacyclin itself.[276] However, Iloprost is orally absorbed and (unlike PGI_2) might be able to produce such cardioprotective effects via the oral route. A potentially serious problem of long-term use of both prostacyclin and its stable mimetics is the eventual increase in platelet aggregation that occurs (see Reference 293).

The area of synthetic prostaglandin analogs and their biological effects are detailed elsewhere in this book (for instance, see the chapter by Muchowski) as are the detailed considerations of all of the many areas touched upon in this chapter. Here, I have tried to merely outline the development and recent progress in this vast eicosanoid field.

ADDENDUM

The following important additional new findings were reported during the time that this volume of the handbook has been "in press":

1.   In the field of lipid mediators, there is now great interest in (Vascular) Endothelium-Derived Relaxing Factor (EDRF). This mediates relaxation of vascular smooth muscle in response to acetylcholine and some other agonists. It is extremely unstable (half life < 6 sec) and there is evidence that it is an eicosanoid metabolite of arachidonic acid. It apparently acts by elevating cyclic GMP in vascular smooth muscle cells (see References 295 to 297).

2.   The guanine nucleotide "G-proteins" have recently been shown to have a dual role in control of adenylate cyclase (thus being implicated in PG actions).[298] They also seem involved in coupling of receptors to phospholipase C in pancreas,[299] pointing to a possibly more widespread role in controlling eicosanoid production and actions.

3.   Several eicosanoid workers are now moving to the study of "Platelet-Activating Factor" (PAF; PAF-Acether, AGEPC). Interestingly, this glycerophospholipid may be formed by the action of a phospholipase A_2 that is arachidonate-specific, so that PAF production is attenuated in cells depleted of arachidonate[302] by EFA deficiency.

4.   The promise that PGs may be become useful therapeutic regimens in their own right now seems to be coming true.

First, there are now several anti-ulcer prostaglandins being introduced into clinical medicine. Misoprostil (Searle) is now on the market in Europe, Canada, and Mexico. Enprostil (Syntex) is on the market in France, Mexico, and New Zealand. Rosaprostil is on the market in Italy, and Rioprostil in Germany.

In addition to the potential in vascular disease, such as peripheral vascular disease, there is now rationale for the possible use of prostacyclin-mimetic drugs to prevent or reverse atherosclerosis per se. This is because PGI_2 and some analogs (exemplified by RS93427 Syntex) potently inhibit release of mitogens from several cells thought to be involved in atherosclerosis, including platelets, macrophages and vascular endothelial cells.[304-307] This is in addition to their ability to inhibit cellular accumulation of cholesteryl esters.[307,308] These effects are extremely potent and the effects of PGI_2 greatly outlast its presence in unmetabolized form.[305,307]

REFERENCES

1. **Willis, A. L. and Stone, K. J.,** Properties of prostaglandins, thromboxanes, their precursors, intermediates, metabolites and analogs — a compendium, in *Handbook of Biochemistry and Molecular Biology,* Vol. 2, 3rd ed., Fasman, G. D., Ed., CRC Press, Cleveland, 1976, 312—423.
2. **Hammarström, S. and Falardeau, P.,** Resolution of prostaglandin endoperoxide synthase and thromboxane synthase of human platelets, *Proc. Natl. Acad. Sci. U.S.A.,* 74, 3691—3695, 1977.
3. **Sun, F. F. and McGuire, J. C.,** Inhibition of human neutrophil arachidonate 5-lipoxygenase by 6,9-deepoxy-6,9-(phenylimino)-$\Delta^{6,8}$-prostaglandin I_1 (4-60257), *Prostaglandins,* 26, 211—221, 1984.
4. **Burr, G. O. and Burr, M. M.,** A new deficiency disease produced by the rigid exclusion of fat from the diet, *J. Biol. Chem.,* 82, 345—367, 1929.
5. **Burr, G. O. and Burr, M. M.,** On the nature and role of the fatty acids essential in nutrition, *J. Biol. Chem.,* 86, 587—621, 1930.
6. **Holman, R. T.,** General introduction to polyunsaturated acids, *Prog. Chem. Fats Other Lipids,* 9, 3—12, 1966.
7. **Hilditch, T. P. and Williams, P. N.,** *The Chemical Constitution of Natural Fats,* 4th ed., John Wiley & Sons, New York, 1964.
8. **Erdmann, E., Bedford, F., and Raspe, F.,** Constitution of linolenic acid, Ber, 42, 1334—1346, 1909.

9. **Heidushka, A. and Lüft, K.**, Fatty oil from the seed of the evening primrose *(Oenothera biennis)*, and a new linolenic acid, *Arch. Pharmacol.*, 257, 33—69, 1919.

10. **Van Dorp, D. A., Beerthuis, R. K., Nugteren, D. H., and Vonkeman, H.**, The biosynthesis of prostaglandins, *Biochim. Biophys. Acta*, 90, 204—297, 1964.

11. **Bergström, S., Danielsson, H., and Samuelsson, B.**, The enzymatic formation of prostaglandin E_2 from arachidonic acid, *Biochim. Biophys. Acta*, 90, 207—210, 1964.

12. **Lands, W. E. M. and Samuelsson, B.**, Phospholipid precursors of prostaglandins, *Biochim. Biophys. Acta*, 164, 426—429, 1968.

13. **Willis, A. L.**, Unanswered questions in EFA and PG research, *Prog. Lipid Res.*, 20, 839—850, 1982.

14. **Haye, B., Champion, S., and Jacquemin, C.**, Existence of two pools of prostaglandins during stimulation of the thyroid by TSH, *FEBS Lett.*, 41, 89—93, 1974.

15. **Carney, J. A., Slinger, S. J., and Walker, B. L.**, Homo-gamma-linolenic acid: a major polyunsaturated fatty acid of swine adrenal cholesteryl esters, *Lipids*, 6, 624, 1971.

16. **Takayasu, K., Okuda, K., and Yoshikawa, I.**, Fatty acid composition of human and bat adrenal lipids: occurrence of ω6 docosatrienoic acid in human adrenal cholesterol ester, *Lipids*, 5, 743—750, 1970.

17. **Vahouny, G. V., Chanderbhan, R., Bisgaier, C., Hodges, V. A., and Naghshineh, S.**, Essential fatty acids and adrenal steroidogenesis, *Prog. Lipid Res.*, 20, 233—240, 1982.

18. **Van Dorp, D.**, Enzymatic conversion of all-cis-polyunsaturated fatty acids into prostaglandins, *Ann. N.Y. Acad. Sci.*, 180, 181—199, 1971.

19. **Sprecher, H., Van Rollins, M., Sun, F., Wyche, A., and Needleman, P.**, Dihomo-prostaglandins and -thromboxane. A prostaglandin family from adrenic acid that may be preferentially synthesized in the kidney, *J. Biol. Chem.*, 257, 3912—3918, 1982.

20. **Struijk, C. B., Beerthuis, R. K., Pabon, H. J. J., and Van Dorp, D. A.**, Specificity in the enzymatic conversion of polyunsaturated fatty acids into prostaglandins, *Recl. Trav. Inst. Ecol. Biogcagr. Acad. Serbe Sci.*, 85, 1233-1252, 1966.

21. **Tobias, L. D., Vane, F. M., and Paulsrud, J. R.**, The biosynthesis of 1a,1b-Dihomo-PGE$_2$ and 1a,1b-Dihomo-PGF$_{2a}$ from acetone pentane powder of sheep vesicular gland microsomes, *Prostaglandins*, 10, 443—468, 1975.

22. **Comai, K., Farber, S. J., and Paulsrud, J. R.**, Analyses of renal medullary lipid droplets from normal, hydronephrotic, and indomethacin treated rabbits, *Lipids*, 10, 555—561, 1975.

23. **Comai, K., Prose, P., Farber, S. J., and Paulsrud, J. R.**, Correlation of renal medullary prostaglandin content and renal interstitial cell lipid droplets, *Prostaglandins*, 6, 375—379, 1974.

24. **Willis, A. L., Hassam, A. G., Crawford, M. A., Stevens, P., and Denton, J. P.**, Relationships between prostaglandins, prostacyclin and EFA precursors in rabbits maintained on EFA-deficient diets, *Prog. Lipid Res.*, 20, 161—167, 1982.

25. **Hassam, A. G., Willis, A. L., Denton, J. P., Stevens, P., and Crawford, M. A.**, The effect of essential fatty acid-deficient diet on the levels of prostaglandins and their fatty acid precursors in the rabbit brain, *Lipids*, 14, 78—80, 1979.

26. **Willis, A. L. and Smith, J. B.**, Some perspectives on platelets and prostaglandins, *Prog. Lipid Res.*, 20, 387—406, 1982.

27. **Weston, P. G. and Johnson, P. V.**, Cerebral prostaglandin synthesis during the dietary and pathological stresses of essential fatty acid deficiency and experimental allergic encephalomyelutus, *Lipids*, 13, 408—414, 1978.

28. **Lagarde, M., Guichardant, M., and Dechavanne, M.**, Human platelet PGE$_1$ and dihomogammalinolienic acid. Comparison to PGE$_2$ and arachidonic acid, *Prog. Lipid Res.*, 20, 439—443, 1982.

29. **von Euler, U. S.**, On the specific vasodilating and plain muscle stimulating substances from accessory genital glands in man and certain animals (prostaglandin and vesiglandin), *J. Physiol. (London)*, 88, 213—234, 1936.

30. **Kurzrok, R. and Lieb, C. C.**, Biochemical studies on human semen. II. The action of semen on the human uterus, *Proc. Soc. Exp. Biol. Med.*, 28, 268—272, 1930.

31. **Goldblatt, M. W.**, A depressor substance in seminal fluid, *J. Soc. Chem. Ind. (London)*, 84, 208—218, 1933.

32. **van Euler, U. S.**, Prostaglandin, historical remarks, *Prog. Lipid Res.*, 20, xxxi—xxxv, 1982.

33. **Bergström, S. and Sjövall, J.**, The isolation of prostaglandin F from sheep prostate glands, *Acta Chem. Scand.*, 14, 1693—1700, 1960.

34. **Bergström, S. and Sjövall, J.**, The isolation of prostaglandin E from sheep prostate glands, *Acta. Scand.*, 14, 1701—1705, 1960.

35. **Bergström, S., Rhyhage, R., Samuelsson, B., and Sjövall, J.**, The structures of prostaglandins E_1, $F_{1\alpha}$ and $F_{1\beta}$, *J. Biol. Chem.*, 238, 3555—3564, 1963.

36. **Hamberg, M.**, A note on nomenclature, *Adv. Biosci.*, 9, 847—850, 1973.

37. **Hamberg, M. and Samuelsson, B.**, New groups of naturally occurring prostaglandins, in *Prostaglandins, Proc. 2nd Nobel Symposium, Stockholm*, Bergström, S. and Samuelsson, B., Eds., Almqvist & Wiksell/ Stockholm Interscience, New York, 1967, 63—70.

38. **Lee, J. B., Covino, B. G., Takman, B. H., and Smith, E. R.**, Renomudullary vasodepressor substance, medullin: isolation, chemical characterization and physiological properties, *Circ. Res.*, 15, 57—77, 1965.

39. **Lee, J. B., Gougoutas, J. Z., Takman, B. H., Daniels, E. G., Grostic, M. F., Pike, J. E., Hinman, J. W., and Muirhead, E. E.**, Vasodepressor and antihypertensive prostaglandins of PGE type with emphasis on the identification of medullin as PGE_{2-217}, *J. Clin. Invest.*, 45, 1036, 1966.

40. **Hamberg, M. and Samuelsson, B.**, Prostaglandins in human seminal plasma, *J. Biol. Chem.*, 241, 257—263, 1966.

41. **Jones, R. L.**, 15-Hydroxy-9-oxoprosta-11,13-dienoic acid as the product of a prostaglandin isomerase, *J. Lipid Res.*, 13, 511—518, 1972.

42. **Jones, R. L.**, Preparation of prostaglandins C: chemical fixation of prostaglandin A isomerase to a gel support and partition chromatography of prostaglandins A, B, and C, *Prostaglandins*, 5, 283—290, 1970.

43. **Jones, R. L. and Cammock, S.**, Purification, properties and biological significance of prostaglandin A isomerase, *Adv. Biosci.*, 9, 61—70, 1973.

44. **Hamberg, M. and Samuelsson, B.**, Detection and isolation of an endoperoxide intermediate in prostaglandin biosynthesis, *Proc. Natl. Acad. Sci. U.S.A.*, 70, 899—903, 1973.

45. **Nugteren, D. H. and Hazelhof, E.**, Isolation and properties of intermediates in prostaglandin biosynthesis, *Biochim. Biophys. Acta*, 236, 448—461, 1973.

46. **Samuelsson, B.**, On the incorporation of oxygen in the conversion of 8,11,14-eicosatrienoic acid to prostaglandin E_1, *J. Am. Chem. Soc.*, 87, 3011—3013, 1965.

47. **Hamberg, M., Svensson, J., Wakabayashi, T., and Samuelsson, B.**, Isolation and structure of two prostaglandin endoperoxides that cause platelet aggregation, *Proc. Natl. Acad. Sci. U.S.A.*, 71, 345—349, 1974.

48. **Willis, A. L., Vane, F. M., Kuhn, D. C., Scott, C. G., and Petrin, M.**, An endoperoxide aggregator (LASS) formed in platelets in response to thrombotic stimuli: purification, identification and, unique biological significance, *Prostaglandins*, 8, 453—507, 1974.

49. **Moncada, S., Gryglewski, R., Bunting, S., and Vane, J. R.**, An enzyme isolated from arteries transforms prostaglandin endoperoxides to an unstable substance that inhibits platelet aggregation, *Nature (London)*, 263, 663—665, 1976.

50. **Johnson, R. A., Morton, D. R., Kinner, J. H., Gorman, R. R., McGuire, J. C., Sun, F. F., Whittaker, N., Bunting, S., Salmon, J., Moncada, S., and Vane, J. R.**, The chemical structure of prostaglandin X (prostacyclin), *Prostaglandins*, 12, 915—928, 1976.

51. **Dusting, G. J., Moncada, S., and Vane, J. R.**, Prostacyclin: its biosynthesis, actions, and clinical potential, in *Prostaglandins and the Cardiovascular System, Advances in Prostaglandin, Thromboxane and Leukotriene Research*, Vol. 10, Oates, J. A., Ed., Raven Press, New York, 1982, 59—106.

52. **Fukushima, M., Kato, T., Ota, K., Arai, Y., and Narumiya, S.**, 9-Deoxy-Δ^9-prostaglandin D_2. A prostaglandin D_2 derivative with potent anti-neoplastic and weak smooth muscle contracting activities, *Biochem. Biophys. Res. Commun.*, 109, 626—632, 1982.

53. **Fitzpatrick, F. A. and Wynalda, M. A.**, Albumin-catalyzed metabolism of prostaglandin D_2-identification of products formed in vitro, *J. Biol. Chem.*, 258, 11713—11718, 1983.

54. **Mahmud, I., Smith, D. L., Willis, A. L., Whyte, M. A., Nelson, J. T., Cho, D., Tokes, L. G., and Alvarez, R.**, On the identification and biological properties of prostaglandin J_2, *Prostaglandins Leukotrienes Med.*, 16, 131—146, 1984.

55. **Rabinowitz, I., Ramwell, P. W., and Davison, P.**, Conformation of prostaglandins, *Nature (London) New Biol.*, 233, 88—90, 1971.

56. **Cahn, R. S., Ingold, C. K., and Prelog, V.**, Specification of asymmetric configuration in organic chemistry, *Experimentia*, 12, 81—94, 1956.

57. **Hamberg, M. and Samuelsson, B. M.**, Prostaglandin endoperoxides. Novel transformations of arachidonic acid in human platelets, *Proc. Natl. Acad. Sci. U.S.A.*, 71, 3400—3404, 1974.

58. **Cram, D. J. and Hammond, G. S.**, *Organic Chemistry*, McGraw-Hill, New York, 1959.

59. **Ojima, M. and Fujita, K.**, Inhibition of platelet aggregation and white thrombus formation by prostaglandin analog ONO-747, in *Advances in Prostaglandin and Thromboxane Research*, Samuelsson, B. and Paoletti, R., Eds., Raven Press, New York, 1976, 781—785.

60. **Andersen, N.**, Program notes on structures and nomenclature, *Ann. N.Y. Acad. Sci.*, 180, 14—23, 1971.

61. **Hamberg, M.**, On the absolute configuration of 19-hydroxy-prostaglandin B_1, *Eur. J. Biochem.*, 6, 147—150, 1968.

62. **Bunting, S., Monada, S., Reed, P., Salmon, J. A., and Vane, J. R.**, PGI_1 synthesis: an anti-serum to 5,6-dihydro prostacyclin (PGI_1) which also binds prostacyclin, *Prostaglandins*, 15, 566—573, 1978.

63. **Levine, L., Alam, I., and Langone, J. J.,** The use of immobilized ligands and [^{125}I] protein A for immunoassays of thromboxane B$_2$, prostaglandin D$_2$, 13,14-dihydro-prostaglandin E$_2$, 5,6-dihydro-prostaglandin I$_2$, 6-keto-prostaglandin F$_{1\alpha}$, 15-hydroxy-9α,11α (epoxymethano) prosta-5,13-dienoic acid and 15-hydroxy-11α,9α (epoxymethano) prosta-5,13-dienoic acid, *Prostaglandins Med.*, 2, 177—189, 1979.

64. **Steer, M. L., MacIntyre, D. E., Levine, L., and Salzman, E. W.,** Is prostacyclin a physiologically important circulating anti-platelet agent?, *Nature (London)*, 283, 194—195, 1980.

65. **Nidy, E. G. and Johnson, R. A.,** Synthesis of prostaglandin I$_3$, *Tetrahedron Lett.*, 27, 2375—2378, 1978.

66. **Needleman, P., Raz, A., Minkes, M. S., Ferrendelli, J. A., and Sprecher, M.,** Triene prostaglandins, prostacyclin and thromboxane biosynthesis and unique biological properties, *Proc. Natl. Acad. Sci. U.S.A.*, 76, 944—948, 1979.

67. **Johnson, R. A., Morton, D. R., and Nelson, N. A.,** Nomenclature for analogs of prostacyclin (PGI$_2$) *Prostaglandins*, 15, 737—750, 1978.

68. **Hamberg, M., Svensson, J., and Samuelsson, B.,** Thromboxanes: a new group of biologically active compounds derived from prostaglandin endoperoxides, *Proc. Natl. Acad. Sci. U.S.A.*, 72, 2994—2998, 1975.

69. **Wlodawer, P. and Hammarström, S.,** Thromboxane synthase from bovine lung-solubilization and partial purification, *Biochem. Biophys. Res. Commun.*, 80, 525—532, 1978.

70. **Smith, J. B., Ingerman, C., and Silver, M. J.,** Persistence of thromboxane A$_2$-like material and platelet release-inducing activity in plasma, *J. Clin. Invest.*, 58, 1119—1122, 1976.

71. **Moncada, S., Needleman, P., Bunting, S., and Vane, J. R.,** Prostaglandin endoperoxide and thromboxane generating systems and their selective inhibition, *Prostaglandins*, 12, 323—325, 1976.

72. **Needleman, P., Minkes, M., and Raz, A.,** Thromboxanes: selective biosynthesis and distinct biological properties, *Science*, 193, 163—165, 1976.

73. **Nugteren, D. H.,** Arachidonate lipoxygenase, in *Prostaglandins in Hematology*, Silver, M. J., Smith, J. B., and Kocsis, J. J., Eds., Spectrum Publications, Jamaica, N.Y., 1977, 11—25.

74. **Falardeau, P., Hamberg, M., and Samuelsson, B.,** Metabolism of 8,11,14-eicosatrienoic acid in human platelets, *Biochim. Biophys. Acta*, 441, 193—200, 1976.

75. **Diczfalusy, U. and Hammerström, S.,** A structural requirement for the conversion of prostaglandin endoperoxides to thromboxanes, *FEBS Lett.*, 105, 291—298, 1979.

76. **Lands, W. E. M., Le Tellier, P. R., Rome, L. H., and Vanderhoek, Y.,** Inhibition of prostaglandin biosynthesis, *Adv. Biosci.*, 9, 15—28, 1973.

77. **Willis, A. L.,** Isolation of a chemical trigger for thrombosis, *Prostaglandins*, 5, 1—25, 1974.

78. **Diczfalusy, U., Falardeau, P., and Hammarström, S.,** Conversion of prostaglandin endoperoxides to C17-hydroxy acids, catalyzed by human platelet thromboxane synthase, *FEBS Lett.*, 84, 271—274, 1977.

79. **Smith, J. B., Ingerman, C. M., and Silver, M. J.,** Malondialdehyde formation as an indicator of prostaglandin production by human platelets, *J. Lab. Clin. Med.*, 88, 167—172, 1976.

80. **Gray, J. I.,** Measurement of lipid oxidation: a review, *J. Am. Oil Chem. Soc.*, 55, 539—546, 1978.

81. **Nugteren, D. H., Vonkeman, H., and Von Dorp, D. A.,** Non-enzymatic conversion of *all cis* 8,11,14-eicosatrienoic acid into prostaglandin E$_1$, *Recl. Trav. Chim. Pays. Belg.*, 86, 1237—1245, 1967.

82. **Gutteridge, J. M. C.,** Identification of malondioldehyde as the TBA-reactant formed by bleomycin-iron free radical damage to DNA, *FEBS Lett.*, 105, 278—282, 1979.

83. **Needleman, P., Whitaker, M. O., Wyche, A., Walters, K., Sprecher, H., and Raz, A.,** Manipulation of platelet aggregation by prostaglandins and their fatty acid precursors: pharmacological basis for a therapeutic approach, *Prostaglandins*, 19, 165—181, 1980.

84. **Hornstra, G. and Nugteren, D. H.,** Fish oil feeding does not result in the endogenous formation of PGI$_3$ in rats, *Prog. Lipid Res.*, 20, 911—912, 1982.

85. **Brocklehurst, W. E.,** The forty year quest of "slow reacting substance of anaphylaxis", *Prog. Lipid Res.*, 20, 709—712, 1982.

86. **Samuelsson, B.,** Leukotrienes: a novel group of compounds including SRS-A, *Prog. Lipid Res.*, 20, 23—30, 1982.

87. **Hammarström, S.,** Leukotriene formation by mastocytoma and basophilic leukemia cells, *Prog. Lipid Res.*, 20, 89—95, 1982.

88. **Hammarström, S.,** Leukotrienes, *Ann. Rev. Biochem.*, 52, 355—377, 1983.

89. **Jubiz, W. and Nolan, G.,** Leukotrienes produced by incubation of dihomo-γ-Linolenic acid with human polymorphonuclear leukocytes, *Biochem. Biophys. Res. Commun.*, 114, 855—862, 1983.

90. **Nugteren, D. H.,** Arachidonate lipoxygenase in blood platelets, *Biochim. Biophys. Acta*, 380, 299—307, 1975.

91. **Borgeat, P., Hamberg, M., and Samuelsson, B.,** Transformation of arachidonic acid and homo-γ-linolenic acid by rabbit polymorphonuclear leukocytes, *J. Biol. Chem.*, 251, 7816—7820, 1976.

92. **Borgeat, P. and Samuelsson, B.,** Transformation of arachidonic acid by rabbit polymorphonuclear leukocytes, *J. Biol. Chem.*, 254, 2643—2646, 1979.

93. **Borgeat, P. and Samuelsson, B.,** Arachidonic acid metabolism in polymorphonuclear leukocytes: effects of ionophore A23187, *Proc. Natl. Acad. Sci. U.S.A.*, 76, 2148—2152, 1979.

94. **Goetzl, E. J.,** Selective feedback inhibition of the 5-lipoxygenation of arachidonic acid in human t-lymphocytes, *Biochem. Biophys. Res. Commun.*, 101, 344—350, 1981.

95. **Bryant, R. W., Bailey, J. M., Schewe, T., and Rapoport, S. M.,** Positional specificity of a reticulocyte lipoxygenase: conversion of arachidonic acid to 15-S-hydroxyperoxy-eicosatetraenoic acid, *J. Biol. Chem.*, 257, 6050—6055, 1982.

96. **Vanderhoek, J. Y., Tare, N. S., Bailey, J. M., Goldstein, A. L., and Pluznic, D.,** New role for 15-hydroxy-eicosatetraenoic acid: activator of leukotriene biosynthesis in PT18 mast/basophil cells, *J. Biol. Chem.*, 257, 12191—12195, 1982.

97. **Vanderhoek, J. Y., Bryant, R. W., and Bailey, J. M.,** 15-hydroxy-5,8,11,13-eicosatetraenoic acid: a potent and selective inhibitor of platelet lipoxygenase, *J. Biol. Chem.*, 255, 5996—5998, 1980.

98. **Vanderhoek, J. Y., Bryant, R. W., and Bailey, J. M.,** Inhibition of leukotriene biosynthesis by the leukocyte product 15-hydroxy-5,8,11-13-eicosatetraenoic acid, *J. Biol. Chem.*, 255, 10064—10066, 1980.

99. **Bailey, J. M., Bryant, R. W., Low, C. E., Pupillo, M. B., and Vanderhoek, J. Y.,** Role of lipoxygenases in regulation of PHA and phorbol ester-induced mitogenesis, in *Leukotrienes and Other Lipoxygenase Products*, Samuelsson, B. and Pavletti, R., Eds., Raven Press, New York, 1982, 341—353.

100. **Borgeat, P., Fruteau De Laclos, B., Picard, S., Vallerand, P., and Sirois, P.,** Double dioxygenation of arachidonic acid in leukocytes by lipoxygenases, in *Leukotrienes and Other Lipoxygenase Products*, Samuelsson, B. and Paoletti, R., Eds., Raven Press, New York, 1982, 45—51.

101. **Bailey, J. M., Bryant, R. W., Whiting, J., and Salata, K.,** Characterization of 11-HETE and 15-HETE, together with prostacyclin, as major products of the cyclo-oxygenase pathway in cultured rat aorta smooth muscle cells, *J. Lipid Res.*, 24, 1419—1428, 1983.

102. **Walker, I. C., Jones, R. L., and Wilson, N. H.,** The identification of an epoxyhydroxy acid as a product from the incubation of arachidonic acid with washed blood platelets, *Prostaglandins*, 18, 173—178, 1979.

103. **Bryant, R. W. and Bailey, J. M.,** Isolation of a new lipoxygenase metabolite of arachidonic acid - 8,11,12-trihydroxy-5,9,14-eicosatrienic acid from human platelets, *Prostaglandins*, 17, 9—18, 1979.

104. **Bild, G. S., Bhat, S. G., and Axelrod, B.,** Inhibition of human platelets by 8,15-dihydroperoxides of 5,9,11,13-eicosatetraenoic acid and 9,11,13-eicosatrienoic acids, *Prostaglandins*, 16, 795—801, 1978.

105. **Bild, G. S., Ramadoss, C. S., Lim, S., and Axelrod, B.,** Double dioxygenation of arachidonic acid by soybean lipoxygenase-1, *Biochem. Biophys. Res. Commun.*, 74, 949—954, 1977.

106. **Bryant, R. W. and Bailey, J. M.,** Isolation of glucose-sensitive platelet lipoxygenase products from arachidonic acid, in *Advances in Prostaglandin and Thromboxane Research*, Vol. 6, Samuelsson, B., Ramwell, P. W., and Paoletti, R., Eds., Raven Press, New York, 1980, 95—99.

107. **Bryant, R. W. and Bailey, J. M.,** Role of selenium-dependent glutathione peroxidase in platelet lipoxygenase metabolism, *Prog. Lipid Res.*, 20, 189—194, 1982.

108. **Lu, A. Y. H. and West, S. B.,** Multiplicity of mammalian microsomal cytochromes P-450, *Pharm. Rev.*, 31, 277—295, 1980.

109. **Falk, J. R. and Manna, S.,** 8,9-Epoxyarachidonic acid: a cytochrome P-450 metabolite, *Tetrahedron Lett.*, 23, 1755—1756, 1982.

110. **Chacos, N., Falck, J. R., Wixtrom, C., and Capdevila, J.,** Novel epoxides formed during liver cytochrome P-450 oxidation of arachidonic acid, *Biochem. Biophys. Res. Commun.*, 104, 916—922, 1982.

111. **Snyder, G. D., Capdevila, J., Chacos, N., Manna, S., and Falk, J. R.,** Action of luteinizing hormone-releasing hormone: involvement of novel arachidonic acid metabolites, *Proc. Natl. Acad. Sci. U.S.A.*, 80, 3504—3507, 1983.

112. **Kutsky, P., Falk, J. R., Weiss, G. B., Manna, S., Chacos, N., and Capdevila, J.,** Effects of newly reported arachidonic acid metabolites on microsomal Ca^{++} binding, uptake and release, *Prostaglandins*, 26, 13—21, 1983.

113. **Oliw, E. M.,** Metabolism of 5(6) oxidoeicosatrienoic acid by ram seminal vesicles. Formation of two stereo isomers of 5-hydroxyprostaglandin I_1, *J. Biol. Chem.*, 259, 2716—2721, 1984.

114. **Serhan, C. N., Hamberg, M., and Samuelsson, B.,** Lipoxins: novel series of biologically active compounds formed from arachidonic acid in human leukocytes, *Proc. Natl. Acad. Sci.*, U.S.A., 81, 5335—5339, 1984.

115. **Marcus, A. J., Weksler, B. B., Jaffe, E. A., and Broekman, M. J.,** Synthesis of prostacyclin from platelet-derived endoperoxides by cultured human endothelial cells, *J. Clin. Invest.*, 66, 979—986, 1980.

116. **Aiken, J. W., Shebuski, R. J., Miller, O. V., and Gorman, R. R.,** Endogenous prostacyclin contributes to the efficacy of a thromboxane synthetase inhibitor for preventing coronary artery thrombosis, *J. Pharmacol. Exp. Ther.*, 219, 299—319, 1981.

117. **Aiken, J. W., Gorman, R. R., and Shebuski, R. J.,** Prostacyclin prevents blockage of partially obstructed coronary arteries, in *Prostacyclin*, Vane, J. R. and Bergström, S., Eds., Raven Press, New York, 1979, 311—321.

118. **Marcus, A. J., Broekman, M. J., Safier, L. B., Ullman, H. L., Islam, N., Serhan, C. N., Rutherford, L. E., Korchak, H. M., and Weissman, G.,** Formation of leukotrienes and other hydroxy acids during platelet-neutrophil interactions in vitro, *Biochem. Biophys. Res. Commun.*, 109, 130—137, 1982.

119. **Marcus, A. J., Safier, L. B., Ullman, H. L., Broekman, M. J., Islam, N., Oglesby, T. D., and Gorman, R. R.,** 12S, 20-dihydroxyicosatetraenoic acid: a new eicosanoid synthesized by neutrophils from 12S-hydroxyicosatetraenoic acid produced by thrombin- or collagen-stimulated platelets, *Proc. Natl. Acad. Sci. U.S.A.*, 81, 903—907, 1984.

120. **Vane, J. R.,** Inhibition of prostaglandin biosynthesis as the mechanism of action of aspirin-like drugs, *Adv. Biosci.*, 9, 395—411, 1973.

121. **Willis, A. L., Davison, P., Ramwell, P. W., Brocklehurst, W. E., and Smith, B.,** Release and actions of prostaglandins in inflammation and fever, in *Prostaglandins in Cellular Biology*, Ramwell, P. W. and Pharriss, B. B., Eds., Plenum Press, New York, 1972, 227—268.

122. **Willis, A. L.,** An enzymatic mechanism for the anti-thrombotic and antihemostatic actions of aspirin, *Science*, 183, 325—327, 1974.

123. **Vargaftig, B. B. and Dao Hai, N.,** Selective inhibition by mepacrine of the release of rabbit aorta contracting substance evoked by the administration of bradykinin, *J. Pharm. Pharmacol.*, 24, 159—161, 1972.

124. **Flower, R. J.,** Prostaglandins and related compounds, in *Inflammation*, Vane, J. R. and Ferreira, S. H., Eds., Springer-Verlag, Berlin, 1978, 374—422.

125. **Greaves, M. W. and McDonald-Gibson, W.,** Prostaglandin biosynthesis by human skin and its inhibition by corticosteroids, *Br. J. Pharmacol.*, 46, 172—175, 1972.

126. **Hong, S.-L. C. and Levine, L.,** Inhibition of arachidonic acid release from cells as the biochemical action of anti-inflammatory corticosteroids, *Proc. Natl. Acad. Sci. U.S.A.*, 73, 1730—1734, 1976.

127. **Gryglewski, R. J., Panczenko, B., Korbut, R., Grodzinska, L., and Ocetkiewicz, A.,** Corticosteroids inhibit prostaglandin release from perfused mesenteric blood vessels of rabbit and from perfused lungs of sensitized guinea pig, *Prostaglandins*, 10, 343—355, 1975.

128. **Blackwell, G. J., Flower, R. J., Nijkamp, F. A., and Vane, J. R.,** Phospholipase A_2 activity of guinea pig isolated perfused lungs. Stimulation and inhibition by anti-inflammatory steroids, *Br. J. Pharmacol.*, 62, 79—89, 1978.

129. **Blackwell, G. J., Carnuccio, R., DiRosa, M., Flower, R. J., Parente, L., and Persico, P.,** Macrocortin: a polypeptide causing the anti-phospholipase effect of glucocorticoids, *Nature (London)*, 287, 147—149, 1980.

130. **Hirata, F., Schiffmann, E., Venkatasubramanian, K., Salomon, D., and Axelrod, J.,** A phospholipase A_2 inhibitory protein in rabbit neutrophils induced by glucocorticoids, *Proc. Natl. Acad. Sci. U.S.A.*, 77, 2533—2536, 1980.

131. **Danon, A. and Assouline, G.,** Inhibition of prostaglandin biosynthesis by corticosteroids requires RNA and protein synthesis, *Nature (London)*, 273, 552—554, 1978.

132. **Vane, J. R.,** Inhibition of prostaglandin synthesis as a mechanism of action for aspirin-like drugs, *Nature (London) New Biol.*, 231, 232—235, 1971.

133. **Ferreira, S. H., Moncada, S., and Vane, J. R.,** Indomethacin and aspirin abolish prostaglandin release from the spleen, *Nature (London) New Biol.*, 231, 237—239, 1971.

134. **Smith, J. B. and Willis, A. L.,** Aspirin selectively inhibits prostaglandin production in human platelets, *Nature (London) New Biol.*, 231, 235—237, 1971.

135. **Moncada, S. and Vane, J. R.,** Pharmacology and endogenous roles of prostaglandin endoperoxides, thromboxane A_2 and prostacyclin, *Pharm. Rev.*, 30, 293—331, 1978.

136. **Kocsis, J. J., Hernandovich, J., Silver, M. J., Smith, J. B., and Ingerman, C.,** Duration of inhibition of platelet prostaglandin formation and aggregation by ingested aspirin or indomethacin, *Prostaglandins*, 3, 141—144, 1973.

137. **Hamberg, M.,** Inhibition of prostaglandin synthesis in man, *Biochem. Biophys. Res. Commun.*, 49, 720—726, 1972.

138. **O'Brien, J. R.,** Effect of anti-inflammatory agents on platelets, *Lancet*, 1, 894—895, 1968.

139. **Willis, A. L.,** Platelet aggregation mechanisms and their implications in haemostasis and inflammatory disease, in *Inflammation*, Vane, J. R. and Ferreira, S. H., Eds., Springer-Verlag, Berlin, 1978, 138—205.

140. **Roth, G. J., Stanford, N., and Majerus, P. W.,** Acetylation of prostaglandin synthase, *Proc. Natl. Acad. Sci. U.S.A.*, 72, 3073—3076, 1975.

141. **Buchanan, M. R., Haas, T. A., Lagarde, M., and Guichardant, M.,** 13-Hydroxyoctadecadienoic acid is the vessel wall chemorepellant factor, LOX, *J. Biol. Chem.*, 260, 16056—16059, 1985.

142. **Flower, R., Gryglewski, R., Herbacynska-Cedro, K., and Vane, J. R.,** Effects of anti-inflammatory drugs on prostaglandin biosynthesis, *Nature (London) New Biol.*, 238, 104—106, 1972.

143. **Flower, R. J. and Vane, J. R.,** Inhibition of prostaglandin synthetase in brain explains the anti-pyretic activity of paracetamol (4-acetamidophenol), *Nature (London)*, 240, 410—411, 1972.

144. **Feldberg, W. and Milton, A. S.,** Prostaglandins and body temperature, in *Inflammation,* Vane, J. R. and Ferreira, S. H., Eds., Springer-Verlag, Berlin, 1978, 617—656.

145. **Stone, K. J., Mather, S. J., and Gibson, P. P.,** Selective inhibition of prostaglandin biosynthesis by gold salts and phenylbutazone, *Prostaglandins,* 10, 241—251, 1975.

146. **Crunkhorn, P. and Willis, A. L.,** Interaction between prostaglandins E and F given intradermally in the rat, *Br. J. Pharmacol.,* 41, 507—512, 1971.

147. **Zurier, R. B. and Quagliata, F.,** Effect of prostaglandin E_1 on adjuvant arthritis, *Nature (London) New Biol.,* 234, 304—305, 1971.

148. **Bonta, I. L., Parnham, M. J., and Adolfs, M. J. P.,** Reduced exudation and increased tissue proliferation during chronic inflammation in cats deprived of exogenous prostaglandin precursors, *Prostaglandins,* 14, 295—307, 1977.

149. **Willis, A. L.,** Essential fatty acids, prostaglandins and related eicosanoids, in *Present Knowledge in Nutrition,* 5th ed., Nutrition Foundation, Inc., Washington, D.C., 1984, 90—116.

150. **Kunkel, S. L. and Zurier, R. B.,** Suppression of chronic inflammation by evening primrose oil, *Prog. Lipid Res.,* 20, 885—888, 1982.

151. **Smith, J. B., Silver, M. J., Ingerman, C. M., and Kocsis, J. J.,** Prostaglandin D_2 inhibits the aggregation of human platelets, *Thromb. Res.,* 5, 291—299, 1974.

152. **Shio, M. and Ramwell, P.,** Effect of prostaglandin E_2 and aspirin on the secondary aggregation of human platelets, *Nature (London) New Biol.,* 236, 45—46, 1972.

153. **Hamberg, M. and Fredholm, B. B.,** Isomerization of prostaglandin H_2 into prostaglandin D_2 in the presence of serum albumin, *Biochim. Biophys. Acta,* 431, 189—193, 1976.

154. **Randall, M. J. and Wilding, R. I. R.,** Acute arterial thrombosis in rabbits. Reduced platelet accumulation after treatment with Dazoxiben (U.K. 37,248-01), *Br. J. Clin. Pharmacol.,* 15, 495—555, 1983.

155. **Defreyn, G., Deckmyn, M., and Vermylen, J.,** A thromboxane synthetase inhibitor reorients endoperoxide metabolism in whole blood towards prostacyclin and prostaglandin E_2, *Thromb. Res.,* 26, 389—400, 1982.

156. **Hanson, S. R. and Harker, L. A.,** Effect of Dazoxiben on arterial graft thrombosis in the baboon, *Br. J. Clin. Pharmacol.,* 15, 575—605, 1983.

157. **Moncada, S., Gryglewski, R. J., Bunting, S., Salmon, J. A., Smith, D. R., Flower, R. J., and Vane, J. R.,** A lipid peroxide inhibits the enzyme in blood vessel microsomes that generates from prostaglandin endoperoxides the substance (prostaglandin X) which prevents platelet aggregation, *Prostaglandins,* 12, 715—737, 1976.

158. **Salmon, J. A., Smith, D. R., Flower, R. J., Moncada, S., and Vane, J. R.,** Further studies on enzymatic conversion of prostaglandin endoperoxide into prostacyclin by porcine aorta microsomes, *Biochim. Biophys. Acta,* 523, 250—262, 1978.

159. **Gryglewski, R. J., Bunting, S., Moncada, S., Flower, R. J., and Vane, J. R.,** Arterial walls are protected against deposition of platelet thrombi by a substance (prostaglandin X) which they make from prostaglandin endoperoxides, *Prostaglandins,* 12, 685—714, 1976.

160. **Hoyns, Ph.F. J. and Van Alphen, G. W. H. M.,** Behavior of IOP and pupil size after topical tranyl-cypromine in the rabbit eye, *Doc. Opthol. (Netherlands),* 51, 225—234, 1981.

161. **Osbond, J. M., Philpott, P. G., and Wickens, J. C.,** Essential fatty acids. Synthesis of linoleic, γ-linolenic, arachidonic, and docosa-4,7,10,13,16-pentaenoic acid, *J. Chem. Soc.,* July, 2779—2787, 1961.

162. **Downing, D. T., Ahern, D. G., and Bachta, M.,** Enzyme inhibition by acetylenic compounds, *Biochem. Biophys. Res. Commun.,* 40, 218—223, 1970.

163. **Downing, D. T.,** Differential inhibition of prostaglandin synthetase and soybean lipoxygenase, *Prostaglandins,* 1, 437—441, 1972.

164. **Willis, A. L., Kuhn, D. C., and Weiss, M. J.,** Acetylenic analog of arachidonate that acts like aspirin on platelets, *Science,* 183, 327—330, 1974.

165. **Shaw, J. E., Jessup, S. J., and Ramwell, P. W.,** Prostaglandin-adenylate cyclase relationships, in *Advances in Cyclic Nucleotide Research,* Vol. 1, Paoletti, R., Ed., Raven Press, New York, 1972, 479—491.

166. **Blackwell, G. J. and Flower, R. J.,** 1-Phenyl-3-pyrazolidone: an inhibitor of arachidonate oxidation in lung and platelets, *Br. J. Pharmacol.,* 63, 360P, 1978.

167. **Higgs, G. A., Flower, R. J., and Vane, J. R.,** A new approach to anti-inflammatory drugs, *Biochem. Pharmacol.,* 28, 1959—1961, 1979.

168. **Tappel, A. L., Lundberg, W. O., and Boyer, P. D.,** Effect of temperature and anti-oxidants upon the lipoxydase-catalysed oxidation of sodium linoleate, *Arch. Biochem. Biophys.,* 42, 293—304, 1953.

169. **Morris, H. R., Piper, P. J., Taylor, G. W., and Tippins, J. R.,** The role of arachidonate lipoxygenase in the release of SRS-A from guinea-pig chopped lung, *Prostaglandins,* 19, 371—383, 1980.

170. **Armour, C. L., Hughes, J. M., Seale, J. P., and Temple, D. M.,** Effects of lipoxygenase inhibitors on release of slow reacting substances from human lung, *Eur. J. Pharmacol.,* 72, 93—96, 1981.

171. **Hammarström, S.,** Selective inhibition of platelet n-8 lipoxygenase by 5,8,11-eicosatriynoic acid, *Biochim. Biophys. Acta,* 487, 517—519, 1977.

172. **Kuehn, M., Holzhuetter, H. G., Schewe, T., Hiebach, C., and Rapoport, S. M.**, The mechanism of inactivation of lipoxygenases by acetylenic fatty acids, *Eur. J. Biochem.*, 139, 577—583, 1984.

173. **Koshihara, Y., Murota, S., and Nicolaou, K. C.**, 5,6-Methano leukotriene A_4: a potent specific inhibitor for 5-lipoxygenase, in *Advances in Prostaglandin, Thromboxane, and Leukotriene Research*, Vol. 11, Samuelsson, B., Paoletti, R., and Ramwell, P., Eds., Raven Press, New York, 1983, 163—172.

174. **Bray, M. A.**, Retinoids are potent inhibitors of the generation of rat leukocyte leukotriene B_4-like activity in vitro, *Eur. J. Pharmacol.*, 98, 61—67, 1984.

175. **Sekiya, K. and Okuda, H.**, Selective inhibition of platelet lipoxygenase by baicalein, *Biochem. Biophys. Res. Commun.*, 105, 1090—1095, 1982.

176. **Egan, R. W., Tischler, A. N., Baptista, E. M., Ham, E. A., Soderman, D. D., and Gale, P. H.**, Specific inhibition and oxidative regulation of 5-lipoxygenase, in *Advances in Prostaglandin, Thromboxane, and Leukotriene Research*, Vol. 11, Samuelsson, B., Paoletti, R., and Ramwell, P., Eds., Raven Press, New York, 1983, 151—157.

177. **Harvey, J., Parish, H., Ho, P. P. K., Boot, J. R., and Dawson, W.**, The preferential inhibition of 5-lipoxygenase product formation by benoxaprofen, *J. Pharm. Pharmacol.*, 35, 44—45, 1983.

178. **Vanderhoek, J. Y. and Bailey, J. M.**, Activation of a 15-lipoxygenase/leukotriene pathway in human polymorphonuclear leukocytes by the anti-inflammatory agent ibuprofen, *J. Biol. Chem.*, 259, 6752—6756, 1984.

179. **Ham, E. A., Soderman, D. D., Zanetti, M. E., Dougherty, H. W., McCauley, E., and Keuhl, F. A., Jr.**, Inhibition by prostaglandins of leukotriene B_4 by release from activated neutrophils, *Proc. Natl. Acad. Sci. U.S.A.*, 80, 4349—4353, 1983.

180. **Bach, M. K., Brashler, J. R., Fitzpatrick, F. A., Griffin, R. L., Iden, S. S., Johnson, H. G., McNee, M. L., McGuire, J. C., Smith, H. W., Smith, R. J., Sun, F. F., and Wasserman, M. A.**, In vivo and in vitro actions of a new selective inhibitor of leukotriene C and D synthesis, in *Advances in Prostaglandin, Thromboxane, and Leukotriene Research*, Vol. 11, Samuelsson, B., Paoletti, R., and Ramwell, P., Eds., Raven Press, New York, 1983, 39—44.

181. **Bito, L. Z.**, Accumulation and apparent active transport of prostaglandins by some rabbit tissue in-vitro, *J. Physiol. (London)*, 221, 371—387, 1972.

182. **Bito, L. Z., Davson, H., and Salvador, E. V.**, Inhibition of in-vitro concentrative prostaglandin accumulation by prostaglandins, prostaglandin analogues and by some inhibitors of organic anion transport, *J. Physiol. (London)*, 256, 257—271, 1976.

183. **Bakhle, Y. S.**, Inhibition by clinically used dyes of prostaglandin inactivation in rat and human lung, *Br. J. Pharmacol.*, 72, 715—721, 1981.

184. **Bito, L. Z. and Baroudy, R. A.**, Inhibition of pulmonary prostaglandin metabolism by inhibitors of prostaglandin biotransport (probenecid and bromcresol green), *Prostaglandins*, 10, 633—639, 1975.

185. **Bito, L. Z.**, Inhibition of renal prostaglandin metabolism and excretion by probenecid, bromcresol green and indomethacin, *Prostaglandins*, 12, 639—646, 1976.

186. **Spagnuold, C., Petroni, A., Blasevich, M., and Galli, C.**, Differential effects of probenecid on the levels of endogeous $PGF_{2\alpha}$ and TXB_2 in brain cortex, *Prostaglandins*, 15, 311—315, 1979.

187. **Petrono, A., Socini, A., and Galli, C.**, Inhibition of fatty acid gastro-intestinal absorption in-vivo in the rat by parenteral administration of probenecid, *Prog. Lipid Res.*, 20, 815—817, 1982.

188. **Uehara, N., Ormstad, K., Orrenius, S., Orning, L., and Hammarström, S.**, Active transport of leukotrienes into rat hepatocytes, in *Advances in Prostaglandin, Thromboxane, and Leukotriene Research*, Vol. 11, Samuelsson, B., Paoletti, R., and Ramwell, P., Eds., Raven Press, New York, 1983, 147—150.

189. **Dawson, C. A., Linehan, J. M., Rickaby, D. A., and Reorig, D. L.**, Influence of plasma protein on the inhibitory effects of indocyanine green and bromcresol green on pulmonary prostaglandin E_1 extraction, *Br. J. Pharmacol.*, 81, 449—455, 1984.

190. **Ferreira, S. H. and Vane, J. R.**, Prostaglandins: their disappearance from and release into the circulation, *Nature (London)*, 216, 868—873, 1967.

191. **Jackson Roberts, L., II, Sweetman, B. J., Maas, R. L., Hubbard, W. C., and Oates, J. A.**, Clinical applications of PG and Tx metabolite quantification, *Prog. Lipid Res.*, 20, 117—121, 1982.

192. **Lands, W. E. M.**, The biosynthesis and metabolism of prostaglandins, *Ann. Rev. Physiol.*, 41, 633—652, 1979.

193. **Samuelsson, B., Granström, E., Green, K., Hamberg, M., and Hammarström, S.**, Prostaglandins, *Ann. Rev. Biochem.*, 44, 669—695, 1975.

194. **Hamberg, M. and Samuelsson, B.**, On the metabolism of prostaglandins E_1 and E_2 in man, *J. Biol. Chem.*, 246, 6713—6721, 1971.

195. **Granström, E. and Samuelsson, B.**, On the metabolism of prostaglandin $F_{2\alpha}$ in female subjects, *J. Biol. Chem.*, 246, 5254—5263, 1971.

196. **Granström, E.**, On the metabolism of prostaglandin $F_{2\alpha}$ in female subjects. Structures of two metabolites in blood, *Eur. J. Biochem.*, 27, 462—469, 1972.

197. **Seyberth, H. W., Oelz, O., Kennedy, T., Sweetman, B. J., Danon, A., Frolich, J. C., Heimberg, M., and Oates, J. A.,** Increased arachidonate in lipids after administration to man: effects on prostaglandin biosynthesis, *Clin. Pharmacol. Ther.*, 18, 521—529, 1975.

198. **Rosenkranz, B., Fischer, C., Reimann, I., Weimer, K. E., Beck, G., and Frölich, J. C.,** Identification of the major metabolite of prostacyclin and 6-keto-PGF$_{1\alpha}$ in man, *Biochim. Biophys. Acta*, 619, 207—213, 1980.

199. **Quilley, C. P., McGiff, J. C., Lee, W. H., Sun, F. F., and Wong, P. Y.-K.,** 6-Keto PGE$_1$: a possible metabolite of prostacyclin having platelet anti-aggregating effects, *Hypertension*, 2, 524—528, 1980.

200. **Wong, P. Y.-K., Lee, K. H., Chao, P. M.-W., Reiss, R. F., and McGiff, J. C.,** Metabolism of prostacyclin by 9-hydroxyprostaglandin dehydrogenase in human platelets, *J. Biol. Chem.*, 255, 9021—9024, 1980.

201. **Hoult, J. R. S., Lofts, F. J., and Moore, P. K.,** Stability of prostacyclin in plasma and its transformation by platelet to a stable spasmogenic product, *Br. J. Pharmacol.*, 73, 2189, 1981.

202. **Gimeno, M. F., Sterin-Borda, L., Borda, S., Lazzari, M., and Gimeno, A. L.,** Human plasma transforms prostacyclin (PGI$_2$) into a platelet anti-aggregatory substance which contracts isolated bovine coronary arteries, *Prostaglandins*, 19, 907—915, 1980.

203. **Lofts, F. J. and Moore, P. K.,** Release of a 6-oxoprostaglandin E$_1$-like substance from human platelets, *Clin. Sci.*, 64, 63—68, 1983.

204. **Lewis, R. A., Lee, C. W., Levine, L., Morgan, R. A., Weiss, J. W., Drazen, J. M., Oh, H., Hoover, D., Corey, E. J., and Austen, K. F.,** Biology of the C-6-sulfidopeptide leukotrienes, in *Advances in Prostaglandin, Thromboxane, and Leukotriene Research*, Vol. 11, Samuelsson, B., Paoletti, R., and Ramwell, P., Eds., Raven Press, New York, 1983, 15—20.

205. **Middleditch, B. S.,** PGA: fact or artifact?, *Prostaglandins*, 9, 409—411, 1975.

206. **Jonsson, H. T., Jr. and Powers, R. E.,** Endogenous prostaglandin As (PGAs): fact or artifact in biological systems, *Prog. Lipid Res.*, 20, 787—790, 1983.

207. **Smith, J. B., Kocsis, J. J., Ingerman, C., and Silver, M. J.,** Inactivation of prostaglandin A$_1$ and A$_2$ by human red cells, *Pharmacologist*, 15, 208, 1973.

208. **Oien, H. G., Ham, E. A., Zanetti, M. E., Ulm, E. H., and Kuehl, T. A., Jr.,** A 15-hydroxy-prostaglandin dehydrogenase specific for prostaglandin A in rabbit kidney, *Proc. Natl. Acad. Sci. U.S.A.*, 73, 1107—1111, 1976.

209. **Paulsrud, J. R. and Miller, O. N.,** Inhibition of 15-OH prostaglandin dehydrogenase by several diuretic drugs, *Fed. Proc.*, 33, 590, 1974.

210. **Thaler-Dao, H., Saintot, M., Baudin, B., Descomps, B., and Crastes de Paulet, A.,** Purification of the human placental 15-hydroxy prostaglandin dehydrogenase properties of the purified enzyme, *FEBS Lett.*, 48, 204—208, 1974.

211. **Stone, K. J. and Hart, M.,** Inhibition of renal PGE$_2$-9-ketoreductase by diuretics, *Prostaglandins*, 12, 197—207, 1976.

212. **Hensby, C. N.,** Distribution studies on the reduction of prostaglandin E$_2$ to prostaglandin F$_{2\alpha}$ by tissue homogenates, *Biochim. Biophys. Acta*, 409, 225—234, 1975.

213. **Kröner, E. E. and Peskar, B. A.,** On the metabolism of prostaglandins by rat brain homogenate, *Experientia*, 32, 1114—1115, 1976.

214. **Ziboh, V. A., Lord, J. T., and Penneys, N. S.,** Alterations of prostaglandin E$_2$-9-ketoreductase activity in proliferating skin, *J. Lipid Res.*, 18, 37—43, 1977.

215. **Stone, K. J. and Hart, M.,** Prostaglandin E$_2$-9-ketoreductase in rabbit kidney, *Prostaglandins*, 10, 273—288, 1975.

216. **Lee, S. C. and Levine, L.,** Purification and regulatory properties of chicken heart prostaglandin E 9-ketoreductase, *J. Biol. Chem.*, 250, 4549—4555, 1975.

217. **Abe, K., Yasujima, M., Chiba, S., Irokawa, N., Ito, T., and Yoshinaga, K.,** Effect of furosemide on urinary excretion of prostaglandin in normal volunteers and patients with essential hypertension, *Prostaglandins*, 14, 513—521, 1977.

218. **Ciabattoni, G., Pugliese, F., Cinotti, G. A., Stirati, C., Ronci, R., Castrucci, G., Pierucci, A., and Patrono, G.,** Characterization of furosemide-induced activation of the renal prostaglandin system, *Eur. J. Pharmacol.*, 60, 181—187, 1979.

219. **Friedman, Z., Demers, L. M., Marks, K. H., Uhrmann, S., and Maisels, M. J.,** Urinary excretion of prostaglandin E following the administration of furosemide and indomethacin to sick low-birth weight infants, *J. Pediatr.*, 93, 512—515, 1978.

220. **Gerber, J. G., Anderson, R. J., Schrier, R. W., and Nies, A. S.,** Prostaglandins and the regulation of renal circulation and function, in *Prostaglandins and the Cardiovascular System*, Oates, J. A., Ed., Raven Press, New York, 1982, 227—254.

221. **Gerber, J. G., Hubbard, W. C., Branch, R. A., and Nies, A. S.,** The lack of an effect of furosemide on uterine prostaglandin metabolism *in vitro*, *Prostaglandins*, 15, 663—670, 1978.

222. **Harris, R. H., Ramwell, P. W., and Gilmer, P. J.,** Cellular mechanisms for prostaglandin action, *Ann. Rev. Physiol.,* 41, 653—668, 1979.
223. **Marx, J. L.,** A new view of receptor action, *Science,* 224, 271—274, 1984.
224. **Nishizuka, Y.,** The role of protein kinase C in cell surface signal transduction and tumour promotion, *Nature (London),* 308, 693—698, 1984.
225. **Eakins, K. E., Miller, J. D., and Karim, S. M.,** The nature of the prostaglandin-blocking activity of polyphloretin phosphate, *J. Pharmacol. Exp. Ther.,* 176, 441—447, 1971.
226. **Sanner, J. H.,** Substances that inhibit the actions of prostaglandins, *Arch. Intern. Med.,* 133, 133—146, 1974.
227. **Eakins, K. E. and Sanner, J. H.,** in the Prostaglandins Progress in Research, Karim, S. M. M., Ed., MTP, Oxford, 1972, 263—292.
228. **Morris, H. R., Taylor, G. W., Jones, C. M., Sculley, N., Piper, P. J., Tippins, J. R., and Samhoun, M. N.,** Structure elucidation and biosynthesis of slow reacting substances and slow reacting substance of anaphylaxis from guinea pig and human lung, *Prog. Lipid Res.,* 20, 719—725, 1981.
229. **Augstein, S., Farmer, J. B., Lee, T. B., Sheard, P., and Tattersall, M. L.,** Selective inhibitor of slow reacting substance of anaphylaxis, *Nature (London) New Biol.,* 245, 215—217, 1973.
230. **Burke, J. A., Levi, R., and Gleason, J. G.,** Antagonism of the cardiac effects of leukotriene C_4 by compound SKF-55046: dissociation of effects on contractility and coronary flow, *J. Cardiovasc. Pharmacol.,* 6, 122—125, 1984.
231. **Nishino, K., Ohkubo, H., Ohashi, M., Hara, S., Kito, J., and Irikura, T.,** KC-404: a potential anti-allergic agent with antagonistic action against slow reacting substance of anaphylaxis, *Jpn. J. Pharmacol.,* 33, 267—278, 1983.
232. **Schrör, K., Smith, E. F., III, Bickerton, M., Smith, J. B., Nicolaou, K. C., Magolda, R., and Lefer, A. M.,** Preservation of ischemic myocardium by pinane thromboxane A_2, *Am. J. Physiol.,* 238, H87—H92, 1980.
233. **Lefer, A. M., Araki, H., Smith, J. B., Nicolaou, K. C., and Magolda, R. L.,** Protective effects of a novel thromboxane analog in lethal traumatic shock, *Prostaglandins Med.,* 3, 139—146, 1979.
234. **Mills, D. C. B. and MacFarlane, D. E.,** in *Prostaglandins in Hematology,* Silver, M. J., Smith, J. B., and Kocsis, J. J., Eds., Spectrum Publications, Jamaica, N.Y., 1977, 219—233.
235. **Amer, M. S. and Kreighbaum, W. E.,** Cyclic nucleotide phosphodiesterases: properties, activators, inhibitors, structure-activity relationships and possible role in drug development, *J. Pharm. Sci.,* 64, 1—37, 1975.
236. **Alvarez, R., Taylor, A., Fazzari, J. J., and Jacobs, J. R.,** Regulation of cyclic AMP metabolism in human platelets. Sequential activation of adenylate cyclase and cyclic AMP phosphodiesterase by prosta-glandins, *Mol. Pharmacol.,* 20, 302—309, 1981.
237. **Moncada, S., Korbut, R., Bunting, S., and Vane, J. R.,** Prostacyclin is a circulating hormone, *Nature (London),* 273, 767—768, 1978.
238. **Moncada, S. and Korbut, R.,** Dipyridamole and other phosphodiesterase inhibitors act as antithrombotic agents by potentiating endogenous prostacyclin, *Lancet,* 1, 1286—1289, 1978.
239. **Pace-Asciak, C. R., Carrara, M. C., and Levine, L.,** PGI_2 is not a circulating vasodepressor hormone, *Prog. Lipid Res.,* 20, 113—116, 1982.
240. **Haslam, R. J. and McClenachan, M. D.,** Measurement of circulating prostacyclin, *Nature (London),* 292, 364—366, 1981.
241. **Fisher, J. M., Willis, A. L., Smith, D. L., and Donegan, D.,** Lack of circulating prostacyclin (PGI_2) in guinea pig and rat, *Thrombosis Hemostasis,* 46, 83, 1981.
242. **Haeusler, G. and Gerold, M.,** Increased levels of prostaglandin-like material in canine blood during arterial hypotension produced by hydralazine, dihydralazine and minoxodil, *Arch. Pharmacol.,* 310, 155—167, 1979.
243. **Rivers, J. P. W., Sinclair, A. J., and Crawford, M. A.,** Inability of the cat to desaturate essential fatty acids, *Nature (London),* 258, 171—173, 1975.
244. **Brenner, R. R.,** Nutritional and hormonal factors influencing desaturation of essential fatty acids, *Prog. Lipid Res.,* 20, 41—47, 1982.
245. **Dyerberg, J. and Bang, H. O.,** Haemostatic function and platelet polyunsaturated fatty acids in Eskimos, *Lancet,* 2, 433—435, 1979.
246. **Bang, H. O., Dyerberg, and Hjorne, N.,** The composition of food consumed by Greenland Eskimos, *Acta Med. Scand.,* 200, 69—73, 1976.
247. **Needleman, P., Whitaker, M. O., Wyche, A., Watters, K., Sprecher, H., and Raz, A.,** Manipulation of platelet aggregation by prostaglandins and their fatty acid precursors: pharmacological basis for a ther-apeutic approach, *Prostaglandins,* 19, 165—181, 1980.
248. **Fischer, S. and Weber, P. C.,** Prostaglandin I_3 is formed in vivo in man after dietary eicosapentaenoic acid, *Nature (London),* 307, 165—168, 1984.

249. **Willis, A. L.,** Nutritional and pharmacological factors in eicosanoid biology, *Nutr. Rev.,* 39, 289—301, 1981.

250. **German, B., Bruckner, G., and Kinsella, J.,** Evidence against a $PGF_{4\alpha}$ prostaglandin structure in trout tissue — a correction, *Prostaglandins,* 26, 207—210, 1983.

251. **Corey, E. J., Shih, C., and Cashman, J. R.,** Docosahexaenoic acid is a strong inhibitor of prostaglandin but not leukotriene biosynthesis, *Proc. Natl. Acad. Sci. U.S.A.,* 80, 3581—3584, 1983.

252. **Iacono, J. M., Judd, J. T., Marshall, M. W., Canary, J. J., Dougherty, R. M., Machlin, J. F., and Weinland, B. T.,** The role of dietary essential fatty acids and prostaglandins in reducing blood pressure, *Prog. Lipid Res.,* 20, 349—364, 1982.

253. **Singer, P., Jaeger, W., Wirth, M., Voigt, S., Naumann, E., Zimontkowski, S., Hajdu, I., and Goedicke, W.,** Lipid and blood pressure lowering effect of mackerel diet in man, *Atherosclerosis,* 49, 99—108, 1983.

254. **Hornstra, G., Lewis, B., Chait, A., Turpeinen, O., Karvonen, M. J., and Vergroesen, A. J.,** Influence of dietary fat on platelet function in man, *Lancet,* 1, 1155—1157, 1973.

255. **Bradlow, B. A., Chetty, N., Van Der Westhuyzen, J., Mendelsohn, D., and Gibson, J. E.,** The effects of a mixed fish diet on platelet function, fatty acids and serum lipids, *Thromb. Res.,* 29, 501—505, 1983.

256. **Sanders, T. B., Vickers, M., and Haines, A. P.,** Effect on blood lipids and hemostasis of a supplement of cod liver oil, rich in eicosapentaenoic and docosahexaenoic acids, in healthy young men, *Clin. Sci.,* 61, 317—324, 1981.

257. **Goodnight, S. H., Harris, W. S., Connor, W. E., and Illingworth, D. R.,** Polyunsaturated fatty acids, hyperlipidemia and thrombosis, *Arteriosclerosis,* 2, 87—113, 1982.

258. **Karim, S. M. M.,** Once a month vaginal administration of prostaglandins E_2 and $F_{2\alpha}$ for fertility control, *Contraception,* 3, 173, 1971.

259. **Karim, S. M. M. and Sharma, S. D.,** Therapeutic abortion and induction of labor by the intravaginal administration of prostaglandins E_2 and $F_{2\alpha}$, *J. Obstet. Gynaecol. Br. Common.,* 78, 294—300, 1971.

260. **Karim, S. M. M., Sharma, S. D., Filshie, G. M., Salmon, J. A., and Ganesan, P. A.,** Termination of pregnancy with prostaglandin analogs, *Adv. Biosci.,* 9, 811—830, 1973.

261. **Herschler, R. C.,** Estrus synchronization and conception rates in beef heifers using fenprostalene in both single and double injection programs, *Agri-Practice,* 4, 28—31, 1983.

262. **Copeland, D. D., Schultz, R. H., and Kemtrup, M. E.,** Induction of abortion in feedlot heifers with cloprostenol (a synthetic analogue of prostaglandin $F_{2\alpha}$): a dose response study, *Can. Vet. J.,* 19, 29—32, 1978.

263. **Friedman, W. F., Printz, M. P., Skidgel, R. A., Benson, L. N., and Zednikova, M.,** Prostaglandins and the ductus arteriosus, in *Prostaglandins and the Cardiovascular System,* Oates, J. A., Ed., Raven Press, New York, 1982, 277—302.

264. **Carlson, L. A. and Olsson, A. G.,** Intravenous prostaglandin E_1 in severe peripheral vascular disease, *Lancet,* 2, 810, 1976.

265. **Szczeklik, A., Nizankowski, R., Shawinsk, S., Szczeklik, J., Gluszko, P., and Gryglewski, R. J.,** Successful therapy of advanced arteriosclerosis obliterans with prostacyclin, *Lancet,* 1, 111, 1979.

266. **Pardy, B. J., Hoare, M. C., Eastcott, H. H. G., Miles, C., Needham, T. N., Harbourne, T., and Ellis, B. W.,** Prostaglandin E_1 in severe Raynaud's phenomenon, *Surgery,* 92, 953—965, 1982.

267. **Dowd, P. M., Martin, M. F. R., Cooke, E. D., Bowcock, S. A., Jones, R., Dieppe, P. A., and Kirby, J. D. T.,** Therapy of Raynaud's phenomenon by intravenous infusion of prostacyclin (PGI_2), *Br. J. Dermatol.,* 106, 81—89, 1982.

268. **Bristol, J. A. and Long, J. F.,** Agents for the treatment of peptic ulcer disease, *Annu. Rep. Med. Chem.,* 16, 83—91, 1981.

269. **Garay, G. L., and Muchowski, J. M.,** Agents for the treatment of peptic ulcer disease, *Annu. Rep. Med. Chem.,* 20, 93—105, 1985.

270. **Birnbaum, J. E., Chan, P. S., Cervoni, P., Dessy, F., and Van Humbeeck, L.,** Cutaneous erythema and blood pressure lowering effects of topically applied 16-vinyl prostaglandins, *Prostaglandins,* 23, 185—199, 1982.

271. **Adaikan, P. G., Lau, L. C., Tai, M. Y., and Karim, S. M. M.,** Inhibition of platelet aggregation with intravenous and oral administration of a carboprostacyclin analogue, 15-cyclopentyl-ω-pentanor-5(E) ω-bacyclin (ONO 41483) in man, *Prostaglandins, Leukotrienes, Med.,* 10, 53—64, 1983.

272. **Tsubui, T., Matano, N., Nakatsuji, K., Fujitani, B., Yoshida, K., Shimizu, M., Kawasaki, A., Sakata, M., and Tsuboshima, M.,** Pharmacological evaluation of OP 1206, a prostaglandin E_1 derivative, as an anti-anginal agent, *Arch. Int. Pharmacodyn.,* 247, 89—102, 1980.

273. **Kottegoda, S. R., Adaikan, P. G., and Karim, S. M. M.,** Reversal of vasopressin-induced coronary vasoconstriction by a PGE_1 analogue (ONO 1206) in primates, *Prostaglandins, Leukotrienes, Med.,* 8, 343—348, 1982.

274. Drugs of the future, 7, 116, 1982.

275. **Smith, E. F., III, Gallenkämper, W., Beckman, R., Thomsen, T., Mannesmann, G., and Schrör, K.,** Early and late administration of a PGI$_2$-analogue, ZK36374 (Iloprost): effects on myocardial preservation, collateral blood flow and infarct size, *Cardiovasc. Res.*, 18, 163—173, 1984.

276. **Lefer, A. M. and Smith, E. F., III,** Protective action of prostacyclin in myocardial ischemia and trauma, in *Prostacyclin*, Vane, J. R. and Bergström, S., Eds., Raven Press, New York, 1979, 339—348.

277. **Pace-Asciak, C. and Wolfe, L. S.,** A novel prostaglandin derivative formed from arachidonic acid by rat stomach homogenates, *Biochemistry*, 10, 3657—3664, 1971.

278. **Pace-Asciak, C.,** Isolation, structure, and biosynthesis of 6-ketoprostaglandin F$_{1\alpha}$ in the rat stomach, *J. Am. Chem. Soc.*, 98, 2348—2349, 1976.

279. **Pace-Asciak, C. R., Granstrom, E., and Samuelsson, B.,** Isolation and structure of two hydroxy-epoxy intermediates in the formation of 8,11,12 and 10,11,12-trihydroxyeicosatrienoic acid acids, *J. Biol. Chem.*, 258, 6835, 1983.

280. **Pace-Asciak, C. R. and Martin, J. M.,** Hepoxilin, a new family of insulin secretogogues formed by intact rat pancreatic islets, *Prostaglandins, Leukotrienes, Med.*, 16, 173—180, 1984.

281. **Pace-Asciak, C. R., Martin, J. M., Corey, E. J., and Su, W. G.,** Endogenous release of hepoxilin A$_3$ from isolated perfused pancreatic islets of langerhans, *Biochem. Biophys. Res. Commun.*, 128, 942—946, 1985.

282. **Holman, R. T.,** Personal communication, 1984.

283. **Bhagwat, S. S., Hamann, P. R., Still, W. C., Bunting, S., and Fitzpatrick, F. A.,** Synthesis and structure of the platelet aggregation factor thromboxane A$_2$, *Nature*, 315, 511—513, 1985.

284. **Samuelsson, B., Hamberg, M., Jackson-Roberts, L., II, Oates, J. A., and Nelson, N. A.,** Nomenclature for thromboxanes, *Prostaglandins*, 16, 857—860, 1978.

285. **Wallner, B. P., Mattaliano, R. J., Hession, C., Cate, R. L., Tizard, R., Sinclair, L. K., Foeller, C., Pingchang Chow, E., Browning, J. L., Ramachandran, K. L., and Pepinsky, R. B.,** Cloning and expression of human lipocortin, a phospholipase A$_2$ inhibitor with potential anti-inflammatory activity, *Nature*, 320, 77—81, 1986.

286. **Ackerman, N. and Willis, A. L.,** Unpublished observations, 1982.

287. **Moore, P. K. and Hoult, J. R. S.,** Selective action of aspirin- and sulphasalazine-like drugs against prostaglandin synthesis and breakdown, *Biochem. Pharmacol.*, 31, 969—971, 1982.

288. **Patrono, C., Ciabattoni, G., Filabozzi, P., Catella, F., Forni, L., Segni, M., Patrignani, P., Pugliese, F., Simonetti, B. M., and Pierucci, A.,** Drugs prostaglandins, and renal function, in *Advances in Prostaglandins, Thromboxanes, and Leukotriene Research*, Vol. 13, Neri Serneri, G. G., McGiff, J. C., Paoletti, R., and Born, G. V. R., Eds., Raven Press, New York, 1985, 131—139.

289. **Musser, J. H., Kreft, A. F., and Lewis, A. J.,** New developments concerning leukotriene antagonists: a review, *Agents and Actions*, 18, 332—341, 1986.

290. **Dunn, M. J. and Grone, H. J.,** The relevance of prostaglandins in human hypertension, in *Advances in Prostaglandin, Thromboxane, and Leukotriene Research*, Vol. 13, Neri Serneri, G. G., McGiff, J. C., Paoletti, R., and Born, G. V. R., Eds., Raven Press, New York, 1985, 179—197.

291. **Herold, P. M. and Kinsella, J. E.,** Fish oil consumption and decreased risk of cardiovascular disease: a comparison of findings from animal and human feeding trials, *Am. J. Clin. Nutr.*, 43, 566—598, 1986.

292. **Schwartz, K., Zaro, B., Burton, P., Hunt, J., and Sevelius, H.,** Effects of enprostil, an oral PGE2 analog on lipoprotein profiles in normocholesteremic volunteers, *Circulation*, 74, Suppl.II,201, Abs.802, 1986.

293. **Yardumian, D. A., Mackie, I. J., Bull, H., and Machin, S. J.,** Platelet responses observed during and after infusions of the prostacyclin analog ZK 36374, in *Advances in Prostaglandin, Thromboxane, and Leukotriene Research*, Vol. 13, Neri Serneri, G. G., McGiff, J. C., Paoletti, R., and Born, G. V. R., Raven Press, New York, 1985, 359—369.

294. **Coleman, R. A., Humphrey, P. P. A., Kennedy, I., and Lumley, P.,** Prostanoid receptors — the development of a working classification, *Trends in Pharmac. Sci.*, 5, 303—306, 1984.

295. **Peach, M. J., Singer, H. A., and Loeb, A. L.,** Mechanisms of endothelium-dependent vascular smooth muscle relaxation, *Biochem. Pharmacol.*, 34, 1867—1874, 1985.

296. **Murad, F.,** Cyclic guanosine monophosphate as a mediator of vasodilatation, *J. Clin. Invest.*, 78, 1—5, 1986.

297. **Ignarro, L. J. and Kadowitz, P. J.,** The pharmacological and physiological role of cyclic GMP in vascular smooth muscle relaxation, *Ann. Rev. Pharmacol. Toxicol.*, 25, 171—191, 1985.

298. **Gilman, A. G.,** G proteins and dual control of adenylate cyclase, *Cell*, 36, 577—579, 1984.

299. **Merritt, J. E., Taylor, C. W., Rubin, R. P., and Putney, J. W., Jr.,** Evidence suggesting that a novel guanine nucleotide regulatory protein couples receptors to phospholipiase C in exocrine pancreas, *Biochem. J.*, 236, 337—343, 1986.

300. **Abdel-Latif, A. A.,** Calcium-mobilizing receptors, polyphosphoinosides, and the generation of second messengers, *Pharmacol. Rev.*, 38, 227—272, 1986.

301. **Benveniste, J.,** Paf-acether (platelet-activating factor), in *Advances in Prostaglandins, Thromboxanes, and Leukotriene Research,* Vol. 13, Neri Serneri, G. G., McGiff, J. C., Paoletti, R., and Born, G. V. R., Raven Press, New York, 1985, 11—18.

302. **Ramesha, C. S. and Pickett, W. C.,** Platelet-activating factor and leukotriene biosynthesis is inhibited in polymorphonuclear leukocytes depleted of arachidonic acid, *J. Biol. Chem.,* 261, 7592—7595, 1986.

303. **Woodward, D. L., Rollins, D. E., Krueger, G., and Harris, B. J.,** A dose-ranging study of viprostol, a topically active synthetic prostaglandin, to assess the effects of cutaneous blood flow, *6th International Conference on Prostaglandins and Related Compounds,* Abstract Book, p. 437, 1986.

304. **Smith, D. L., Willis, A. L., Nguyen, N., Dave, S., Yih, R., and Nakamura, C.,** Anti-atherosclerotic properties of prostacyclins on release of mitogens from platelets and endothelial cells, in *Atherosclerosis VII: Excerpta Medica Int. Cong. Ser.,* Vol. 696, (Edited by Fidge, N. H. and Nestel, P. J., Eds., Elsevier, Amsterdam, 1986, 453—456.

305. **Willis, A. L., Smith, D. L., Vigo, C., and Kluge, A. F.,** Effects of prostacyclin and orally active mimetic agent RS93427-007 on basic mechanisms of atherosclerosis, *Lancet,* ii, 682—683, 1986.

306. **Willis, A. L., Smith, D. L., Vigo, C., Kluge, A., O'Yang, C., Kertesz, D., and Wu, H.,** The orally active prostacyclin-mimetic, RS93427: therapeutic potential in vascular occlusive disease associated with atherosclerosis, in *Prostaglandins and Related Compounds,* Samuelsson, B., Ed., Advances in Prostaglandin, Thromboxane and Leukotriene Research, Vol. 17, 1987, 254—265.

307. **Willis, A. L., Smith, D. L., and Vigo, C.,** Suppression of principal atherosclerotic mechanisms by prostacyclins and other eicosanoids, *Prog. Lipid Res.,* 25, 645, 1986.

308. **Hajjar, D. P.,** Prostaglandins and cyclic nucleotides. Modulators of arterial cholesterol metabolism, *Biochem. Pharmacol.,* 34, 295—300, 1985.

CYCLOOXYGENASE AND LIPOXYGENASE PRODUCTS: A COMPENDIUM

Donald L. Smith

The figures presented here review the chemical structure of known or presumed cyclooxygenase and lipoxygenase products of arachidonic acid (20:4ω6; 5-*cis*,8-*cis*,11-*cis*,14-*cis*-eicosatetraenoic acid) (Figures 1 and 6); dihomo-γ-linolenic acid (DGLA; 20:3ω6; 8-*cis*,11-*cis*,14-*cis*-eicosatrienoic acid) (Figures 2 and 7); adrenic acid (22:4ω6; 7-*cis*-10-*cis*,13-*cis*-16-*cis*-docosatetraenoic acid) (Figures 3 and 8); eicosapentaenoic acid (EPA; 20:5ω3; 5-*cis*,8-*cis*,11-*cis*,14-*cis*,17-*cis*-eicosapentaenoic acid) (Figures 4 and 9); Mead acid (20:3ω9; 5-*cis*,8-*cis*,11-*cis*-eicosatrienoic acid) (Figures 5 and 10) and docosahexaenoic acid (22:6ω3; 4-*cis*,7-*cis*,10-*cis*,13-*cis*,16-*cis*,19-*cis*-docosahexaenoic acid) (Figure 11). The pathways shown are also summarized in Figure 3 in the first chapter of this volume.

Below each structure is shown an abbreviated name. Wherever possible, a systematic approach has been used in assigning abbreviated names. For prostaglandins, standard "PG" notation and A through J ring designation are already widely used, as is the subscript denoting number of double bonds in the side chains. "TXA" and "TXB" notations for thromboxanes and "LT" A through F sulfidopeptide designations for leukotrienes are similarly well established.

For some other products shown here, names prefixed with positional number* followed by "H", "Hp", or "Ep" denote (respectively) a hydroxy, hydroperoxy, or epoxy group. Similarly, "DH" denotes dihydroxy, "TH" denotes trihydroxy, etc. The middle part of the abbreviated name denotes chain length and number of double bonds. For example "ET" is an abbreviated version of "eicosatetraenoic", "DT" means "docosatetraenoic", "ETr" means "eicosatrienoic", "DH" means "docosahexaenoic", "ED" means "eicosadienoic", "HTr" means "heptadecatrienoic", "NTr" means "nonadecatrienoic" etc. The final "E" in the name is the usual "enoic acid" designation. Newer classes such as the "hepoxilins", "trioxilins", "lipoxins", and "lipoxenes" are shown here with names reflecting these new designations. In many cases, alternative names or previously used names are given in the figure legends.

Compounds that have not been actually isolated but are strongly presumed to exist by analogy to products of closely related substrates (such as monooxygenase derived epoxides of DGLA or EPA) are denoted by an asterisk.

Parentheses around a structure denotes uncertainty in double bond geometry and structures in square brackets are presumed biosynthetic intermediates to other products shown. A dashed line is used to denote pathways that are uncertain or where essential intermediates have not been determined.

For purposes of organization of products into separate figures, the categories of lipoxygenase enzymes (5-lipoxygenase, 12-lipoxygenase, 15-lipoxygenase etc.) refer to the enzymes that catalyze at least the first step in a given pathway (e.g. arachidonic acid to 15-HpETE). Subsequent steps may be catalyzed by different enzymes. In the case of products that can be formed by alternative routes (e.g. 5,15-DHpETE from initial catalysis by either 5- or 15-lipoxygenase) reference to both pathways is noted. As another example, it has recently been found that after initial formation of 15-HpETE, action by a 12-lipoxygenase can produce 14,15-LTA₄ and 14,15-DHpETE (by acting as a 14-lipoxygenase) and 8,15-DHpETE (by acting as an 8-lipoxygenase).[140,141,174] Also, recent evidence suggests that cellular 14,15-DHETE might not derive from the epoxide.[141]

---

* Position of hydroxy (and other functional groups) are denoted by single numbers, counting from the carboxyl end, separated by commas (e.g., 5-15-DHETE). Where there is a bridge between two carbons (e.g., an epoxy group), the number of the second carbon is shown in parentheses (e.g., the monooxygenase product 5(6)-EpETrE).

An exception to the formation of eicosanoids by cyclooxygenase and lipoxygenase pathways is the NADPH-dependent formation of the four regiospecific *cis* epoxytrienoic acids (EpETrEs) from arachidonic acid, catalyzed by the cytochrome P-450 monooxygenase (epoxygenase) pathway.[34,35] It is of interest that 8(9)- and 14(15)-EpETrEs are cyclooxygenase inhibitors.[175]

Although this chapter is devoted to currently known products of oxygenation of these six polyunsaturated fatty acids, it has recently been reported that cyclooxygenase or lipoxygenase products of other polyunsaturated fatty acids can be produced. These include products of 6,9,12-octadecatrienoic acid ($18:3\omega6$),[123] $\omega3$-eicosatetraenoic acid ($20:4\omega3$);[124] $\omega3$-docosapentaenoic acid ($22:5\omega3$);[125] $\omega6$-docosapentaenoic acid ($22:5\omega6$);[126] columbinic acid ($18:3\omega5$),[127] and linoleic acid ($18:2\omega6$).[165-169] Significance of these findings is further discussed in the chapter in this volume by Mead and Willis.

As a further addendum to the products shown in these figures, several novel eicosanoids have only very recently been described. Among these are eicosanoids that appear to be unique to psoriatic lesions, such as 2,9-HETE[173] and 12(R)-HETE.[152] It is also of interest that, in addition to the prostaglandins formed by mammalian systems, some unique eicosanoids have been found in marine organisms. Examples include the clavulones and punaglandins from octacoral[170] and a hatching factor from barnacles.[108]

Other recent studies have suggested that in addition to the dihydroxytetraenoic acids (DHETEs) formed in human platelet mixtures via the 15-lipoxygenase reaction, other DHETEs are formed in platelets via the 12-lipoxygenase pathway.[142] These included four 5,12- and two 11,12-DHETEs with conjugated triene structures.

The wide variety of biological activity expressed by several of the products reviewed here can be seen by a sampling of the references given (particularly the more recent ones) and in several other chapters of this series of volumes. It is very likely that some more eicosanoids will be discovered in the next few years, although it does seem that there is diminishing room for major new discoveries, particularly with regards to those with important biological activity.

It could now also be argued that with the plethora of eicosanoids now known and the numerous possibilities for opposing or synergistic biological actions, a return to biological assays of crude extracts and *then* subsequent identification of the active principals is in order. This is classically the way that all of the important eicosanoids have, in fact, been discovered and seems preferable to initial biochemistry followed by intense efforts to find some "use" for the products identified.

ACKNOWLEDGMENTS

I would like to acknowledge Dr. A. L. Willis and Dr. D. V. K. Murthy for critical review of the information presented here. I would also like to thank Ngoc Nguyen, Elizabeth Jacobs, and Sue Matheson for assistance with layouts and design.

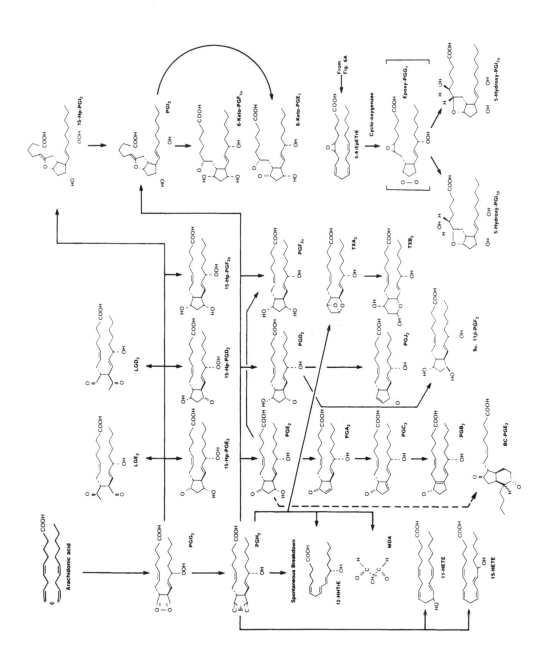

FIGURE 1. Cyclooxygenase products of arachidonic acid. Prostaglandin G$_2$ (PGG$_2$): 15α-Hydroperoxy-9α,11α-peroxidoprosta-5-cis,13-trans-dienoic acid.[47,50] Prostaglandin H$_2$ (PGH$_2$; PGR$_2$; "LASS"): 15α-Hydroxy-9α,11α-peroxidoprosta-5-cis,13-trans-dienoic acid.[47,50] 12-HHTrE (previously called "HHT"): 12(S)-Hydroxy-5-cis,8-trans,10-trans-heptadecatrienoic acid.[73,111] MDA: Malon-dialdehyde.[73] 11-HETE: 11-Hydroxy-5-cis,8-cis,12-trans,14-cis-eicosatetraenoic acid.[3,111,131] 15-HETE: 15-Hydroxy-5-cis,8-cis,11-cis,13-trans-eicosatetraenoic acid.[3] 15-Hydroperoxy-prostaglandin E$_2$ (15-Hp-PGE$_2$): 11α-Hydroxy-15α-hydroperoxy-9-ketoprosta-5-cis,13-trans-dienoic acid.[49] Prostaglandin E$_2$ (PGE$_2$): 11α,15α-Dihydroxy-9-ketoprosta-5-cis,13-trans-dienoic acid.[53] Prostaglandin A$_2$ (PGA$_2$; "medullin"): 15α-Hydroxy-9-ketoprosta-5-cis,10-13-trans-trienoic acid.[55,56] Prostaglandin C$_2$ (PGC$_2$): 15α-Hydroxy-9-ketoprosta-5-cis,11-13-trans-trienoic acid.[58,59] Prostaglandin B$_2$ (PGB$_2$): 15α-Hydroxy-9-ketoprosta-5-cis-8(12),13-trans-trienoic acid.[55,57,99] 15-Hydroperoxy-prostaglandin D$_2$ (15-Hp-PGD$_2$): 9α-Hydroxy-15α-hydroperoxy-11-ketoprosta-5-cis,13-trans-dienoic acid.[49] Prostaglandin D$_2$ (PGD$_2$); once called 11-dehydro-PGF$_{2α}$): 9α,15α-Dihydroxy-11-ketoprosta-5-cis,13-trans-dienoic acid.[47,48] Prostaglandin J$_2$ (PGJ$_2$; 9-Deoxy-Δ^9-PGD$_2$): 15α-Hydroxy-11-ketoprosta-5-cis,9-13-trans-trienoic acid.[63] 15-Hydroperoxy-prostaglandin F$_{2α}$ (15-Hp-PGF$_{2α}$): 9α,11α-Dihydroxy-15α-hydroperoxyprosta-5-cis,13-trans-dienoic acid.[49] Prostaglandin F$_{2α}$ (PGF$_{2α}$): 9α,11α,15α-Trihydroxyprosta-5-cis,13-trans-dienoic acid.[52] Thromboxane A$_2$ (TXA$_2$): 7{[β-(3α-Hydroxy-1-trans-octenyl)]-2,7-dioxabicyclo-(3.1.1)-heptan-4α-yl}-5-cis-heptenoic acid.[67,68] Thromboxane B$_2$ (TXB$_2$; PHD; "polar hydroxylated derivative"): 7{[4α,6β-Dihydroxy-2β-(3α-hydroxy-1-trans-octenyl)]-3α-tetrahydropyranyl}-5-cis-heptenoic acid.[67,68] 15-Hydroperoxyprostacyclin (15-Hp-PGI$_2$): 6(9)-Oxy-11-α-hydroxy-15-α-hydroperoxy-5-trans,13-trans-dienoic acid.[111] Prostacyclin (PGI$_2$; eoprostenol; 5Z-9-Deoxy-6,9α-epoxy-Δ^5-prostaglandin F$_{1α}$): 6,9α-Oxy-11α,15α-dihydroxyprosta-5-cis,13-trans-dienoic acid.[60,62] 6-Keto-prostaglandin F$_{1α}$ (6-keto-PGF$_{1α}$): 9α,11α,15α-Trihydroxy-6-ketoprost-13-trans-enoic acid.[60-62,111] 6-Keto-prostaglandin E$_1$(6-keto-PGE$_1$): 11α,15α-Dihydroxy-6,9-diketoprost-13-trans-enoic acid.[79] 5(6)-EpETrE (5(6)Oxido-20:3): 5(6)-Epoxy-8-cis,11-cis,14-cis-eicosatrienoic acid.[36] 5-Hydroxy-PGI$_{1β}$ and 5-Hydroxy-PGI$_{1α}$: 5(R),6(R)- and 5(S),6(9)-Oxy-5,11,15-Trihydroxyprosta-13-trans-enoic acid.[36] 9α,11β-Prostaglandin F$_2$ (9α,11β-PGF$_2$): 9α,11β,15(S)-Trihydroxyprosta-5-cis,13-trans-dienoic acid.[15c,157] BC-PGE$_2$ (Bicyclo-PGE$_2$): 11-Deoxy-13,14-dih-ydro-15-keto-11β,16-cycloprostaglandin E$_2$.[176,177] LGE$_2$ (Levuglandin E$_2$): 8(R)-Acetyl-9(R)-formyl-12(S)-hydroxy-5-cis,10-trans-heptadecadienoic acid.[135] LGD$_2$ (Levuglandin D$_2$): 8-Formyl,9-acetyl-12(S)-hydroxy-5-cis,10-trans-heptadecadienoic acid.[135] 6,15-Diketo-prostaglandin F$_{1α}$ (6,15-Diketo PGF$_{1α}$): 9α,11α-Dihydroxy-6,15-ketoprost-13-trans-enoic acid (not shown).[111]

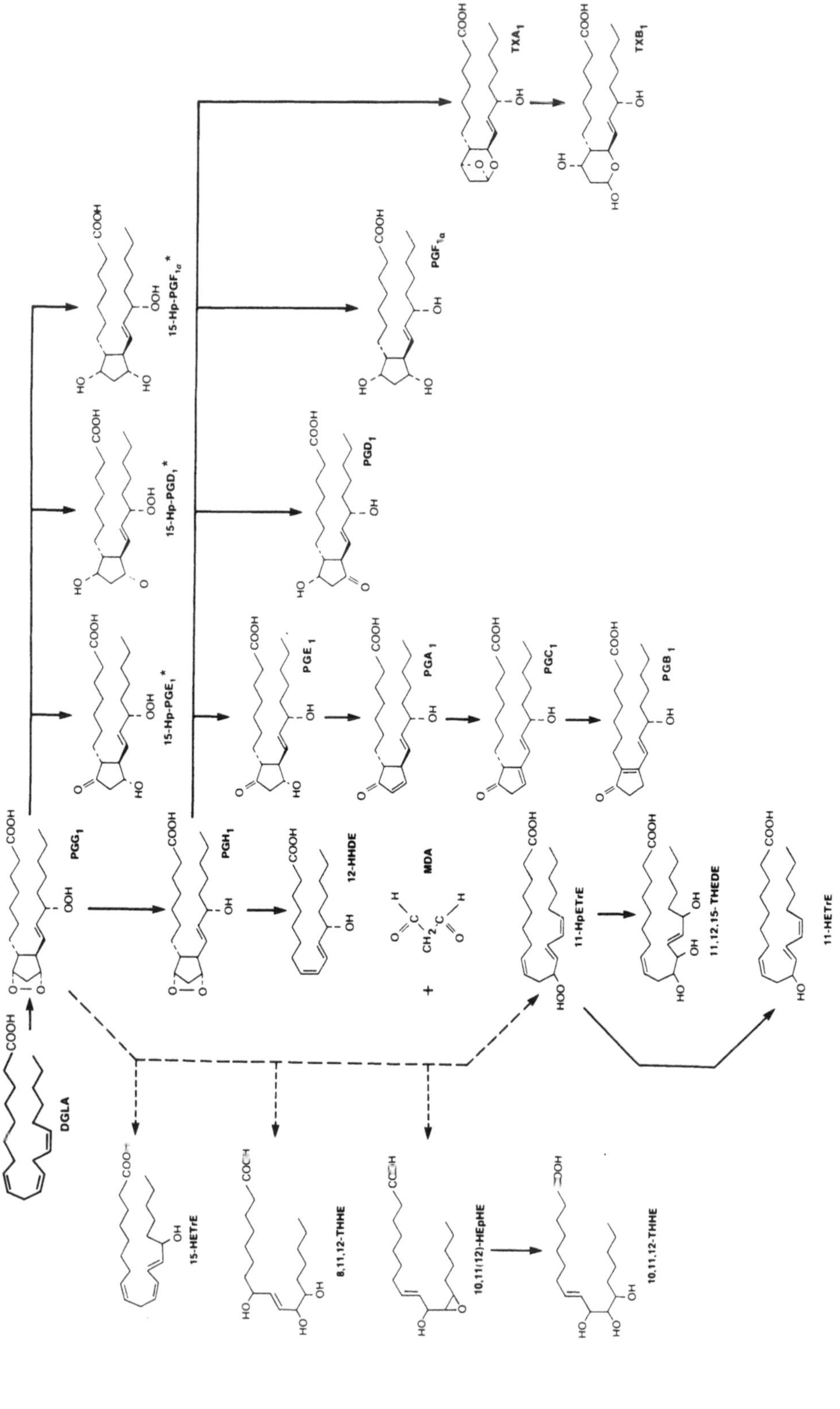
DGLA
PGG1
PGH1
12-HHDE
MDA
11-HpETrE
11,12,15-THEDE
11-HETrE
15-HETrE
8,11,12-THHE
10,11(12)-HEpHE
10,11,12-THHE
15-Hp-PGE1 *
15-Hp-PGD1 *
15-Hp-PGF1α *
PGE1
PGA1
PGC1
PGB1
PGD1
PGF1α
TXA1
TXB1

FIGURE 2. Cyclooxygenase products of dihomo-γ-linolenic acid. Prostaglandin G_1 (PGG_1): 15α-Hydroperoxy-9α,11α-peroxidoprost-13-*trans*-enoic acid.[32,48,49] Prostaglandin H_1 (PGH_1; PGR_1): 15α-Hydroxy-9α,11α-peroxidoprost-13-*trans*-enoic acid.[32,48,50,75] 12-HHDE (previously called "HHD"): 12(S)-Hydroxy-8-*cis*,10-*trans*-heptadecadienoic acid.[32,44,66,73] MDA: Malondialdehyde.[32,66,73] 11-HpETrE (11-Hydroperoxy-20:3): 11-Hydroperoxy-8-*cis*,12-*trans*,14-*cis*-eicosatrienoic acid.[44] 11-HETrE (11-Hydroxy-20:3): 11-Hydroxy-8-*cis*,12-*trans*,14-*cis*-eicosatrienoic acid.[44] 11,12,15-THEDE: 11,12,15-Trihydroxy-8-*cis*,13-*trans*-eicosadienoic acid.[44] 15-Hydroperoxy-prostaglandin E_1 (15-Hp-PGE_1): 11α-Hydroxy-15α-hydroperoxy-9-ketoprost-13-*trans*-enoic acid. Prostaglandin E_1 (PGE_1): 11α,15α-Dihydroxy-9-ketoprost-13-*trans*-enoic acid.[32,51,54,76] Prostaglandin A_1 (PGA_1): 15α-Hydroxy-9-ketoprosta-10-13-*trans*-dienoic acid.[76,77] Prostaglandin C_1 (PGC_1): 15α-Hydroxy-9-ketoprosta-11-13-*trans*-dienoic acid.[78] Prostaglandin B_1 (PGB_1): 15α-Hydroxy-9-ketoprosta-8(12),13-*trans*-dienoic acid.[76,77] 15-Hydroperoxy-prostaglandin D_1 (15-Hp-PGD_1): 9α-Hydroxy-15α-hydroperoxy-11-ketoprost-13-*trans*-enoic acid. Prostaglandin D_1 (PGD_1): 9α,15α-Dihydroxy-11-ketoprost-13-*trans*-enoic acid.[32] 15-Hydroperoxy-prostaglandin $F_{1α}$ (15-Hp-$PGF_{1α}$): 9α,11α,15α-Trihydroxy-prost-13-*trans*-enoic acid. Prostaglandin $F_{1α}$ ($PGF_{1α}$): 9α,11α-Dihydroxy-15-α-hydroperoxy-prost-13-*trans*-enoic acid. Thromboxane A_1 (TXA_1): 7[[3β-(3α-Hydroxy-1-*trans*-octenyl)]-2,7-dioxabicyclo-(3.1.1)-heptan-4α-yl]-heptanoic acid.[69,70] Thromboxane B_1 (TXB_1): 7[[4α,6β-Dihydroxy-2β-(3α-hydroxy-1-*trans*-octenyl)]-3α-tetrahydropyranyl]-heptanoic acid.[32,69,70] 15-HETrE (15-OH-20:3): 15-Hydroxy-8-*cis*,11-*cis*,13-*trans*-eicosatrienoic acid.[44] 8,11,12-THHE (8,11,12-Trihydroxy-9-*trans*-heptadecenoic acid.[44] 10,11(12)-HEpHE (10-OH-11 (12)epoxy-17:1): 10-Hydroxy-11(12)-epoxy-8-*trans*-heptadecenoic acid.[44,112] 10,11,12-THHE (10,11,12-Trihydroxy-17:1): 10,11,12-Trihydroxy-8-*trans*-heptadecenoic acid.[44,112]

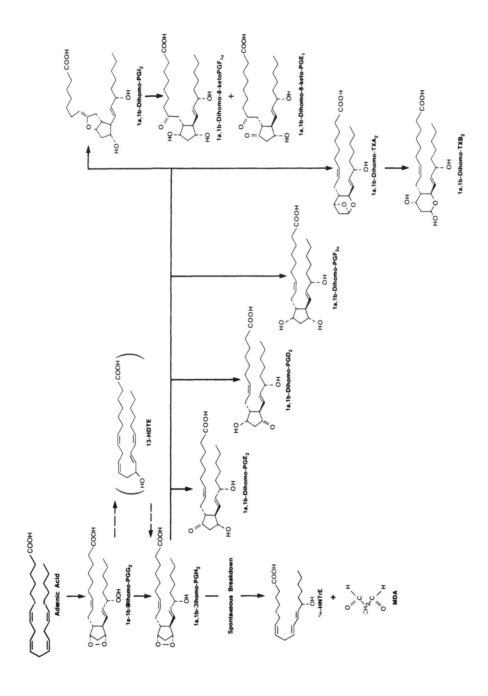

FIGURE 3. Cyclooxygenase products of adrenic acid. 1a,1b-Dihomo-PGG$_2$: 17α-Hydroperoxy-11α,13α-peroxido-1a,1b-dihomoprosta-7-cis,15-trans-dienoic acid.[45] 1a,1b-Dihomo-PGH$_2$: 17α-Hydroxy-11α,13α-peroxido-1a,1b-dihomoprosta-7-cis,15-trans-dienoic acid.[45] 14-HNTrE (1a,1b-Dihomo-12-HHTrE: 14(S)-Hydroxy-7-cis,10-trans,12-trans-monadecatrienoic acid.[45,122] MDA: Malondialdehyde.[45] 1a,1b-Dihomo-PGE$_2$: 13α,17α-Dihydroxy-11-keto 1a,1b-dihomoprosta-7-cis,15-trans-dienoic acid.[45,171,178] 1a,1b-Dihomo-PGD$_2$: 11α,17α-Dihydroxy-13-keto-1a,1b-dihomoprosta-7-cis,15-trans-dienoic acid.[45] 1a,1b-Dihomo-PGF$_{2α}$: 11α,13α,17α-Trihydroxy-1a,1b-dihomoprosta-7-cis,15-trans-dienoic acid.[45,178] 1a,1b-Dihomo-TXA$_2$: 9{[3β-(3α-Hydroxy-1-trans-octenyl)]-2,7-dioxabicyclo-(3.1.1)-heptan-4α-yl}-7-cis-nonenoic acid.[45,122] 1a,1b-Dihomo-TXB$_2$: 9{[4α,6β-Dihydroxy-2β-(3α-hydroxy-1-trans-octenyl)]-3-α-tetrahydropyranyl}-7-cis-nonenoic acid.[45,122] 1a,1b-Dihomo-PGI$_2$ (7Z-11-Deoxy-8,11α-epoxy-Δ^7-1a,1b-dihomoprostaglandin F$_{1α}$): 8(11)-Oxy-13α,17α-dihydroxyprosta-7-cis,15-trans-dienoic acid.[45] 1a,1b-Dihomo-8-keto-PGF$_{1α}$: 11α,13α,17α-Trihydroxy-8-keto-1a,1b-dihomoprost-15-trans-enoic acid.[45,171] 1a,1b-Dihomo-8-keto-PGE$_1$: 13α,17α-Dihydroxy-8,11-diketo-1a,1b-dihomoprost-15-trans-enoic acid.[45] 13-HDTE: 13-Hydroxy-7,10,14,16-docosatetraenoic acid.[45,122]

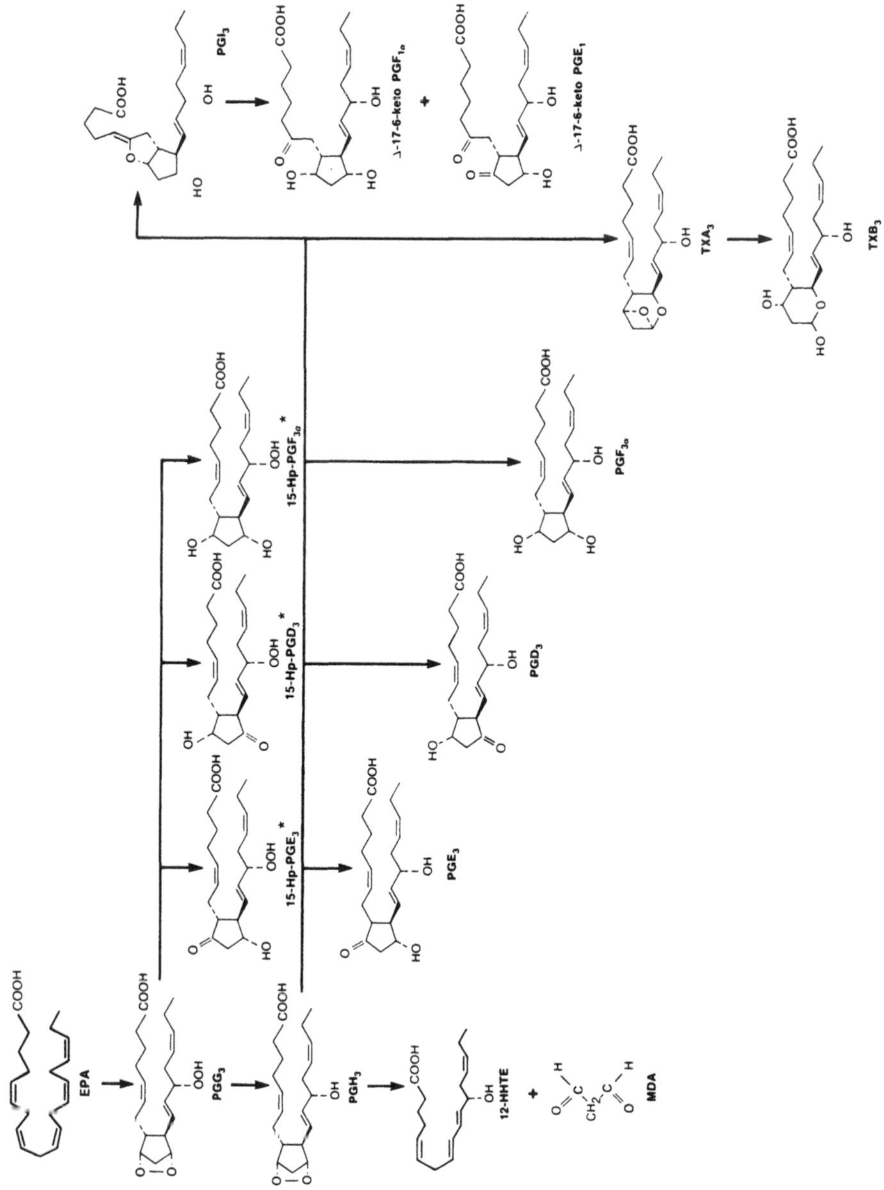
EPA
PGG3
PGH3
12-HHTE
MDA
15-Hp-PGE3 *
15-Hp-PGD3 *
15-Hp-PGF3α *
PGE3
PGD3
PGF3α
TXA3
TXB3
PGI3
Δ-17-6-keto PGF1α
Δ-17-6-keto PGE1

FIGURE 4. Cyclooxygenase products of eicosapentaenoic acid. Prostaglandin G$_3$ (PGG$_3$): 15α-Hydroperoxy-9α,11α-peroxidoprosta-5-*cis*,13-*trans*,17-*cis*-trienoic acid.[71] Prostaglandin H$_3$ (PGH$_3$): 15α-Hydroxy-9α,11α-peroxidoprosta-5-*cis*,13-*trans*,17-*cis*-trienoic acid.[64,69,71,74] 12-HHTE: 12(S)-Hydroxy-5-*cis*,8-*trans*,10-*trans*,14-*cis*-heptadecatetraenoic acid.[73,104,172] MDA: Malondialdehyde.[73] 15-Hydroperoxy-prostaglandin E$_3$ (15-Hp-PGE$_3$): 11α-Hydroxy-15α-hydroperoxy-9-ketoprosta-5-*cis*,13-*trans*,17-*cis*-trienoic acid. Prostaglandin E$_3$ (PGE$_3$): 11α,15α-Dihydroxy-9-ketoprosta-5-*cis*,13-*trans*,17-*cis*-trienoic acid.[64,109] 15-Hydroperoxy-prostaglandin D$_3$ (15-Hp-PGD$_3$): 9α-Hydroxy-15α-hydroperoxy-11-ketoprosta-5-*cis*,13-*trans*,17-*cis*-trienoic acid. Prostaglandin D$_3$ (PGD$_3$): 9α,15α-Dihydroxy-11-ketoprosta-5-*cis*,13-*trans*,17-*cis*-trienoic acid.[64,109] 15-Hydroxy-prostaglandin F$_{3α}$ (15-Hp-PGF$_{3α}$): 9α,11α-Dihydroxy-15α-hydroperoxy-prosta-5-*cis*,13-*trans*,17-*cis*-trienoic acid. Prostaglandin F$_{3α}$ (PGF$_{3α}$): 9α,11α,15α-Trihydroxy-prosta-5-*cis*,13-*trans*,17-*cis*-trienoic acid.[64,109] Thromboxane A$_3$ (TXA$_3$): 7{[3β-(3α-Hydroxy-1-*trans*,5-*cis*-octadienyl)]-2,7-dioxabicyclo-(3.1.1)-heptan-4α-yl}-5-*cis*-heptenoic acid.[64,69,74] Thromboxane B$_3$ (TXB$_3$): 7{[4α,6β-Dihydroxy-2β-(3α-Hydroxy-1-*trans*,5-*cis*-octadienyl)]-3α-tetrahydropyranyl}-5-*cis*-heptenoic acid.[64,69,74,104,109] Prostaglandin I$_3$ (PGI$_3$; 5Z-9-Deoxy-6,9α-epoxy-Δ^{5,17}-prostaglandin F$_{1α}$): 6(9)-Oxy-11α,15α-dihydroxyprosta-5-*cis*,13-*trans*,17-*cis*-trienoic acid.[64,72,74,109,172] Δ^{17}-6-Keto-prostaglandin F$_{1α}$: 9α,11α,15α-Trihydroxy-6-keto-prost-14-*trans*,17-*cis*-dienoic acid.[64,74,109,172] β-Hydroxy isomer of Δ^{17}-6-keto-PGF$_{1α}$: 9α,11α,13-Trihydroxy-6-keto-prost-13-*trans*,17-*cis*-dienoic acid (not shown).[172] Δ^{17}-6-Keto-prostaglandin E$_1$ (Δ^{17}-6-keto-PGE$_1$): 11α,15α-Dihydroxy-6,9-diketoprost-13-*trans*,17-*cis*-dienoic acid.[64] 11-HEPE: 11-Hydroxy-5,8,12,14,17-eicosapentaenoic acid (not shown).[172] 12-HEPE: 12-Hydroxy-5,8,10,14,17-eicosapentaenoic acid (not shown).[64] 14-HEPE: 14-Hydroxy-5,8,11,15,17-eicosapentaenoic acid (not shown).[172] 15-HEPE: 15-Hydroxy-5,8,11,13,17-eicosapentaenoic acid (not shown).[172] 18-HEPE: 18-Hydroxy-5,8,11,14,16-eicosapentaenoic acid (not shown).[172]

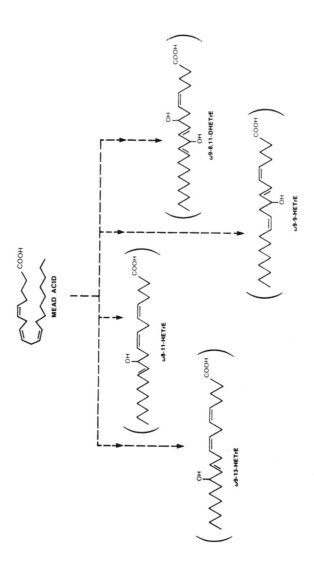

FIGURE 5. Cyclooxygenase products of Mead acid. ω9-13-HETrE: 13-Hydroxy-5,8,11-eicosatrienoic acid.[113] ω8-11-HETrE: 11-Hydroxy-5,8,12-eicosatrienoic acid.[113] ω9-9-HETrE: 9-Hydroxy-5,7,11-eicosatrienoic acid.[113] ω9-8,11-DHETrE: 8,11-Dihydroxy-5,9,12-eicosatrienoic acid.[113]

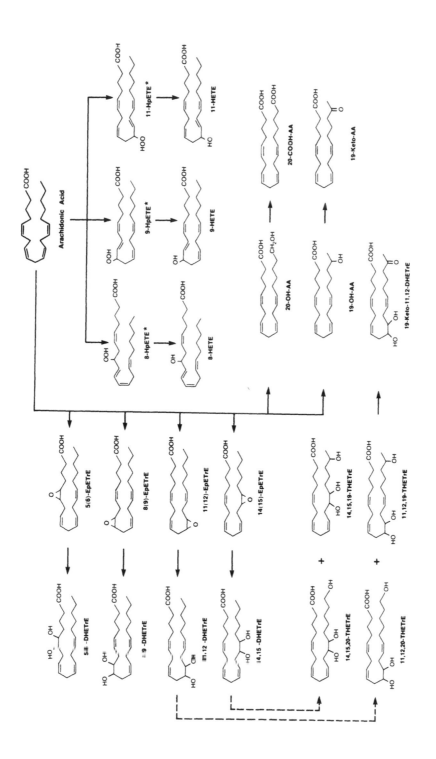

FIGURE 6A.   Lipoxygenase products of arachidonic acid. A. Monoxygenase, 8-9-, and 11-lipoxygenase products. 5(6)-EpETrE (5(6)-Oxido-20:3): 5(6)-Epoxy-8-*cis*-11-*cis*,14-*cis*-eicosatrienoic acid.[34,35,159] 8(9)-EpETrE (8(9)-Oxido-20:3): 8(9)-Epoxy-5-*cis*,11-*cis*,14-*cis*-eicosatrienoic acid.[34,35] 11(12)-EpETrE (11(12)-Oxido-20:3): 11(12)-Epoxy-5-*cis*,8-*cis*,14-*cis*-eicosatrienoic acid.[34,35] 14(15)-EpETrE (14(15)-Oxido-20:3): 14(15)-Epoxy-5-*cis*,8-*cis*,11-*cis*-eicos-atrienoic acid.[34,35] 8(S)-HpETE: 8(S)-Hydroperoxy-5-*cis*,9-*trans*,11-*cis*,14-*cis*-eicosatetraenoic acid.[46] 8(S)-HETE: 8(S)-Hydroxy-5-*cis*,9-*trans*,11-*cis*,14-*cis*-eicosatetraenoic acid.[46,81,94,153,155] 9-HpETE: 9-Hydroperoxy-5-*cis*,7-*trans*,11-*cis*,14-*cis*-eicosatetraenoic acid.[46] 9-HETE: 9-Hydroxy-5-*cis*,7-*trans*,11-*cis*,14-*cis*-eicosatetraenoic acid.[46,81,155] 11-HpETE: 11-Hydroperoxy-5-*cis*,8-*cis*,12-*trans*,14-*cis*-eicosatetraenoic acid.[46] 11-HETE: 11-Hydroxy-5-*cis*,8-*cis*,12-*trans*,14-*cis*-eicosatetraenoic acid.[46,81,155] 20-OH-AA (20-Hydroxy-arachidonic acid): 20-Hydroxy-5-*cis*,8-*cis*,11-*cis*,14-*cis*-eicosatetraenoic acid.[137,138] 20-COOH-AA (20-carboxy-arachidonic acid): 5-*cis*-8,*cis*-11-*cis*,14-*cis*-Eicosatetraen-1,20-dioic acid.[137,138] 19-OH-AA (19-Hydroxy-arachidonic acid): 19-Hydroxy-5-*cis*,8,*cis*-11-*cis*,14-*cis*-eicosatetraenoic acid.[137,138,160] 19-Keto-AA (19-keto-arachidonic acid): 19-Keto-5-*cis*,8-*cis*,11-*cis*,14-*cis*-eicosatetraenoic acid.[138] 8,9-DHETrE: 8,9-Dihydroxy-5-*cis*,11-*cis*,14-*cis*-eicosatrienoic acid.[138] 5,6-DHETrE: 5,6-Dihydroxy-8-*cis*,11-*cis*,14-*cis*-eicosatrienoic acid.[138,160] 14,15-DHETrE: 14,15-Dihydroxy-5-*cis*,8-*cis*,11-*cis*-eicosatrienoic acid.[138,160] 11,12-DHETrE: 11,12-Dihydroxy-5-*cis*,8-*cis*,14-*cis*-eicosatrienoic acid.[138,160] 14,15,20-THETrE: 14,15,20-Trihydroxy-5-*cis*,8-*cis*,11-*cis*-eicosatrienoic acid.[160] 14,15,19-THETrE: 14,15,19-Trihydroxy-5-*cis*,8-*cis*,11-*cis*-eicosatrienoic acid.[160] 11,12,20-THETrE: 11,12,20-Trihydroxy-5-*cis*,8-*cis*,14-*cis*-eicosatrienoic acid.[160] 11,12,19-THETrE: 11,12,19-Trihydroxy-5-*cis*,8-*cis*,14-*cis*-ei-cosatrienoic acid.[160] 19-Keto-11,12-DHETrE: 11,12-Dihydroxy-19-keto-5-*cis*,8-*cis*,14-*cis*-eicosatrienoic acid.[160]

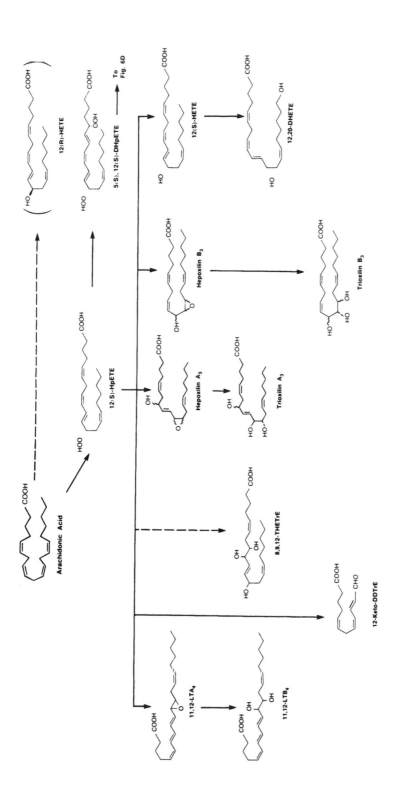

FIGURE 6B. 12-Lipoxygenase products. 12(S)-HpETE: 12(S)-Hydroperoxy-5-*cis*,8-*cis*,10-*trans*,14-*cis*-eicosatetaenoic acid.[1,15,20-22,27] 12(S)-HETE: 12(S)-Hydroxy-5-*cis*,8-*cis*,10-*trans*,14-*cis*-eicosatetraenoic acid.[1,15,20-23,46,94,111] 12,20-DiHETE (12,20-DiHETE): 12(S)-12,20-Dihydroxy-5-*cis*,8-*cis*,10-*trans*,14-*cis*-eicosatetraenoic acid.[38,39] Hepoxilin A₃ (8-Hydroxy-11(12)-Leukotriene A₃; previously called "8,11,12-HEPA"): 8-Hydroxy-11(12)-epoxy-5-*cis*,9-*trans*,14-*cis*-eicosatrienoic acid.[29,46,98,99,100] Hepoxilin B₃ (10-Hydroxy-11(12)-Leukotriene A₃; previously called "10,11,12-HEPA" or "EPHETA"): 10-Hydroxy-11(12)-epoxy-5-*cis*,8-*cis*,14-*cis*-eicosatrienoic acid.[28,46,80,98,99,100] 8,9,12-THETrE (previously called "8,9,12-THETA"): 8,9,12-Trihydroxy-5-*cis*,10-*trans*,14-*cis*-eicosatrienoic acid.[30,46,80] Trioxilin A₃ (8,11,12-THETrE): 8,11,12-Trihydroxy-5-*cis*,9-*trans*,14-*cis*-eicosatrienoic acid.[29,30,41,46,97,98,100] Trioxilin B₃ (10,11,12-THETrE): 10,11,12-Trihydroxy-5-*cis*,8-*cis*,14-*cis*-eicosatrienoic acid.[41,98,100] 11(12)-Leukotriene A₄ (11(12)-LTA₄): 11(12)-Epoxy-5-*cis*,7-*trans*,9-*trans*,14-*cis*-eicosatetraenoic acid.[1,43] 11,12-Leukotriene B₄ (11,12-LTB₄): 11,12-Dihydroxy-5-*cis*,7-*trans*,9-*trans*,14-*cis*-eicosatetraenoic acid.[1,43,142] 12-Keto-DDTrE: 12-Keto-5-*cis*,8-*cis*,10-*trans*-dodecatrienoic acid.[136,143] 12(R)-HETE: 12(R)-Hydroxy-5,8,10,14-eicosatetraenoic acid.[152] 5(S),12(S)-DHpETE: 5(S),12(S)-Dihydroperoxy-6-*trans*,8-*cis*,10-*trans*,14-cis-eicosatetraenoic acid.[89,140]

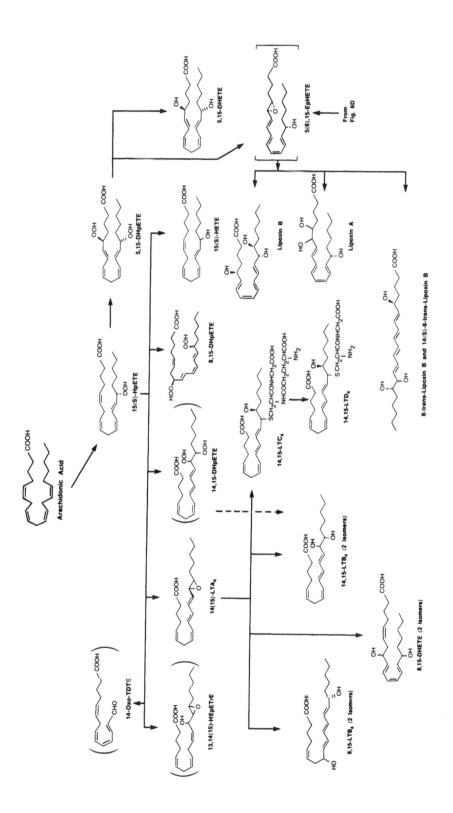

FIGURE 6C. 15-Lipoxygenase products. 15(S)-HpETE: 15(S)-Hydroperoxy-5-*cis*,8-*cis*,11-*cis*,13-*trans*-eicosatetraenoic acid.[23,26,27,46,91,92,130] 15(S)-HETE: 15(S)-Hydroxy-5-*cis*,8-*cis*,11-*cis*,13-*trans*-eicosatetraenoic acid.[31,92,132,140] 5,15-DHpETE (previously called "5,15-DiHPETE"): 5(S),15(S)-Dihydroperoxy-5-*cis*,9-*trans*,11-*cis*,13-*trans*-eicosatetraenoic acid. 8,15-DHpETE: 8(S),15(S)-Dihydroperoxy-6-*trans*,8-*cis*,11-*cis*,13-*trans*-eicosatetraenoic acid.[89,92] 5,15-DHETE (previously called "5,15-DiHETE"): 5(S),15(S)-Dihydroxy-6-*trans*,8-*cis*,11-*cis*,13-*trans*-eicosatetraenoic acid.[5,9,89,91] Lipoxin B (LXB): 5(S),14(R),15(S)-Trihydroxy-6-*trans*,8-*cis*,10-*trans*,12-*trans*-eicosatetraenoic acid.[37,101,102,134] Lipoxin A (LXA): 5,6,15-Trihydroxy-7-*trans*,9-*trans*,11-*cis*,13-*trans*-eicosatetraenoic acid.[37,101,102,134] 14(15)-Leukotriene A₄ (14(15)-LTA₄): 14(15)-Epoxy-5-*cis*,8-*cis*,10-*trans*,12-*trans*-eicosatetraenoic acid.[1,5,43,46,9,92] 14,15-Leukotriene B₄ (14,15-LTB₄; previously called "Erythro-14,15-LTB₄" and "14,15-DHETE"): 14(R),15(S)- and 14(S),15(S)-Dihydroxy-5-*cis*,8-*cis*,10-*trans*,12-*trans*-eicosatetraenoic acid.[1,6-9,43,46,91,93,94,149] 14,15-Leukotriene B₄ (8,15-LTB₄): 8(R),15(S)- and 8(S),15(S)-Dihydroxy-5-*cis*,9-*trans*,11-*trans*,13-*trans*-eicosatetraenoic acid.[1,6-9,43,46,91-95,117,140] 14,15-Leukotriene C₄ (14,15-LTC₄): 15-Hydroxy-14-S-glutathionyl-5-*cis*,8-*cis*,10-*trans*,12-*trans*-eicosatetraenoic acid.[1,5,43,46] 14,15-Leukotriene D₄ (14,15-LTD₄): 15-Hydroxy-14-S-cysteinylglycyl-5-*cis*,8,*cis*,10-*trans*,12-*trans*-eicosatetraenoic acid.[8,91,94,95,117,132] 14,15-DHpETE: 14,15-Dihydroperoxy-5,8,10,12-eicosatetraenoic acid.[140,141] 8-*trans*-Lipoxin B and 14(S)-8-*trans*-Lipoxin B: 5(S),14(R),15(S) and 5(S),14(S),15(S)-Trihydroxy-6-*trans*,8-*trans*,10-*trans*,12-*trans*-e cosatetraenoic acid.[103] 13,14(15)-HEPETrE(13-OH-14(15)-EPETrE; previously called "15α-HEPA"): 13-Hydroxy-14(15)-epoxy-5,8,11-eicosatrienoic acid.[92] 14-Oxc-TDTE: 14-Keto-5,8,10,12-tetradecatetraenoic acid.[143] 8,15-DHETE (previously called "8,15-DiHETE"): 8(R),15(S)- and 8(S),15(S)-5-*cis*,9-*trans*,11-*cis*,13-*trans*-eicosatetraenoic acid.[1,5,46]

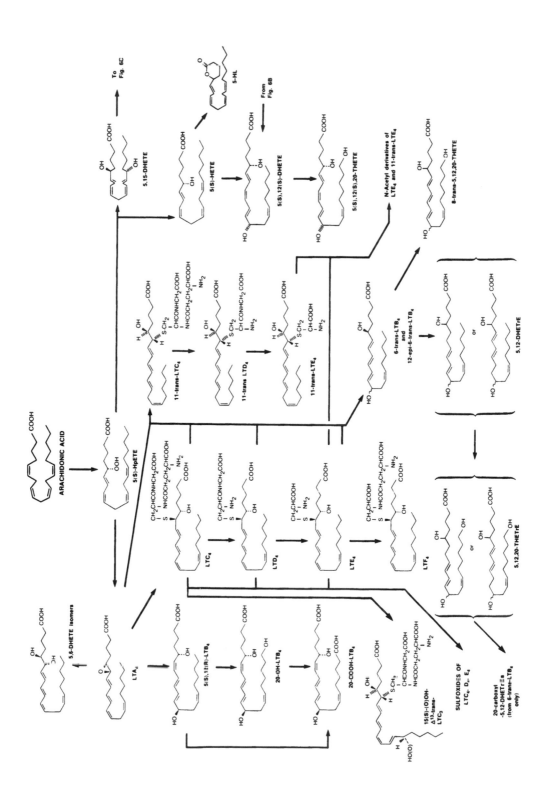

FIGURE 6D. 5-Lipoxygenase products. 5(S)-HpETE: 5(S)-Hydroperoxy-6-*trans*,8-*cis*,11-*cis*,14-*cis*-eicosatetraenoic acid.[1,2,15,18,23-25,46,91,105,140] 5(S)-HETE: 5(S)-Hydroxy-6-*trans*,8-*cis*,11-*cis*,14-*cis*-eicosatetraenoic acid.[1,2,15,18,23-25,46,91,105] 5(S),12(S)-DHETE: 5(S),12(S)-Dihydroxy-6-*trans*,8-*cis*,10-*trans*,14-*cis*-eicosatetraenoic acid.[1,14,15,46,82,89,91,129,142] 5(S),12(S),20-THETE: 5(S),12(S), 20-Trihydroxy-6-*trans*,8-*cis*,10-*trans*,14-*cis*-eicosatetraenoic acid.[5,9,89,91] 5,15 DHETE (previously called "5,15-DiHETE"): 5(S),15(S)-Dihydroxy-6-*trans*,8-*cis*,11-*cis*,13-*trans*-eicosatetraenoic acid.[1,2,15,18,46,90,91,133,140] Leukotriene C₄ (LTC₄): 5(S)-Hydroxy-6(R)-S-glutathionyl-7-*trans*,9-*trans*,11-*cis*,14-*cis*-eicosatetraenoic acid.[1,2,18,43,46,84,87,91,105,144-146] Leukotriene D₄ (LTD₄): 5(S)-Hydroxy-6(R)-S-cysteinylglycyl-7-*trans*,9-*trans*,11-*cis*,14-*cis*-eicosatetraenoic acid.[1,2,18,43,46,86,87,91,144-146] Leukotriene E₄ (LTE₄): 5(S)-Hydroxy-6(R)-S-cysteinyl-7-*trans*,9-*trans*,11-*cis*,14-*cis*-eicosatetraenoic acid.[1,11,18,43,46,85,91,145] Leukotriene F₄ (LTF₄): 5(S)-Hydroxy-6(R)-S-cysteinyl-γ-glutamyl-7-*trans*,9-*trans*,11-*cis*,14-*cis*-eicosatetraenoic acid.[1,43-46] 5(S),12(R)-Leukotriene B₄ (5(S),12(R)-LTB₄; commonly designated "LTB₄"): 5(S),12(R)-Dihydroxy-6-*cis*,8-*trans*,10-*trans*,14-*cis*-eicosatetraenoic acid.[1,15,18,24,46,82,90,91,105,142,147,148,164] 20-Hydroxy-leukotriene B₄ (20-OH-LTB₄; ω-Hydroxy-LTB₄): 5(S),12(R),20-Trihydroxy-6-*cis*,8-*trans*,10-*trans*,14-*cis*-eicosatetraenoic acid.[1,15,46,91,128] 20-Carboxy-leukotriene B₄ (20-COOH-LTB₄): 5(S),12(R)-Dihydroxy-6-*cis*,8-*trans*,10-*trans*,14-*cis*-eicosatetraen-1,20-dioic acid.[1,15,16,46] 6-*trans*-Leukotriene B₄ (6-*trans*-LTB₄ and 12-epi-6-*trans*-LTB₄): 5(S),12(R)- and 5(S),12(S)-Dihydroxy-6-*trans*,8-*trans*,10-*trans*,14-*cis*-eicosatetraenoic acid.[15,17,82,83,88-91,105,110,140] 11-*trans*-Leukotriene C₄ (11-*trans*-LTC₄): 5(S)-Hydroxy-6(R)-S-glutathionyl-7-*trans*,9-*trans*,11-*trans*,14-*cis*-eicosatetraenoic acid.[17,18,9,144] 11-*trans*-Leukotriene D₄ (11-*trans*-LTD₄): 5(S)-Hydroxy-6(R)-S-cysteinyl-7-*trans*,9-*trans*,11-*trans*,14-*cis*-eicosatetraenoic acid.[11,85,139] 11-*trans*-Leukotriene E₄ (11-*trans*-LTE₄): 5(S)-Hydroxy-6(R)-S-cysteinyl-7-*trans*,9-*trans*,11-*trans*,14-*cis*-eicosatetraenoic acid.[88-91,140,148] Sulfoxides of LTC₄, D₄, E₄.[17,83] 5,6-DHETE (previously called "5,6-Di-HETE"): 5,6-Dihydroxy-7-*trans*,9-*trans*,11-*cis*,14-*cis*-eicosatetraenoic acid.[118] 5,12-DHETrE: 5,12-Dihydroxy-6-*trans*,8-*trans*,14-*cis*-eicosatrienoic acid or 5,12-Dihydroxy-8-*trans*,10-*trans*,14-*cis*-eicosatrienoic acid.[118] 5,12,20-THETrE: 5,12,20-Trihydroxy-6-*trans*,8-*trans*,14-*cis*-eicosatrienoic acid or 5,12,20-Trihydroxy-8-*trans*,10-*trans*,14-*cis*-eicosatrienoic acid.[118] 20-COOH-5,12-DHETrE (20-Carboxy-5,12-DHETrE): 5,12-Dihydroxy-6-*trans*,8-*trans*,14-*cis*-eicosatrien-1,20-dioic acid or 5,12-Dihydroxy-8-*trans*,10-*trans*,14-*cis*-eicosatrien-1,20-dioic acid.[118] 5-HL (5-HETE lactone):5-Hydroxy-6-*trans*,8-*cis*,11-*cis*,14-*cis*-eicosatetraenoic acid-delta lactone.[154] 15(S)-Hydroxy-Δ¹³-*trans*-leukotriene C₃ (15(S)-OH-Δ¹³-*trans*-LTC₃): 5(S),15(S)-Dihydroxy-6(R)-S-glutathionyl-7-*trans*,9-*trans*,11-*cis*,13-*trans*-eicosatetraenoic acid.[65] N-Acetyl-LTE₄: 5-Hydroxy-6-S-(2-acetamido-3-thiopropionyl)-7-*trans*,9-*trans*,11-*cis*,14-*cis*-eicosatetraenoic acid.[158,162,163] N-Acetyl-11-*trans*-LTE₄: 5-Hydroxy-6-S-(2-acetamido-3-thiopropionyl)-7-*trans*,9-*trans*,11-*trans*,14-*cis*-eicosatetraenoic acid.[158,162]

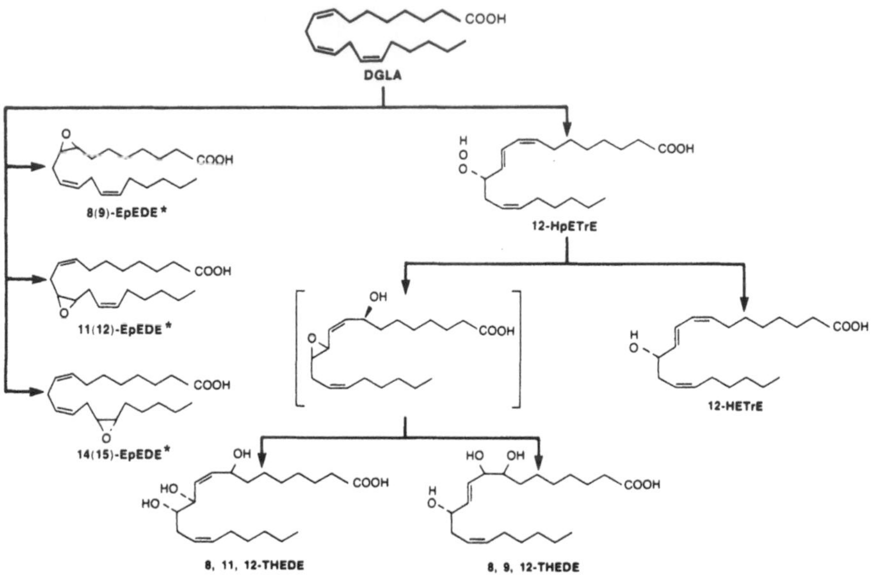

A

FIGURE 7. Lipoxygenase products of dihomo-γ-linolenic acid. (A) Monooxygenase and 12-li-poxygenase products. 8(9)-EpEDE (8(9)-Oxido-20:2) 8(9)-Epoxy-11-*cis*-14-*cis*-eicosadienoic acid. 11(12)-EpEDE (11(12)-Oxido-20:2): 11(12)-Epoxy-8-*cis*,14-*cis*-eicosadienoic acid. 14(15)-EpEDE (14(15)-Oxido-20:2): 14(15)-Epoxy-8-*cis*,11-*cis*-eicosadienoic acid. 12-HpETrE: 12(S)Hydroperoxy-8-*cis*,10-*trans*,14-*cis*-eicosatrienoic acid.[32] 12-HETrE: 12(S)-Hydroxy-8-*cis*,10-*trans*,14-*cis*-eicosatrienoic acid.[32] 8,11,12-THEDE: 8,11,12-Trihydroxy-9-*cis*,14-*cis*-eicosadienoic acid.[32] 8,9,12-THEDE: 8,9,12-Trihydroxy-10-*trans*,14-*cis*-eicosadienoic acid.[32] (B) 15- and 8-lipoxygenase products. 15-HpETrE: 15(S)-Hydroperoxy-8-*cis*,11-*cis*,13-*trans*-eicosatrienoic acid.[27,33] 15-HETrE: 15(S)-Hydroxy-8-*cis*,11-*cis*,13-*trans*-eicosatrienoic acid.[33] 8,15-DHpETrE: 8,15-Dihydroperoxy-9-*cis*,11-*cis*,13-*trans*-eicosatrienoic acid.[31,33] 8,15-DHETrE: 8,15-Dihydroxy-9-*cis*,11-*cis*,13-*trans*-eicosatrienoic acid.[32] 14,15-Leukotriene B₃ (14,15-LTB₃): 14,15-Dihydroxy-8-*cis*,10-*cis*-12-*trans*-eicosatrienoic acid.[32] 8,15-Leukotriene B₃ (8,15-LTB₃): 8,15-Dihydroxy-9-*trans*,11-*trans*,13-*cis*-eicosatrienoic acid.[19] 8-HpETrE: 8-Hydroperoxy-9-*trans*,11-*cis*,14-*cis*-eicosatrienoic acid.[1,23] 8-HETrE: 8-Hydroxy-9-*trans*,11-*cis*,14-*cis*-eicosatrienoic acid.[1,23,81] 8(9)-Leukotriene A₃ (8(9)-LTA₃): 8(9)-Epoxy-10-*trans*,12-*trans*,14-*cis*-eicosatrienoic acid.[1,4,43] 8,9-Leukotriene B₃ (8,9-LTB₃): 8(R),9(S)-Dihydroxy-10-*trans*,12-*trans*,14-*cis*-eicosatrienoic acid. 8,9-Leukotriene C₃ (8,9-LTC₃): 8(R)-Hydroxy-9(S)-S-glutathionyl-10-*trans*,12-*trans*,14-*cis*-eicosatrienoic acid.[1,4,43]

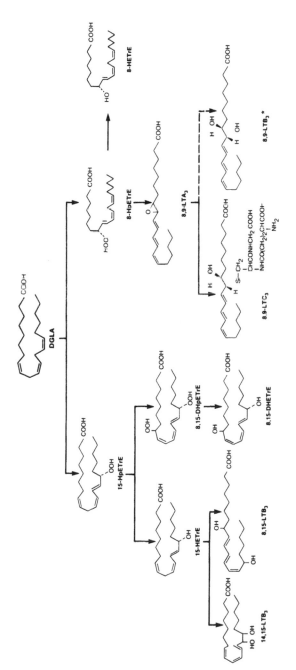

FIGURE 7B.

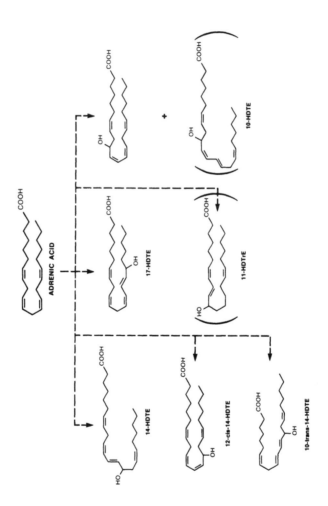

FIGURE 8. Lipoxygenase products of adrenic acid. 14-HDTE: 14-Hydroxy-7-*cis*,10-*cis*,12-*trans*,16-*cis*-docosatetraenoic acid.[122] 12-*cis*-14-HDTE: 14-Hydroxy-7-*cis*,10-*cis*,12-*cis*,16-*cis*-docosatetraenoic acid.[122] 10-*trans*-14-HDTE: 14-Hydroxy-7-*cis*,10-*trans*,12-*trans*,16-*cis*-docosatetraenoic acid.[122] 17-HDTE: 17-Hydroxy-7-*cis*,10-*cis*,13-*cis*,15-*trans*-docosatetraenoic acid.[122] 11-HDTrE: 11-Hydroxy-7,9,16-do-cosatrienoic acid.[122] 10-HDTE: 10-Hydroxy-7,11,13,16-docosatetraenoic acid.[122]

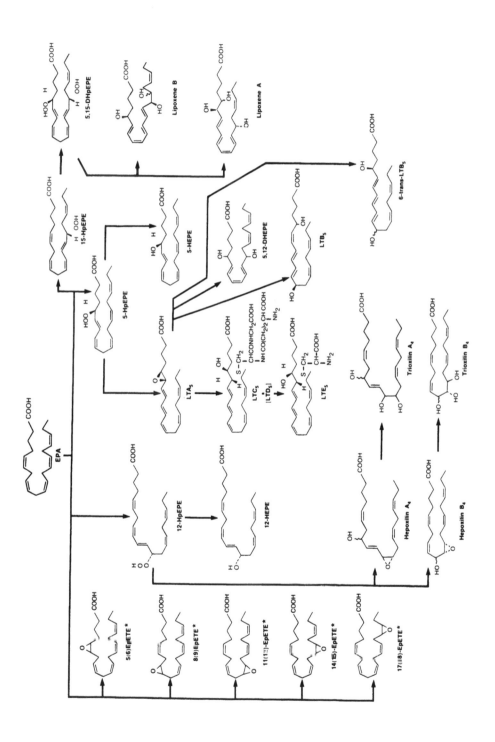

FIGURE 9. Lipoxygenase products of eicosapentaenoic acid. 5(6)-EpETE (5(6)Oxido 20:4): 5(6)-Epoxy-8-*cis*,11-*cis*,14-*cis*,17-*cis*-eicosatetraenoic acid. 8(9)-EpETE (8(9)Oxido 20:4): 8(9)-Epoxy-5-*cis*,11-*cis*,14-*cis*,17-*cis*-eicosatetraenoic acid. 11(12)-EpETE (11(12)Oxido-20:4): 11(12)-Epoxy-5-*cis*,8-*cis*,14-*cis*,17-*cis*-eicosatetraenoic acid. 14(15)-EpETE (14(15)Oxido-20:4): 14(15)-Epoxy-5-*cis*,8-*cis*,11-*cis*,17-*cis*,-eicosatetraenoic acid. 17(18)-EpETE (17(18)Oxido-20:4): 17(18)-Epoxy-5-*cis*,8-*cis*,11-*cis*,14-*cis*-eicosatetraenoic acid. 12-HpEPE: 12(S)-Hydroperoxy-5-*cis*,8-*cis*,10-*trans*,14-*cis*,17-*cis*-eicosapentaenoic acid.[21] 12-HEPE: 12(S)-Hydroxy-5-*cis*,8-*cis*,10-*trans*,14-*cis*,17-*cis*-eicosapentaenoic acid.[21,104,115] 5-HpEPE: 5(S)-Hydroperoxy-6-*trans*,8-*cis*,11-*cis*,14-*cis*,17-*cis*-eicosapentaenoic acid.[42,109] 5-HEPE: 5(S)-Hydroxy-6-*trans*,8-*cis*,11-*cis*,14-*cis*,17-*cis*-eicosapentaenoic acid.[42,105,109] 5,12-DHEPE (6-*trans*,8-*cis*-LTB_5): 5(S),12(S)-Dihydroxy-6-*trans*,8-*cis*,10-*trans*,14-*cis*,17-*cis*-eicosapentaenoic acid.[42,82,105,109] Leukotriene A_5 (LTA_5): 5(S),6(S)-5(6)-Epoxy-7-*trans*,9-*trans*,11-*cis*,14-*cis*,17-*cis*-eicosapentaenoic acid.[1,2,42] Leukotriene C_5 (LTC_5): 5(S)-Hydroxy-6(R)-S-glutathionyl-7-*trans*,9-*trans*,11-*cis*,14-*cis*,17-*cis*-eicosapentaenoic acid.[1,2,105,110,1[illegible]] Leukotriene E_5 (LTE_5): 5(S)-Hydroxy-6(R)-S-cysteinyl-7-*trans*,9-*trans*,-11-*cis*,14-*cis*,17-*cis*-eicosapentaenoic acid.[11] Leukotriene B_5 (LTB_5): 5(S),12(R)-Dihydroxy-6-*cis*,8-*trans*,10-*trans*,14-*cis*,17-*cis*-eicosapentaenoic acid.[82,105,110,161] 6-*trans*-LTB_5: 5(S),12(R)- and 5(S),12(S)-Dihydroxy-6-*trans*,8-*trans*,10-*trans*,14-*cis*,17-*cis*-eicosapentaenoic acid.[82,85,110] 15-HpEPE: 15(S)-Hydroperoxy-5-*cis*,8-*cis*,11-*cis*,13-*trans*,17-*cis*-eicosapentaenoic acid.[151] 5,15-DHpEPE: 5(S),15(S)-Dihydroperoxy-6-*trans*,8-*cis*,11-*cis*,13-*trans*,17-*cis*-eicosapentaenoic acid.[151] Lipoxene B: 5(S),14(S),15(S)-Trihydroxy-6-*trans*,8-*cis*,10-*trans*,12-*trans*,17-*cis*-eicosapentaenoic acid.[151] Lipoxene A: 5,6,15-Trihydroxy-7-*trans*,9-*trans*,11-*cis*,13-*trans*,17-*cis*-eicosapentaenoic acid.[151] Hepoxilin A_4 (8-OH-11(12)-EpETE): 8-Hydroxy-11(12)-epoxy-5-*cis*,9-*trans*,14-*cis*,17-*cis*-eicosatetraenoic acid.[96] Hepoxilin B_4 (10-OH-11(12)-EpETE): 10-Hydroxy-11(12)-epoxy-5-*cis*,8-*cis*,14-*cis*,17-*cis*-eicosatetraenoic acid.[96] Trioxilin A_4: (previously called "8,11,12-THETE" and "8,11,12-THETA"): 8,11,12-Trihydroxy-5-*cis*,9-*trans*,14-*cis*,17-*cis*-eicosatetraenoic acid.[40,96] Trioxilin B_4 (previously called "10,11,12-THETE" and "10,11,12-THETA"): 10,11,12-Trihydroxy-5-*cis*,8-*cis*,14-*cis*,17-*cis*-eicosatetraenoic acid.[96,108]

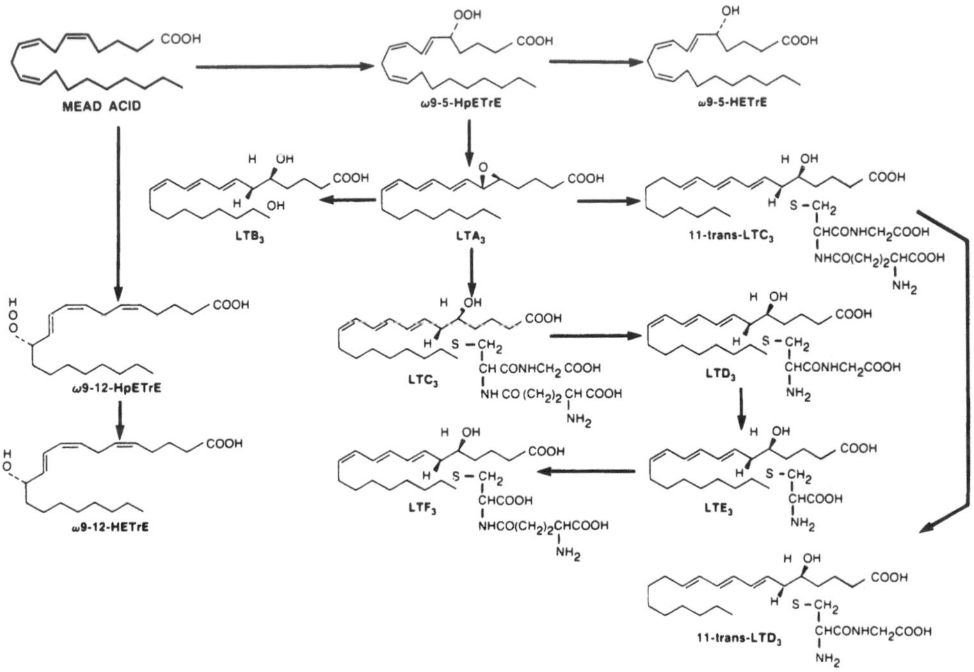

FIGURE 10. Lipoxygenase products of Mead acid. ω9-12-HpETrE: 12(S)-Hydroperoxy-5-*cis*,8-*cis*,10-*trans*-eicosatrienoic acid.[21] ω9-12-HETrE: 12(S)-Hydroperoxy-5-*cis*,8-*cis*,10-*trans*-eicosatrienoic acid.[21] ω9-12-HETrE: 12(S)-Hydroxy-5-*cis*,8-*cis*,10-*trans*-eicosatrienoic acid.[21,120] ω9-5-HpETrE: 5(S)-Hydroperoxy-6-*trans*,8-*cis*,11-*cis*-eicosatrienoic acid.[1,27] ω9-5-HETrE: 5(S)-Hydroxy-6-*trans*,8-*cis*,11-*cis*-eicosatrienoic acid.[1,27] Leukotriene A₃ (LTA₃): 5(S),6(S)-5(6)-Epoxy-7-*trans*,9-*trans*,11-*cis*-eicosatrienoic acid.[1,115,121] Leukotriene B₃ (LTB₃): 5(S), 6(R)-Dihydroxy-7-*trans*,9-*trans*,11-*cis*-eicosatrienoic acid.[115,116,121] Leukotriene C₃ (LTC₃): 5(S)-Hydroxy-6(R)-S-glutathionyl-7-*trans*,9-*trans*,11-*cis*-eicosatrienoic acid.[1,2,114] Leukotriene D₃ (LTD₃): 5(S)-Hydroxy-6(R)-S-cysteinylglycyl-7-*trans*,9-*trans*,11-*cis*-eicosatrienoic acid.[1,10,114] Leukotriene E₃ (LTE₃): 5(S)-Hydroxy-6(R)-S-cysteinyl-7-*trans*,9-*trans*,11-*cis*-eicosatrienoic acid.[1,10,11] Leukotriene F₃ (LTF₃): 5(S)-Hydroxy-6(R)-S-cysteinyl-γ-glutamyl-7-*trans*,9-*trans*,11-*cis*-eicosatrienoic acid.[1,12,13] 11-*trans*-Leukotriene C₃ (11-*trans*-LTC₃): 5(S)-Hydroxy-6(R)-S-glutathionyl-7-*trans*,9-*trans*,11-*trans*-eicosatrienoic acid.[1,2,114] 11-*trans*-Leukotriene D₃ (11-*trans*-LTD₃): 5(S)-Hydroxy-6(R)-S-cysteinylglycyl-7-*trans*,9-*trans*,11-*trans*-eicosatrienoic acid.[114]

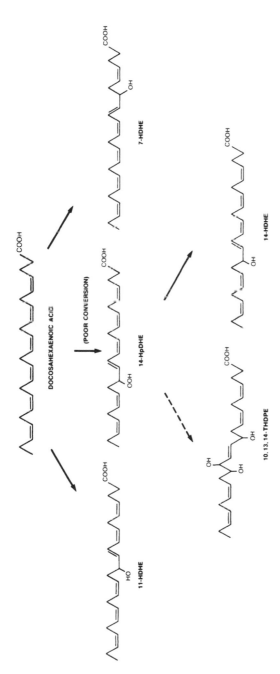

FIGURE 11. Lipoxygenase products of docosahexaenoic acid. 14-HpDHE: 14-Hydroperoxy-4-*cis*,7-*cis*,10-*cis*,12-*trans*,16-*cis*,19-*cis*-docosahexaenoic acid.[21] 14-HDHE: 14-Hydroxy-4-*cis*,7-*cis*,10-*cis*,12-*trans*,16-*cis*,19-*cis*-docosahexaenoic acid.[21,106,107] 10,13,14-THDPE: 10,13,14-Trihydroxy-4-*cis*,7-*cis*,11-*trans*,16-*cis*,19-*cis*-docosapentaenoic acid.[40] 7-HDHE: 7-Hydroxy-4-*cis*,8-*trans*,10-*cis*,13-*cis*,16-*cis*,19-*cis*-docosahexaenoic acid.[106,107] 11-HDHE: 11-Hydroxy-4-*cis*,7-*cis*,9-*trans*,13-*cis*,16-*cis*,19-*cis*-docosahexaenoic acid.[105,107]

REFERENCES

1. **Hammarström, S.**, Leukotrienes, *Ann. Rev. Biochem.*, 52, 355—377, 1983.
2. **Hammarström, S.**, Leukotriene formation by mastocytoma and basophilic leukemia cells, *Prog. Lipid Res.*, 20, 89—95, 1982.
3. **Bailey, J. M., Bryant, R. W., Whiting, J., and Salata, K.**, Characterization of 11-HETE and 15-HETE, together with prostacyclin, as major products of the cyclooxygenase pathway in cultured rat aorta smooth muscle cells, *J. Lipid Res.*, 24, 1419—1428, 1983.
4. **Hammarström, S.**, Conversion of dihomo-γ-linolenic acid to an isomer of leukotriene C_3, oxygenated at C-8, *J. Biol. Chem.*, 256, 7712—7714, 1981.
5. **Sok, D.-E., Han, C.-O., Shieh, W.-R., Zhou, B.-N., and Sih, C. J.**, Enzymatic formation of 14,15-leukotriene A and C(14)-sulfur-linked peptides, *Biochem. Biophys. Res. Commun.*, 104, 1363—1370, 1982.
6. **Lundberg, U., Radmark, O., Malmsten, C., and Samuelsson, B.**, Transformation of 15-hydroperoxy-5,9,11,13-eicosatetraenoic acid into novel leukotrienes, *FEBS Lett.*, 126, 127—132, 1981.
7. **Jubiz, W., Radmark, O., Lindgren, J. A., Malmsten, C., and Samuelsson, B.**, Novel leukotrienes: products formed by initial oxygenation of arachidonic acid at C-15, *Biochem. Biophys. Res. Commun.*, 99, 976—986, 1981.
8. **Maas, R. L., Brash, A. R., and Oates, J. A.**, A second pathway of leukotriene biosynthesis in porcine leukocytes, *Proc. Natl. Acad. Sci. U.S.A.*, 78, 5523—5527, 1981.
9. **Turk, J., Maas, R. L., Brash, A. R., Roberts, L. J., and Oates, J. A.**, Arachidonic acid 15-lipoxygenase products for human eosinophils, *J. Biol. Chem.*, 257, 7068—7076, 1982.
10. **Hammarström, S.**, Metabolism of leukotriene C_3 in the guinea pig. Identification of metabolites formed by lung, liver, and kidney, *J. Biol. Chem.*, 256, 9573—9578, 1981.
11. **Bernström, K. and Hammarström, S.**, Metabolism of leukotriene D by porcine kidney, *J. Biol. Chem.*, 256, 9579—9582, 1981.
12. **Anderson, M. E., Allison, R. D., and Meister, A.**, Interconversion of leukotrienes catalyzed by purified gamma-glutamyl transpeptidase: concomitant formation of leukotriene D_4 and gamma-glutamyl amino acids, *Proc. Natl. Acad. Sci. U.S.A.*, 79, 1088—1091, 1982.
13. **Bernström, K. and Hammarström, S.**, A novel leukotriene formed by transpeptidation of leukotriene E, *Biochem. Biophys. Res. Commun.*, 109, 800—804, 1982.
14. **Lindgren, J. A., Hansson, G., and Samuelsson, B.**, Formation of hydroxylated eicosatetranoic acids in preparations of human polymorphonuclear leukocytes, *FEBS Lett.*, 128, 329—335, 1981.
15. **Hansson, G., Lindgren, J. A., Dahlen, S. E., Hedqvist, P., and Samuelsson, B.**, Identification and biological activity of novel ω-oxidized metabolites of leukotriene B_4 from human leukocytes, *FEBS Lett.*, 130, 107—112, 1981.
16. **Jubiz, W., Radmark, O., Malmsten, C., Hansson, G. H., Lindgren, J. A., Palmblad, J., Uden, A. M., and Samuelsson, B.**, A novel leukotriene produced by stimulation of leukocytes with formylmethionylleucylphenylalanine, *J. Biol. Chem.*, 257, 6106—6110, 1982.
17. **Lewis, R. A., Lee, C. W., Levine, L., Morgan, R. A., Weiss, J. W., Drazen, J. M., Oh, M., Hoover, D., Corey, E. J., and Austen, K. F.**, Biology of the C-6 sulfidopeptide leukotrienes, *Adv. Prostaglandin, Thromboxane, Leukotriene Res.*, 11, 15—25, 1983.
18. **Samuelsson, B.**, Leukotrienes: a novel group of compounds including SRS-A, *Prog. Lipid Res.*, 20, 23—30, 1982.
19. **Jubiz, W. and Nolan, G.**, Leukotriene produced by incubation of dihomo-γ-linolenic acid with human polymorphonuclear leukocytes, *Biochem. Biophys. Res. Commun.*, 114, 855—862, 1983.
20. **Hamberg, M. and Samuelsson, B. M.**, Prostaglandin endoperoxides. Novel transformations of arachidonic acid in human platelets, *Proc. Natl. Acad. Sci. U.S.A.*, 71, 3400—3404, 1974.
21. **Nugteren, D. H.**, Arachidonic lipoxygenase, in *Prostaglandins in Hematology*, Silver, M. J., Smith, J. B., and Kocsis, J. J., Eds., Spectrum Publications, Jamaica, N.Y., 1977, 11—25.
22. **Nugteren, D. H.**, Arachidonate lipoxygenase in blood platelets, *Biochim. Biophys. Acta*, 380, 299—307, 1975.
23. **Borgeat, P., Hamberg, M., and Samuelsson, B.**, Transformation of arachidonic acid and homo-γ-linolenic acid by rabbit polymorphonuclear leukocytes, *J. Biol. Chem.*, 251, 7816—7820, 1976.
24. **Borgeat, P. and Samuelsson, B.**, Transformation of arachidonic acid by rabbit polymorphonuclear leukocytes, *J. Biol. Chem.*, 254, 2643—2646, 1979.
25. **Borgeat, P. and Samuelsson, B.**, Arachidonic acid metabolism in polymorphonuclear leukocytes: effects of ionophore A23187, *Proc. Natl. Acad. Sci. U.S.A.*, 76, 2148—2152, 1979.
26. **Goetzl, E. J.**, Selective feed-back inhibition of the 5-lipoxygenase of arachidonic acid in human T-lymphocytes, *Biochem. Biophys. Res. Commun.*, 101, 344—350, 1981.
27. **Bryant, R. W., Bailey, J. M., Schewe, T., and Rapoport, S. M.**, Positional specificity of a reticulocyte lipoxygenase: conversion of arachidonic acid to 15-S-hydroperoxyeicosatetraneoic acid, *J. Biol. Chem.*, 257, 6050—6055, 1982.

28. **Walker, I. C., Jones, R. L., and Wilson, N. H.**, The identification of an epoxyhydroxy acid as a product from the incubation of arachidonic acid with washed blood platelets, *Prostaglandins*, 18, 173—178, 1979.

29. **Bryant, R. W. and Bailey, J. M.**, Isolation of a new lipoxygenase metabolite of arachidonic acid-8,11,12-trihydroxy-5,9,14-eicosatrienoic acid from human platelets, *Prostaglandins*, 17, 9—18, 1979.

30. **Jones, R. L., Kerry, P. J., Poyser, N. L., Walker, I. C., and Wilson, N. H.**, The identification of trihydroxyeicosatrienoic acids as products from the incubation of arachidonic acid with washed blood platelets, *Prostaglandins*, 16, 583—589, 1978.

31. **Bild, G. S., Bhat, S. G., Axelrod, B., and Iatridis, P. G.**, Inhibition of aggregation of human platelets by 8,15-dihydroperoxides of 5,9,11,13 eicosatetraenoic and 9,11,13-eicosatrienoic acids, *Prostaglandins*, 16, 795—801, 1978.

32. **Falardeau, P., Hamberg, M., and Samuelsson, B.**, Metabolism of 8,11,14-eicosatrienoic acid in human platelets, *Biochim. Biophys. Acta*, 441, 193—200, 1976.

33. **Bild, G. S., Ramadoss, C. S., Lim, S., and Axelrod, B.**, Double dioxygenation of arachidonic acid by soybean lipoxygenase-1, *Biochem. Biophys. Res. Commun.*, 74, 949—954, 1977.

34. **Falck, J. R. and Manna, S.**, 8,9-Epoxyarachidonic acid: a cytochrome P-450 metabolite, *Tetrahedron Lett.*, 23, 1755—1756, 1982.

35. **Chacos, N., Falck, J. R., Wixtrom, C., and Capdevila, J.**, Novel epoxides formed during the liver cytochrome P-450 oxidation of arachidonic acid, *Biochem. Biophys. Res. Commun.*, 104, 916—922, 1982.

36. **Oliw, E. M.**, Metabolism of 5(6) oxidoeicosatrienoic acid by ram seminal vessels. Formation of two stereoisomers of 5-hydroxyprostaglandin I_1, *J. Biol. Chem.*, 259, 2716—2721, 1984.

37. **Serhan, C. N., Hamberg, M., and Samuelsson, B.**, in *Proceedings Prostaglandins and Leukotrienes '84: Their Biochemistry, Mechanism of Action and Clinical Applications*, Washington, D.C., Bailey, J. M., Ed., Abstract volume 1984, 31.

38. **Marcus, A. J., Brockman, M. J., Safier, L. B., Ullman, H. L., Islam, N., Serhan, C. N., Rutherford, L. E., Korchak, H. M., and Weissman, G.**, Formation of leukotrienes and other hydroxy acids during platelet-neutrophil interactions in vitro, *Biochem. Biophys. Res. Commun.*, 109, 130—137, 1982.

39. **Marcus, A. J., Safier, L. B., Ullman, H. L., Broekman, M. J., Islam, N., Oglesby, T. D., and Gorman, R. R.**, 12S,20-Dihydroxyeicosatetraenoic acid: a new eicosanoid synthesized by neutrophils from 12S-hydroxyeicosatetraenoic acid produced by thrombin- or collagen-stimulated platelets, *Proc. Natl. Acad. Sci. U.S.A.*, 81, 903—907, 1984.

40. **German, B., Bruckner, G., and Kinsella, J.**, Evidence against a $PGF_{4\alpha}$ prostaglandin structure in trout tissue — a correction, *Prostaglandins*, 26, 207—210, 1983.

41. **Pace-Asciak, C. R., Mizumo, K., Yamamoto, S., Granstrom, E., and Samuelsson, B.**, Oxygenation of arachidonic acid into 8,11,12- and 10,11,12-trihydroxyeicosatrienoic acid by rat lung, *Adv. Prostaglandin, Thromboxane, Leukotriene Res.*, 11, 133—139, 1983.

42. **Prescott, S. M.**, The effect of eicosapentaenoic acid on leukotriene B production by human neutrophils, *J. Biol. Chem.*, 259, 7615—7621, 1984.

43. **Samuelsson, B.**, Leukotrienes: a new class of mediators of immediate hypersensitivity reaction and inflammation, *Adv. Prostaglandin, Thromboxane, Leukotriene Res.*, 11, 1—26, 1983.

44. **Powell, W. S. and Funk, C. D.**, Metabolism of 8,11,14-eicosatrienoic acid by fetal calf aorta, *Adv. Prostaglandin, Thromboxane, Leukotriene Res.*, 11, 111—117, 1983.

45. **Sprecher, H., Van Rollins, M., Sun, F., Wyche, A., and Needleman, P.**, Dihomo prostaglandins and thromboxane, *J. Biol. Chem.*, 257, 3912—3918, 1982.

46. **Atkinson, J. G. and Rokach, J.**, Synthesis of the leukotrienes and other lipoxygenase-derived products, in *CRC Handbook of Eicosanoids: Prostaglandins and Related Lipids*, Vol. 1, Willis, A. L., Ed., CRC Press, Inc., Boca Raton, Fla., 1987.

47. **Hamberg, M. and Samuelsson, B.**, Detection and isolation of an endoperoxide intermediate in prostaglandin biosynthesis, *Proc. Natl. Acad. Sci. U.S.A.*, 70, 899—903, 1973.

48. **Nugteren, D. H. and Hazelhof, E.**, Isolation and properties of intermediates in prostaglandin biosynthesis, *Biochim. Biophys. Acta*, 326, 448—461, 1973.

49. **Hamberg, M., Svenssen, J., Wakabayashi, T., and Samuelsson, B.**, Isolation and structure of two prostaglandin endoperoxides that cause platelet aggregation, *Proc. Natl. Acad. Sci. U.S.A.*, 71, 345—349, 1974.

50. **Willis, A. L., Vane, F. M., Kuhn, D. C., Scott, C. G., and Petrin, M.**, An endoperoxide aggregator (LASS) formed in platelets in response to thrombotic stimuli: purification, identification, and unique biological significance, *Prostaglandins*, 8, 453—507, 1974.

51. **Samuelsson, B.**, On the incorporation of oxygen in the conversion of 8,11,14-eicosatrienoic acid to prostaglandin E_1, *J. Am. Chem. Soc.*, 87, 3011—3013, 1965.

52. **Bergstrom, S. and Sjovall, J.**, The isolation of prostaglandin F from sheep prostate glands, *Acta Chem. Scand.*, 14, 1693—1700, 1960.

53. **Bergstrom, S. and Sjovall, J.**, The isolation of prostaglandin E from sheep prostate glands, *Acta Chem. Scand.*, 14, 1701—1705, 1960.

54. **Bergstrom, S., Rhyhage, R., Samuelsson, B., and Sjovall, J.,** The structures of prostaglandin E_1, $F_{1\alpha}$, and $F_{1\beta}$, *J. Biol. Chem.*, 238, 3555—3564, 1963.

55. **Hamberg, M. and Samuelsson, B.,** New groups of naturally occurring prostaglandins, in *Prostaglandins, Proc. 2nd Nobel Symposium, Stockholm,* Bergstrom, S. and Samuelsson, B., Eds., Almqvist & Wicksell, Stockholm and Interscience, New York, 1967, 63—70.

56. **Lee, J. B., Covino, B. G., Takman, B. H., and Smith, E. R.,** Renomedullary vasodepressor substance medullin: isolation, chemical characterization, and physiological properties, *Circ. Res.*, 15, 57—77, 1965.

57. **Jones, R. L.,** 15-Hydroxy-9-oxoprosta-11,13-dienoic acid as the product of a prostaglandin isomerase, *J. Lipid Res.*, 13, 511—518, 1972.

58. **Jones, R. L.,** Preparation of prostaglandins C. Chemical fixation of prostaglandin A isomerase to a gel support and partition chromatography of prostaglandins A, B, and C, *Prostaglandins*, 5, 283—290, 1970.

59. **Jones, R. L. and Cammock, S.,** Purification, properties, and biological significance of prostaglandin A isomerase, *Adv. Biosci.*, 9, 61—70, 1973.

60. **Moncada, S., Gryglewski, R., Bunting, S., and Vane, J. R.,** An enzyme isolated from arteries transforms prostaglandin endoperoxides to an unstable substance that inhibits platelet aggregation, *Nature (London)*, 263, 663—665, 1976.

61. **Johnson, R. A., Morton, D. R., Kinner, J. H., Gorman, R. R., McGuire, J. C., Sun, F. F., Whittaker, N., Bunting, S., Salmon, J., Moncada, S., and Vane, J. R.,** The chemical structure of prostaglandin X (prostacyclin), *Prostaglandins*, 12, 915—928, 1976.

62. **Dusting, G. J., Moncada, S., and Vane, J. R.,** Prostacyclin: its biosynthesis, actions, and clinical potential, in *Prostaglandins and the Cardiovascular System,* Oates, J. A., Ed., Raven Press, New York, 1982, 59—106.

63. **Fukushima, M., Kato, T., Ota, K., Arai, Y., and Narumiya, S.,** 9-Deoxy-Δ^9-prostaglandin D_2. A prostaglandin D_2 derivative with potent anti-neoplastic and weak smooth muscle contracting activities, *Biochem. Biophys. Res. Commun.*, 109, 626—633, 1982.

64. **Needleman, P., Raz, A., Minkes, M. S., Ferrendelli, J. A., and Sprecher, M.,** Triene prostaglandins, prostacyclin, and thromboxane biosynthesis and unique biological properties, *Proc. Natl. Acad. Sci. U.S.A.*, 76, 944—948, 1979.

65. **Örning, L. and Hammarström, S.,** Isolation and characterization of 15-hydroxylated metabolites of leukotriene C_4, *FEBS Lett.*, 153, 253—256, 1983.

66. **Needleman, P., Whitaker, M. O., Wyche, A., Walters, K., Sprecher, H., and Raz, A.,** Manipulation of platelet aggregation by prostaglandins and their fatty acid precursor: pharmacological basis for a therapeutic approach, *Prostaglandins*, 19, 165—181, 1980.

67. **Hamberg, M., Svensson, J., and Samuelsson, B.,** Thromboxane: a new group of biologically active compounds derived from prostaglandin endoperoxides, *Proc. Natl. Acad. Sci. U.S.A.*, 72, 2944—2998, 1975.

68. **Wlodawer, P. and Hammarström, S.,** Thromboxane synthase from bovine lung-solubilization and partial purification, *Biochem. Biophys. Res. Commun.*, 80, 525—532, 1978.

69. **Needleman, P., Minkes, M., and Raz, A.,** Thromboxanes: selective biosynthesis and distinct biological properties, *Science*, 193, 163—165, 1976.

70. **Falardeau, P., Hamberg, M., and Samuelsson, B.,** Metabolism of 8,11,14-eicosatrienoic acid in human platelets, *Biochim. Biophys. Acta*, 441, 193—200, 1973.

71. **Lands, W. E. M., LeTellier, P. R., Rome, L. H., and Vanderhoek, Y.,** Inhibition of prostaglandin biosynthesis, *Adv. Biosci.*, 9, 15—28, 1973.

72. **Nidy, E. G. and Johnson, R. A.,** Synthesis of prostaglandin I_3, *Tetrahedron Lett.*, 27, 2375—2378, 1978.

73. **Diczfalusy, U., Falardeau, P., and Hammarström, S.,** Conversion of prostaglandin endoperoxides to C17-hydroxy acids, catalyzed by human platelet thromboxane synthase, *FEBS Lett.*, 84, 271—274, 1977.

74. **Needleman, P., Sprecher, H., Whitaker, M. O., and Wyche, A.,** Mechanism underlying the inhibition of platelet aggregation by eicosapentaenoic acid and its metabolites, in *Advances in Prostaglandin and Thromboxane Research,* Vol. 6, Samuelsson, B., Ramwell, P. W., and Paoletti, R., Eds., Raven Press, New York, 1980, 61—68.

75. **Willis, A. L., Comai, K., Kuhn, D. C., and Paulsrud, J.,** Dihomo-gamma-linolenate suppresses platelet aggregation when administered in vitro or in vivo, *Prostaglandins*, 8, 509—519, 1974.

76. **Bergstrom, S., Carlson, L. A., and Weeks, J. R.,** The prostaglandins: a family of biologically active lipids, *Pharmacol. Rev.*, 20, 1—48, 1968.

77. **Von Euler, U. S. and Eliasson, R.,** *Prostaglandins, Medical Chemistry Series of Manuscripts,* Vol. 8, Academic Press, New York, London, 1967.

78. **Jones, R. L. and Cammock, S.,** Purification, properties and biological significance of prostaglandin A isomerase, *Adv. Biosci.*, 9, 61—70, 1973.

79. **Wong, P. Y.-K., Lee, W. H., Chao, P. H.-W., Reiss, R. F., and McGiff, J. C.,** Metabolism of prostacyclin by 9-hydroxyprostaglandin dehydrogenase in human platelets, *J. Biol. Chem.*, 255, 9021—9024, 1980.

80. **Bryant, R. W. and Bailey, J. M.,** Isolation of glucose-sensitive platelet lipoxygenase products from arachidonic acid, in *Advances in Prostaglandin and Thromboxane Research,* Vol. 6, Samuelsson, B., Ramwell, P. W., and Paoletti, R., Eds., Raven Press, New York, 1980, 95—99.
81. **Goetzl, E. J. and Sun, F. F.,** Generation of unique monohydroxyeicosatetraenoic acids from arachidonic acid by human neutrophils, *J. Exp. Med.,* 150, 406—411, 1979.
82. **Lee, T. H., Mencia-Huerta, J.-M., Shih, C., Corey, E. J., Lewis, R. A., and Austen, K. F.,** Characterization and biologic properties of 5,12-dihydroxy derivatives of eicosapentaenoic acid, including leukotriene B₅ and the double lipoxygenase product, *J. Biol. Chem.,* 259, 2383—2389, 1984.
83. **Lee, C. W., Lewis, R. A., Tauber, A. I., Mehrota, M., Corey, E. J., and Austen, K. F.,** The myeloperoxidase-dependent metabolism of leukotrienes C₄, D₄, and E₄ to 6-*trans*-leukotriene B₄ diastereomers and the subclass-specific S-diastereomeric sulfoxides, *J. Biol. Chem.,* 258, 15004—15010, 1983.
84. **Lewis, R. A., Austen, K. F., Drazen, J. M., Clark, D. A., Marfat, A., and Corey, E. J.,** Slow reacting substances of anaphylaxis: identification of leukotriene C-1 and D from human and rat sources, *Proc. Natl. Acad. Sci. U.S.A.,* 77, 3710—3714, 1980.
85. **Lewis, R. A., Drazen, J. M., Austen, K. F., Clark, D. A., and Corey, E. J.,** Identification of the C(6)-S-conjugate of leukotriene A with cysteine as a naturally-occurring slow reacting substance of anaphylaxis (SRS-A). Importance of the 11-*cis* geometry for biological activity, *Biochem. Biophys. Res. Commun.,* 96, 271—277, 1980.
86. **Morris, H. A., Taylor, G. W., Piper, P. J., and Tippins, J. R.,** Structure of slow-reacting substance of anaphylaxis from guinea pig lung, *Nature (London),* 285, 104—105, 1980.
87. **Lewis, R. A., Austen, K. F., Drazen, J. M., Clark, D. A., and Corey, E. J.,** Slow reacting substance of anaphylaxis: identification of leukotrienes C-1 and D *in vivo* and *in vitro, Proc. Natl. Acad. Sci. U.S.A.,* 77, 4354—4358, 1980.
88. **Borgeat, P. and Samuelsson, B.,** Metabolism of arachidonic acid in polymorphonuclear leukocytes. Structural analyses of novel hydroxylated compounds, *J. Biol. Chem.,* 254, 7865—7869, 1979.
89. **Ueda, N., Kaneko, S., Yoshimoto, T., and Yamamoto, S.,** Purification of arachidonate 5-lipoxygenase from porcine leukocytes and its reactivity with hydroperoxyeicosatetraenoic acids, *J. Biol. Chem.,* 261, 7982—7988, 1986.
90. **Borgeat, P. and Samuelsson, B.,** Arachidonic acid metabolism in polymorphonuclear leukocytes: unstable intermediate in formation of dihydroxy acids, *Proc. Natl. Acad. Sci. U.S.A.,* 76, 3213—3217, 1979.
91. **Borgeat, P.,** Biochemistry of the Leukotrienes, in *CRC Handbook of Eicosanoids: Prostaglandins and Related Lipids,* Vol. I, Willis, A. L., Ed., CRC Press, Boca Raton, Fla., 1987.
92. **Bryant, R. W., Schewe, J., Rapoport, S. M., and Bailey, J. M.,** Leukocyte formation by a purified reticulocyte lipoxygenase enzyme. Conversion of arachidonic acid and 15-hydroxyeicosatetraenoic acid to 14,15-leukotriene A₄, *J. Biol. Chem.,* 260, 3548—3555, 1985.
93. **Wong, P.Y-K., Westlund, P., Hamberg, M., Granström, E., Chao, P.H-W., and Samuelsson, B.,** 15-Lipoxygenase in human platelets, *J. Biol. Chem.,* 260, 9162—9165, 1985.
94. **Hunter, J. A., Finkbeiner, W. E., Nadel, J. A., Goetzl, E. J., and Holtzman, M. J.,** Predominant generation of 15-lipoxygenase metabolites of arachidonic acid by epithelial cells from human trachea, *Proc. Natl. Acad. Sci. U.S.A.,* 82, 4633—4637, 1985.
95. **Hopkins, N. K., Oglesby, T. D., Bundy, G. L., and Gorman, R. R.,** Biosynthesis and metabolism of 15-hydroperoxy-5,8,11,13-eicosatetraenoic acid by human umbilical vein endothelial cells, *J. Biol. Chem.,* 259, 14048—14053, 1984.
96. **Pace-Asciak, C. R.,** Formation of hepoxilin A4,B4 and the corresponding trioxilins from 12(S)-hydroperoxy-5,8,10,14,17-eicosapentaenoic acid, *Prostaglandins Leukotrienes Med.,* 22, 1—9, 1986.
97. **Pace-Asciak, C. R., Mizuno, K., and Yamamoto, S.,** Formation of 8,11,12-trihydroxyeicosatrienoic acid by a rat lung high speed supernatant fraction, *Biochim. Biophys. Acta,* 665, 352—354, 1981.
98. **Pace-Asciak, C. R., Granström, E., and Samuelsson, B.,** Arachidonic acid epoxides. Isolation and structure of two hydroxy epoxide intermediates in the formation of 8,11,12- and 10,11,12-trihydroxyeicosatrienoic acids, *J. Biol. Chem.,* 258, 6835—6840, 1983.
99. **Pace-Asciak, C. R. and Martin, J. M.,** Hepoxilin, a new family of insulin secretagogues formed by intact pancreatic islets, *Prostaglandins Leukotrienes Med.,* 16, 173—180, 1984.
100. **Pace-Asciak, C. R., Martin, J. M., Corey, E. J., and Su, W.-G.,** Endogenous release of hepoxilin A₃ from isolated perfused pancreatic islets of langerhans, *Biochem. Biophys. Res. Commun.,* 128, 942—946, 1985.
101. **Serhan, C. N., Hamberg, M., and Samuelsson, B.,** Lipoxins: novel series of biologically active compounds formed from arachidonic acid in human leukocytes, *Proc. Natl. Acad. Sci. U.S.A.,* 81, 5335—5339, 1984.
102. **Kühn, H., Wiesner, R., and Stender, H.,** The formation of products containing a conjugated tetraenoic system by pure reticulocyte lipoxygenase, *FEBS Lett.,* 177, 255—259, 1984.
103. **Serhan, C. N., Hamberg, M., Samuelsson, B., Morris, J., and Wishka, D. G.,** On the stereochemistry and biosynthesis of lipoxin B, *Proc. Natl. Acad. Sci. U.S.A.,* 83, 1983—1987, 1986.

104. **Hamberg, M.,** Transformations of 5,8,11,14,17-eicosapentaenoic acid in human platelets, *Biochim. Biophys. Acta,* 618, 389—398, 1980.

105. **Lee, T. H., Mencia-Huerta, J-M., Shih, C., Corey, E. J., Lewis, R. A., and Austen, K. F.,** Effects of exogenous arachidonic, eicosapentaenoic, and docosahexaenoic acid on the generation of 5-lipoxygenase pathway products by ionophore-activated human neutrophils, *J. Clin. Invest.,* 74, 1922—1933, 1984.

106. **Aveldano, M. I. and Sprecher, H.,** Synthesis of hydroxy fatty acids from 4,7,10,13,16,19-[1-^{14}C] docosahexaenoic acid by human platelets, *J. Biol. Chem.,* 258, 9339—9343, 1983.

107. **Fischer, S., Schacky, C. V., Siess, W., Strasser, T., and Weber, P. C.,** Uptake, release and metabolism of docosahexaenoic acid (DHA $C_{22}\omega$ 613) in human platelets and neutrophils, *Biochem. Biophys. Res. Commun.,* 120, 907—918, 1984.

108. **Holland, D. L., East, J., Gibson, K. H., Clayton, E., and Oldfield, A.,** Identification of the hatching factor of the barnacle *balanus balanoides* as the novel eicosanoid 10,11,12-trihydroxy-5,8,14,17-eicosatetraenoic acid, *Prostaglandins,* 29, 1021—1029, 1985.

109. **Kulkarni, P. S. and Srinivasan, B. D.,** Eicosapentaenoic acid metabolism in human and rabbit anterior uvea, *Prostaglandins,* 31, 1159—1164, 1986.

110. **Terano, T., Salmon, J. A., and Moncada, S.,** Anti-inflammatory effects of eicosapentaenoic acid: relevance to icosanoid formation, in *Advances in Prostaglandin, Thromboxane and Leukotriene Research,* Vol. 15, Hayaishi, O. and Yamamoto, S., Eds., Raven Press, New York, 1985, 253—256.

111. **Powell, W. S.,** Formation of 6-oxoprostaglandin $F_{1\alpha}$, 6,15-dioxoprostaglandin $F_{1\alpha}$ and monohydroxy fatty acids from arachidonic acid by fetal calf aorta and ductus arteriosus, *J. Biol. Chem.,* 257, 9457—9464, 1982.

112. **Funk, C. D. and Powell, W. S.,** Conversion of 8,11,14-eicosatrienoic acid to 11,12-epoxy-10-hydroxy-8-heptadecenoic acid by aorta, *Prostaglandins,* 25, 299—309, 1983.

113. **Elliott, W. J., Morrison, A. R., Sprecher, H., and Needleman, P.,** Calcium-dependent oxidation of 5,8,11-icosatrienoic acid by the cyclooxygenase enzyme system, *J. Biol. Chem.,* 261, 6719—6724, 1986.

114. **Hammarström, S.,** Conversion of 5,8,11-eicosatrienoic acid to leukotrienes C_3 and D_3, *J. Biol. Chem.,* 256, 2275—2279, 1981.

115. **Jakschik, B. A., Morrison, A. R., and Sprecher, H.,** Products derived from 5,8,11-eicosatrienoic acid by the 5-lipoxygenase-leukotriene pathway, *J. Biol. Chem.,* 258, 12797—12800, 1983.

116. **Evans, J. F., Nathaniel, D. J., Zamboni, R. J., and Ford-Hutchinson, A. W.,** Leukotriene A_3: a poor substrate but a potent inhibitor of rat and human neutrophil leukotriene A_4 hydrolase, *J. Biol. Chem.,* 260, 10966—10970, 1985.

117. **Levine, J. D., Lam, D., Taiwo, Y. O., Donatoni, P., and Goetzl, E. J.,** Hyperalgesic properties of 15-lipoxygenase products of arachidonic acid, *Proc. Natl. Acad. Sci. U.S.A.,* 83, 5331—5334, 1986.

118. **Powell, W. S.,** Novel pathway for the metabolism of 6-trans-leukotriene B_4 by human polymorphonuclear leukocytes, *Biochem. Biophys. Res. Commun.,* 136, 707—712, 1986.

119. **Hammarström, S.,** Leukotriene C_5: a slow reacting substance derived from eicosapentaenoic acid, *J. Biol. Chem.,* 255, 7093—7094, 1980.

120. **Lagarde, M., Burtin, M., Rigaud, M., Sprecher, H., Dechavanne, M., and Renaud, S.,** Prostaglandin E_2-like activity of 20:3ω9 platelet lipoxygenase end product, *FEBS Lett.,* 181, 53—56, 1985.

121. **Stenson, W. F., Prescott, S. M., and Sprecher, H.,** Leukotriene B formation by neutrophils from essential fatty acid deficient rats, *J. Biol. Chem.,* 259, 11784—11789, 1984.

122. **Van Rollins, M., Horrocks, L., and Sprecher, H.,** Metabolism of 7,10,13,16-docosatetraenoic acid to dihomo-thromboxane, 14-hydroxy-7,10,12-nonadecatrienoic acid and hydroxy fatty acids by human platelets, *Biochim. Biophys. Acta,* 833, 272—280, 1985.

123. **Hamberg, M.,** ω6-Oxidation of 6,9,12-Octadecatrienoic acid in human platelets, *Biochem. Biophys. Res. Commun.,* 117, 593—600, 1983.

124. **Sprecher, H. and Careaga, M. M.,** Metabolism of (n-6) and (n-3) polyunsaturated fatty acids by human platelets, *Prostaglandins Leukotrienes Med.,* 23, 129—134, 1986.

125. **Careaga, M. M. and Sprecher, H.,** Synthesis of two hydroxy fatty acids from 7,10,13,16,19-docosapentaenoic acid by human platelets, *J. Biol. Chem.,* 259, 14413—14417, 1984.

126. **Milks, M. M. and Sprecher, H.,** Metabolism of 4,7,10,13,16-docosapentaenoic acid by human platelet cyclooxygenase and lipoxygenase, *Biochim. Biophys. Acta,* 835, 29—35, 1985.

127. **Elliott, W. J., Morrison, A. R., Sprecher, H. W., and Needleman, P.,** The metabolic transformation of columbinic acid and the effect of topical application of the major metabolites on rat skin, *J. Biol. Chem.,* 260, 987—992, 1985.

128. **Powell, W. S.,** Properties of leukotriene B_4 20-hydroxylase from polymorphonuclear leukocytes, *J. Biol. Chem.,* 259, 3082—3089, 1984.

129. **Borgeat, P., Picard, S., Vallerand, P., and Sirois, S.,** Transformation of arachidonic acid in leukocytes. Isolation and structural analysis of a novel dihydroxy derivative, *Prostaglandins and Medicine,* 6, 557—570, 1981.

130. **Deorge, D. R. and Corbett, M. D.,** An arachidonic acid specific lipoxygenase from the gorgonian coral *Pseudoplexaura porosa, Experientia,* 38, 901—902, 1982.

131. **Setty, B. N. Y., Stuart, M. J., and Walenga, R. W.,** Formation of 11-hydroxy-eicosatetraenoic acid in umbilical arteries is catalyzed by cyclooxygenase, *Biochim. Biophys. Acta,* 833, 484—494, 1985.

132. **Schewe, T., Kühn, M., and Rapoport, S. M.,** Positional specificity of lipoxygenases and their suitability for testing potential drugs, *Prostaglandins Leukotrienes Med.,* 23, 155—160, 1986.

133. **Shimizu, T., Rädmark, O., and Samuelsson, B.,** Enzyme with dual lipoxygenase activities catalyzes leukotriene A$_4$ synthesis from arachidonic acid, *Proc. Natl. Acad. Sci. U.S.A.,* 81, 689—693, 1984.

134. **Serhan, C., Fahlstadius, P., Dahlen, S. E., Hamberg, M., and Samuelsson, B.,** Biosynthesis and biological activities of lipoxins, in *Advances in Prostaglandin Thromboxane and Leukotriene Research,* Vol. 15, Hayaishi, O. and Yamamoto, S., Eds., Raven Press, New York, 1985, 163—166.

135. **Salomon, R. G. and Miller, D. B.,** Levuglandins: isolation, characterization and total synthesis of new secoprostanoid products from prostaglandin endoperoxides, in *Advances in Prostaglandin Thromboxane and Leukotriene Research,* Vol. 15, Hayaishi O. and Yamamoto, S., Eds., Raven Press, New York, 1985, 323—326.

136. **Glasgow, W. C., Harris, T. M., and Brash, A. R.,** A short-chain aldehyde is a major lipoxygenase product in arachidonic acid stimulated porcine leukocytes, *J. Biol. Chem.,* 261, 200—204, 1986.

137. **Morrison, A. R., and Pascoe, N.,** Metabolism of arachidonate through NADPH-dependent oxygenase of renal cortex, *Proc. Natl. Acad. Sci. U.S.A.,* 78, 7375—7378, 1981.

138. **Oliw, E. H., Guengerich, F. P., and Oates, J. A.,** Oxygenation of arachidonic acid by hepatic mono-oxygenases. Isolation and metabolism of four epoxide intermediates, *J. Biol. Chem.,* 257, 3771—3781, 1982.

139. **Lewis, R. A. and Austen, K. F.,** Molecular determinants for functional responses to the sulfidopeptide leukotrienes: metabolism and receptor subclasses, *J. Allergy Clin. Immunol.,* 74, (3-Pt.2), 369—372.

140. **Yamamoto, S., Ueda, N., Yokoyama, C., Kaneko, S., Shinjo, F., Yoshimoto, T., Oates, J. A., and Brash, A. R.,** Dioxygenase and leukotriene A synthase activities of arachidonate 5- and 12-lipoxygenase purified from porcine leukocytes, in *Proceedings, Sixth International Conference on Prostaglandins and Related Compounds,* Florence, Italy, 1986, p. 6.

141. **Brash, A. R., Oates, J. A., Yokoyama, C., and Yamamoto, S.,** Mechanism of transformation of 15(S)-HPETE to 15-series leukotrienes by purified porcine leukocyte 12-lipoxygenase, in *Proceedings, Sixth International Conference on Prostaglandins and Related Compounds,* Florence, Italy, 1986, p. 8.

142. **Westlund, P.,** Formation of dihydroxy leukotrienes in human platelets catalyzed via the 12-lipoxygenase pathway, in *Proceedings, Sixth International Conference on Prostaglandins and Related Compounds,* Florence, Italy, 1986, p. 89.

143. **Glasgow, W. C. and Brash, A. R.,** Specific short-chain aldehydes are cleavage products of the transformation of arachidonic acid to HPETEs and leukotrienes, in *Proceedings, Sixth International Conference on Prostaglandins and Related Compounds,* Florence, Italy, 1986, p. 90.

144. **Feinmark, S. J. and Cannon, P. J.,** Endothelial cell-neutrophil interactions lead to endothelial cell LTC$_4$ synthesis, in *Proceedings, Sixth International Conference on Prostaglandins and Related Compounds,* Florence, Italy, 1986, p. 91a.

145. **Claesson, H-E. and Haeggström, J.,** Metabolism of leukotriene A$_4$ by human endothelial cells, in *Proceedings, Sixth International Conference on Prostaglandins and Related Compounds,* Florence, Italy, 1986, p. 91b.

146. **Klein, J., Pace-Asciak, C. R., and Speilberg, S. P.,** Biosynthesis and metabolism of tritiated LTC$_4$ by human platelets, in *Proceedings, Sixth International Conference on Prostaglandins and Related Compounds,* Florence, Italy, 1986, p. 92.

147. **Fitzpatrick, F. A., Haeggström, J., Granström, E., and Samuelsson, B.,** Metabolism of leukotriene A$_4$ by an enzyme in blood plasma: a possible leukotactic mechanism, *Proc. Natl. Acad. Sci. U.S.A.,* 80, 5425—5429, 1983.

148. **Haeggström, J., Meijer, J., and Rädmark, O.,** Leukotriene A$_4$. Enzymatic conversion into 5,6-dihydroxy-7,9,11,14-eicosatetraenoic acid by mouse liver cytosolic epoxide hydrolase, *J. Biol. Chem.,* 261, 6332—6337, 1986.

149. **Wetterholm, A., Haeggström, J., Meijer, J., and Rädmark, O.,** Metabolism of 14,15-leukotriene A$_4$ by cytosolic epoxide hydrolase, in *Proceedings, Sixth International Conference in Prostaglandins and Related Compounds,* Florence, Italy, 1986, p. 185.

150. **Puustinen, T., Serhan, C. N., and Samuelsson, B.,** Evidence for a 5(6)-epoxy-tetraene intermediate in the biosynthesis of lipoxin B in human leukocytes, in *Proceedings, Sixth International Conference on Prostaglandins and Related Compounds,* Florence, Italy, 1986, p. 187.

151. **Wong, P.Y.-K., Hughes, R., and Lam, B.,** Lipoxene: a new group of trihydroxy pentaenes of eicosapentaenoic acid derived from porcine leukocytes, *Biochem. Biophys. Res. Commun.,* 126, 763—772, 1985.

152. **Woolard, P. M.,** Stereochemical difference between 12-hydroxy-5,8,10,14-eicosatetraenoic acid in platelets and psoriatic lesions, *Biochem. Biophys. Res. Commun.,* 136, 169—176, 1986.

153. **Furstenberger, G., Gschwendt, M., Hagedorn, H., and Marks, F.,** On the role of lipoxygenase production in the hyperplastic response of mouse epidermis to exogenous stimuli, in *Proceedings, Sixth International Conference on Prostaglandins and Related Compounds*, Florence, Italy, 1986, p. 434.

154. **Schröder, J. M. and Christophers, E.,** 5-HETE-lactone, a natural chemotaxis-antagonist and marker for 5-lipoxygenase activation, in *Proceedings, Sixth International Conference on Prostaglandins and Related Compounds*, Florence, Italy, 1986, p. 438.

155. **Lianos, E. A., Rahman, M. A., and Dunn, M. J.,** Glomerular arachidonate lipoxygenation in rat nephrotoxic serum nephritis, *J. Clin. Invest.*, 76, 1355—1359, 1985.

156. **Liston, T. E. and Roberts, L. J., II.,** Transformation of prostaglandin D_2 to $9\alpha,11\beta$-15(S)-trihydroxy prosta-(5Z,13E)-dienoic acid ($9\alpha,11\beta$-prostaglandin F_2): a unique biologically active prostaglandin produced enzymatically in vivo in humans, *Proc. Natl. Acad. Sci. U.S.A.*, 82, 6030—6034, 1985.

157. **Liston, T. E., Oates, J. A., and Roberts, L. J., II,** Prostaglandin D_2 is metabolized in humans to 9α, 11β-prostaglandin F_2, a novel biologically active prostaglandin, in *Advances in Prostaglandin, Thromboxane, and Leukotriene Research*, Vol. 15, Hayaishi, O. and Yamamoto, S., Eds., Raven Press, New York, 1985, 365—367.

158 **Hammarström, S., Örning, L., Bernström, K., Gustafsson, B., Norin, E., and Kaijser, L.,** Metabolism of leukotriene C_4 in rats and humans, in *Advances in Prostaglandin, Thromboxane, and Leukotriene Research*, Vol. 15, Hayaishi, O. and Yamamoto, S., Eds., Raven Press, New York, 1985, pp. 185—188.

159. **Snyder, G., Lattanzio, F., Yadagiri, P., Falck, J. R. and Capdevila, J.,** 5,6-Epoxyeicosatrienoic acid mobilizes Ca^{2+} in anterior pituitary cells, *Biochem. Biophys. Res. Commun.*, 139, 1188—1194, 1986.

160. **Oliw, E. H. and Oates, J. A.,** Oxygenation of arachidonic acid by hepatic microsomes of the rabbit, *Biochim. Biophys. Acta*, 666, 327—340, 1981.

161. **Prescott, S. M., Zimmerman, G. A., and Morrison, A. R.,** The effects of a diet rich in fish oil on human neutrophils: identification of leukotriene B_5 as a metabolite, *Prostaglandins*, 30, 209—227, 1985.

162. **Örning, L., Norin, E., Gustafsson, B., and Hammarström, S.,** In vivo metabolism of leukotriene C_4 in germ free and conventional rats. Fecal excretion of N-acetylleukotriene E_4, *J. Biol. Chem.*, 261, 766—771, 1986.

163. **Bernström, K. and Hammarström, S.,** Metabolism of leukotriene E_4 by rat tissues: formation of N-acetylleukotriene E_4, *Arch. Biochem. Biophys.*, 244, 486—491, 1986.

164. **Haeggström, J., Rädmark, O., and Fitzpatrick, F. A.,** Leukotriene A_4-hydrolase activity in guinea pig and human liver, *Biochim. Biophys. Acta*, 835, 378—384, 1985.

165. **Claeys, M., Kivits, G. A. A., Christ-Hazelhof, E., and Nugteren, D. H.,** Metabolic profile of linoleic acid in porcine leukocytes through the lipoxygenase pathway, *Biochim. Biophys. Acta*, 837, 35—51, 1985.

166. **Claeys, M., Coene, M.-C., Herman, A. G., Jouvenaz, G. H., and Nugteren, D. H.,** Characterization of monohydroxylated lipoxygenase metabolites of arachidonic and linoleic acids in rabbit peritoneal tissue, *Biochim. Biophys. Acta*, 713, 160—169, 1982.

167. **Coene, M.-C., Bult, H., Claeys, M., Laekeman, G. M., and Herman, A. G.,** Inhibition of rabbit platelet activation by lipoxygenase products of arachidonic and linoleic acid, *Thrombosis Res.*, 42, 205—214, 1986.

168. **Funk, C. D. and Powell, W. S.,** Metabolism of linoleic acid by prostaglandin endoperoxide synthase from adult and fetal blood vessels, *Biochim. Biophys. Acta*, 754, 57—71, 1983.

169. **Funk, C. D. and Powell, W. S.,** Release of prostaglandins and monohydroxy and trihydroxy metabolites of linoleic and arachidonic acids by adult and fetal aorta and ductus arteriosis, *J. Biol. Chem.*, 260, 7481—7488, 1985.

170. **Fukushima, M. and Kato, T.,** Antitumor marine icosanoids: clavulones and punaglandins, in *Advances in Prostaglandin, Thromboxane, and Leukotriene Research*, Vol. 15, Hayaishi, O. and Yamamoto, S., Eds., Raven Press, New York, 1985, 415—418.

171. **Campbell, W. B., Falck, J. R., Okita, J. R., Johnson, A. R., and Callahan , K. S.,** Synthesis of dihomoprostaglandins from adrenic acid (7,10,13,16-docosatetraenoic acid) by human endothelial cells, *Biochim. Biophys. Acta*, 837, 67—76, 1985.

172. **Powell, W. S. and Gravelle, F.,** Metabolism of eicosapentaenoic acid by aorta: formation of a novel 13-hydroxylated prostaglandin, *Biochim. Biophys. Acta*, 835, 201—211, 1985.

173. **Camp, R. D. R., Cunningham, F. M., Woolard, P. M., Fincham, N. J., Mallet, A. I., Black, A. K., and Greaves, M. W.,** Novel biologically active monohydroxy fatty acid-like lipid in psoriatic skin lesions, in *Proceedings, Sixth International Conference on Prostaglandins and Related Compounds*, Florence, Italy, 1986, p. 435.

174. **Shimizu, T., Izumi, T., Ohishi, N., Seyama, Y., Rädmark, O., and Samuelsson, B.,** Biosynthesis and further transformation of leukotriene A_4, in *Proceedings, Sixth International Conference on Prostaglandins and Related Compounds*, Florence, Italy, 1986, p. 7.

175. **Fitzpatrick, F. A., Ennis, M. D., Baze, M. E., Wynalda, M. A., McGee, J. E., and Liggett, W. F.,** Inhibition of cyclooxygenase activity and platelet aggregation by epoxyeicosatetraenoic acids. Influence of stereochemistry, *J. Biol. Chem.*, 261, 15334—15338, 1986.

176. **Granström, E. and Kindahl, H.**, Radioimmunologic determination of 15-keto-13,14-dihydro-PGE_2; a method for its stable degradation product 11-deoxy-13,14-dihydro PGE_2, in *Advances in Prostaglandin and Thromboxane Research*, Vol. 6, Samuelsson, B., Paoletti, R., and Ramwell, P., Eds., Raven Press, New York, 1980, 181—182.

177. **Bothwell, W., Verburg, M., Wynalda, M., Daniels, E. G., and Fitzpatrick, F. A.**, A radioimmunoassay for the unstable pulmonary metabolites of prostaglandins E_1 and E_2: an indirect index of their *in vivo* deposition and pharmacokinetics, *J. Pharm. Exp. Therap.*, 220, 229—235, 1982.

178. **Tobias, L. D., Vane, F. M., and Paulsrud, J. R.**, The biosynthesis of 1a,1b-dihomo-PGE_2 and 1a,1b-dihomo-$PGF_{2\alpha}$ from 7,10,13,16-docosatetraenoic acid by an acetone-pentane powder of sheep vesicular gland microsomes, *Prostaglandins*, 10, 443—468, 1975.

Eicosanoid Precursors

THE ESSENTIAL FATTY ACIDS: THEIR DERIVATION AND ROLE

James F. Mead and Anthony L. Willis

It is certainly true that the ultimate source of the precursors of the essential fatty acids (EFA) is a group of polyunsaturated fatty acids produced only in plants (largely linoleic, 18:2ω6, and α-linolenic, 18:3ω3). However, the same cannot be said with accuracy for the eicosanoids. With only a few minor exceptions, the eicosapolyenoic acids neccessary as eicosanoid precursors are synthesized only in the animal organism. Thus, for the purpose of the present discussion, it is necessary only to consider those biosynthetic pathways common to animals. It will also be of importance to have some idea of the relative amounts of these precursors likely to be present in different dietary sources.

OCCURRENCE OF EICOSAPOLYENOIC ACIDS IN FOOD SOURCES

Table 1 shows fatty acid composition of the depot fat of bullocks as a composite from different locations.[1] It can be seen that the immediate precursors of the prostanoids are virtually absent from the depot fat of ruminants. In nonruminant herbivores, such as the pig, they may represent as much as 2% of the total; in the organ lipids of herbivores, particularly in the phospholipid fraction, they may amount to over 20% of the total. At the opposite extreme is the body fat of marine fish. Table 2 lists the percent composition of the fatty acids of the body fat of kelp bass.[2] It can readily be seen that the eicosapolyenoic acid content is considerably higher in this source. However, the 20:5 of fish lipids is almost entirely of the ω3 family (derived from α-linolenic acid). The implications of this structure are discussed below.

THE CONVERSION OF PRECURSORS TO POLYUNSATURATED FATTY ACID PRODUCTS

In Vivo Conversions Demonstrated

For several years, interconversions of the polyunsaturated fatty acids remained obscure. Then, experiments were initiated in which labeled precursors and suspected intermediates were fed to rats and labeled products isolated and identified.[3,4] These studies revealed that the precursor fatty acids are alternately desaturated and elongated to form the principal polyunsaturated fatty acids (PUFA) found in animal tissues. These processes produce discrete families of PUFA but no interconversion between these families can occur. The ω3 family of PUFA are derived from α-linolenic acid (18:3ω3), the ω6 family from linoleic acid (18:2ω6), the ω9 family from oleic acid (18:1ω9), and the ω7 family from palmitoleic acid (16:1ω7). As depicted in Figure 1, in each family, principal products are formed and incorporated into the various lipids. Other members of the families are usually present only in small amounts.

Since the first step in each pathway is Δ6 desaturation, it is evident that this rate-controlling step represents a point at which competition could occur. Experiments were done in which large amounts of oleic acid in a diet minimal in linoleic acid actually produced EFA deficiency symptoms in guinea pigs.[5] Thus, evidence was provided that such competitive inhibition does take place in vivo. Extension of these findings[6,7] revealed that in competition for the Δ6 desaturase, 18:3ω3, appears to have the greatest affinity, followed by 18:2ω6, which has a much greater affinity than the monoenes. For this reason, the formation of 20:3ω7 and 20:3ω9 is suppressed as long as dietary 18:3ω3 or 18:2ω6 are available. In their absence, however, as with a fat-deficient diet, this inhibition is relieved and the formation of 20:3ω9

Table 1
FATTY ACID COMPOSITION OF THE DEPOT FAT OF BULLOCKS[1]

Fatty acid	Total (by GLC) (%)
12:0	0.3
14:0	3.1
15:0	0.9
16:0	27.7
17:0	1.9
18:0	23.9
14:1	1.1
16:1	5.9
16:2	1.6
18:1	31.1
18:2	2.0
18:3	0.3
20:1	0.3

Table 2
FATTY ACID COMPOSITION OF THE BODY FAT OF KELP BASS (*PARALABLAX CLATHRATUS*)[2]

Fatty acid	Total (%)
14:0	7.7
14:1	Trace
16:0	27.3
16:1	9.8
16:2	Trace
18:0	3.4
18:1	14.5
18:2	1.6
18:3	1.2
18:4 (20:1)	2.5
20:2	0.2
20:3	0.1
20:4	1.3, 1.0
20:5	9.9
22:4	0.6
22:5	0.2, 1.8
22:6	16.9

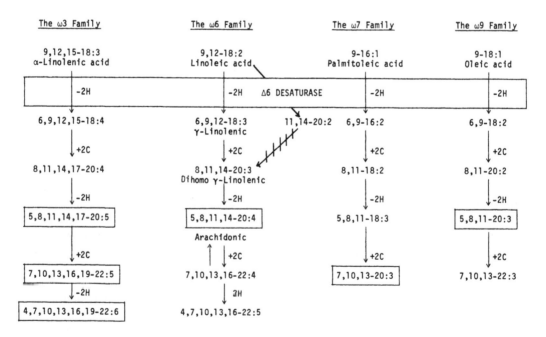

FIGURE 1.   Pathways of interconversion of the polyunsaturated fatty acids. Elongation is designated as +2C, desaturation by −2H. Principle products in each family are enclosed in boxes.

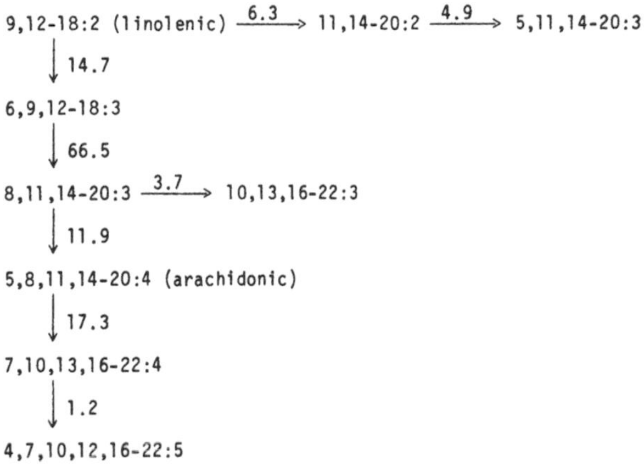

9,12-18:2 (linolenic) $\xrightarrow{6.3}$ 11,14-20:2 $\xrightarrow{4.9}$ 5,11,14-20:3

$\downarrow$ 14.7

6,9,12-18:3

$\downarrow$ 66.5

8,11,14-20:3 $\xrightarrow{3.7}$ 10,13,16-22:3

$\downarrow$ 11.9

5,8,11,14-20:4 (arachidonic)

$\downarrow$ 17.3

7,10,13,16-22:4

$\downarrow$ 1.2

4,7,10,12,16-22:5

FIGURE 2.   Rates of desaturation and elongation steps in in vitro conversion of linoleic to arachidonic acid.[14] Rates are expressed as nanomoles of product formed during a 3-min incubation of the substrate fatty acid (150 nmol of carboxylabeled precursor) with 5 mg of microsomal protein and adequate amounts of required cofactors.[14]

proceeds. In fact, the relative amount of 20:3ω9 (compared to 20:4ω6) is often used as a measure of EFA deficiency.[8]

In Vitro Experiments and Conditions

Bloomfield and Bloch[9] showed that desaturation of stearic to oleic acid could be carried out with a liver microsomal preparation in the presence of oxygen and NADPH. Subsequent reports from other laboratories indicated that the same conditions could be used for the desaturation of the PUFA. Moreover, with the addition of malonyl-CoA the same system could anaerobically carry out the elongation reactions.[10-12] Thus, both in vivo and in vitro experiments showed that the alternating desaturation-elongation pathway was the principal route by which the precursors were converted to PUFA products. It had once been thought that 11,14-20:2ω6 might be an intermediate in the ω6 pathway because it could often be found in systems carrying out the conversion of 18:2ω6 to 20:4ω6. However, Ullman and Sprecher[13] showed that the 20:3 produced from this fatty acid is actually 5,11,14 eicosatrienoic acid (5,11,14-20:3), a dead-end, rather than the intermediate 8,11,14-20:3 (20:3ω6; see Figure 1). This reaction has proved to be quite general in that any fatty acid with a double bond in the Δ11-position is desaturated only in the Δ5 position. There is no Δ8 desaturase and fatty acids with a double bond in the 8-position must be formed by elongation of fatty acids with a double bond in the 6-position. This includes, of course, all the derived EFA precursors of the prostaglandins. Such alternate pathways for production of arachidonic acid are described in more detail in the following chapter by Brenner.

It should also be noted that the elongation reaction, (but not desaturation) is inhibited by concentrations of saturated fatty acids normally present in the tissues. This type of inhibition should also favor the pathway from 18:2ω6 to 18:3ω6 rather than that to 20:2ω6. Several unexplained findings from the in vivo work have been clarified by subsequent studies. Sprecher and co-workers[14] have carried out experiments to determine the rates of the individual steps in the pathways of PUFA transformations. For 18:2ω6, the rates are given in Figure 2. It can readily be seen why the desaturation of 18:2ω6 is preferred over the elongation step and why little 18:3ω3 accumulates in the tissues. What is not so apparent, however, is why arachidonic acid (20:4ω6) is found to accumulate rapidly when it can be converted to adrenic acid (22:4ω6). There are two possible reasons. First, each step in the pathway is in competition with a transferase that transfers the acyl group to a glyceride. Moreover, the longer, more unsaturated fatty acids are subject to a reaction known as retroconversion,[15] in which they are shortened by two carbons. This reaction becomes more important as the PUFA chain length increases and, as a result, adrenic acid (22:4ω6) is converted to arachidonic acid rather than being further desaturated or incorporated into lipids.[16]

Such considerations cannot yet predict the fatty acid composition of the tissue lipids (see the section "Anomalies and Exceptions in EFA Metabolism" later in this chapter); however, they can aid in the interpretation of the analytical findings. It may be noted that of the two members of the ω6 family that serve as prostanoid precursors, 20:4ω6 is generally much more available than 20:3ω6. The availability of 20:5ω3, of course, may depend on the presence of 18:3ω3 in substantial amounts in the diet and/or of its occurrence in ingested fish oil (as in Eskimos).

Returning to a consideration mentioned in the first section of this chapter, it is evident that the eicosanoid precursors can be derived from two major sources. First, they can be obtained in the diet of flesh-eating animals. In the case of a land-based diet, arachidonic acid is a far more likely component than its immediate precursor, 20:3ω6. The availability of the ω3 fatty acids is usually limited for human consumption, being mainly in marine animals.

Second, they can be obtained by conversion from precursors within the organism. In this case as well, it is evident that arachidonic acid is likely to be by far the most prevalent substrate except perhaps in the very rare case in which large quantities of α-linolenic acid are consumed. It is not surprising, therefore, that 20:4ω6 is the precursor of most of the known physiologically active prostanoids.

OTHER PUFA WITH EFA ACTIVITY

The Odd-Chain Fatty Acids

In addition to the EFA already discussed, there are several fatty acids with EFA activity that do not fall within the usual definition of members of the ω6 or, possibly, the ω3 family. One such group consists of polyunsaturated odd-chain fatty acids of the ω5 and ω7 families and, possibly, the ω2 family. Schlenk[17] fed a mixture of 17-carbon PUFA derived from mullet oil to rats and noted conversions similar to those of the even carbon ω6 and ω3 fatty acids. In particular, 5,8,11,14-19:4 (19:4ω5), an analog of arachidonic acid (5,8,11,14-20:4; 20:4ω6), was formed and probably accounted for the EFA activity, which approached that of arachidonate. The ω7 odd-chain fatty acids, which occur in small amounts in mullet oil, were synthesized by Beerthuis and co-workers[18] and were tested for EFA activity and converted to prostaglandins. Both the 19:4ω5 and the 21:4ω7 had EFA activities and rates of conversion to eicosanoids about half those of arachidonic acid.

From these data it would appear that the requirement for essentiality in a fatty acid is a chain length of 19, 20, or 21 carbons and a system of *cis* double bonds in the 5,8,11,14-positions. The definition of EFA would also, of course, include those fatty acids that are readily converted to these products in the animal body. The ω3 even chain acids and, probably, the ω2 odd chain acids are in a special class and their exact essentiality for higher animals and their function have not yet been satisfactorily established.

Columbinic Acid

A very significant finding bearing on the function of EFA is that reported by Houtsmuller[19] on the EFA activity of a fatty acid isolated from the seed oil of the columbine. This fatty acid, designated as columbinic acid, has the structure: 5-*trans*,9-*cis*,12-*cis*-18:3. It thus has the structure of linoleic acid with an additional *trans* double bond at the 5-position. Metabolic

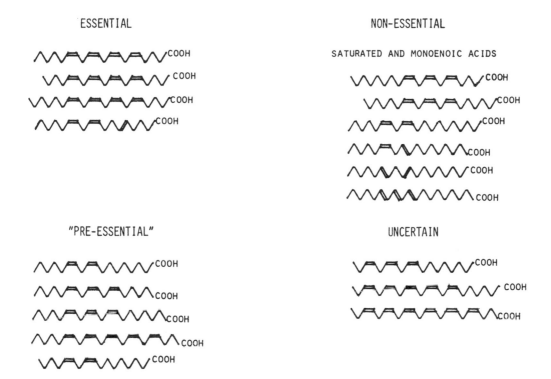

FIGURE 3. Structures of some essential and nonessential fatty acids. By "pre-essential" is meant those fatty acids known to be readily converted to EFA although they may function as EFA themselves. The uncertain classification includes members of the ω3 family fatty acids which have important, though not well-understood functions in the animal body, but may not be able to replace the known EFA.

studies in the rat revealed that columbinic acid is incorporated into lipids unchanged and is not elongated to a C_{20} homolog and, indeed, that dihomo columbinic acid is not an EFA. Moreover, columbinic acid is not converted to a prostaglandin-like compound and appears to inhibit the formation of prostaglandins from arachidonate. This finding, therefore, may afford the opportunity of distinguishing between EFA and eicosanoids functions (see Addendum).

In the case of the functions most often used to characterize the EFA-growth rate and skin permeability, columbinic is almost as efficient as is linoleic acid. Certain functions associated with prostaglandins, however, are inhibited by columbinic acid. It suppresses the inflammatory process, does not support adequate uterine contractions in labor, and does not prevent (but enhances) the hematuria produced in EFA deficiency by lesions in the papillary region of the kidneys.

Although there is undoubtedly some overlap of EFA and prostaglandin functions when cell membrane integrity is concerned with other reactions, this finding seems to present the opportunity not only to distinguish the structural from the precursor functions of the EFA but also to permit further consideration of the nature of the structural function.

STRUCTURE-FUNCTION RELATIONSHIPS OF THE EFA

Structures and Biosynthesis of EFA

Before considering the biosynthesis of the EFA, it is important to determine just which structures should be considered. Figure 3 presents a list of those structures with known EFA activity as well as some that are obviously not essential and some about which there is some doubt. Those structures listed as essential consist of the 19-, 20-, and 21-carbon 5,8,11,14-tetraenoic acids, as discussed above, and strangely, columbinic acid. These structures all have high EFA activity without the obvious necessity of further elongation or desaturation. Those structures listed under "pre-essential" include those that are precursors of the known EFA and could be expanded by addition of precursors of the odd-chain EFA. It is uncertain whether these fatty acids are active by virtue of their conversion into their 19—21 carbon homologues or whether, like columbinic acid, they are active in their own right. The evidence on this point is equivocal. For example, in experiments with HeLa cells in culture,[20] arachidonic acid added to the medium almost corrected the inhibition in growth rate, oxidative phosphorylation, and respiratory control resulting from a fat-free medium. Linoleic acid had $^1/_3$ to $^1/_2$ the activity of arachidonic and, in these cells, was not converted to it. Rivers et al.,[21,22] on the other hand, have reported that strict carnivores, such as the lion or cat, lack the Δ6 desaturase and cannot convert 18:2ω6 to 20:4ω6. In these animals, arachidonic acid must be obtained in the diet and it has been reported that linoleic acid does not replace it.[21,22] Evidence to the contrary has been presented by Houtsmuller and Van Der Beek[23] who have reported that skin symptoms in EFA-deficient rats are cured by topical application of linoleic acid in the absence of its conversion to arachidonate. In the cat, also, in which no conversion of 18:2ω6 to 20:4ω6 could be shown, linoleate was found to cure the skin and liver symptoms of EFA deficiency.[24] In the absence of further evidence, it will be assumed that although the 18-carbon PUFA may have some EFA activity in their own right, their major function is as precursors of the 20-carbon PUFA listed as EFA.

In the "uncertain" group are listed the ω3 PUFA. The fatty acids of this family have been found to promote growth of young rats but do not cure or prevent the skin symptoms associated with fat deficiency. Moreover, in experiments in which linolenic acid is effectively eliminated from a diet containing adequate linoleic acid, no obvious symptoms of deficiency appear.[25] However, the ω3 PUFA do appear to have some special function in the brain[26,27] and are present in high concentration in the retina,[28,29] where 22:6ω3 seems necessary for proper visual activity in primates.[29] They also serve as prime EFA in some fish.[30]

Within the limitations considered here, the biosynthesis of the EFA has already been discussed. EFA are formed, in the animal, by a process involving desaturation and elongation of the shorter, less unsaturated members of each family derived from plant sources. That these precursors are not synthesized *de novo* in animals has been demonstrated many times and the persistent reports of incorporation of ^{14}C from acetate into linoleic or linolenic acids are probably the result of the inability to separate synthetic 18:2 or 18:3 isomers or of the known 2-carbon exchange reaction involving oxidative removal and replacement of carbons 1 and 2 of the carboxylic acids.

Anomalies and Exceptions in EFA Metabolism

As reviewed by Brenner,[30a] desaturation of EFA may be controlled by hormonal and other influences. However, there are other factors that also determine the EFA/eicosanoid precursor levels in animal tissues. These are described below.

As reviewed above, the studies of Mead[3] and Klenk[4] on linoleic acid (18:2ω6) conversion to arachidonic acid (20:4ω6) were done in the rat. However, it is largely overlooked that there are considerable species differences in hepatic desaturase activity. Indeed, the rat is almost unique in its avid conversion of linoleate (18:2ω6) to arachidonate (20:4ω6). This explains why plasma arachidonate levels are so high in comparison with other species.[31,32] In the cat, both Δ6 and Δ5 desaturases seem lacking.[21,22] Since this animal must of necessity obtain its eicosanoid precursor fatty acids from meat or fish constituents of the diet, it has been termed an "obligate carnivore". Eskimos have also been proposed to be obligate carnivores.[33]

According to Willis,[32] the lack of desaturase activity in cats may explain why some years ago, one group was able to study the release of PGE$_2$ from the cat spleen[34] while under the same conditions another group was unable to observe any prostaglandin production.[35]

Perhaps cats of the latter group received a mainly fish diet, rich in 20:5ω3 and 22:6ω3, which tends to inhibit cyclooxygenase[36,37] rather than act as substrates.

Workers at Hoffman La Roche[32,38] showed that hepatic Δ^5-desaturase activity is extremely low in most species, the rat and mouse being exceptions. This was shown both for hepatic microsomal preparations in vitro and in feeding experiments with pure dihomo-γ-linolenic acid (20:3ω6) the substrate for Δ^5 desaturation to arachidonic acid (20:4ω6). Such negligible Δ^5-desaturase activity was seen[32,38] in rabbits, guinea pigs, and man (Figures 4 and 5).

In rats, metabolism of ω3 fatty acids is similar to that of the ω6 series,[3] indeed signs of linoleate (EFA) deficiency can be produced in this species by administering excess 18:3ω3 in the diet.[39] Nevertheless, elevation of 20:5ω3 reportedly was not produced in human subjects through ingestion of large amounts of dietary α-linolenic acid (18:3ω3).[40,40a] This is further confirmation of the poor Δ^5 desaturation in man.[32,38]

What are the implications of the low hepatic Δ^5 desaturation seen in man? Clearly, arachidonate may be derived directly from dietary meat products, although dihomo-γ-linolenate is likely to be derived in greater amounts from dietary linoleate. Correspondingly, the ratio of 20:3ω6 to 20:4ω6 in human plasma is over twice that of most ingested meat products.[32]

There are two additional considerations: one is the elevated levels of 20:3ω6 in serum of the elephant.[42] Is this due to low Δ^5 desaturase activity and high linolenate intake by this herbivore? The second is the apparent selective uptake of fatty acids into different tissues. Thus, in human platelets, ratios of 20:3ω6/20:4ω6 are much higher than the 20:3ω6/20:4ω6 ratio in plasma.[32,38] Is there selective uptake of 20:4ω6 or platelet Δ^5 desaturation?

The opposite situation is seen in the adrenal cortex and vesicular gland in which amounts of 20:3ω6 are much greater than in the plasma or indeed anywhere else in the body. Is there selective uptake of 20:3ω6 or its precursors? Similarly, why does 22:6ω3 accumulate in brain[42] and retinal rods?[28] Is there also a selective uptake mechanism for this fatty acid?

One factor that might be involved in these anomalies is receptor-mediated uptake of low density lipoproteins (LDL) by cells.[43] This process has been extensively studied with regards to cholesterol transport and metabolism. It is overlooked, however, that cholesterol esters and phospholipid constituents of the lipoproteins have EFA constituents that will also be transported into the cell. The cholesterol ester of LDL is rich in linoleate (18:2ω6), while the phospholipid will be rich in arachidonate or other EFAs which is largely determined by dietary and metabolite sources.

Although high density lipoprotein (HDL) seems largely to function in transporting cholesterol away from peripheral cells and into the plasma, it does transport cholesterol esters to the adrenal cortex.[44] Could this fact be related to the unique fatty acid profile of adrenal cholesteryl esters?

The Nature of Essentiality

The discovery that certain of the EFAs are obligatory precursors of the prostaglandins and related compounds quickly led to the conclusions that herein lay the reason for their essentiality. Further reflection, however, reveals that this cannot be their sole or even major function. First, the symptoms associated with EFA deficiency are not those normally associated with an action of prostaglandins. Moreover, the knowledge that aspirin inhibits the cyclooxygenase would lead to the conclusion that toxicology studies with aspirin-type drugs might produce the signs and symptoms of EFA deficiency; this is not the case. Finally, the EFA activity of columbinic acid, which is not only incapable of conversion to a prostaglandin but even inhibits their formation, would confirm the supposition that some other action must be sought. As discussed above, columbinic acid actually aids in separating those functions of EFA as prostaglandin precursors and the major apparent functions.

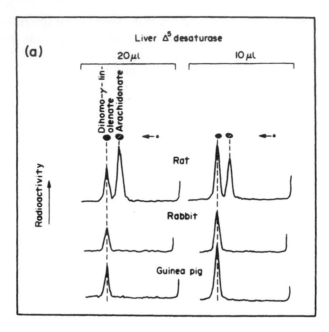

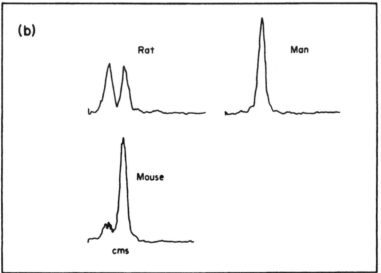

FIGURE 4.   Species variation of in vitro hepatic Δ^5-desaturase activity. ^{14}C-labeled dihomo-γ-linolenic acid (20:3ω6) was incubated with microsomes of liver that had been freshly excised and then either used immediately or stored frozen for up to 21 days at $-80°C$. In one experiment, microsomes were prepared from a homogenate of normal human liver obtained from a road accident victim. This was provided courtesy of the Liver Unit, Kings College Hospital, London. The tracings show radiochromatograms of the esterified products of 5 min incubations at 37°C. Methodology was similar to that of Marcel et al. and details are given in Reference 32. The greatest conversion of [14]C-20:3ω6 to [14]C-20:4ω6 by hepatic Δ^5-desaturase is seen in the mouse and in the rat. In rabbit, guinea pig, and man, Δ^5-desaturase activity appeared to be very low. These in vitro results are paralleled by in vivo experiments in which dihomo-γ-linolenic acid or its ethyl ester were orally administered (see Figure 5 and References 32 and 38). From Willis, A. L., *Prog. Lipid Res.*, 20, 834—850, 1982 and Stone, K. J., et al., *Lipids*, 14, 174—180, 1979. With permission.)

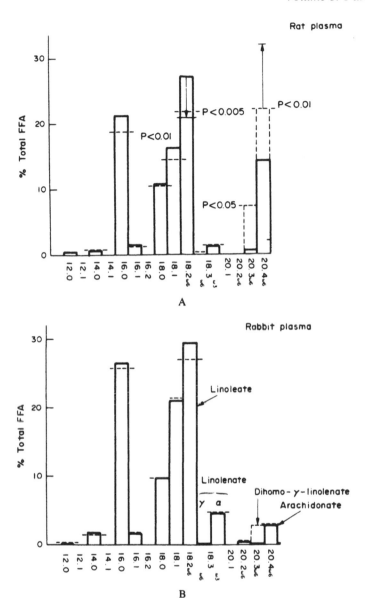

FIGURE 5. Fatty acid profiles in various species and the influence of orally administered dihomo-γ-linolenate (20:3ω6). (A) The histogram shows mean plasma total fatty acids in untreated control rats (solid line, n = 6) compared with those treated with dihomo-γ-linolenic acid at 400 mg/kg given perorally (p.o.) for 8 consecutive days (broken line, n = 7). There was a marked increase in 20:3ω6 and also a significant ($p < 0.01$) increase in arachidonate (20:4ω6), probably arising via Δ^5-desaturation of the administered 20:3ω6. Also shown (arrows, n = 7) are the values seen after treating the rats orally with arachidonic acid (400 mg/kg, p.o. for 8 days). Although increases in 20:4ω6 were observed, the mean ratio of 20:3ω6/20:4ω6 was not significantly different from the untreated controls ($p > 0.05$). Statistical significance of other changes in the fatty acid profile induced by administration of 20:3ω6 are indicated on the histogram. In rats treated with either 20:3ω6 or 20:4ω6 there was a decrease in linoleate (18:2ω6) similar to that reported by others in rabbit or man. (From Willis, A. L., *Prog. Lipid Res.*, 20, 834—850, 1982. With permission.) (B) Mean values in untreated control rabbits (n = 9) are shown by a solid line. The broken line indicates mean values in rabbits treated with dihomo-γ-linolenate ethyl ester administered p.o. at 100 mg/kg for 8 days (n = 10). Modified from Willis.[32]

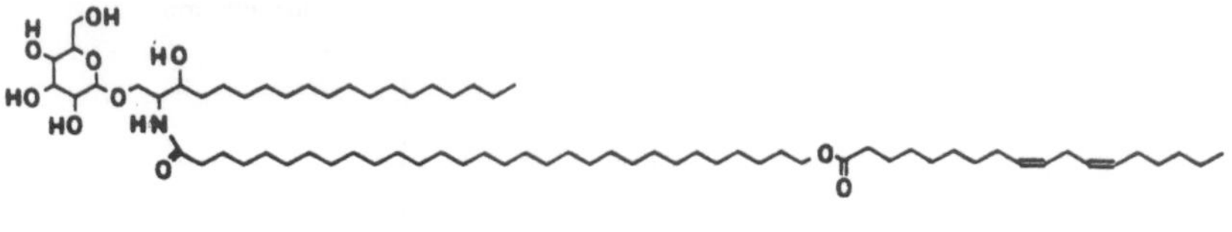

FIGURE 6.   Representative structures of the acylglucosylceramides of pig epidermis. From Wertz and Downing[47] as modified by them in a subsequent personal communication to A. L. Willis. See also Reference 51.

Stemming from the observed symptoms of EFA deficiency, it can be concluded that the obvious EFA action concerns membrane formation and integrity. The skin permeability and other symptoms strongly suggest such a function and the fact that in the absence of EFA, cell division ceases, at least in vivo, indicating that they are required for membrane formation. Unfortunately, the exact nature of such a function has so far eluded the investigators. It is tempting to consider that EFAs are involved in regulation of the viscotropic properties of the membrane lipid bilayer, but if so, this function can be performed equally well by low-melting fatty acids that are not EFA.

It is possible that they are involved in some much more specific relationship with the membrane proteins, particularly with one or more transport proteins. Such a relationship might explain the water permeability changes in the skin of EFA-deficient animals[45] as well as other symptoms of deficiency. The boundary lipid in contact with transmembrane proteins does not have the average composition of the lipid bilayer and may be rich in EFAs and have a "gating" function for transport of water and some salts and small molecules. Indeed, according to a recent paper, specific linoleate-containing glycolipids (Figure 6) may be involved in maintaining normal water barrier function of the skin.[46-48,51]

This contrasts with the possible involvement of derived EFA (prostaglandin precursors) important in human patients in whom there is a lack of Δ^6 desaturase and thus formation of 20:4ω6 in various clinical syndromes,[49,50] including diabetes mellitus, atopic eczema, hepatic cirrhosis, and cystic fibrosis (see also Brenner chapter). It seems obvious that the final chapter in the story of functions of the EFA has not been written and that much research remains to be done.

ADDENUM

Since this chapter was prepared, some further items of interest in the EFA area have come to our attention mostly via an excellent review by Lefkowith et al.[52] These items are as follows:

1.   The claim that columbinic acid has EFA activity independent of conversion to prostaglandins now has to be qualified in light of its recently described conversion to oxygenated metabolites.[53] Columbinic acid is converted a 9-hydroxyoctadecatrienoic acid by cyclo-oxygenase and to a 13-hydroxyoctadecatrienoic acid by lipoxygenase. The latter product is the principal epidermal metabolite and, like columbinic acid itself, is able to reverse the scaly dermatitis of EFA deficiency in rats when applied topically (the 9-hydroxy metabolite is inactive).[53] However, only columbinic acid itself is able to reverse the increase in transepidermal water loss.[54] On the basis of these findings Lefkowith et al.[52] suggested that lipoxygenase metabolites of linoleic acid might play a similar role in epidermal function.

2.   Although Mead Acid (20:3ω9) cannot be cyclized into prostaglandin-like molecules by cyclo-oxygenase, it can be converted by this enzyme into a variety of hydroxy-eicosanoids, the major product being 13-hydroxy-5,8,11-eicosatrienoic acid.[55] An in-

teresting aspect of these conversions are that they are dependent upon calcium ions and are suppressed by indomethacin.[55] Mead acid is also readily converted via leukotriene A_3 and the action of glutathione-S-transferase to leukotriene C_3 and its metabolites.[56-58] However, the hydrolysis of LTA_3 to LTB_3 is insignificant[57] so the enzyme requires not only three double bonds for formation of the triene, but also the double bond at C-14 to efficiently convert LTA to LTB.[58] Furthermore, the LTA_3 is a potent inhibitor of LTA_4 conversion to LTB_4.[59] These new insights into the biosynthesis of leukotrienes, explain why LTB_4 production in EFA-deficient rats was reduced by almost 90%, although arachidonate content of phospholipids was reduced by only 34% (Reference 60). These findings are also of considerable relevance to the diminished inflammatory response, observed during EFA deficiency.[61,62] Thus LTB_4 (whose production is lacking in EFA-deficiency) is a potent stimulator of neutrophil chemotaxis, adherence, aggregation and lysosomal enzyme release (see References 63 to 67 and Chapter by Ford-Hutchinson in Volume III of this Handbook). Furthermore, production of PGE_2, a well-established mediator of inflammation[65] would also be lacking. It is also possible (as suggested by Lefkowith et al.[52] that the C- and D-type leukotrienes derived from Mead acid are less potent in producing the increased vascular permeability observed[69] for the LTC_4 and LTD_4 derived from arachidonic acid.

3. Previous reports that EFA-deficient diets could lead to diminished formation of prostaglandins in unstimulated tissues, despite no reduction in phosphilipid arachidonate was attributed to reduced metabolic pools of precursor trickling down from linoleate (see Reference 70, and Chapter by Willis in this Volume). Such findings have now been extended to show that there are organ and phospholipid differences in ability to retain arachidonate during EFA deficiency.[71] When mice are maintained on an EFA-deficient diet, the liver is readily depleted of arachidonate. By contrast, phosphilipids of the renal cortex tenaciously retain arachidonate (in all phospholipids except phosphatidylinositol) and the heart actually doubles its content of arachidonate! Even more surprisingly, the accumulation of cardiac arachidonate occurred only in phosphatidylethanolamine (PE), in which there was a fourfold increase in arachidonate). Uptake into PE involved preferential incorporation of arachidonate rather than the Mead acid that had accumulated in the liver and plasma.[71] Phosphatidylinositol (PI) is unique in that it is the only phospholipid depleted of arachidonate in both renal cortex and heart.[71] Experiments with the perfused heart and kidney from these animals showed that receptor-mediated (angiotensin II-stimulated) prostaglandin production by the heart was diminished and therefore might have involved release from PI (since that is the only one depleted of arachidonate). It had previously been shown[72] that there is a hormone-selective lipase that selectively releases arachidonate from heart phospholipids. This more recent data suggested that (at least in the case of angiotensin II stimulation), this lipase acts selectively on PI. However the nonspecific stimulus of ischemia caused undiminished PG production which therefore had probably also derived from the arachidonate of other phospholipids[71] and via a less-specific lipase.[72]

4. EFA-deficiency has recently been shown to potentiate the effects of volatile anesthetics, an effect reversed by parental supplementation with linoleate but not omega 3 fatty acids and is therefore not merely a function of membrane unsaturation. It was shown that administration of linoleate resulted in a preferential normalization of the content of arachidonate in brain PI. Turnover of PI (containing arachidonate) may thus be implicated in the mode of action of volatile anesthetics.[73]

5. Wound healing is reduced in EFA deficiency in mice, rats, and infants. It is restored to normal by topical administration of arachidonate but results with PGE_2 or $PGF_{2\alpha}$ were equivocal (see Reference 74). In light of our present knowledge concerning

lipoxygenase products (see above), it is possible that LTB-mediated leukocyte function (perhaps release of mitogens from macrophages) might be involved in wound healing (see Reference 75) and is consequently lacking in EFA-deficiency. The role of EFA in skin is dealt with in more detail by Ziboh, in Volume III of this Handbook.

REFERENCES

1. **Hilditch, T. P. and Williams, P. N.,** *The Chemical Constitution of the Natural Fats,* 4th ed., John Wiley & Sons, New York, 1964.
2. **Kayama, M., Tsuchiya, Y., Nevenzel, J. C., Fulco, A. J., and Mead, J. F.,** Incorporation of linoleic-1-C^{14} acid into eicosapentaenoic and docosahexaenoic acids in fish, *J. Am. Oil Chem. Soc.,* 40, 499—502, 1963.
3. **Mead, J. F.,** The metabolism of polyunsaturated fatty acids, in *Progress in the Chemistry of Fats and Other Lipids,* Vol. 9, Holman, R. T., Ed., Pergamon Press, Oxford, 1971, 161—189.
4. **Klenk, E.,** Chemie und Stoffwechsel der Polyenfettsauren, *Experientia,* 17, 199—204, 1961.
5. **Dhopeshwarkar, G. A. and Mead, J. F.,** Role of oleic acid in the metabolism of the essential fatty acids, *J. Am. Oil Chem. Soc.,* 38, 297—301, 1961.
6. **Holman, R. T.,** Nutritional and metabolic interrelationships between fatty acids, *Fed. Proc.,* 23, 1062—1067, 1964.
7. **Brenner, R. R.,** The desaturation step in the animal biosynthesis of polyunsaturated fatty acids, *Lipids,* 6, 567—571, 1971.
8. **Holman, R. T.,** The ratio of trienoic: tetraenoic acids in tissue lipids as a measure of essential fatty acid requirement, *J. Nutr.,* 70, 405—410, 1960.
9. **Bloomfield, D. K. and Bloch, K.,** The formation of Δ9 unsaturated fatty acids, *J. Biol. Chem.,* 235, 337—345, 1960.
10. **Nugteren, D. H.,** Conversion *in vitro* of linoleic acid into γ-linolenic acid by rat liver enzymes, *Biochim. Biophys. Acta,* 60, 656—657, 1962.
11. **Stoffel, W.,** Enzymatic studies on polyunsaturated fatty acid metabolism, *Wiss. Veroeff. Deutsch. Ges. Ernaehr.,* 22, 12—34, 1971.
12. **Brenner, R. R., DeTomas, M. E., and Peluffo, R. O.,** Effect of poly-unsaturated fatty acids on the desaturation *in vitro* of linoleic to γ-linolenic acid, *Biochim. Biophys. Acta,* 106, 640—642, 1965.
13. **Ullman, D. and Sprecher, H.,** *In vitro* and *in vivo* study of the conversion of eicosa-11,14-dienoic acid to eicosa-5,11,14-trienoic acid and of the conversion of eicosa-11-enoic acid to eicosa-5,11-dienoic acid in the rat, *Biochim. Biophys. Acta,* 248, 186—197, 1971.
14. **Bernent, J. T., Jr. and Sprecher, H.,** Studies to determine the role rates of chain elongation and desaturation play in regulating the unsaturated fatty acid composition of rat liver lipids, *Biochim. Biophys. Acta,* 398, 354—363, 1975.
15. **Schlenk, H., Sand, D. M., and Gellerman, J. L.,** Retroconversion of docosahexaenoic acid in the rat, *Biochim. Biophys. Acta,* 187, 201—207, 1969.
16. **Ayala, S., Gaspar, G., Brenner, R. R., Peluffo, R. O., and Kunau, W.,** Fate of linolenic, arachidonic and docosa-7,10-13,16-tetraenoic acids in rat testicles, *J. Lipid Res.,* 14, 296—305, 1973.
17. **Schlenk, H.,** Odd numbered polyunsaturated fatty acids, in *Progress in the Chemistry of Fats and Other Lipids,* Vol. 9, Holman, R. T., Ed., Pergamon Press, Oxford, 1971, 589—605.
18. **Beerthuis, R. K., Nugteren, D. H., Pabon, H. J. J., and Van Dorp, D. A.,** Biologically active prostaglandins from some new odd-numbered essential fatty acids, *Rec. Trav. Chim.,* 87, 461—480, 1968.
19. **Houtsmuller, U. M. T.,** Columbinic acid, a new type of essential fatty acid, in *Progress in Lipid Research,* Vol. 20, Holman, R. T., Ed., Pergamon Press, Oxford, 1982, 889—896.
20. **Gerschenson, L. E., Mead, J. F., Harary, I., and Haggerty, D. R., Jr.,** Studies on the effects of essential fatty acids on growth rate, fatty acid composition, oxidative phosphorylation, and respiratory control of HeLa cells in culture, *Biochim. Biophys. Acta,* 131, 42—49, 1967.
21. **Rivers, J. P. W., Sinclair, A. J., and Crawford, M. A.,** Inability of the cat to desaturate essential fatty acids, *Nature (London),* 258, 171—173, 1975.
22. **Rivers, J. P. W., Hassam, A. G., Crawford, M. A., and Brambell, M. R.,** The inability of the lion, *Panthera leo,* to desaturate linolenic acid, *FEBS Lett.,* 67, 269—270, 1976.
23. **Houtsmuller, U. M. T. and Van Der Beek, A.,** Effects of topical application of fatty acids, in *Progress in Lipid Res.,* Vol. 20, Holman, R. T., Ed., Pergamon Press, Oxford, 1982, 219—224.

24. **MacDonald, M. L., Rogers, Q. R., and Morris, J. G.,** Role of linoleate as an essential fatty acid for the cat, independent of arachidonic synthesis, *J. Nutr.*, 113, 1422, 1983.

25. **Tinoco, J., Williams, M. A., Hincenbergs, I., and Lyman, R. L.,** Evidence for nonessentiality of linolenic acid in the diet of the rat, *J. Nutr.*, 101, 937—945, 1971.

26. **Crawford, M. A. and Sinclair, A. J.,** The limitations of whole tissue analysis to define linoleic acid deficiency, *J. Nutr.*, 102, 1315—1321, 1972.

27. **Lamptey, M. S. and Walker, B. L.,** A possible essential role for linolenic acid in the development of the young rat, *J. Nutr.*, 106, 86—93, 1976.

28. **Bazan, N. G., Avaldano De Caldironi, M. I., Guisto, N. M., and Rodriguez De Turco, E. B.,** Phosphatidic acid in the central nervous system, *Prog. Lipid Res.*, 20, 307—313, 1982.

29. **Neuringer, M., Connor, W. E., Van Petten, C., and Barstad, L.,** Dietary omega-3 fatty acid deficiency and visual loss in infant rhesus monkeys, *J. Clin. Invest.*, 73, 272—276, 1984.

30. **Yu, T. C. and Sinnhuber, R. O.,** Effect of dietary linolenic acid and docosahexaenoic acid on growth and fatty acid compositon of rainbow trout *(Salmo gairdneri), Lipids*, 7, 450—454, 1972.

30a. **Brenner, R. R.,** Nutritional and hormonal factors influencing desaturation of essential fatty acids, *Prog. Lipid Res.*, 20, 41—47, 1982.

31. **Horrobin, D. F., Huang, Y.-S., Cunnane, S. C., and Manku, M. S.,** Essential fatty acids in plasma, red blood cells and liver phospholipids in common laboratory animals as compared to humans, *Lipids*, 19, 806—811, 1984.

32. **Willis, A. L.,** Unanswered questions in EFA and PG research, *Prog. Lipid Res.*, 20, 834—850, 1982.

33. **Gibson, R. A. and Sinclair, A. J.,** Are Eskimos obligate carnivores?, *Lancet*, 1, 1110, 1981.

34. **Gilmore, N., Vane, J. R., and Wyllie, J. R.,** Prostaglandin released by the spleen, *Nature (London)*, 218, 1135—1140, 1968.

35. **Bedwani, J. R. and Millar, G. C.,** Prostaglandin release from cat and dog spleen, *Br. J. Pharmacol.*, 54, 499—505, 1975.

36. **Lands, W. E. M.,** The biosynthesis and metabolism of prostaglandins, *Ann. Rev. Physiol.*, 41, 633—652, 1979.

37. **Corey, E. J., Shih, C., and Cashman, J. R.,** Docosahexaenoic acid is a strong inhibitor of prostaglandin but not leukotriene biosynthesis, *Proc. Natl. Acad. Sci. U.S.A.*, 80, 3551—3584, 1983.

38. **Stone, K. J., Willis, A. L., Hart, M., Kirtland, S. J., Kernoff, P. B. A., and McNicol, G. P.,** *Lipids*, 14, 174—180, 1979.

39. **Morhauer, H. and Holman, R. T.,** *J. Lipid Res.*, 4, 151, 1963.

40. **Dyerberg, J., Bang, H. O., and Aagaard, O.,** α-linolenic acid and eicosapentaenoic acid, *Lancet*, i, 199, 1980.

40a. **Singer, P., Berger, I., Wirth, M., Godicke, W., Jaeger, W., and Voight, S.,** Slow desaturation and elongation of linoleic and a-linolenic acids as a rationale of eicosapentaenoic acid-rich diet to lower blood pressure and serum lipids in normal, hypertensive and hyperlipidemic subjects, *Prostaglandins, Leukotrienes, and Medicine*, 24, 173—193, 1986.

41. **Moore, J. H. and Sykes, S. K.,** The serum and adrenal lipids of the African elephant, *Loxodonta africana, Comp. Biochem. Physiol.*, 20, 779—792, 1967.

42. **Dhopeshwarker, G. A. and Mead, J. F.,** Uptake and transport of fatty acids into the brain and the role of the blood-brain barrier system, *Adv. Lipid Res.*, 2, 109—142, 1973.

43. **Brown, M. S., Kovanen, P. T., and Goldstein, J. L.,** Regulation of plasma cholesterol by lipoprotein receptors, *Science*, 212, 628—635, 1981.

44. **Gwynne, J. T. and Hess, B.,** Role of high density lipoproteins in rat adrenal cholesterol metabolism and steroidogenesis, *J. Biol. Chem.*, 255, 10875—10883, 1980.

45. **Basnayake, V. and Sinclair, H. M.,** The effect of deficiency of essential fatty acid upon the skin, in *Biochemical Problems of Lipids*, Popjak, J. and Breton, E. L., Eds., Butterworths, London, 1956, 476—478.

46. **Wertz, P. W. and Downing, D. T.,** Glycolipids in mammalian epidermis: structure and function in the water barrier, *Science*, 217, 1261—1262, 1982.

47. **Wertz, P. W. and Downing, D. T.,** Acyglucosylceramides of pig epidermis: structure determination, *J. Lipid Res.*, 24, 753—758, 1983.

48. **Wertz, P. W. and Downing, D. T.,** Ceramides of pig epidermis: structure determination, *J. Lipid Res.*, 24, 759—765, 1983.

49. **Holman, R. T. and Johnson, S.,** Changes in essential fatty acid profile of serum phospholipids in human disease, *Prog. Lipid Res.*, 20, 67—73, 1982.

50. **Horrobin, D. F.,** The regulation of prostaglandin biosynthesis by the manipulation of essential fatty acid metabolism, *Rev. Pure Appl. Pharmacol. Sci.*, 4, 339—343, 1983.

51. **Abraham, W., Wertz, P. W., and Downing, D. T.,** Linoleate-rich acetylglucosylceramide of pig epidermis: structure determination by proton magnetic resonance, *J. Lipid Res.*, 26, 761—766, 1985.

52. **Lefkowith, J. B., Evers, A. S., Elliot, W. J., and Needleman, P.,** Essential fatty acid deficiency: a new look at an old problem, *Prostaglandins, Leukotrienes and Medicine,* 23, 123—127, 1986.

53. **Elliott, W. J., Morrison, A. R., Sprecher, H., Needleman, P.,** The metabolic transformations of columbinic acid and the effect of topical applications of the major metabolites on rat skin, *J. Biol. Chem.,* 260, 987—992, 1985.

54. **Elliott, W. J., Sprecher, H., Needleman, P.,** Physiologic effects of columbinic acid and its metabolites on rat skin, *Biochim. Biophys. Acta,* 835, 158—160, 1985.

55. **Elliott, W. J., Morrison, A. R., Sprecher, H., and Needleman, P.,** Calcium-dependent oxidation of 5,8,11-eicosatrienoic acid by the cyclo-oxygenase enzyme system, *J. Biol. Chem.,* 261, 6719—6724, 1986.

56. **Jakschik, B. A., Sams, A. R., Sprecher, H. and Needleman, P.,** Fatty acid structural requirements for leukotriene biosynthesis, *Prostaglandins,* 20, 401—410, 1980.

57. **Hammarstrom, S.,** Conversion of 5,8,11-eicosatrienoic acid to leukotrienes C3 and D3, *J. Biol. Chem.,* 256, 2275—2279, 1981.

58. **Jakschik, B. A., Morrison, A. R., and Sprecher, H.,** Products derived from 5,8,11-icosatrienoic acid by the 5-lipoxygenase-leukotriene pathway, *J. Biol. Chem.,* 258, 12797—12800, 1983.

59. **Evans, J. F., Nathaniel, D. J., Zamboni, R. J., and Ford-Hutchinson, A. W.,** Leukotriene A_4. A poor substrate but a potent inhibitor of rat and human neutrophil leukotriene A_4 hydrolase, *J. Biol. Chem.,* 260, 10966—10970, 1985.

60. **Stenson, W. F., Prescott, S. M., and Sprecher, H.,** Leukotriene B formation by neutrophils from essential fatty acid-deficient rats, *J. Biol. Chem.,* 259, 11784—11789, 1984.

61. **Bonta, I. L. and Parnham, M. J.,** Prostaglandins, essential fatty acids and cell-tissue interactions in immune inflammation, *Progr. Lipid Res.,* 20, 617—623, 1982.

62. **Hurd, E. R., Johnston, J. M., Okita, J. R., MacDonald, P. C., and Ziff, M.,** Prevention of glomerulonephritis and prolonged survival in New Zealand black/New Zealand white F1 hybrid mice fed an essential fatty acid-deficient diet, *J. Clin. Invest.,* 67, 476—485, 1981.

63. **Ford-Hutchinson, A. W., Bray, M. A., Doig, M. V., Shipley, M. E., and Smith, M. J. H.,** Leukotriene B4, a potent chemokinetic and aggregating substance released from polymorphonuclear leukocytes, *Nature,* 286, 264—265, 1980.

64. **Goezl, E. J. and Picket, W. C.,** The human PMN leukocyte chemotactic activity of complex hydroxy-eicosatetraenoic acids (HETEs), *J. Immunol.,* 125, 1789—1791, 1980.

65. **Palmblad, J., Malmsten, C. L., Uden, A. M., Radmark, O., Engstedt, L., and Samuelsson, B.,** Leukotriene B4 is a potent and stereospecific stimulator of neutrophil chemotaxis and adherence, *Blood,* 58, 658—661, 1981.

66. **Hafstrom, I., Palmblad, J., Malmsten, C., Radmark, O., and Samuelsson, B.,** Leukotriene B4 — A Stereospecific stimulatory for release of lysosomal enzymes from neutrophils, *FEBS Lett.,* 130, 146—148, 1981.

67. **Palmblad, J., Hafstrom, I., Malmsten, C. L., Uden, A. M., Radmark, O., Engstedt, L., and Samuelsson, B.,** Effects of leukotrienes on in vitro neutrophil functions, *Advances in Prostaglandin, Thromboxane and Leukotriene Research,* 9, 293—299, 1982.

68. **Willis, A. L., Davidson, P., Ramwell, P. W., Brocklehurst, W. E., and Smith, J. B.,** Release and actions of prostaglandins during inflammation and fever: inhibition by anti-inflammatory and anti-pyretic drugs, in *Prostaglandins in Cellular Biology and the Inflammatory Process,* Ramwell, P. W. and Pharriss, B. B., Eds., Plenum Press, New York, 1972, 227—268.

69. **Hedqvist, P., Dahlen, S. E., Gustafsson, L., Hammarstrom, S. A., and Samuelsson, B.,** Biological profile of leukotrienes C_4 and D_4, *Acta Physiol. Scand.,* 110, 331—333, 1980.

70. **Willis, A. L., Hassam, A. G., Crawford, M. A., Stevens, P., and Denton, J. P.,** Relationships between prostaglandins, prostacyclins and EFA precursors in rabbits maintained on EFA-deficient diets, *Progress in Lipid Research,* 20, 161—167, 1982.

71. **Lefkowith, J. B., Flippo, V., Sprecher, H., and Needleman, P.,** Paradoxical conservation of cardiac and renal arachidonate in essential fatty acid deficiency, *J. Biol. Chem.,* 260, 15736—15744, 1985.

72. **Hsueh, W., Isaksun, P. C., and Needleman, P.,** Hormone-selective lipase activation in the isolated rabbit heart, *Prostaglandins,* 13, 1073—1091, 1977.

73. **Evers, A. S., Elliot, W. J., Lefkowith, J. B., and Needleman, P.,** Manipulation of rat brain fatty acid composition alters volatile anesthetic potency, *J. Clin. Invest.,* 77, 1028—1033, 1986.

74. **Caffrey, B. B. and Jonsson, H. T., Jr.,** Role of essential fatty acids in cutaneous wound healing in rats, *Progress in Lipid Research,* 20, 641—647, 1982.

75. **Leibovich, S. J. and Ross, R.,** A macrophage-dependent factor that stimulates the proliferation of fibroblasts in vitro, *Am. J. Pathol.,* 84, 501—513, 1976.

BIOSYNTHESIS AND INTERCONVERSION OF THE ESSENTIAL FATTY ACIDS

Rodolfo R. Brenner

DERIVATION OF ESSENTIAL FATTY ACID PRECURSORS AND THEIR POSITION AS ESSENTIAL DIETARY COMPONENTS

Natural essential fatty acids (EFA) of the animal kingdom are traditionally regarded as the unsaturated fatty acids linoleic and α-linolenic that cannot be synthesized *de novo* by animals and the fatty acids derived from them. That is, plants synthesize *de novo* linoleic and α-linolenic acids and animals convert them to fatty acids that are more unsaturated and of longer chain length. However, there are exceptions to these general rules. The sequence of reactions involved in the biosynthesis of long-chain unsaturated fatty acids derived from linoleate and α-linolenate was shown in experiments done both in vivo and in vitro. They are the result of coordination between two types of reactions: desaturation at specific positions of the hydrocarbon chain and elongation by addition of units of two carbon atoms. The reactions are catalyzed by enzymes which are integral parts of the endoplasmic reticulum membrane.[1-5] The same enzymes are involved in the biosynthesis of fatty acids derived from linoleate, α-linolenate, oleate, and palmitoleate.[5] The fatty acids derived from each one of these precursors constitutes a family. Each family is independent and there is no direct crossover. Fatty acids of the linoleic acid (ω6) family cure the typical symptoms produced in the rat and other animals by a fat-free diet.[6] The major symptoms produced in the rat by EFA deficiency are dermatosis and increased water permeability of the skin, decrease in weight, decreased capillary resistance, cholesterol accumulation in lung, impaired reproduction and lactation in females and degeneration of seminiferous tubules of males, alteration of endocrine glands, changes in the fatty acid composition of tissues, modification of cholesterol levels in organs and plasma, swelling of mitochondria, and increased triacylglycerol synthesis by the liver. α-Linolenic acid is not able to cure the skin signs.

In the plant kingdom, linoleic acid is synthesized by desaturation of oleic acid[7,8] and the new double bond is formed between carbons 12-13; α-linolenic acid is formed from linoleic acid.[8] The essentiality of linoleic acid in the animal kingdom is determined by the absence of the enzyme that desaturates oleate in carbons localized between the 9-10 double bond and the methyl end. α-Linolenic acid requires an additional desaturation between carbons 15-16 that is not produced in animals. On the contrary, the desaturases that produce double bonds between a preexisting double bond and the carboxyl group are considered animal desaturases. However, these differences between both kingdoms may disappear when going down the evolutionary scale. Thus, some unicellular algae, in addition to linoleic and α-linolenic acids, *de novo* synthesize polyunsaturated acids considered of typical animal structure. Some protista, like *Ochromonas danica*, synthesize linoleic and γ-linolenic acids.[9-11] Some animals low in the evolutionary scale like the ciliated protozoan *Tetrahymena pyriformis*[12] not only desaturates acyl-2-oleoyl-*sn*-glycerol-3-phosphorylcholine to the linoleophosphatidylcholine,[13] but also produces γ-linolenic acid.[14] It desaturates linoleate, α-linolenate, and *cis*-vaccenate between carbons 6-7.[14] *Acanthamoeba castellanii*[15] and Trypanosomae, specifically *T. cruzi*, have been also shown to synthesize linoleic acid *de novo*.[16]

It has also been claimed that other animals in higher steps of the evolutionary scale may also synthesize linoleate *de novo*. Evidence has been presented[17] that the cockroach *Periplaneta americana* synthesizes linoleate.

In addition, it has been shown[18] that the livers of hen, goat, and pig possess an enzyme that desaturates cis-12-octadecenoic acid between carbons 9-10 synthesizing linoleic acid.

This reaction does not take place in liver of hamster or rat.[18,19] Therefore, the unnatural substrate cis-2-octenoic acid is converted to linoleic by the hen[20,21] after elongation to cis-12-octadecenoic acid, but not by the rat.[22]

The other situation represented by the absence of enzymes that convert linoleate into arachidonate may be found in some animals. There is a general consensus that insects are unable to produce these reactions.[23] Besides, the cat[24] and other felines are also reportedly unable to desaturate linoleic acid. Transformed cells easily lose or decrease their capacity to desaturate linoleic acid[25] whereas the capacity to desaturate α-linolenic acid[26] and 8,11,14-eicosatrienoic acid[25-27] is more firmly preserved.

THE CONVERSION OF PRECURSORS TO EFA PRODUCTS

In Vivo Demonstration of Interconversions

The Linoleic (ω6) Family

The pioneering work of Burr and Burr[28] established the concept of animal EFAs. From these results, a relationship between linoleic and arachidonic acids was recognized based upon nutritional studies and their effects on EFA-deficient animals and the changes of fatty acid composition.[29-31] An EFA-deficient diet evokes, in the tissues of young animals, a rapid decline, first of linoleic acid and then of arachidonic acid with corresponding increases of oleic and 5,8,11-eicosatrienoic acids. These effects of EFA deficiency and the corresponding biological symptoms reported in early work have been reviewed by Holman.[32] The decline of linoleic and arachidonic acids is seen very early in the endoplasmic reticulum of liver.[33] Moreover, dietary linoleic acid incorporation in EFA-deficient rats and arachidonic acid increase follow sequential curves.[34] These results and the similarity of the double bond position in relation to the methyl end of the hydrocarbon chain of the acids are indicative of a metabolic relation. Both acids are ω6, following Thomasson's nomenclature.[35] The relative biological effect measured by Thomasson in rats by using tests based on the disturbances in water metabolism showed that it was high for linoleic, γ-linolenic, eicosa-8,11,14-trienoic, and arachidonic acids, stressing a possible biological and biochemical relation among these acids.[35]

The in vivo conversion of linoleic acid (18:2ω6) into arachidonic acid (20:4ω6) was definitively shown by Mead's group[36,37] by administration of labeled acetate and ^{14}C-linoleic acid to rats. They also demonstrated[38] that linoleic acid was not converted to eicosa-5,8,11-trienoic acid by the rat, and that this last acid was synthesized from oleic acid. However, γ-linolenic (18:3ω6)[39] and 8,11,14-eicosatrienoic acid (20:3ω6)[40] were readily transformed into arachidonic acid. The conversion of linoleic into arachidonic acid is seen in the liver and other organs. In addition, arachidonic acid is converted to 7,10,13,16-docosatetraenoic acid and 4,7,10,13,16-docosapentaenoic acid. These reactions have been studied preferentially in rat adrenals and testes since the aforementioned acids are (respectively) relevant components of these tissues.[41] The in vivo conversion of ^{14}C arachidonate to 9,12,15,18-tetracosa tetraenoate and 6,9,12,15,18-tetracosa pentaenoate has been also demonstrated in rat testes.[42]

Labeled linoleic acid is also elongated in the animal tissues to 11,14-eicosadienoic acid.[43] However, feeding 1-[14]C-11,14-eicosadienoic acid to fat deficient rats demonstrated that the acid is further desaturated not to the 8,11,14-eicosatrienoic acid, but to 5,11,14-eicosatrienoic acid, that is *not* converted into arachidonic acid.[44]

The same type of reactions have been also shown in HTC cells (ascites form of a rat-carried Morris hepatoma 7288 C) when they were incubated with 1-^{14}C linoleic acid.[45] However, some of these reactions are generally absent in other transformed cells.

All these results indicate that the members of linoleic acid family are synthesized in the animal by the main route outlined in Figure 1.

Linoleic Acid family (ω6)

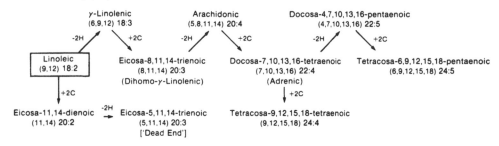

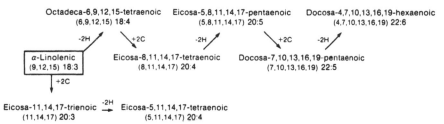

α-Linolenic Acid family (ω3)

FIGURE 1. Schemes of the EFA families of linoleic (ω6) and α-linolenic (ω3) acids recognized by in vivo experiments.

The α-Linolenic (ω3) Family

The members of the α-linolenic acid family are characterized by the ω3 double bond. The first indication of the existence of this family also came from the comparison of the structure of fatty acids and also from nutritional studies.[46] Feeding α-linolenic acid to rats demonstrated an increase of 5,8,11,14,17-eicosapentaenoic, 7,10,13,16,19-docosapentaenoic, and 4,7,10,13,16,19-docosahexaenoic acids in the tissues. Experiments carried out with labeled 1-[14]C α-linolenic acid followed by analyses of labeling distribution confirmed the conversion of α-linolenic acid to 5,8,11,14,17-docosapentaenoic acid.[49] Furthermore, similar data were obtained in fish, whose tissues are rich in ω3 fatty acids.[50] Moreover, Klenk and Mohrhauer[43] demonstrated the following conversions when including 2-^{14}C labeled acids in the diet of rats:

6,9,12,15-Octadecatetraenoic acid	→	5,8,11,14,17-Eicosapentaenoic acid
11,14,17-Eicosatrienoic acid	→	5,8,11,14,17-Eicosapentaenoic acid
8,11,14,17-Eicosatetraenoic acid	→	5,8,11,14,17-Eicosapentaenoic acid
7,10,13,16,19-Docosapentaenoic acid	→	4,7,10,13,16,19-Docosahexaenoic acid

From these in vivo experiments, it was postulated that α-linolenic acid is converted to docosa-4,7,10,13,16,19-hexaenoic acid, as is shown in Figure 1.

The same series of reactions were confirmed by incubation of HTC cells with 1-[14]C α-linolenic acid.[26,51] However, the incubation also showed that similarly to linoleic acid, 1-[14]C α-linolenic acid was also elongated to 11,14,17-eicosatrienoic acid and desaturated to 5,11,14,17-eicosatetraenoic acid, an acid in which the Δ8 double bond has been skipped[51,52] (Figure 1). The yield of these reactions depends on the composition of the incubation medium and it increases in parallel with a decrease of the Δ6 desaturase pathway.

The Oleic and Palmitoleic Family and the Idea of Competitive Inhibition Leading to the Formation of 20:3ω9

In the rat, an EFA-deficient diet evokes a rapid decrease of linoleic and arachidonic acids

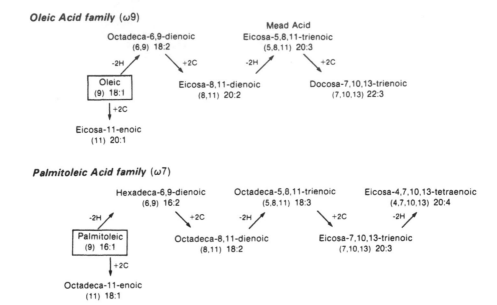

FIGURE 2. Schemes of the non-EFA families of oleic (ω9) and palmitoleic (ω7) acids recognized by in vivo experiments.

in the lipids of the tissues and the increase of oleic, palmitoleic, and 5,8,11-eicosatrienoic acids.[32,53] An increase of 7,10-13-docosatrienoic acid has also been shown.[54] Oleic and palmitoleic acids are not EFAs and are synthesized *de novo* from acetate via stearic and palmitic acid desaturation, respectively.[55-57] The formal demonstration that oleic acid was a precursor of 5,8,11-eicosatrienoic acid was shown in in vivo studies with ^{14}C labeled acids.[38] By administration of labeled oleic acid to rats raised on a fat-free diet, it was also shown that small amounts of 6,9-octadecadienoic acid are formed.

Therefore, the structural similarity of the aforementioned acids and in vivo experiments done preferentially with EFA-deficient animals indicate that oleic acid is precursor of an independent family constituted by ω9 acids converted by similar reactions of desaturation and elongation to the ω6 and ω3 families and perhaps by the same enzymes (Figure 2).

Palmitoleic (16:1ω7) acid is the precursor of the ω7 family as suggested originally by Klenk. It is produced by reactions of desaturation and elongation similar to the ω6, ω3, and ω9 series but with small yields. The main pathways[53] would be the one of Figure 2, mainly based on the following information.

It has been shown that in liver phospholipids of fat-deficient rats, 30% of the 18:1 acid belongs to 18:1ω7.[59] Budny and Sprecher [60] have shown that 1-[14]C 18:2ω7 and 1-[14]C 20:3ω7 injected to rats are converted to 18:3ω7 and 20:4ω7, respectively, with small yields. A small increase of 20:4ω7 is also shown when rats raised on a fat-free diet are fed 18:3ω7 or 20:3ω7.[61] However, 1-[14]C 20:2ω7 is a poor precursor for 20:3ω7 and 20:4ω7 and injected 20:2ω7 acid is degraded to 18:2ω7.[60]

Since the linoleic (ω6), α-linolenic (ω3), oleic (ω9), and palmitoleic acid (ω7) families are built by the same type of reactions, it was probable that there would be competition between them.

To support this notion, a rise of palmitoleic, oleic, and 5,8,11-eicosatrienoic was produced in the rat as a consequence of EFA deficiency;[29-34] these reactions were reversed by administration of linoleic,[34] arachidonic,[34] α-linolenic,[48] or 4,7,10,13,16,19-docosahexaenoic acid,[48] indicative of competitive reactions. Moreover, upon administering linoleate to EFA-deficient rats the kinetics of oleic and 5,8,11-eicosatrienoic acids' decrease and linoleic and arachidonic

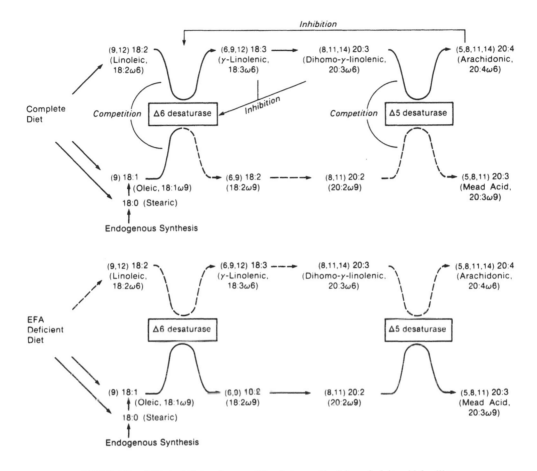

FIGURE 3.   Effect of diet and competition between linoleic and oleic acid families.

acids' increase[34] were found to fit well with the theoretical curves calculated by Lindstrom and Tinsley,[62] considering mutual competitive inhibitions of both families. They confirm the results obtained by Holman[63] in feeding experiments. It was also shown by Dopeshwarkar and Mead[64] that oleic acid may compete with linoleic acid, since they found that an excess of oleic acid aggravated EFA deficiency. However, an absolute demonstration of the competition between the different families was obtained in vitro experimentation as described under the section "In Vitro Experiments and Conditions" (Figure 3).

In Vitro Experiments and Conditions
Reactions of Desaturation and Elongation

The mechanism of desaturation and elongation reactions involved in the animal biosynthesis of polyunsaturated fatty acids was recognized by in vitro experiments.

The first in vitro demonstrations of the desaturation reaction and the experimental conditions and subcellular particles involved in 8,11,14-eicosatrienoic acid conversion to arachidonic acid, linoleic acid conversion to α-linolenic acid, oleic acid conversion to 6,9-octadecadienoic acid and α-linolenic acid conversion to 6,9,12,15-octadecatetraenoic acid, were performed by Stoffel,[1] Nugteren,[2] Holloway et al.,[65] and by Brenner and Peluffo,[66] respectively. The reactions were produced on acyl CoA derivatives in the presence of oxygen and NADH or NADPH as electron donors. The endoplasmic reticulum membrane (microsome) is the obligate cellular subfraction that produces desaturation. The mechanism is similar to the one recognized in the steroyl CoA desaturase.[67-69] The electrons of the NADH

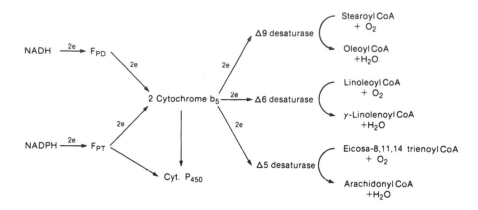

FIGURE 4.   Electron flow from NADH or NADPH and cytochrome b_5 to desaturases.

are transported to the cytochrome $(b)_5$ by the NADH-ferricytochrome b_5 oxidoreductase (E.C. 1.6.2.2) and then to the desaturase[5,70-72] (Figure 4).

All the components of the desaturation reaction are integral proteins of the membrane that can so far only be separated by detergents.[73-75] They are amphipathic proteins with a nonpolar tail deeply embedded in the lipid bilayer of the membrane. The stearoyl-CoA desaturase has been isolated and purified by Strittmatter's group[76] and has also been shown to be an amphipathic protein. Recently, Okayasu et al.[77] claimed to have isolated the linoleyl-CoA desaturase. It has a molecular weight in its native form of 65,000 to 68,000 and this also is embedded in the membrane.

All the components of the electron system are apparently distributed at random in the outer surface of the membrane.[78,79] The comparative rates of the electron transport by the NADH-cytochrome b_5 reductase-cytochrome b_5 system and the activity of the stearoyl-CoA desaturation to oleyl CoA[79] and linoleoyl-CoA desaturation to α-linoleoyl-CoA[80] have been specifically measured. It was shown that the rate of the NADH-cytochrome b_5 transport is many times faster than the desaturase activity. Therefore, there would generally be enough electron provision in the endoplasmic reticulum to fulfill the requirements of the desaturases which could then be regarded as the "pacemakers" of fatty acid desaturation.

The desaturases are inhibited by different substances. Table 1 summarizes the results obtained by Stoffel[81] with Δ5 desaturase.

The disclosure of the sequence of reactions involved in the elongation reaction of EFA is due to Stoffel et al.,[3,81] Nugteren,[4] Seubert and Podack,[82] and Bernert and Sprecher[83] and their colleagues and is illustrated in Figure 5. It is produced on unsaturated acyl-CoA of 16, 18, 20, 22, and 24 carbons by enzymes localized in the endoplasmic reticulum of cells and requires malonyl-CoA and NADH. It does not require oxygen and so it is not inhibited by CN^- as with the fatty acid desaturases. Thus, elongation is easily examined in the absence of desaturation by addition of CN^- and performing all incubations under nitrogen.

Seubert and Podack[82] showed that the 2-trans-enoyl-CoA reductase is the leading reaction of the enzymatic complex with the highest energy change (-140 kcal/mol) and is specially active with hexenoyl-CoA and α-linolenyl-CoA derivatives. However, Bernert et al.[85,86] have proved that the rates of the β-hydroxyacyl-CoA dehydrase and 2-transenoyl-CoA reductase are faster than condensation or overall chain elongation, suggesting that the condensation was the rate-limiting step.

Confirmation of In Vivo Findings

The biosynthetic pathways of polyunsaturated fatty acids, outlined by in vivo experiments have been confirmed by in vitro experiments, once the incubation conditions necessary for

Table 1
INHIBITORS IN VITRO OF EICOSA-8,11,14-TRIENOIC ACID DESATURATION TO ARACHIDONIC ACID BY RAT LIVER MICROSOMES

Inhibitor	Conc.(M)	Yield (%)
FAD	10^{-5}	89
Atebrin	10^{-4}	44
Vitamin K_3	10^{-4}	100
Ubiquinone (0)	10^{-4}	100
Methylene blue	10^{-5}	44
Dichlorophenolindophenol	10^{-5}	73
Tetrahydrofolic acid	5.10^{-4}	100
Dihydrobiopterin	10^{-4}	100
Ascorbic acid	10^{-3}	32
FeSO$_4$	10^{-3}	28
Ascorbic acid + FeSO$_4$ (each)	10^{-3}	43
KCN	10^{-3}	25
NaN$_3$	10^{-3}	28
O-Phenanthroline	10^{-2}	17
Ethylenediaminetetraacetate	10^{-3}	16
p-Chloromercuribenzoate	10^{-3}	14

Note: The composition of the solution was: 0.2 μM eicosa-8,11,14-trienoate; 10 μM ATP; 10 μM MgCl$_2$; 0.2 μM CoA; 10 μM cysteine; 10 mg microsomal protein; 200 μM potassium phosphate buffer, pH 7.4; 2 μM NADP; 20 μM glucose-6-phosphate; 3 I.U. glucose-6-phosphate dehydrogenase. Total vol: 2 mℓ.

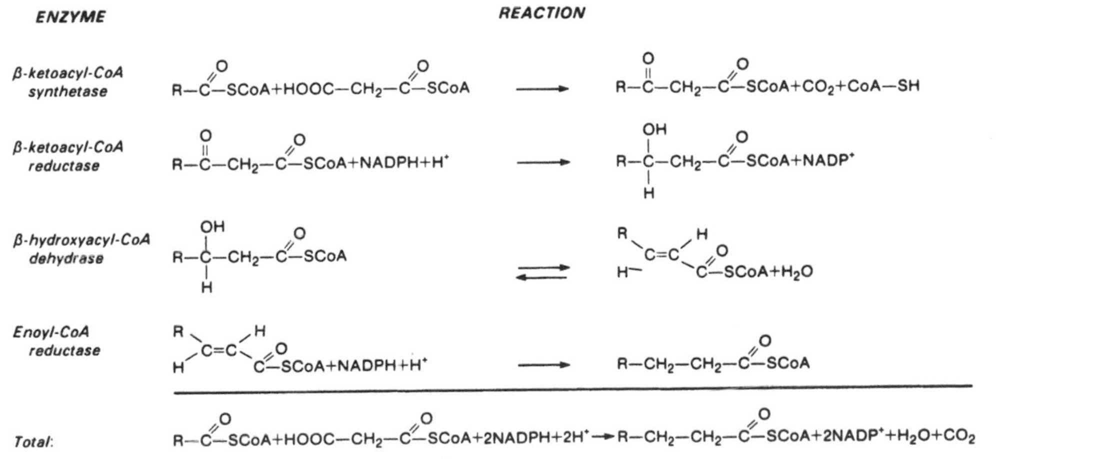

FIGURE 5. Sequence of reactions in the microsomal chain elongation of unsaturated acids.

the desaturation and elongation enzymes were each determined (see previous section). The principal desaturation and elongation steps of the different families have been tested and measured using the corresponding ^{14}C-labeled fatty acids that were totally synthesized by chemical procedures[87,88] in the laboratories of Stoffel, Sprecher, Holman, Kunau, Gunstone, and Brenner (Tables 2 and 3).

Table 2
OXIDATIVE DESATURATION REACTIONS

Family	Substrate	Desaturase	Product	Ref.
ω9	(9)18:1	Δ6	(6,9)18:2	65
ω6	(9-12)18:2	Δ6	(6,9,12)18:3	2
ω3	(9-12,15)18:3	Δ6	(6,9,12,15)18:4	66
ω9	(8,11)20:2	Δ5	(5,8,11)20:3	89, 90
ω6	(8,11,14)20:3	Δ5	(5,8,11,14)20:4	1
ω6	(7,10,13,16)22:4	Δ4	(4,7,10,13,16)20:5	91

Table 3
ELONGATION REACTIONS

Family	Substrate	Product	Ref.
ω9	(9)18:1	(11)20:1	81
ω6	(9,12)18:2	(11,12)20:2	81
ω3	(9,12,15)18:3	(11,14,17)20:3	81
ω9	(6,9)18:2	(8,11)20:2	81
ω6	(6,9,12)18:3	(8,11,14)20:3	81
ω3	(6,9,12,15)18:4	(8,11,14,17)20:4	81
ω9	(5,8,11)20:3	(7,10,13)22:3	81
ω6	(5,8,11,14)20:4	(7,10,13,16)22:4	81

All the unsaturated fatty acid families have in common (Figures 1 and 2) desaturation reactions at Δ6, Δ5, and Δ4 carbons. The desaturation in Δ6 is the first reaction of the series and the desaturation at Δ5 converts the unsaturated acids of 20 carbons such as eicosa-8,11,14-trienoic acid to arachidonic. It has been shown, as suggested by Brenner and Peluffo[66] that the desaturation at Δ9, Δ6, and Δ5 is produced by independent enzymes[5,92] and these results have been confirmed under different experimental conditions.[5,25,26]

A single Δ6 desaturase is capable of accepting oleic, linoleic, and α-linolenic acids,[5,93] and Okayasu et al.[77] have recently purified an enzyme with specific Δ6 desaturase activity. The Δ6 desaturation of linoleic and α-linolenic in human liver microsomes has been measured.[94] However, there may be at least two types of Δ5 desaturases; in addition to the enzyme that desaturates 8,11,14-eicosatrienoyl-CoA, Pugh and Kates[95,96] demonstrated the existence of another enzyme that desaturates the phospholipid-bound eicosatrienoic acid. The Δ4 desaturase that produces an additional double bond in the unsaturated fatty acids of 22 carbons has little activity.[5,91,97,98]

The different steps of desaturation/elongation of linoleic acid were identified by incubation of rat liver or testes microsomes with 1-^{14}C linoleate in simultaneous desaturating and elongating conditions in the presence of oxygen, NADH, CoA, ATP, and malonyl-CoA.[97] The label was transferred to 18:3ω6, 20:2ω6, 20:3ω6, 20:4ω6, 22:2ω6, and 22:5ω6. These results confirmed and extended the earlier work of Stoffel and Ach[98] in which the label from 1-^{14}C linoleic acid was found only in 20:2ω6 and 20:4ω6. However, the same authors showed that the incubation of 1-[14]C 18:3ω6 produced labeled 20:3ω6, whereas 1-[14]C 20:3ω6 was converted to 20:4ω6, 22:3ω6, and 22:4ω6; 1-[14]C 20:4ω6 was transferred to 22:4ω6 and 22:5ω6.

Thus, linoleic acid (18:2ω6) is converted to arachidonic by the following sequence of reactions: Δ6 desaturation to (6,9,12)18:3 acid (γ-linolenic, 18:3ω6), then an elongation to (8,12,14)20:3 (dihomo-γ-linolenic, 20:3ω6), and finally by a Δ5 desaturation to arachidonic acid (20:4ω6). However, it could also be produced alternatively by an elongation to (11,14)20:2

followed by a Δ8 desaturation to (8,11,14)20:3 and a Δ5 desaturation to arachidonic acid.[98] However, the last pathway was shown to be inoperative in vivo (section "The Linoleic (ω6) Family") and in other in vitro tests,[45] since the (11,14)20:2ω6 acid could not be a substrate for a Δ8 desaturase in rat liver[44] or brain.[99] However, Albert and Coniglio[100] presented evidence that this enzyme is active in rat testes. Holman's group also showed that the preferred pathway for linoleic acid conversion to arachidonic in rat liver microsomes is the one going through γ-linolenic acid.[101] An excellent discussion on the biosynthetic pathways of polyunsaturated fatty acids and the absence of Δ8 desaturase in tissues has been published by Sprecher.[102] Holman and Mahfouz[104] and Pollard et al.[105] have shown, using labeled acids, that the substrate specificity of Δ6 and Δ5 desaturases is relatively broad, at least in vitro. The Δ6 desaturase may desaturate, with different efficiencies, octadecenoic acids of different efficiencies with double bonds in positions different from Δ9,[103,104] *cis-trans* dienoic acid,[103] and even saturated acids, but with very small yield.[103] However, these reactions do not, in general, produce acids with EFA properties. All *trans* linoleic acid is not a substrate for the Δ6 desaturase,[5] and *cis-trans* isomers are not converted through to arachidonic acid.

Competitive Inhibition in In Vitro Studies

The first in vitro experiment showing the existence of competition among different unsaturated fatty acids for rat liver Δ6 desaturase was described in 1966,[66] confirming the results of in vivo tests. It was shown that palmitic or stearic saturated fatty acids did not inhibit the Δ6 desaturation of oleic, linoleic, and α-linolenic acids, whereas the three aforementioned unsaturated acids did compete among themselves for the enzyme. The ability to compete increased with the number of double bonds. These results demonstrated that ω6, ω3, and ω9 fatty acid families compete at the Δ6 desaturation step. Thus, the increase of 20:3ω9 acid produced in EFA-deficient animals is due to unsuppressed desaturation of oleic acid (Figure 3). By contrast, oleic acid is a poor inhibitor of linoleic acid desaturation;[66] such results are consistent with the in vivo work of Dopeschwarkar and Mead.[64]

A similarly small decrease in the Δ6 desaturation of lineolate is also produced by other 18:1 acids as vaccenic, petroselinic, elaidic, and by all *trans* linoleic acids[84] in addition to other unnatural *cis* and *trans* monoenoic acids.[106,107]

The polyethylenic acids 20:4ω6, 20:5ω3, and 22:4ω6 may appear to activate in vitro Δ6 desaturation of linoleate.[66] However, this effect is apparently produced by a simultaneous competition with the linoleate substrate for phospholipid transacylation in the microsomes.[108] Under other experimental conditions, it was shown that 18:3ω6,[109] 20:3ω6, 20:4ω6,[109] 22:5ω6,[111] and 22:6ω3[111,112] inhibited Δ6 desaturation of linoleate. The inhibitory effect of 20:3ω6, 20:4ω6, 22:4ω6 and 22:5ω6 (linoleic family) and 20:4ω3, 20:5ω3, and 22:6ω3 (α-linolenic family) on α-linolenic Δ6 desaturation was also shown.[111-113] These results indicate that intermediate members and the last member of a family may inhibit the first desaturation enzyme of the same or different series as was shown in in vivo experiments (see section entitled "The Oleic and Palmitoleic Family and the Idea of Competitive Inhibition Leading to the Formation of 20:3ω9"). These results have been confirmed repeatedly.[106,113]

Competitive effect of fatty acids at the Δ5 desaturase step has been studied by Sprecher's group.[89] They studied the Δ5 desaturation of 20:2ω9 to 20:3ω9 and found that whereas 18:2ω6 was a poor inhibitor, 20:3ω6 and 20:4ω6 compete actively for the enzyme, with 20:3ω6 being the most active. α-Linoleic acid also inhibited the reaction. The biosynthesis of 20:3ω9 can thus be seen also to be reduced by competition of fatty acids of both the ω6 and ω3 families at the Δ5 desaturase step. The competitive effect of a whole series of *cis* and *trans* monoenoic acids of 18 carbons with different double bond positions in the chain has been demonstrated by Mahfouz et al.[106,107]

Chain elongation of linoleic acid is inhibited by other fatty acids.[114] This rat liver microsomal enzyme was inhibited by saturated fatty acids from 8 to 20 carbons with myristic acid

Ⓐ *Linoleic family (ω6)*

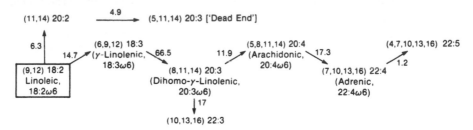

Ⓑ *Oleic family (ω9)*

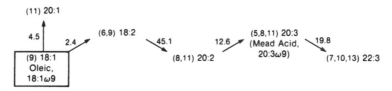

Ⓒ *Palmitoleic family (ω7)*

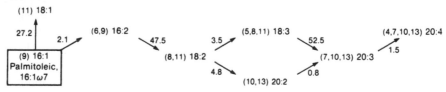

FIGURE 6.   Rates of desaturation and elongation for fatty acids of: (A) linoleate sequence; (B) oleate sequence; (C) palmitoleate sequence.[91] Rates are in nanomoles of product produced with 5 mg of rat microsomal protein during 3 min incubation.

being the most active acid. The unsaturated acids 18:1ω9, 18:3ω3, 18:3ω6, 20:4ω6, 20:5ω3, and 22:6ω3 were all inhibitors of the elongation. Sprecher[115] also showed that 18:2ω9 inhibited the elongation of 18:3ω6, and that conversely, 18:3ω6 inhibited 18:2ω9 elongation.

Rates of Interconversion

Knowledge of the rates of the individual reactions in the fatty acid families are very relevant to understanding their dynamic interactions. The rates of the Δ6 desaturation of the fatty acids α-linolenic, linoleic, and oleic acids have been shown[66] to decrease markedly following the order 18:3ω3 > 18:2ω6 > 18:1ω9. These relative rates were shown to be maintained in species as different as rat and fish although the absolutely highest Δ6 desaturase specific activity was found in fish.[92] This difference of Δ6 desaturation rate between the three acids as well as competition of the corresponding substrates for the Δ6 and Δ5 de-saturases of the ω3, ω6, and ω9 families can explain the absence of higher ω9 fatty acids when a normal diet is ingested by animals or man; 20:3ω9 only appears during EFA defi-ciency. Since the Δ5 desaturation of 20:2ω9 is similar to that of 20:3ω6,[92] Δ6 desaturase must be the main factor controlling the relative biosynthesis of ω6 and ω9 fatty acids. The regulatory functions of the Δ6 desaturase have been extensively investigated by Brenner.[116]

The relative speeds of desaturation and elongation of the fatty acid of the linoleic, oleic, and palmitoleic families has been measured by incubation of the corresponding labeled acids synthesized ad hoc by Sprecher's group. The results are summarized in Figure 6. They show again that the most striking difference between ω6 and ω9 fatty acid metabolism rates is seen for Δ6 desaturation and not for Δ5 desaturation.[66,92] The speeds of the elongation reaction are high, especially for 18:3ω6, 18:2ω9, and 18:2ω7. Correspondingly they do not accumulate in vivo.

The Δ8 Desaturases and "Dead End" Products

It has been shown by in vivo and in vitro experiments that rat liver and brain are unable to desaturate (11,14)20:2 to (8,11,14)20:3, but that this acid *is* a substrate for the Δ5 desaturase synthesizing (5,11,14)20:3.[44,99] Similarly,[47,115] the metabolite of the oleic acid series (11)20:1 is desaturated to (5,11)20:2 and not to (8,11)20:2. It has also been shown that (11,14,17)20:3, the analogous metabolite of the α-linolenic series, is Δ5 desaturated to (5,11,14,17)20:4.[51,52,117]

However, the (5,11,14)-20:3 derived from linoleic acid (18:2ω6) *cannot* be converted by Δ8 desaturase to arachidonic acid (5,8,11,14-20:4; 20:4ω6). Similarly, the (5,11,14,17)-20:4 derivative from α-linolenic acid (18:3ω3) cannot be converted to an eicosanoic precursor, 5,8,11,14,17-eicosapentaenoic acid (20:5ω3). It is the same for the ω9 series fatty acids. Thus, (5,11)-20:2 is not a substrate for Δ8 desaturase and therefore cannot be converted to 5,8,11-eicosatrienoic acid (Mead acid, 20:3ω9).

These alternate routes of eicosanoid precursor synthesis have been examined in rat liver and brain and in some lines of cultured cells.[51,52,119]

Retroconversion Reactions

The pathways of polyunsaturated fatty biosynthesis illustrated in Figures 2 and 3 show desaturation and elongation reactions that evoke the conversion of "lower" to "higher" fatty acids. In addition to these pathways, a reversed metabolic sequence has been demonstrated in which a decrease in the number of double bonds and chain length is produced in acids from 20 to 24 carbons. Such retroconversions are more pronounced in normal than in EFA-deficient rats.

The first retroconversion reaction recognized in the rat was the conversion of 22:5ω6 to 20:4ω6 as reported by Verdino et al.,[120] Schlenk et al.,[121] and Kunau.[122] These results were extended to the α-linolenic series and it was found that uniformly ^{14}C-labeled 22:6ω3 when fed to rats was retroconverted to 22:5ω3 and 20:5ω3.[123] The acids produced were incorporated into the phospholipids.

In addition to the aforementioned acids, a long list of even and odd numbered unsaturated fatty acids of 20 to 24 carbons is now known to be retroconverted (Table 4). The process may involve either the loss of a single two-carbon fragment or both a double bond and two or four carbons. Stoffel et al.[129] showed that this process occurs in the mitochondria.

Neither linoleate,[128] 5,8,11-eicosatrienoate, arachidonate, nor 5,8,11,14,17-eicosapentaenoate are substrates for retroconversion. Such results and the retroconversions illustrated in Table 5 indicate that 22 carbon acids derived from oleate (ω9), linoleate (ω6), or α-linolenate (ω3) families may be used as substrates in an optional route in production of 20 carbon polyunsaturated acids in which the first double bond is in the Δ5 position.

POLYUNSATURATED ACIDS WITH EFA ACTIVITY

Odd Chain Fatty Acids

In addition to natural EFAs of the linoleic (ω6) and α-linolenic acid (ω3) families, some odd chain unsaturated fatty acids have also been reported to cure EFA deficiency symptoms.[132-135] Table 5 summarizes some of the acids that possess EFA activity measured by Thomasson's[136] test based on the disturbance of water metabolism. It shows that polyunsaturated acids of odd chain length with 17, 19, and 21 carbons and double bonds 5,8,11,14 may give rise to 5,8,11,14-19:4 and 21:4, structures that possess EFA activity. They have ω5 or ω7 unsaturation, since ω6 structure would not correspond to a Δ5,8,11,14, double

Table 4
RETROCONVERSION OF *cis* UNSATURATED ACIDS

Substrate	Products	Ref.
(11)20:1	(9)18:1	124
(10,13)20:2	(8,11)18:2	60
(11,14)20:2	(9,12)18:2	125
(7,10,13)22:3	(5,8,11)20:3	126, 127
(10,13,16)22:3	(8,11,14)20:3	128, 129
(6,9,12,15)21:4	(4,7,10,13)19:4 (5,8,11,14,17)23:5	130
(4,7,10,13)22:4	(7,10,13)22:3 (5,8,11)20:3	126, 127
(6,9,12,15)22:4	(4,7,10,13)20:4	128
(7,10,13,16)22:4	(5,8,11,14)20:4	131
(4,7,10,13,16)22:5	(7,10,13,16)22:4 (5,8,11,14)20:4	120, 121, 123, 126, 127
(7,10,13,16,19)22:5	(5,8,11,14,17)20:5	126, 127
(6,9,12,15,18)24:5	(4,7,10,13,16)22:5 (7,10,13,16)22:4 (5,8,11,14)20:4	126, 127
(4,7,10,13,16,19)22:6	(7,10,13,16,19)22:5 (5,8,11,14,17)20:3	123, 126, 127
(6,9,12,15,18,21)24:6	(4,7,10,13,16,19)22:6 (7,10,13,16,19)22:5 (5,8,11,14,17)20:5	126, 127

Table 5
NATURAL AND SYNTHEIC UNNATURAL POLYUNSATURATED FATTY ACIDS WITH EFA ACTIVITY

Fatty acid	Double bond position (Δ)	EFA potency U/g	Prostaglandin formation
17:2	9,12	100	+
18:2	9,12	100	+
γ18:3	6,9,12	115	+
18:4	6,9,12,15	34	+
19:2	10,13	9	+
19:3	8,11,14	22—33	+
19:4	5,8,11,14	19	+
20:2	11,14	46	+
20:3	8,11,14	100	+
20:4	5,8,11,14	139	+
21:3	8,11,14	56	+
21:4	5,8,11,14	62,68	+
22:5	4,7,10,13,16	139	+

bond sequence. It has also been shown that ω6 odd chain fatty acids such as (10,13)19:2 and (7,10,13)19:3 are inactive.[135,137] It is remarkable that all of the EFAs mentioned are capable of being converted into prostaglandins.[133-135,138] Van Dorp[138,139] postulated two requirements for a fatty acid to have EFA activity: "First, the animal must be able to convert it into a prostaglandin; second, the prostaglandin thus formed, must have biological activity." The structural requirements necessary to meet these two demands includes a fixed structure of the carboxyl end of the fatty acids and the presence of *cis* double bonds in the positions 8,11,14 or 5,8,11,14.

That the essentiality of fatty acids is determined by their conversion into physiologically active prostaglandins is still a matter of some controversy, even though scaliness of skin (typical of EFA deficiency) has, on one occasion, been ascribed to insufficient synthesis of

Table 6
STRUCTURE AND SOME BIOLOGICAL PROPERTIES OF COLUMBINIC AND
bis-HOMO COLUMBINIC ACIDS[142]

	Biological effects						
	Potency			Fertility			Platelet function
Structure	Body wt (%)	Skin permeability (%)	Mitochondria swelling (%)	Pregnancy	Newborn alive	Newborn dead	Obturation time aortic loop (hr)
Columbinic acid (5-*trans*,9-*cis*,12-*cis*)18:3	80	100	100	5/5	29	8	188
Bis-homo columbinic acid (7-*trans*,11-*cis*,14-*cis*)20:3	25	85	25	—	—	—	—
Linoleic acid (9-*cis*,12-*cis*)18:2	100	100	100	3/5	32	0	117

* Time it takes an extracorporeal loop to be obstructed by thrombus formation.

prostaglandins.[140] Nevertheless, as remarked by Lands et al.,[141] the polyunsaturated acids also have activity in their own right in maintaining cell membrane integrity and in producing specific 1-chain protein interactions. Such concepts are discussed in more detail in the preceding chapter by Mead and Willis.

Columbinic Acid

Columbinic acid is 5-*trans*,9-*cis*,12-*cis*-octadecatrienoic acid and can be designated *trans*-5-linoleic acid. It is abundant in the seed oil of the columbine *(Aquilegia vulgaris)*,[142] and Houtsmuller[143] showed that in rats it induces a growth rate comparable to that found by linoleic acid and also cures skin and fertility symptoms of EFA deficiency; it does not give rise to prostaglandins (Table 6). Prefeeding of columbinic acid enhances the conversion of trace amounts of linoleic acid to arachidonic, but large doses aggravate the EFA deficiency.[145] Therefore, it is possible that the effect of columbinic acid may be produced by mobilization of linoleic acid from depots. Columbinic acid is converted neither into linoleic acid nor into arachidonic acid in the animal tissues, and it is incorporated without transformation into such complex lipids as cholesteryl esters and phospholipids, replacing arachidonic acid.

Both columbinic and linoleic acids shared the ability of several ω6 fatty acids (18:3ω6, 20:3ω6, 20:4ω6, and 22:4ω6) to activate the Δ5 desaturation of 1-^{14}C-20:3ω6 to 1-^{14}C-20:4ω6 in HTC cells.[144] Among ω6 fatty acids, 4,7,10,13,16-docosapentaenoic acid alone suppressed Δ5 desaturation. Therefore, it seems that activity in this system of both linoleic and columbinic acids depends upon their own structure and not upon prior conversion to higher homologs. Columbinic acid is thus a magnificent tool for study of EFA function. Its use has already illuminated the relationship existing between structure and disparate physiological effects of EFAs.

REFERENCES

1. **Stoffel, W.,** Biosynthesis of polyenoic fatty acids, *Biochem. Biophys. Res. Commun.*, 6, 270—273, 1961.
2. **Nugteren, D. H.,** Conversion *in vitro* of linoleic acid into γ-linolenic acid by rat-liver enzymes, *Biochim. Biophys. Acta*, 60, 656—657, 1962.
3. **Stoffel, W. and Ach, K. L.,** Der Stoffwechsel der ungesättigten Fettsäure II Eigenshafter des kettenver-längernden. Enzyms zur frage der Biohydrogenierung der ungesättigten Fettsäuren, *Z. Physiol. Chem.*, 337, 123—132, 1964.
4. **Nugteren, D. H.,** The enzymatic chain elongation of fatty acids by rat-liver microsomes, *Biochim. Biophys. Acta*, 106, 280—290, 1965.
5. **Brenner, R. R.,** The oxidative desaturation of unsaturated fatty acids in animals, *Mol. Cell. Biochem.*, 3, 41—52, 1974.
6. **Holman, R. T.,** Essential fatty acid deficiency, in *Progress in the Chemistry of Fat and Other Lipids*, Vol. 9, Part II, Holman, R. T., Ed., Pergamon Press, Oxford, 1970, 279—348.
7. **McMahon, V. and Stumpf, P. K.,** Synthesis of linoleic acid by particulate system from safflower seed, *Biochim. Biophys. Acta*, 84, 359—361, 1964.
8. **Stumpf, P. K.,** Biosynthesis of saturated and unsaturated fatty acids, in *The Biochemistry of Plants*, Vol. 4, *Lipids: Structure and Function*, Stumpf, P. K., Ed., Academic Press, New York, 1980, 177—204.
9. **Haines, T. H., Aaronson, S., Gellerman, L., and Schlenk, H.,** Occurrence of arachidonic and related acids in the protozoon *Ochromonas danica*, *Nature (London)*, 194, 1282—1283, 1962.
10. **Gellerman, J. L. and Schlenk, H.,** Preparation of fatty acids labeled with ^{14}C from *Ochromonas danica*, *J. Protozool.*, 12, 178—189, 1965.
11. **Pollero, R. J., Brenner, R. R., and de Gómez Dumm, C. G.,** Comparative biosynthesis of polyethylenic fatty acids in *Acanthamoeba castellanii* and in *Ochromonas danica*, *Acta Physiol. Lat. Am.*, 25, 412—424, 1975.
12. **Kasai, R., Kitajima, Y., Martin, C. E., Nozawa, Y., Skriver, L., and Thompson, G. A., Jr.,** Molecular control of membrane properties during temperature acclimation. Membrane fluidity regulation of fatty acid desaturase activity, *Biochemistry*, 15, 5228—5233, 1976.
13. **Kameyama, Y., Yoshioka, S., and Nozawa, Y.,** The occurrence of direct desaturation of phospholipid acyl chain in *Tetrahymena pyriformis*. Thermal adaptation of membrane phospholipids, *Biochim. Biophys. Acta*, 618, 214—222, 1980.
14. **Neville, M., Jr., Min, T., and Ferguson, K. A.,** Differential biosynthesis of polyunsaturated fatty acids by *Tetrahymena* supplemented with ergosterol, *Biochim. Biophys. Acta*, 573, 201—206, 1979.
15. **Korn, E. D.,** Biosynthesis of unsaturated fatty acids in *Acanthamoeba* sp., *J. Biol. Chem.*, 239, 396—400, 1964.
16. **Aeberhard, E. A., de Lema, M. G., and Bromia, D. I. M.,** Biosynthesis of fatty acids by *Trypanosoma cruzi*, *Lipids*, 16, 623—625, 1981.
17. **Dwyer, L. A. and Blomquist, G.,** Biosynthesis of linoleic acid in the American cockroach, in *Essential Fatty Acids and Prostaglandins: Progress in Lipid Research*, Vol. 20, Holman, R. T., Ed., Pergamon Press, New York, 1981, 215—218.
18. **Curr, M. I., Robinson, M. P., James, A. T., Morris, J., and Howling, D.,** The substrate specificity of desaturases: the conversion of *cis* 12-octadecenoic acid into linoleic acid in different animal and plant species, *Biochim. Biophys. Acta*, 280, 415—421, 1972.
19. **Fulco, A. J. and Mead, J. F.,** Metabolism of essential fatty acids. IX. The biosynthesis of the octade-cadienoic acids of the rat, *J. Biol. Chem.*, 235, 3379—3384, 1960.
20. **Murty, N. L., Rakoff, H., and Reiser, R.,** The conversion of cis-2-octenoic acid to linoleic acid, *Biochem. Biophys. Res. Commun.*, 8, 372—376, 1962.
21. **Reiser, R., Murty, N. L., and Rakoff, H.,** Biosynthesis of linoleic acid from 1-^{14}C cis-2-octenoic acid by the laying hen, *J. Lipid Res.*, 3, 56—59, 1962.
22. **Brenner, R. R., Mercuri, O., and de Tomás, M. E.,** Influence dietary medium chain fatty acids in the rat lipid composition. I. Cis-2-octenoic acid as precursor of linoleic acids, *J. Nutr.*, 77, 203—209, 1962.
23. **Rees, H. H.,** *Insect Biochemistry*, Chapman & Hall, London, 1977, 64.
24. **Rivers, J. P. W., Sinclair, A. J., and Crawford, M. A.,** Inability of the cat to desaturate essential fatty acids, *Nature (London)*, 258, 171—173, 1975.
25. **Dunbar, L. M. and Bailey, J. M.,** Enzyme deletions and essential fatty acid metabolism in cultured cells, *J. Biol. Chem.*, 250, 1152—1153, 1975.
26. **de Alaniz, M. J. T., Ponz, G., and Brenner, R. R.,** Biosynthesis of unsaturated fatty acids in cultured minimal deviation hepatoma 7288 C cells, *Acta Physiol. Lat. Am.*, 25, 1—11, 1976.
27. **Gaspar, G., de Alaniz, M. J. T., and Brenner, R. R.,** Uptake and metabolism of exogenous eicosa-8,11,14-trienoic acid in minimal deviation hepatoma 7288 C cells, *Lipids*, 10, 726—731, 1975.
28. **Burr, G. O. and Burr, M. M.,** A new deficiency disease produced by the rigid exclusion of fat from the diet, *J. Biol. Chem.*, 82, 345—367, 1929.

29. **Nunn, L. C. A. and Smedley-Maclean, I.,** The nature of the fatty acids stored by the liver in the fat deficiency disease of rats, *Biochem. J.,* 32, 2179—2184, 1938.

30. **Smedley-MacLean, I. and Hume, E. M.,** Fat deficiency disease of rats. The storage of fat in the fat-starved rat, *Biochem. J.,* 35, 975—990, 1942.

31. **Holman, R. T. and Taylor, T. S.,** Polyethenoid fatty acid metabolism. III. Arachidonate supplementation, *Arch. Biochem.,* 29, 295—301, 1950.

32. **Holman, R. T.,** Essential Fatty Acid deficiency, in *Progress in the Chemistry of Fats and Other Lipids,* Vol. 9, Part I, Holman, R. T., Ed., Pergamon Press, Oxford, 1968, 279—348.

33. **Brenner, R. R., Garda, H., de Gómez Dumm, I. N. T., Pezzano, H.,** Early effects of essential fatty acid deficiency on the structure in enzymatic activity of rat liver microsomes, in *Essential Fatty Acids and Prostaglandins, Progress in Lipid Research,* Vol. 20, Holman, R. T., Ed., Pergamon Press, New York, 1981, 315—321.

34. **Brenner, R. R. and Nervi, A. M.,** Kinetics of linoleic and arachidonic acid incorporation and eicosatrienoic acid depletion in the lipids of fat-deficient rats fed methyl linoleate arachidonate, *J. Lipid Res.,* 6, 363—368, 1965.

35. **Thomasson, H. J.,** Essential fatty acid, *Nature (London),* 194, 973, 1962.

36. **Mead, J. F., Steinberg, G., and Howton, D. R.,** Metabolism of essential fatty acids. Incorporation of acetate into arachidonic acid, *J. Biol. Chem.,* 205, 683—689, 1953.

37. **Steinberg, G., Slaton, W. H., Jr., Howton, D. R., and Mead, J. F.,** Metabolism of essential fatty acids. IV. Incorporation of linoleate into arachidonic acid, *J. Biol. Chem.,* 220, 257—264, 1956.

38. **Fulco, A. T. and Mead, J. F.,** Metabolism of essential fatty acids. VIII. Origin of 5,8,11-eicosatrienoic acid in the fat deficient rat, *J. Biol. Chem.,* 234, 1411—1416, 1959.

39. **Mead, J. F. and Howton, D. R.,** Metabolism of essential fatty acids. VII. Conversion of γ-linolenic acid into arachidonic acid, *J. Biol. Chem.,* 229, 575—582, 1957.

40. **Howton, D. R. and Mead, J. F.,** Metabolism of essential fatty acids. X. Conversion of 8,11,14-eicosatrienoic acid to arachidonic acid in the rat, *J. Biol. Chem.,* 235, 3385—3386, 1960.

41. **Davis, J. T. and Coniglio, J. G.,** The biosynthesis of docosapentaenoic and other fatty acids by rat testes, *J. Biol. Chem.,* 241, 610—612, 1966.

42. **Bridges, R. B. and Coniglio, J. G.,** The biosynthesis of Δ9,12,15,18-tetracosatetraenoic and of 6,9,12,15,18-tetracosapentaenoic acid by rat testes, *J. Biol. Chem.,* 245, 46—49, 1970.

43. **Klenk, E. and Mohrhauer, H.,** Untersuchungen uber den Stoffwechsel der Polyenfettsauren bei der Ratte, *Z. Physiol. Chem.,* 320, 218—232, 1960.

44. **Ullman, D. and Sprecher, H.,** An *in vitro* and *in vivo* study of the conversion of 11,14-eicosa-dienoic acid to 5,11,14-eicosa-trienoic acid and of the conversion of 11-eicos-enoic acid to 5,11-eicosa dienoic acid in the rat, *Biochim. Biophys. Acta,* 248, 186—197, 1971.

45. **de Alaniz, M. J. T., de Gómez Dumm, I. N. T., and Brenner, R. R.,** Biosynthesis of polyunsaturated acids from the linoleic acid family in cultured cells, in *Function and Biosynthesis of Lipids, Advances in Experimental Medicine and Biology,* Bazán, N. G., Brenner, R. R., and Giusto, N. M., Eds., Plenum Press, New York, 1979, 617—624.

46. **Widmer, C., Jr. and Holman, R. T.,** Polyethenoid fatty acid metabolism. II. Deposition of polyunsaturated fatty acids in fat-deficient rats upon single fatty acid supplementation, *Arch. Biochem.,* 25, 1—12, 1950.

47. **Klenk, E. and Oette, K.,** Uber die Natur der in den Leberphosphatiden auftretenden C_{20} und C_{22}-Polyen-säuren bei Verabrunchung von Linol- und Linolensäure an fettfrei ernabute Ratten, *Z. Physiol. Chem.,* 318, 86—99, 1960.

48. **Brenner, R. R. and José, P.,** Action of linolenic and docosahexaenoic acid upon the eicosatrienoic acid level in rat lipids, *J. Nutr.,* 85, 196—204, 1965.

49. **Steinberg, G., Slaton, W. H., Jr., Howton, D. R., and Mead, J. W.,** Metabolism of essential fatty acids. V. Metabolic pathway of linolenic acid, *J. Biol. Chem.,* 224, 841—849, 1957.

50. **Kayama, M., Tsuchiya, Y., Nevenzel, J. C., Fulco, A. J., and Mead, J. F.,** Incorporation of linolenic 1-^{14}C acid into eicosapentaenoic and docosahexaenoic acids in fish, *J. Am. Oil Chem. Soc.,* 40, 499—502, 1963.

51. **de Alaniz, M. J. T., de Gómez Dumm, I. N. T., and Brenner, R. R.,** The action of insulin and dibutyryl cyclic AMP on the biosynthesis of polyunsaturated acids of α-linolenic acid family in HTC cells, *Mol. Cell. Biochem.,* 12, 3—8, 1976.

52. **de Alaniz, M. J. T. and Brenner, R. R.,** Effect of different carbon sources on the biosynthesis of polyunsaturated fatty acids of α-linolenic acid family in culture of minimal deviation hepatoma 7288 C cells, *Mol. Cell. Biochem.,* 12, 81—87, 1976.

53. **Peluffo, R. O., Brenner, R. R., and Mercuri, O.,** Action of linoleic and arachidonic acids upon eicos-atrienoic acid level in rat heart and liver, *J. Nutr.,* 81, 110—116, 1963.

54. **Nervi, A. M. and Brenner, R. R.,** Identification and origin of docosatrienoic acid of rat phospholipids, *Biochim. Biophys. Acta,* 106, 205—207, 1965.

55. **Schoenheimer, R. and Rittenberg, D.,** Deuterium as an indicator in the study of intermediary metabolism. VI. Synthesis and destruction of fatty acids in the organism, *J. Biol. Chem.,* 114, 381—396, 1936.

56. **Anker, H. S.,** On the mechanism of fatty acid synthesis *in vivo, J. Biol. Chem.,* 194, 177, 1952.

57. **Dauben, W. G., Hoerger, E., and Petersen, J. W.,** Distribution of acetic acid carbon in high fatty acids synthesized from acetic acid by the intact mouse, *J. Am. Chem. Soc.,* 75, 2347—2351, 1953.

58. **Mead, J. F.,** Metabolism of polyunsaturated fatty acids, *Am. J. Clin. Nutr.,* 8, 55—61, 1961.

59. **Klenk, E.,** The metabolism of polyenoic fatty acids, in *Advances in Lipid Research,* Vol. 3, Paoletti, R. and Kritchevsky, D., Eds., Academic Press, New York, 1965, 1.

60. **Budny, J. and Sprecher, H.,** A study of some of the factors involved in regulating the conversion of octadeca-8,11-dienoate to eicosa-4,7,10,13-tetraenoate in the rat, *Biochim. Biophys. Acta,* 239, 190—207, 1971.

61. **Sprecher, H.,** The total synthesis and metabolism of octadeca-5,8,11-trienoate,eicosa-10,13-dienoate and eicosa-7,10,13-trienoate in the fat deficient rat, *Biochim. Biophys. Acta,* 231, 122—130, 1971.

62. **Lindstrom, T. and Tinsley, I. J.,** Interactions in the metabolism of polyunsaturated fatty acids; analysis by a simple mathematical model, *J. Lipid Res.,* 7, 758—762, 1966.

63. **Holman, R. T.,** Nutritional and metabolic interrelationships between fatty acids, *Fed. Proc.,* 23, 1062—1067, 1964.

64. **Dopeshwarkar, G. A. and Mead, J. F.,** Role of oleic acid in the metabolism of essential fatty acids, *J. Am. Oil Chem. Soc.,* 38, 297—301, 1961.

65. **Holloway, P. W., Peluffo, R. O., and Wakil, S. J.,** On the biosynthesis of dienoic fatty acids by animal tissues, *Biochem. Biophys. Res. Commun.,* 12, 300—304, 1963.

66. **Brenner, R. R. and Peluffo, R. O.,** Effect of saturated and unsaturated fatty acids on the desaturation *in vitro* of palmitic, stearic, oleic, linoleic and linolenic acids, *J. Biol. Chem.,* 241, 5213—5219, 1966.

67. **Oshino, N., Imai, Y., and Sato, R.,** Electron transfer mechanism associated with fatty acid desaturation catalyzed by liver microsomes, *Biochim. Biophys. Acta,* 128, 13—28, 1966.

68. **Holloway, P. W. and Wakil, S. J.,** Requirement for reduced diphosphopyridine nucleotide-cytochrome b_5 reductase in stearyl coenzyme A desaturation, *J. Biol. Chem.,* 245, 1862—1865, 1970.

69. **Enoch, H. G., Catalá, A., and Strittmatter, P.,** Mechanism of rat liver microsomal stearyl-CoA desaturase. Studies of the substrate specificity, enzyme-substrate interactions and the function of lipid, *J. Biol. Chem.,* 251, 5095—5103, 1976.

70. **Okayasu, T., Ono, T., Shinojima, K., and Imai, J.,** Involvement of cytochrome b_5 in the oxidative desaturation of linoleic acid to γ-linolenic acid in rat liver microsomes, *Lipids,* 12, 267—271, 1977.

71. **Lee, T. C., Baker, R. C., Stephens, N., and Snyder, F.,** Evidence of participation of cytochrome b_5 in microsomal Δ6 desaturation of fatty acids, *Biochim. Biophys. Acta,* 489, 25—31, 1977.

72. **Keyes, S. R., Alfano, J. A., Jansson, I., and Cinti, D. L.,** Rat liver microsomal elongation of fatty acids: possible involvement of cytochrome b_5, *J. Biol. Chem.,* 254, 7778—7784, 1979.

73. **Ito, A. and Sato, R.,** Purification by means of detergents and properties of cytochrome b_5 from liver microsomes, *J. Cell. Biol.,* 40, 179, 1969.

74. **Spatz, L. and Strittmatter, P.,** Form of cytochrome b_5 that contains an additional hydrophobic sequence of 40 amino acid residues, *Proc. Natl. Acad. Sci. U.S.A.,* 68, 1042—1046, 1971.

75. **Spatz, L. and Strittmatter, P.,** Form of reduced nicotinamide adenine dinucleotide containing both the catalytic site and an additional hydrophobic membrane binding segment, *J. Biol. Chem.,* 248, 793—799, 1973.

76. **Strittmatter, P., Spatz, L., Corcoran, D., Rogers, M. J., Setlow, B., and Redline, R.,** Purification and properties of rat liver microsomal stearyl coenzyme A desaturase, *Proc. Natl. Acad. Sci. U.S.A.,* 71, 4565—4569, 1974.

77. **Okayasu, T., Nagao, M., Ishibashi, T., and Imai, Y.,** Purification and partial characterization of linoleyl-CoA desaturase from rat liver microsomes, *Biochem. Biophys. Res. Commun.,* 206, 21—28, 1981.

78. **Rogers, M. J. and Strittmatter, P.,** Evidence of random distribution and translational movement of cytochrome b_5 in endoplasmic reticulum, *J. Biol. Chem.,* 249, 895—900, 1974.

79. **Strittmatter, P. and Rogers, M. J.,** Apparent dependence of interactions between cytochrome b_5 and cytochrome b_5 reductase upon translational diffusion in dimyristoyl lecithin liposomes, *Proc. Natl. Acad. Sci. U.S.A.,* 72, 2658—2661, 1975.

80. **Brenner, R. R.,** Nutritional and hormonal factors influencing desaturation of essential fatty acids, in *Essential Fatty Acids and Prostaglandins, Progress in Lipid Research,* Vol. 20, Holman, R. T., Ed., Pergamon Press, New York, 1981, 41—47.

81. **Stoffel, W.,** Uber Biosynthese und biologischen Abban hochungesattigter Fettsauren, *Naturwissenschaften,* 53, 621—630, 1966.

82. **Seubert, W. and Podack, E. R.,** Mechanisms and physiological roles of fatty acid chain elongation in microsomes and mitochondria, *Mol. Cell. Biochem.,* 1, 29—40, 1973.

83. **Bernert, T. J., Jr. and Sprecher, H.,** The isolation of acyl-CoA derivatives as products of partial reactions in the microsomal chain elongation of fatty acids, *Biochim. Biophys. Acta,* 573, 436—442, 1979.

84. **Brenner, R. R. and Peluffo, R. O.**, Regulation of unsaturated fatty acid biosynthesis. I. Effects of unsaturated fatty acid of 18 carbons on the microsomal desaturation of linoleic acid into γ-linolenic acid, *Biochim. Biophys. Acta*, 176, 471—479, 1969.

85. **Bernert, T. J., Jr. and Sprecher, H.**, An analysis of partial reactions in the overall chain. Elongation of saturated and unsaturated fatty acids by rat liver microsomes, *J. Biol. Chem.*, 252, 6736—6744, 1977.

86. **Bernert, J. T., Jr., Bourre, J. M., Baumann, N. H., and Sprecher, H.**, The activity of partial reactions in the chain elongation of palmitoyl CoA and stearoyl CoA by mouse brain microsomes, *J. Neurochem.*, 32, 85—90, 1979.

87. **Sprecher, H.**, The organic synthesis of polyunsaturated fatty acids, in *Polyunsaturated Fatty Acids*, Kunau, W. H. and Holman, R. T., Eds., American Oil Chemists Society, Champaign, Ill., 1977, 69—79.

88. **Osbond, J. M.**, The synthesis of naturally-occurring and labelled 1,4-polyunsaturated fatty acids, in *Progress in the Chemistry of Fats and Other Lipids*, Vol. 9, Part I, Holman, R. T., Ed., Pergamon Press, Oxford, 1966, 119—157.

89. **Ullman, D. and Sprecher, H.**, An *in vitro* study of the effects of linoleic, 8,11,14-eicosatrienoic acid and arachidonic acids on the desaturation of stearic, oleic and 8,11-eicosadienoic acids, *Biochim. Biophys. Acta*, 243, 61—70, 1971.

90. **Castuma, J. C., Catalá, A., and Brenner, R. R.**, Oxidative desaturation of eicosa-8,11-dienoic acid to eicosa-5,8,11-trienoic acid: comparison of different diets on oxidative desaturation at the 5,6 and 6,7 positions, *J. Lipid Res.*, 13, 783—789, 1972.

91. **Bernert, J. T., Jr. and Sprecher, H.**, Studies to determine the role rates of chain elongation and desaturation play in regulating the unsaturated fatty acid composition of rat liver lipids, *Biochim. Biophys. Acta*, 398, 354—363, 1975.

92. **Ninno, R. E., de Torrengo, M. P., Castuma, J. C., and Brenner, R. R.**, Specificity of 5- and 6- fatty acid desaturases in rat and fish, *Biochim. Biophys. Acta*, 360, 124—133, 1974.

93. **Brenner, R. R.**, The desaturation step in the animal biosynthesis of polyunsaturated fatty acids, *Lipids*, 6, 567—575, 1971.

94. **de Gómez Dumm, I. N. T. and Brenner, R. R.**, Oxidative desaturation of α-linolenic, linoleic and stearic acids by human liver microsomes, *Lipids*, 10, 315—317, 1975.

95. **Pugh, E. L. and Kates, M.**, Direct desaturation of eicosatrienoyl lecithin to arachidonyl lecithin by rat liver microsomes, *J. Biol. Chem.*, 252, 68—73, 1977.

96. **Pugh, E. L. and Kates, M.**, Membrane-bound phospholipid desaturases, *Lipids*, 14, 159—165, 1979.

97. **Ayala, S., Gaspar, G., Brenner, R. R., Peluffo, R. O., and Kunau, W.**, Fate of linoleic, arachidonic and docosa-7,10,13,16-tetraenoic acids in rat testicles, *J. Lipid Res.*, 14, 296—305, 1973.

98. **Stoffel, W. and Ach, K. L.**, Der Stoffwechsel der ungesättigten Fettesauren. II. Eigesischaften de Rettenverlangernden Enzyms. Zur frage der Biohydrogenierung der ungesättigten Fettsauren, *Z. Physiol. Chem.*, 337, 123—132, 1964.

99. **Dhopeshwarkar, G. A. and Subramanian, G. J.**, Intracranial conversion of linoleic acid to arachidonic acid; evidence for lack of Δ8 desaturase in the brain, *J. Neurochem.*, 26, 1175—1179, 1976.

100. **Albert, D. H. and Coniglio, J. G.**, Metabolism of eicosa-11,14-dienoic acid in rat testes. Evidence for Δ8-desaturase activity, *Biochim. Biophys. Acta*, 489, 390—396, 1977.

101. **Marcel, Y. L., Christiansen, K., and Holman, R. T.**, The preferred metabolic pathway from linoleic acid to arachidonic acid *in vitro*, *Biochim. Biophys. Acta*, 164, 25—34, 1968.

102. **Sprecher, H.**, Biosynthetic pathways of polyunsaturated fatty acids, in *Function and Biosynthesis of Lipids, Advances in Experimental Medicine and Biology*, Bazán, N. G., Brenner, R. R., and Giusto, N. M., Eds., Plenum Press, New York, 1977, 35.

103. **Sprecher, H. and James, A. T.**, Biosynthesis of long chain fatty acids in mammalian systems, in *Geometrical and Positional Fatty Acids' Isomers*, Emken, E. A. and Dutton, H. J., Eds., American Oil Chemists Society, Champaign, Ill., 1979, 303—338.

104. **Holman, R. T. and Mahfouz, M. M.**, *Cis* and *trans* octadecenoic acids as precursors of polyunsaturated acids, in *Essential Fatty Acids and Prostaglandins, Progress in Lipid Research*, Vol. 20, Holman, R. T., Ed., Pergamon Press, New York, 1981, 151—156.

105. **Pollard, M. R., Gunstone, F. D., James, R. T., and Morris, L. J.**, Desaturation of positional and geometrical isomers of monoenoic fatty acids by microsomal preparation from rat liver, *Lipids*, 15, 306—314, 1980.

106. **Mahfouz, M., Johnson, S., and Holman, R. T.**, Inhibition of desaturation of palmitic, linoleic and eicosa-8,11,14-trienoic acids *in vitro* by isomeric *cis* octadecenoic acids, *Biochim. Biophys. Acta*, 663, 58—68, 1981.

107. **Mahfouz, M., Johnson, S., and Holman, R. T.**, The effect of isomeric *trans* 18:1 acids on the desaturation of palmitic, linoleic and eicosa-8,11,14-trienoic acids by rat liver microsomes, *Lipids*, 15, 100—107, 1980.

108. **Nervi, A. M., Brenner, R. R., and Peluffo, R. O.**, Effect of arachidonic acid on the microsomal desaturation of linoleic into γ-linolenic acid and their simultaneous incorporation into the phospholipids, *Biochim. Biophys. Acta*, 152, 539—551, 1968.

109. **Brenner, R. R., Peluffo, R. O., Nervi, A. M., and de Tomás, M. E.,** Competitive effect of α and γ linolenyl CoA and arachidonyl-CoA in linoleyl-CoA desaturation to γ-linolenyl CoA, *Biochim. Biophys. Acta,* 176, 420—422, 1969.

110. **Brenner, R. R.,** Reciprocal interactions in the desaturation of linoleic acid into γ-linolenic acid and eicosa-8,11,14-trienoic into arachidonic, *Lipids,* 4, 621—623, 1969.

111. **Actis Dato, S. M. and Brenner, R. R.,** Comparative effects of docosa-4,7,10,13,16-pentaenoic acid and docosa, 4,7,10,13,16,19-hexaenoic acid on the desaturation of linoleic acid and γ-linolenic acid, *Lipids,* 5, 1020—1022, 1970.

112. **Brenner, R. R. and Peluffo, R. O.,** Inhibitory effect of docosa-4,7,10,13,16,19-hexaenoic acid upon the oxidative desaturation of linoleic into γ-linolenic acid and α-linolenic into octadeca-6,9,12,15-tetraenoic acid, *Biochim. Biophys. Acta,* 137, 184—186, 1967.

113. **Castuma, J. C., Brenner, R. R., and Kanau, W.,** Specificity of Δ6 desaturase-effect of chain length and number of double bonds, in: *Function and Biosynthesis of Lipids, Advances in Experimental Medicine and Biology,* Vol. 83, Bazán, N. G., Brenner, R. R., and Giusto, N. M., Eds., Plenum Press, New York, 1977, 127—134.

114. **Mohrhauer, H., Christiansen, K., Gan, M. V., Deubig, M., and Holman, R. T.,** Chain elongation of linoleic acid and its inhibition by other fatty acids *in vitro, J. Biol. Chem.,* 242, 4507—4514, 1967.

115. **Sprecher, H.,** The influence of dietary alterations, fasting and competitive interactions on the microsomal chain elongation of fatty acids, *Biochim. Biophys. Acta,* 360, 113—123, 1974.

116. **Brenner, R. R.,** Regulatory function of Δ6 desaturase — key enzyme of polyunsaturated fatty acid synthesis, in *Function and Biosynthesis of Lipids, Advances in Experimental Medicine and Biology,* Bazán, N. G., Brenner, R. R., and Giusto, R. R., Eds., Plenum Press, New York, 1977, 85—101.

117. **Sprecher, H. and Lee, C.,** The absence of an Δ8-desaturase in rat liver: a reevaluation of optional pathways for the metabolism of linoleic and linolenic acids, *Biochim. Biophys. Acta,* 388, 113—125, 1975.

118. **Schlenk, H., Sand, D. M., and Gellerman, J. L.,** Non conversion of 8,11,14-eicosatrienoic into arachidonic acid by rats, *Lipids,* 5, 575—577, 1970.

119. **Maeda, M., Doi, O., and Akamatsu, Y.,** Metabolic conversion of polyunsaturated fatty acids in mammalian cultured cells, *Biochim. Biophys. Acta,* 530, 153—164, 1978.

120. **Verdino, B., Blank, M. L., Privett, O. S., and Lundberg, W. O.,** Metabolism of 4,7,10,13,16-docosa pentaenoic acid in the essential fatty acid deficient rat, *J. Nutr.,* 83, 234—238, 1964.

121. **Schlenk, H., Gellerman, J. L., and Sand, D. M.,** Retroconversion of polyunsaturated fatty acids *in vivo* by partial degradation and hydrogenation, *Biochim. Biophys. Acta,* 137, 420—426, 1967.

122. **Kunau, W. H.,** Synthesis of 4,7,10,13,16-docosapentaenoic acid with all double bonds tritium-labeled and its conversion to 5,8,11,14-eicosatetraenoic acid in rats on a fat-free diet, *Hoppe-Seyler's Z. Physiol. Chem.,* 349, 333—338, 1968.

123. **Schlenk, H., Sand, D. M., and Gellerman, J. L.,** Retroconversion of docosahexaenoic acid in the rat, *Biochim. Biophys. Acta,* 187, 201—207, 1969.

124. **Sprecher, H.,** Regulation of polyunsaturated fatty acid biosynthesis in the rat, *Fed. Proc.,* 31, 1451—1457, 1972.

125. **Stearns, E. M., Rysavy, J. A., and Privett, O. S.,** Metabolism of *cis-*11,*cis-*14- and *trans-*11,*trans-*14-eicosadienoic acids in the rat, *J. Nutr.,* 93, 485—490, 1967.

126. **Kunau, W. H. and Bartnik, F.,** Studies on the partial degradation of polyunsaturated fatty acids in rat-liver mitochondria, *Eur. J. Biochem.,* 98, 311—318, 1974.

127. **Kunau, W. H.,** Oxidation of polyunsaturated fatty acids, in *Polyunsaturated Fatty Acids,* Kunau, W. H. and Holman, R. T., Eds., American Oil Chemists Society, Champaign, Ill., 1977, 51.

128. **Sprecher, H.,** The synthesis and metabolism of hexadeca-4,7,10-trienoate, eicosa-8,11,14-trienoate, docosa-10,13-16-trienoate and docosa-6,9,12,15-tetraenoate in the rat, *Biochim. Biophys. Acta,* 152, 519—530, 1968.

129. **Stoffel, W., Ecker, W., Assad, H., and Sprecher, H.,** Enzymatic studies on the mechanism of the retroconversion of C_{22} polyenoic fatty acids to their C_{20} homologues, *Hoppe-Seyler's Z. Physiol. Chem.,* 351, 1545—1554, 1970.

130. **Schlenk, H., Gerson, T., and Sand, D. M.,** Conversion of non-biological polyunsaturated fatty acids in rat liver, *Biochim. Biophys. Acta,* 176, 740—747, 1969.

131. **Sprecher, H.,** The total synthesis and metabolism of 7,10,13,16-docosatetraenoate in the rat, *Biochim. Biophys. Acta,* 144, 295—304, 1967.

132. **Schlenk, H. and Sand, D. M.,** A new group of essential fatty acids and their comparison with other polyenoic fatty acids, *Fed. Proc.,* 26, 412, 1967.

133. **Beerthuis, R. K., Nugteren, D. H., Pabon, H. J. J., and Van Dorp, D. A.,** Biologically active prostaglandins from some new odd-numbered essential fatty acids, *Rec. Trav. Chem., Pays-Bas,* 87, 461—480, 1968.

134. **Beerthuis, R. K., Nugteren, D. H., Pabon, H. J. J., Steenhock, A., and Van Dorp, D. A.,** Synthesis of a series of polyunsaturated fatty acids, their potencies as essential fatty acids and as precursors of prostaglandins, *Rec. Trav. Chim.,* 90, 943—960, 1971.

135. **Schlenk, H.,** Odd numbered and new essential fatty acid, *Fed. Proc.,* 31, 1430—1435, 1972.

136. **Thomasson, H. J.,** Biological standardization of essential fatty acids, *Int. Rev. Vitamin Res.,* 25, 62, 1953.

137. **Rahm, J. J. and Holman, R. T.,** The relation of single dietary polyunsaturated fatty acids to fatty acid composition of lipids from subcellular particles of liver, *J. Lipid Res.,* 5, 169—176, 1964.

138. **Van Dorp, D. A.,** Essential fatty acids and prostaglandins, *Acta Biol. Germ.,* 35, 1041, 1976.

139. **Van Dorp, D. A.,** Recent developments in the biosynthesis and the analyses of prostaglandins, *Ann. N.Y. Acad. Sci.,* 180, 181—199, 1971.

140. **Ziboh, V. A. and Hsia, S. L.,** Effects of prostaglandin E_2 on rat skin. Inhibition of sterol ester biosynthesis and clearing of scaly lesions in essential fatty acid deficiency, *J. Lipid Res.,* 13, 458—467, 1972.

141. **Lands, W. E. M., Hemler, M. E., and Crawford, C. G.,** Functions of polyunsaturated fatty acids: biosynthesis of prostaglandins, in *Polyunsaturated Fatty Acids,* Kunau, W. H. and Holman, R. T., Eds., American Oil Chemists Society, Champaign, Ill., 1977, 193.

142. **Kaufmann, H. P. and Barve, J.,** Uber das Samenöl von *Aquilegia vulgaris, Fette Seifen. Anstrichmittel,* 67, 14—16, 1965.

143. **Houtsmuller, U. M. T.,** Columbinic acid, a new type of essential fatty acids, *Essential Fatty Acid and Prostaglandins: Progress in Lipid Research,* Vol. 20, Holman, R. T., Ed., Pergamon Press, New York, 1981, 889—896.

144. **de Alaniz, M. J. T., de Gómez Dumm, I. N. T., and Brenner, R. R.,** Effect of fatty acids of ω6 series on the biosynthesis of arachidonic acid in HTC cells, *Mol. Cell. Biochem.,* 64, 31—37, 1984.

145. **Houtsmuller, U. M. T.,** Personal communication.

UPTAKE AND RELEASE OF EICOSANOID PRECURSORS FROM PHOSPHOLIPIDS AND OTHER POOLS: PLATELETS

Holm Holmsen

Platelets are anucleate cells that are shed from the megakaryocytes in the bone marrow and live in the circulation for 7 to 9 days in man. Their principal physiological role is in the hemostatic process in which they rapidly adhere to each other and to exposed subendothelial structures when a blood vessel is damaged. They eventually form a "platelet plug" that causes arrest of blood loss. Formation of such platelet deposits also occurs pathologically and can cause obstruction of vessels that provide blood to vital organs, with severe or fatal results (thrombosis). The mechanisms that activate platelets in hemostasis and thrombosis are generally thought to be similar and to involve production of platelet-stimulating prostaglandins and thromboxanes within the platelets. These active substances are synthesized from eicosanoid precursors within the platelets, precursors that are covalently bound in the phospholipids of resting platelets and set free upon platelet stimulation.

The metabolism of *added* eicosatri-, -tetra-, and -pentaenoic acid by the cyclooxygenase and lipoxygenase pathways in platelets is well described.[1-4] However, since the platelet-active oxygenation products that operate in hemostasis and thrombosis are derived from eicosanoid precursors esterified in platelet phospholipids, considerable interest has recently been shown in the mechanisms for uptake and release of these precursors by platelets. Despite this interest, our understanding of the mechanisms underlying uptake and release is quite obscure. The uptake mechanisms have not been studied to the same extent as those of release, and five well-supported but separate mechanisms have been suggested for the latter. Therefore, it is not possible to make a clear picture of these mechanisms in the present review. I will, nevertheless, attempt to describe the main experimental evidence that has been reported and suggest causes for controversial findings and interpretations.

UPTAKE OF EICOSANOID PRECURSORS

Eicosatri-, -tetra-, and -pentaenoic acids have been demonstrated in all major glycerophospholipids in platelets (PC, PE, PS, and PI) with eicosatetraenoate as the major fatty acid (Table 1). Eicosapentaenoate is more predominant in platelets from individuals on a diet rich in this fatty acid (cod liver oil) and from inhabitants of arctic regions than from those of the rest of the world (Table 1). It should be noted that per individual phospholipid PI has the highest percentage of eicosanoid precursors, with up to 45% of total fatty acid as eicosatetraenoate. When investigated, the eicosanoid precursors are predominantly esterified in the 2-position of the glycerol backbone.

Platelets are capable of both *de novo* fatty acid synthesis (cytoplasmic pathway) and fatty acid elongation (mitochondrial pathway). Degradation studies of platelet phospholipid fatty acids have revealed that myristic and palmitic acids (14:0 and 16:0) are normally made by *de novo* synthesis, while fatty acids containing 18 or more carbons are made by chain elongation of preformed fatty acid.[13,14] Newly synthesized myristic and palmitic acids (from radioactive acetate) are found primarily in nonesterified fatty acids, while the radioactive fatty acids containing 18 or more carbons are found predominantly in phospholipids,[15] glycerides, and ceramides.[13,15] The lipid radioactivity is found primarily in palmitic (16:0), myristic (14:0), and arachidic (20:0) acids.[13,14,16]

None of the eicosanoid precursors are formed by these pathways in platelets, and their presence in the cells' glycerophospholipids therefore suggests that they are directly incorporated into the phospholipids from the respective intact species taken up from the platelets'

Table 1
DISTRIBUTION (MOL%) OF EICOSANOID PRECURSORS IN PLATELET PHOSPHOLIPIDS

Diet	20:3	20:4	20:5	Ref.	Diet	20:3	20:4	20:5	Ref.
Phosphatidylethanolamine					**Phosphatidylserine + phosphatidylinositol**				
	1.1	37.4	0.5	6, 7	Control	2.0	28.6	0.5	11
Control	—	32.0	—	8	CLO[b]	2.3	28.2	0.5	
	1.4	35.8	—	9[a]					
Control	1.4	35.4	1.6	11	**Phosphatidylinositol**				
CLO[b]	1.6	32.4	5.2				40.1	—	8
Phosphatidylcholine					Control		36.4	—	9[a]
	—	11.9	—	6, 7					
Control	1.2	12.2	—	8	**Total phospholipid**				
	1.4	5.9	—	9	D[c]	1.0	25.2	0.5	11, 12
Control	2.2	13.6	0.9	11[c]	E[c]	—	28.5	8.0	12
CLO[b]	1.4	11.2	2.4						
Phosphatidylserine									
	0.9	24.9	—	6, 7					
Control	1.4	22.0	—	8					
	1.6	9.9	—	9[a]					

[a]   Guinea pig platelets; other numbers refer to human platelets.
[b]   Cod liver oil.
[c]   D = Controls of Danish population; E = Eskimos.

environment. This suggestion is, for example, supported by the significant increase in the eicosapentaenoid acid content in PC, PE, and PS + PI of platelets from man fed cod liver oil, which is rich in this fatty acid.[11] Further support comes from the results of the widespread use of radioactive arachidonic acid to label platelet phospholipids in vitro in order to study the liberation of this eicosanoid precursor during platelet stimulation. No systematic study of the mechanisms underlying uptake and incorporation into phospholipids of eicosatetraenoic acid (or other eicosanoid precursors) by platelets seems to have been undertaken. The behavior of eicosatetraenoic acid, however, appears to follow that of fatty acids with 16 or more carbon atoms. These long-chain fatty acids in plasma dissociate from albumin and are taken up by platelets within which they equilibrate with the nonesterified fatty acid pool,[17,18] from which they are mainly incorporated into phospholipids. Two mechanisms are operative, *de novo* synthesis and remodeling. The *de novo* synthesis pathway[19-29] is shown schematically in Figure 1 (solid lines) which also denotes the enzymes involved, all of which have been demonstrated in platelets. A fatty acid is esterified, as acyl-CoA, with an glycerol-3-phosphate to yield lysophosphatidic acid. In turn, lysophosphatidic acid is acylated by a second acyl-CoA to yield phosphatidic acid (PA), which is dephosphorylated to diglyceride, the common precursor for the diacylglycerol moiety in phosphatidylcholine (PC), phosphatidylethanolamine (PE), and phosphatidylserine (PS). PA is directly converted (i.e., without dephosphorylation) to the phosphoinositides through activation by CTP in combination with inositol and successive phosphorylations. For all phospholipids it is evident from Figure 1 that their fatty acid composition obtained by *de novo* synthesis is determined by the two first acylation steps. However, it is not clear whether there exists individual acyl transferase for each acyl CoA species.

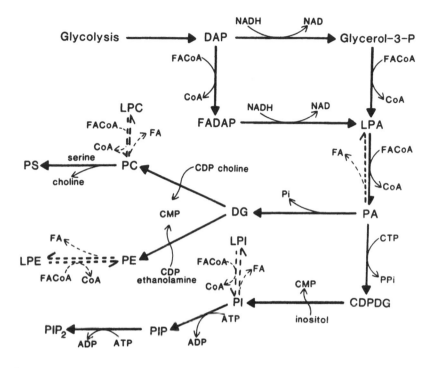

FIGURE 1. The *de novo* and remodeling pathway of platelet phospholipid synthesis. The *de novo* pathways are indicated by solid lines and the remodeling pathways are shown by broken lines. Abbreviations used: DAP = dihydroxyacetone phosphate; CoA = coenzyme A (reduced form); FA = fatty acid; FACoA = fatty acid coenzyme A (acyl coenzyme A); NADH and NAD = nicotineamide adenine dinucleotide, reduced and oxidized forms, respectively; FADAP = fatty acid dihydroxyacetone phosphate; LPA = lysophosphatidic acid; PA = phosphatidic acid; CTO = cytidine triphosphate; PP_i = inorganic pyrophosphate; CDPDG = cytidine triphosphate diglyceride; CMP = cytidine monophosphate; PI = phosphatidylinositol-4-phosphate; PIP_2 = phosphatidylinositol-4,5-bisphosphate; DG = diglyceride (diacyl glycerol); PE = phosphatidylethanolamine; LPE = lysophosphatidylethanolamine; CDP = cytidine diphosphate; PC = phosphatidylcholine; LPC = lysophosphatidylcholine; PS = phosphatidylserine.

In the remodeling pathway,[24,28-33] preexisting PC, PE, and PA are deacylated by phospholipase A_1 and/or A_2 to the corresponding lysophospholipids, which in turn are reacylated by acyltransferases of varying specificity (Figure 1, broken lines). It is possible that the deacylation step can be carried out by the acyl transferase alone, as in some other tissues.[34,35] It is not known how these deacylation and reacylation processes are controlled or how the *de novo* synthesis and remodeling pathways interact in the resting platelet. An acyl-CoA:1-acyl-*sn*-glycero-3-phosphocholine acyltransferase has been demonstrated in platelets by McKean et al.[33] that catalyzes transfer of eicosatrienoate markedly faster than eicosatetraenoate from the respective acyl-CoA. A similar transferase with high activity for n-3 polyunsaturated fatty acids was later reported by Iritani et al.[36] Strict control of the acyl-CoA species entering phosphatidyl inositol (PI) must exist, since this phospholipid contains almost exclusively stearyl (18:0) in the 1-position and arachidonyl (20:4) in the 2-position.[27,37-40] Another pathway (of uncertain significance) for conversion of PS to PC through decarboxylation to PE followed by three successive methylations has been demonstrated in human platelets.[41,42] Methylation of arachidonyl-rich PE species to arachidonyl-rich PC may be a significant pathway in rabbit platelets.[43] Besides acyl-CoA synthetase(s), platelets also contain acyl-CoA hydrolases which are present both in cytoplasm and mitochondria;[44,45] the

Table 2
DISTRIBUTION (%) OF
RADIOACTIVITY OR
ARACHIDONATE MASS IN
PLATELET PHOSPHOLIPIDS AFTER
INCUBATION IN VITRO WITH
RADIOACTIVE ARACHIDONATE

Endogenous arachidonate	PC	PE	PS	PI	Ref.
Nonradioactive	26	49	13	12	47
Radioactive	69	12	6	12	26
(t < 1 hr[a])	55	14	14	16	46
	53	19	17	16	48
	64	12	12	11	49
Radioactive	53	28	10	8	47
(t > 10 hr[a])	56	28	3	13	50

[a] Incubation time with arachidonate at 0.01 μM initial concentration of arachidonate added to platelet-rich plasma. The platelets were thereafter gel-filtered before extraction.

latter group of enzymes may also be of importance in selecting the acyl-CoA species to be incorporated in platelet phospholipids. There are evidently multiple potential steps in the various routes of phospholipid interconversions at which control of introduction of fatty acids can be exercised.

The presence of extracellular albumin appears to be essential for the direction of arachidonic acid metabolism. In the absence of albumin 93% of the added acid is oxygenated through the cyclooxygenase and lipooxygenase pathways, while the presence of 0.2% or more albumin directs metabolism to incorporation of more than 95% into phospholipids.[5,46] Radiolabeled arachidonate taken up by platelets is distributed among the phospholipids in proportions that are fairly consistent from laboratory to laboratory, but which depends on the time of incubation with arachidonate (Table 2). In particular, the content of radioactive arachidonate in PE (both in DPE and PPE) increases with time. However, the proportions of radioactive arachidonate in PC is twice and that in PE is 20% (1 hr incubation) of the respective proportions of endogenous arachidonate (Table 2). These discrepancies suggest that only a small pool of PC and PE exchanges or is formed with radioactive arachidonate. Different subcellular localization of the exchangeable/newly formed pools could not be demonstrated,[47,50] and it is presently unclear what this metabolic compartmentalization of PC and PE means. For PI it has been demonstrated that after labeling with [³H]arachidonate the PI radioactivity and PI mass (P content) decreases in parallel when the platelets are treated with thrombin,[46] suggesting homogenous labeling of the entire PI pool.

Radioactive eicosapentaenoate is also taken up and incorporated into platelet phospholipids.[51,52] The distribution of this eicosaenoate in human platelets is approximately similar to that shown for radioactive arachidonate in Table 2, except for markedly higher proportion of pentaenoate than arachidonate in PI.[52] In murine platelets the reverse appears to be the case.[51]

Preliminary experiments in our laboratory indicate that the distribution among phospholipids of the [³H]arachidonate taken up by platelets also depends on the concentration of [³H]arachidonate (Table 3). In particular, the proportion of [³H]arachidonate in PI is markedly higher with micromolar concentrations of arachidonate than in the (commonly used, Table

Table 3
DISTRIBUTION OF [³H] IN PHOSPHOLIPIDS AFTER INCUBATION OF GEL-FILTERED PLATELETS WITH DIFFERENT CONCENTRATIONS OF [³H]ARACHIDONATE

[³H]Arachidonate (μM)	[³H], of total PL[a] radioactivity				[³H]Arachidonate consumed[b]	[³H]Arachidonate in PL[c]
	PC	PE	PS	PI		
(0.01)[d]	(55)	(14)	(14)	(16)	—	—
5	42	13	12	33	36	31
10	41	15	13	30	32	37
50	39	20	12	28	28	23
100	25	39	22	14	92	30

Note: The platelets were gel filtered into a Ca²⁺-free Tyrode's solution containing 0.2% albumin (parentheses) and incubated for 60 min at 37°. The results are from preliminary experiments in our laboratory.[109]

[a] PL = Phospholipid.
[b] Percent of [³H]arachidonate at the beginning of incubation.
[c] Percent of the [³H]arachidonate consumed.
[d] Percent distribution in platelets gel-filtered after incubation in plasma with 0.01 μM [³H]arachidonate.[46]

2) nanomolar range. By increasing the arachidonate concentration from 5 to 100 μM, PE in particular increases its proportion of [³H]arachidonate. At 100 μM arachidonate and 0.2% albumin cell lysis has reached values of 20%, which may have contributed to the much higher proportion of arachidonate consumed (both incorporation in phospholipids and oxygenation) at this than at lower concentrations of arachidonate (Table 3). Thus, incubation time, concentration of arachidonate, concentration of albumin, and the intactness of the platelets are all factors that determine the uptake and metabolism of this prostanoid precursor. These factors — and most likely other unknown factors — have to be meticulously studied in order to obtain a clear picture of eicosanoid precursor uptake mechanisms in platelets.

Arachidonate *esterified* in the 2-position of PC in extracellular HDL rapidly exchanges with a pool of PC *within* the platelets and the arachidonate in this PC pool is transferred to platelet PI.[53] Conversely, platelets having incorporated radioactive arachidonate in their phospholipids rapidly release the labeled arachidonate to plasma.[54] In both cases arachidonate was exchanged markedly faster than oleate or linoleate. Whether this exchange consists in the above-described deacylation-reacylation (remodeling) pathway or a direct transacylation is not clear. Spin label studies indicated that extracellular PC first exchanges with PC in the outer leaflet of the plasma membrane and then is internalized by flip-flop;[54] thus, the radiolabeled arachidonate is apparently not released from the extracellular PC and then reincorporated intracellularly. In a recent study with double-labeled fatty acids (free ¹⁴C-labeled, esterified ³H-labeled FA), it was elegantly shown that both exchange and acylation pathways exist for phospholipid labeling incorporation and produce distinctly different labeling patterns.[55] This type of approach seems to be very promising, and will most likely yield important, basic information about the various mechanisms of uptake of eicosanoid precursors into platelet phospholipids.

The molecular species composition of individual diacylglycerophospholipids has been determined and the 1-stearoyl 2-arachidonoyl species are the most predominant of the arachidonate-containing species.[56] However, a considerable proportion of platelet glycerophospholipids, PC and PE in particular, consists of 1-alkyl-2-acyl and 1-alkenyl-2-acyl species[57,58] with 2-arachidonyl as the major species.[58] Besides the diacylglycerophospholi-

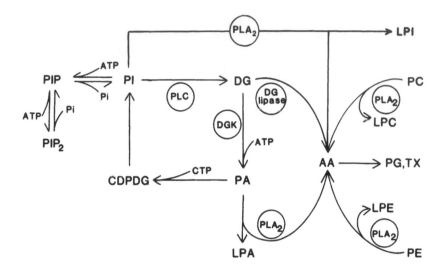

FIGURE 2.   Possible mechanisms of liberation of arachidonic acid in stimulated platelets. Most of the abbreviations used here are given in Figure 1. Exceptions are PLC = phospholipase C; DG lipase = diglyceride lipase; PLA_2 = phospholipase A_2; PG = prostaglandin; TX = thromboxane; LPI = lyso PI.

pids, therefore, the 1-alkenyl- or 1-alkyl-glycerophospholipids are potential sources for the free arachidonate liberated during platelet activation.

RELEASE OF EICOSANOID PRECURSORS

Although eicosatrienoic and eicosapentanoic acids are present in platelet phospholipids (Table 1), the study of the release of eicosanoid precursors from these phospholipids by stimulated platelets has concentrated on the release of eicosatetraenoate, the most abundant precursor (Table 1). Sufficient platelet stimulation causes liberation of arachidonate esterified in phospholipids, and the freed arachidonate is oxygenated as the added acid.[1-4] However, our understanding of the mechanisms for this liberation is controversial, and the five alternate pathways claimed are shown in Figure 2. The controversy is probably due to differences in methodology and a combination of unevenness of arachidonate distribution in the platelet membranes[59] and different phospholipase-susceptible domains in the membrane.[60]

A frequently used approach is to incubate platelets with radioactive arachidonate which becomes esterified in PC, PE, PS, and PI as shown in Table 2. Treatment of such labeled platelets with thrombin causes loss of radioactivity from PI (60% decrease) and PC (40% decrease) with a corresponding accumulation of radioactive PA and free arachidonate and its oxygenation products.[26,28,30,61-71] Due to the specificities of platelet phospholipase A_2 (PLA_2) and phospholipase C (PLC) these results suggest that platelet stimulation is associated with activation of PLA_2 which causes liberation of arachidonate from PC and activation of PLC which cases breakdown of PI, although without liberation of arachidonate. This labeling technique has also shown that thrombin causes incorporation of radioactive arachidonate into platelet PE plasmalogen,[72] a process that is poorly understood, although a CoA-independent transacylase from human platelets[73] and a CoA-dependent transacylase from rat platelets[74] have been characterized and found to be specific for arachidonate.

Lyso-PA is formed during thrombin stimulation;[68] inhibition of PLA_2 with calmodulin inhibitors abolishes both liberation of radioactive arachidonate[68,75,76] and the formation of lyso-PA, suggesting that PA is a source of arachidonate through activation of a PA-specific

PLA$_2$. in studies where the mass of individual phospholipids have been determined,[30,40,77,79] thrombin causes the same loss in PI mass as its content of radioactive arachidonate, while there was virtually no decrease in PC mass and an increase in lyso-PE. Phospholipase C, which becomes activated upon platelet stimulation, acts only on PI and produces diglyceride (DG), and its concentration only increases transiently.[70] DG is a substrate for both diglyceride kinase and a diglyceride lipase activity which has been described in platelets.[77,80,81] Since PI, and thereby DG, contains almost exclusively arachidonate in the 2-position,[27,38-40] the hydrolysis of DG by the lipase will produce free arachidonate, and the PLC-DG lipase pathway is an apparent potential route for production of free arachidonate. Actually, there is a preferential utilization of 1-acyl-2-arachidonyl species of PI during the thrombin-platelet interaction.[82]

PI may also be a source for arachidonate through direct deacylation, since lyso-PI is formed in platelets stimulated with collagen.[83] Based on the difference in PI breakdown caused in intact platelets by thrombin and A23187[84,85] and the formation of lyso-PI,[86] a PI-specific phospholipase A$_2$ has also been suggested, in sharp contrast to previous findings and conclusions by the same authors.[28,65] PI may therefore be degraded by both phospholipase A$_2$ and phospholipase C to yield free arachidonate depending on the nature of platelet activation.[85] However, an acyl transferase specific for 1-acylglycerophosphorylinositol and with high activity for arachidonyl CoA has been demonstrated,[87] and may rapidly reacylate the lyso-PI formed by phospholipase A$_2$.

Taken together, the approaches with arachidonate labeling and measurement of mass appear to agree only with respect to PI breakdown, and the PLC-DG lipase pathway may be quantitatively important. This is supported by the finding that serine esterase blockers, which inhibit PLC, inhibit arachidonate liberation in thrombin- and collagen-treated platelets.[64] However, since neither calcium nor calmodulin are necessary for the DG lipase,[77,80,81] the almost complete inhibition of arachidonate liberation caused by calmodulin inhibitors[66,75,76] does not agree with a major role of the PLC-DG lipase pathway. Also, the finding that almost all [^{3}H]-arachidonate-labeled DG formed is subsequently recovered in PA[88] speaks against the PLC-DG lipase pathway as a major route for liberation of arachidonate. Finally, arachidonate liberation was reported to be blocked by the diglyceride lipase-specific blocked RHC 80267;[89] however, these studies are inconclusive since too high a concentration of inhibitor, giving nonspecific effects, seems to have been used.[90]

The discrepancies between the labeling and mass methods suggest that mass-undetectable and rapidly labeled pools of PC and PA and a mass-detectable, unlabeled pool of PE are potential sources of arachidonate and through PLA$_2$ action on the platelet phospholipids. Recent studies with simultaneous measurement of mass and radioactivity showed that the specific radioactivity of the arachidonate set free by thrombin from platelets prelabeled with radioactive arachidonate was between that of arachidonate in PC and PI, but six times greater than that of arachidonate in PE.[91] Thus, PC and PI are clearly major sources of the liberated arachidonate while PE is not. These studies also showed that the specific radioactivity of arachidonate in PC and PI was the same, respectively, before and after thrombin-induced arachidonate liberation.[91] Evidently, rapidly labeled pools of PC and PI in platelets are not preferred sources of the freed arachidonate. However, until more is known about the re-modeling pathway in resting and stimulated platelets (Figure 1), and until mass and labeling techniques are used together in the same experiment, it is difficult to determine the quantitative importance of the potential arachidonate-liberating pathways (Figure 2).

In platelets prelabeled with radioactive eicosapentaenoate thrombin caused liberation of radioactive pentaenoate from PI and PC, but not from PE or PC.[52]

Nevertheless, it is clear that phospholipase A$_2$ is involved in arachidonate liberation and that its regulation is of central importance for the modulation of platelet response through prostaglandins and thromboxanes. The enzyme is membrane-bound, stimulated by Ca^{2+} and

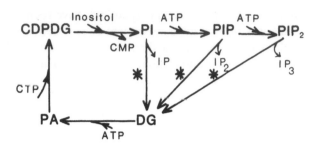

FIGURE 3.   Three possible routes of DG formation in activated platelets. The abbreviations are defined in Figure 1, except IP (inositol-1,4-bis-phosphate) and IP_3 (inositol-1,4,5-trisphosphate). The asterisks indicate phosphodiesterases (PLC-like enzymes) that may be controlled by receptor occupancy.

detergents,[28,80,81,86,92] the calcium activation is enhanced by calmodulin,[29] it does not hydrolyze PI, but uses PC, PE, and PS as substrates;[28] hydrolysis depends on the physical state of the substrate[81] and on its conformation in its native milieu, the platelet membrane,[53] and is inhibited by certain nonsteroidal anti-inflammatory drugs, i.e., indomethacin, meclofenamate, and flufenamate.[93,94] Recently, a lipid-soluble, endogenous inhibitor of phospholipase A_2 was demonstrated in platelets, suggesting that the enzyme activity can be regulated by other means than Ca^{2+} and calmodulin.[95] A possible activation of phospholipase A_2 by DG could not be demonstrated.[96]

The less likely enzyme to be involved in arachidonate liberation, PLC, is specific for PI and its phosphates. It is present in the cytosol and has an absolute requirement for Ca^{2+} at an optimal pH 6.5 to 7.5.[28,62,97,98] The PI-specific phospholipase C from platelets is powerfully inhibited by spermine and spermidine.[99] It is, however, unlikely that these naturally occurring polyamines play any role in the regulation of phospholipase C activity in platelets, a regulation that currently is poorly understood. The production of DG in platelets in response to thrombin appears to take place with minimal increase in the cytoplasmic Ca^{2+} concentration, suggesting regulation of phospholipase C by Ca^{2+}-independent mechanisms.[100] Also, a possible self-potentiation of the phospholipase C reaction by the product DG does not take place with the platelet enzyme.[96] The calmodulin inhibitors trifluoperazine and quinacrine do not affect thrombin-induced PI breakdown under conditions where phospholipase A_2 is markedly inhibited,[66,73,75] indicating that phospholipase C activation by calcium may be independent of calmodulin.

Liberation of arachidonate from phospholipids during platelet stimulation is inhibited by reduction in the ATP availability in the cells.[101] This can happen for several reasons; for example, because energy is needed for the mobilization of Ca^{2+} necessary for activation of phospholipase A_2, or because PA (the formation of which requires ATP) is a major substrate for the enzyme. Supporting the latter possibility is the finding that gradual reduction of the ATP availability by incubating the cells with metabolic inhibitors caused parallel decay of thrombin-stimulated PA formation and arachidonate liberation. The concomitant thrombin-induced reduction in the PI level was also inhibited by ATP deprivation, even more strongly than PA formation.[102] This inhibition of agonist-controlled PI disappearance suggests ATP involvement in the activation of phospholipase C or in making PI available for the enzyme. The mechanism(s) for such involvement of energy is not known; however, the finding that thrombin causes a transient fall in PI-bisphosphate followed by an increase in this polyphosphoinositide[102-105] suggests that PI may not disappear by hydrolysis, but by successive phosphorylations to PIP_2, which is continuously being hydrolyzed by a phospholipase C-like enzyme to DG and inositol-1,4,5-trisphosphate (Figure 3). Such a mechanism would fully explain the sensitivity of PI disappearance to ATP deprivation. Recent studies have demonstrated that platelet phospholipase C can hydrolyze PI and PIP but not PIP_2, which is a potent inhibitor of PI hydrolysis.[106]

Irrespective of mechanism, the liberation of (prelabeled) arachidonate from platelet phospholipids by thrombin is significantly enhanced by increase in the cholesterol in the membrane; this may have resulted from an effect by cholesterol on the organization of PC and PI in the membrane which renders them more available to phospholipases A_2 and C, respectively.[107] In contrast, incubation of human platelet phospholipids with linoleic acid caused replacement of phospholipid-bound arachidoniate with linoleate; distinctly less arachidonate was liberated from the lineolate-enriched platelet than normal platelets, and linoleate was exclusively released from PI.[108]

REFERENCES

1. **Marcus, A. J.,** The role of prostaglandins in platelet function, in *Progress in Hematology,* Vol. 5, Spaet, T. H., Ed., Grune & Stratton, New York, 1979, 147—171.
2. **Smith, J. B. and Silver, M. J.** Prostaglandins synthesis by platelets and its biological significance, in *Platelets in Biology and Pathology,* Vol. 1, Gordon, J. L., Ed., North-Holland, Amsterdam, 1976, 331—352.
3. **Needleman, P., Raz, A., Minkes, M. S., Ferrendell, J. A., and Sprecher, H.,** Triene prostaglandins: prostacyclin and thromboxane biosynthesis and unique biological properties, *Proc. Natl. Acad. Sci. U.S.A.,* 76, 944—948, 1979.
4. **Samuelsson, B.,** Prostaglandin endoperoxides and thromboxanes. Role in platelets and vascular and respiratory smooth muscle, *Acta Biol. Med. Ger.,* 35, 1055—1063, 1976.
5. **Stuart, M. J., Gerrard, J. M., and White, J. G.,** The influence of albumin and calcium on human platelet arachidonic acid metabolism, *Blood,* 55, 418—423, 1980.
6. **Marcus, A. J., Ullman, H. L., Safier, L. B., and Ballard, H. S.,** Platelet phosphatides. Their fatty acid and aldehyde composition and activity in different clotting systems, *J. Clin. Invest.,* 41, 2198—2212, 1962.
7. **Marcus, A. J., Ullman, H. L., and Safier, L. B.,** Lipid composition of subcellular particles of human platelets, *J. Lipid Res.,* 10, 108—114, 1969.
8. **Imai, A., Yano, K., Kaneyama, Y., and Nozawa, Y.,** Reversible thrombin-induced modification of positional distribution of fatty acids in platelet phospholipids: involvement of deacylation-reacylation, *Biochem. Biophys. Res. Commun.,* 103, 1092—1099, 1981.
9. **Schick, B., Schick, P., and Chase, P. R.,** Lipid composition of guinea pig platelets and megakaryocytes. The megakaryocyte as a probable source of platelet lipids, *Biochim. Biophys. Acta,* 663, 239—248, 1981.
10. **Saduska, B., Tacconi, M. T., di Minno, G., Roncaglioni, M. C., and Pagrazzi, J., Donati, M. B., Bizzi, A., and Silver, M. J.,** Plasma and platelet lipid composition and platelet aggregation by arachidonic acid in women on the pill, *Thromb. Haemostas.,* 45, 150—153, 1981.
11. **Brox, J. H., Killie, J. E., Gunnes, S., and Nordøy, A.,** The effect of cod liver oil and corn oil on platelet and vessel wall in man, *Thromb. Haemostas.,* 46, 604—611, 1981.
12. **Dyerberg, J. and Bang, H. O.,** Haemostatic function and platelet unsaturated fatty acids in eskimos, *Lancet,* 2, 433—435, 1979.
13. **Deykin, D. and Desser, R. K.,** The incorporation of acetate and palmitate into lipids by human platelets, *J. Clin. Invest.,* 47, 1590—1602, 1968.
14. **Hennes, A. R. and Awai, K.,** Studies of incorporation of radioactivity into lipids by human blood. IV. Abnormal incorporation of acetate 1-C-14 into fatty acids by whole blood and platelets from insulin-independent diabetics, *Diabetes,* 14, 709—715, 1965.
15. **Hennes, A. R., Awai, K., and Hammerstrand, K.,** Studies of incorporation of radioactivity into lipids by human blood. V. Pattern of fatty acid radioactivity in lipid fractions of platelets, *Biochim. Biophys. Acta,* 84, 610—612, 1964.
16. **Awai, K. and Hennes, A. R.,** Studies of incorporation of radioactivity into lipids by human blood. II. Pattern of incorporation of radioactivity into fatty acids by platelets from normal subjects and patients in diabetic acidosis, *Diabetes,* 13, 592—599, 1964.

17. **Spector, A. A., Hoak, J. C., Warner, E. D., and Fry, G. L.,** Utilization of long-chain free fatty acids by human platelets, *J. Clin. Invest.*, 49, 1489—1496, 1970.
18. **Deykin, D. and Desser, R. K.,** The incorporation of acetate and palmitate into lipids by human platelets, *J. Clin. Invest.*, 47, 1590—1602, 1968.
19. **Okuma, M., Yamashita, S., and Numa, S.,** Enzymatic studies on phosphatidic acid synthesis in human platelets, *Blood*, 41, 379—389, 1973.
20. **Call, F. L., II and Williams, W. J.,** Biosynthesis of cytidine diphosphate diglyceride by human platelets, *J. Clin. Invest.*, 49, 392—399, 1970.
21. **Lucas, C. T., Call, F. L., II, and Williams, W. J.,** The biosynthesis of phosphatidylinositol by human platelets, *J. Clin. Invest.*, 49, 1949—1955, 1970.
22. **Call, F. L., II and Williams, W. J.,** Phosphatidate phosphatase in human platelets, *J. Lab. Clin. Med.*, 82, 663—673, 1973.
23. **Call, F. L. and Rubert, M.,** Diglyceride kinase in human platelets, *J. Lipid Res.*, 14, 466—474, 1973.
24. **Elsbach, P., Pettis, P., and Marcus, A.,** Lysolecithin metabolism by human platelets, *Blood*, 37, 675—683, 1971.
25. **Call, F. L. and Rupert, M.,** Synthesis of ethanolamine phosphoglycerides by human platelets, *J. Lipid Res.*, 16, 352—359, 1975.
26. **Bills, T. K., Smith, J. B., and Silver, M. J.,** Metabolism of ^{14}C-arachidonic acid by human platelets, *Biochim. Biophys. Acta*, 424, 303—314, 1976.
27. **Cohen, P., Broekman, M. J., Verkley, A., Lismann, J. W. W., and Derksen, A. L.,** Quantification of human platelet inositides and the influence of ionic environment on their incorporation of orthophosphate-^{32}P, *J. Clin. Invest.*, 50, 762—772, 1971.
28. **Billah, M. M., Lapetina, E. G., and Cuatrecasas, P.,** Phospholipase A_2 and phospholipase C activities of platelets. Differential substrate specificity, Ca^{2+} requirement, pH dependence and cellular localization, *J. Biol. Chem.*, 255, 10227—10231, 1980.
29. **Wong, P. Y.-K. and Cheung, W. Y.,** Calmodulin stimulates human platelet phospholipase A_2, *Biochem. Biophys. Res. Commun.*, 90, 473—480, 1979.
30. **Rittenhouse-Simmons, S. and Deykin, D.,** The activation by Ca^{2+} of platelet phospholipase A_2. Effects of dibutyryl cyclic adenosine monophosphate and 8(*N,N*-diethyl amino)-octyl-3,4,5-trimetoxybenzoate, *Biochim. Biophys. Acta*, 543, 409—422, 1978.
31. **Silk, S. T., Wong, K. T., and Marcus, A. J.,** Arachidonic acid releasing activity in platelet membranes: effects of sulfhydryl-modifying agents, *Biochemistry*, 20, 391—397, 1981.
32. **Lagarde, M., Menashi, S., and Crawford, N.,** Localization of phospholipase A_2 and diglyceride lipase activities in human platelet intracellular membranes, *FEBS Lett.*, 124, 23—26, 1981.
33. **McKean, M. L., Smith, J. B., and Silver, M. J.,** Phospholipid biosynthesis in human platelets: formation of phosphatidylcholine from L-acyl lysophosphotidylcholine by acyl CoA: 1-acyl-*sn*-glycerol-3-phospho-choline acyltransferase, *J. Biol. Chem.*, 257, 11278—11283, 1982.
34. **Trotter, J., Flesch, I., Schmidt, B., and Ferler, E.,** Acyltransferase-catalyzed cleavage of arachidonic acid from phospholipids and transfer to lysophosphatides in lymphocytes and macrophages, *J. Biol. Chem.*, 257, 1816—1823, 1982.
35. **Gross, R. W. and Sobel, B. E.,** Lysophosphatidylcholine metabolism in the rabbit heart. Characterization of metabolic pathways and partial purification of myocardial phospholipase-transacylase, *J. Biol. Chem.*, 257, 6702—6708, 1982.
36. **Iritani, N., Ikeda, Y., and Kajitani, H.,** Selectivities of 1-acylglycerophosphorylcholine acyltransferase and acyl-CoA synthase for n-3 polyunsaturated fatty acids in platelets and liver microsomes, *Biochim. Biophys. Acta*, 793, 416—422, 1984.
37. **Marcus, A. J., Ullman, H. L., and Safier, L. B.,** Lipid composition of subcellular particles of human blood platelets, *J. Lipid Res.*, 10, 108—114, 1969.
38. **Broekman, M. J., Handin, R. I., Derksen, A., and Cohen, P.,** Distribution of phospholipid, fatty acids, and platelet factor 3 activity among subcellular fractions of human platelets, *Blood*, 47, 963—971, 1976.
39. **Prescott, S. M. and Majerus, P. W.,** The fatty acid composition of phosphatidyl inositol from thrombin-stimulated human platelets, *J. Biol. Chem.*, 256, 579—582, 1981.
40. **Broekman, M. J., Ward, J. W., and Marcus, A. J.,** Fatty acid composition of phosphatidyl inositol and phosphatidic acid in stimulated platelets: persistence of arachidonyl-stearyl structure, *J. Biol. Chem.*, 256, 8271, 1981.
41. **Hotchkiss, A., Jordan, J. V., Hirata, F., Shulman, N. R., and Axelrod, J.,** Phospholipid methylation and human platelet function, *Biochem. Pharmacol.*, 30, 3089—3095, 1981.
42. **Shattil, S. J., McDonough, M., and Burch, J. W.,** Inhibition of platelet phospholipid methylation during platelet secretion, *Blood*, 57, 537—544, 1981.
43. **Kannagi, R., Koizumi, K., Hata-Tanoue, S., and Masuda, T.,** Mobilization of arachidonic acid from phosphatidyl-ethanolamine fraction to phosphatidylcholine fraction in platelets, *Biochem. Biophys. Res. Commun.*, 96, 711—718, 1980.

44. **Berge, R. K., Vollset, S. E., and Farstad, M.**, Intracellular localization of palmitoyl CoA hydrolase and synthetase in human blood platelets and liver, *Scand. J. Clin. Lab. Med.*, 40, 271—278, 1980.

45. **Berge, R. K., Hagen, L. E., and Farstad, M.**, Isolation of palmitoyl CoA hydrolase from human blood platelets, *Biochem. J.*, 199, 639—647, 1981.

46. **Holmsen, H.**, Unpublished observations.

47. **Ritterhouse-Simmons, S. and Deykin, D.**, Release and metabolism of arachidonate in human platelets, in *Platelets in Biology and Pathology*, Vol. 2, Gordon, J. L., Ed., North-Holland, Amsterdam, 1981, 349—372.

48. **Tangney, C. C. and Driskell, J. A.**, Effects of vitamin E deficiency on the relative incorporation of ^{14}C-arachidonate into platelet lipids of rabbits, *J. Nutr.*, 111, 1839—1847, 1981.

49. **Rittenhouse-Simmons, S., Russell, F. A., and Deykin, D.**, Mobilization of arachidonic acid in human platelets. Kinetics and Ca^{2+} dependency, *Biochim. Biophys. Acta*, 488, 370—380, 1977.

50. **Lagarde, M., Guichardant, M., Menashi, S., and Crawford, N.**, The phospholipid and fatty acid composition of human platelet surface and intracellular membranes isolated by high voltage free flow electrophoresis, *J. Biol. Chem.*, 257, 3100—3014, 1982.

51. **Morita, J., Saito, Y., Chang, W. C., and Murota, S.**, Effects of purified eicosapentaenoic acid on arachidonic acid metabolism in cultured murine aortic smooth muscle cells, vessel walls and platelets, *Lipids*, 18, 42—49, 1983.

52. **Ahmed, A. A., Mahadevappa, V. G., and Holub, B. J.**, The turnover of 1-^{14}C eicosapetaenoic acid among individual phospholipids of human platelets, *Nutr. Res.*, 3, 673, 1983.

53. **Béréziat, G., Chambaz, J., Trugnan, G., Pépin, D., and Polonovska, J.**, Turnover of phospholipid linoleic and arachidonic acids in human platelets from plasma lecithins, *J. Lipid Res.*, 19, 495—500, 1978.

54. **Chambaz, J., Wolf, C., Pépin, D., and Béréziat, G.**, Phospholipid and fatty acid exchange between human platelet and plasma, *Biol. Cell.*, 37, 223, 1980.

55. **Pantavid, M., Perret, B. P., Chap, H., Simon, M.-F., and Douste-Blazy, L.**, Asymmetry of arachidonic acid metabolism in the phospholipids of the human platelet membrane as studied with purified phospholipases, *Biochim. Biophys. Acta*, 693, 450—460, 1982.

56. **Mahadevappa, V. G. and Holub, B. J.**, The molecular species composition of individual diacyl phospholipids in human platelets, *Biochim. Biophys. Acta*, 713, 73—79, 1982.

57. **Natarajan, V., Zuzarte-Augustin, M., Schmid, H. H. O., and Graff, G.**, The alkylacyl and alkenylacyl glycerophospholipids of human platelet, *Thromb. Res.*, 30, 119—125, 1982.

58. **Mueller, H. A., Purdon, A. D., Smith, J. B., and Wykle, R. C.**, 1-0-alkyl-linked phosphoglycerides of human platelets: distribution of arachidonate and other acyl residues in the ether-linked and diacyl species, *Lipids*, 18, 814—819, 1983.

59. **Perret, B., Chap, J. H., and Douste-Blazy, L.**, Asymmetric distribution of arachidonic acid in the plasma membrane of human platelets. A determination using purified phospholipases and rapid method for membrane isolation, *Biochim. Biophys. Acta*, 556, 434—436, 1979.

60. **Kannagi, R., Koizumi, K., and Masuda, T.**, Limited hydrolysis of platelet membrane phospholipids, On the proposed phospholipase susceptible domain in platelet membranes, *J. Biol. Chem.*, 256, 1177—1184, 1981.

61. **Holmsen, H., Dangelmaier, C. A., and Holmsen, H. K.**, Thrombin-induced platelet responses differ in requirement for receptor occupancy, Evidence for tight coupling occupancy and compartmentalized phosphatidic acid formation, *J. Biol. Chem.*, 256, 9393-9396, 1981.

62. **Mauco, G., Chap, H. M.-F., and Douste-Blazy, L.**, Phosphatidic and lysophosphatidic acid production in phospholipase C- and thrombin-treated platelets. Possible involvement of a platelet lipase, *Biochemistry*, 60, 653—661, 1978.

63. **Lapetina, E. G. and Cuatrecasas, P.**, Ionophore A23187- and thrombin-induced platelet aggregation: independence from cyclooxygenase products, *Proc. Natl. Acad. Sci. U.S.A.*, 75, 818—822, 1978.

64. **Walenga, R., Vanderhoek, J. Y., and Feinstein, M. B.**, Serine esterase inhibitors block stimulus-induced mobilization of arachidonic acid and phosphatidylinositide-specific phospholipase C activity in platelets, *J. Biol. Chem.*, 255, 6024—6027, 1980.

65. **Lapetina, E. G., Billah, M. M., and Cuatrecasas, P.**, The phosphatidylinositol cycle and the regulation of arachidonic acid production, *Nature (London)*, 292, 367—369, 1981.

66. **Lapetina, E. G., Billah, M. M., and Cuatrecasas, P.**, The initial action of thrombin on platelets. Conversion of phosphatidylinositol to phosphatidic acid preceding the production of arachidonic acid, *J. Biol. Chem.*, 256, 5037—5040, 1981.

67. **Lapetina, E. G., Chandrabose, K. A., and Cuatrecasas, P.**, Ionophore A23187 and thrombin-induced platelet aggregation: independence from cyclooxygenase products, *Proc. Natl. Acad. Sci. U.S.A.*, 75, 818—822, 1978.

68. **Billah, M. M., Lapetina, E. G., and Cuatrecasas, P.**, Phospholipases A_2 activity specific for phosphatidic acid. A possible mechanism for the production of arachidonic acid in platelets, *J. Biol. Chem.*, 256, 5399—5403, 1981.

69. **Billah, M. M., Lapetina, E. G., and Cuatrecasas, P.,** Phosphatidylinositol-specific phospholipase C of platelets: association with 1,2-diacylglycerolkinase and inhibition by cyclic-AMP, *Biochem. Biophys. Res. Commun.,* 90, 92—98, 1979.

70. **Rittenhouse-Simmons, S.,** Production of diglyceride from phosphatidylinositol in activated human platelets, *J. Clin. Invest.,* 63, 580—597, 1979.

71. **Bills, T. K., Smith, J. B., and Silver, M. J.,** Selective release of arachidonic acid from the phospholipids in response to thrombin, *J. Clin. Invest.,* 60, 1—6, 1977.

72. **Rittenhouse-Simmons, S., Russel, F. A., and Deykin, D.,** Transfer of arachidonic acid to human platelet plasmalogen in response to thrombin, *Biochem. Biophys. Res. Commun.,* 70, 295—301, 1976.

73. **Kramer, R. M. and Deykin, D.,** Arachidonyl transacylase in human platelets, *J. Biol. Chem.,* 258, 13806—13811, 1983.

74. **Colard, O., Breton, M., and Bereziat, G.,** Induction by lysophospholipids of CoA-dependent arachidonyl transfer between phospholipids in rat platelet homogenates, *Biochim. Biophys. Acta,* 793, 42—48, 1984.

75. **Walenga, R. W., Opas, E. E., and Feinstein, M. B.,** Differential effect of calmodulin antagonists on phospholipases A_2 and C in thrombin-stimulated platelets, *J. Biol. Chem.,* 256, 12523—12528, 1981.

76. **Holmsen, H., Daniel, J. L., Dangelmaier, C. A., Molish, I., Rigmaiden, M., and Smith, J. B.,** Differential effects of trifluoperazine on arachidonate liberation, secretion and myosin phosphorylation in intact platelets, *Thromb. Res.,* 36, 419—428, 1984.

77. **Bell, R. L., Kennerly, D. A., Stanford, N., Majerus, P. W., Constrini, N. V., and Bradshaw, R. A.,** Diglyceride lipase pathway for arachidonate release from human platelets, *Proc. Natl. Acad. Sci. U.S.A.,* 76, 3238—3241, 1979.

78. **Broekman, M. J., Ward, J. W., and Marcus, A. J.,** Phospholipid metabolism in stimulated human platelets. Changes in phosphatidylinositol, phosphatidic acid and lysophospholipids, *J. Clin. Invest.,* 66, 275—283, 1980.

79. **Bell, R. L. and Majerus, P. W.,** Thrombin-induced hydrolysis of phosphatidylinositol in human platelets, *J. Biol. Chem.,* 255, 1790—1792, 1980.

80. **Chau, L. Y. and Tai, H. H.,** Release of arachidonate from diglyceride in human platelets requires the sequential action of a diglyceride lipase and a monoglyceride lipase, *Biochem. Biophys. Res. Commun.,* 100, 1688—1695, 1981.

81. **Kannagi, R. and Koizumi, K.,** Effect of different physical states of phospholipid substrates in partially purified platelet phospholipase A_2 activity, *Biochim. Biophys. Acta,* 556, 423—433, 1979.

82. **Mahadavappa, V. G. and Holub, B. J.,** Degradation of different molecular species of phosphatidylinositol in thrombin-stimulated human platelets, *J. Biol. Chem.,* 258, 5337—5339, 1983.

83. **Rittenhouse, S. E. and Allen, C. L.,** Synergistic activation by collagen and 15-hydroxy 9α, 11α-peroxidoprosta-5,13-dienoic acid (PGH_2) of phosphatidylinositol metabolism and arachidonic acid release in human platelet, *J. Clin. Invest.,* 70, 1216—1224, 1982.

84. **Rittenhouse-Simmons, S.,** Differential activation of platelet phospholipases by thrombin and ionophore A23187, *J. Biol. Chem.,* 256, 4153—4155, 1981.

85. **Billah, M. M. and Lapetina, E. G.,** Evidence for multiple metabolic pools of phosphatidylinositol in stimulated platelets, *J. Biol. Chem.,* 257, 11856—11859, 1982.

86. **Billah, M. M. and Lapetina, E. G.,** Formation of lysophosphatidylinositol in platelets stimulated with thrombin or ionophore A23187, *J. Biol. Chem.,* 257, 5196—5200, 1982.

87. **Kameyama, Y., Yoshioka, S., Imai, A., and Nozawa, Y.,** Possible involvement of 1-acyl-glycoerophosphorylinositol acetyltransferase in arachidonate enrichment of phosphatidylinositol in human platelets, *Biochim. Biophys. Acta,* 752, 244—250, 1983.

88. **Imai, A., Yano, K., Kameyama, Y., and Nozawa, Y.,** Evidence for predominance of phospholipase A_2 in release of arachidonic acid in thrombin-activated platelets: phosphatidylinositol-specific phospholipase C may play a minor role in arachidonate liberation, *Jpn. J. Exp. Med.,* 52, 99—105, 1982.

89. **Sutherland, C. A. and Amin, D.,** Relative activities of rat and dog platelet phospholipase A_2 and diglyceride lipase, *J. Biol. Chem.,* 257, 14006—14010, 1982.

90. **Oglesby, T. D. and Gorman, R. R.,** The inhibition of arachidonic acid metabolism by RHC 80267, a diacylglycerol lipase inhibitor, *Biochim. Biophys. Acta,* 793, 269—277, 1984.

91. **Newfeld, E. J. and Majerus, P. W.,** Arachidonate release and phosphatidic acid turnover in stimulated human platelets, *J. Biol. Chem.,* 258, 2461—2467, 1983.

92. **Apitz-Castro, R. J., Mas, M. A., Cruz, M. R., and Jain, M. K.,** Isolation of homogenous phospholipase A_2 from human platelets, *Biochem. Biophys. Res. Commun.,* 91, 63—71, 1979.

93. **Jesse, R. L. and Franson, R. C.,** Modulation of purified phospholipase A_2 activity from human platelets by calcium and indomethacin, *Biochim. Biophys. Acta,* 575, 467—470, 1979.

94. **Franson, R. C., Eisen, D., Jesse, R., and Lanni, C.,** Inhibition of highly purified mammalian phospholipases A_2 by non-steroidal antiinflammatory agents. Modulation by calcium ions, *Biochem. J.,* 186, 633—636, 1980.

95. **Ballou, L. R. and Cheung, W. Y.**, Marked increase of human platelet phospholipase A_2 activity in vitro and demonstration of an endogeneous inhibitor, *Proc. Natl. Acad. Sci. U.S.A.*, 80, 5203—5207, 1983.

96. **Watson, S. P., Ganong, B. R., Bell, R. M., and Lapetina, E. G.**, 1,2-Diacylglycerols do not potentiate the actions of phospholipids A_2 and C in human platelets, *Biochem. Biophys. Res. Commun.*, 121, 386—391, 1984.

97. **Mauco, G., Chap, H., and Douste-Blazy, L.**, Characterization and properties of a phosphatidylinositol phosphodiesterase (phospholipase C) from platelet cytosol, *FEBS Lett.*, 100, 367—370, 1979.

98. **Siess, W. and Lapetina, E. G.**, Properties and distribution of phosphatidylinositol-specific phospholipase C in human and horse platelets, *Biochim. Biophys. Acta*, 752, 328—338, 1983.

99. **Nahas, N. and Graff, G.**, Inhibitory activity of polyamines on phospholipase C from human platelets, *Biochem. Biophys. Res. Commun.*, 109, 1035—1040, 1982.

100. **Simon, M.-F., Chap, H., and Douste-Blazy, L.**, Activation of phospholipase C in thrombin-stimulated platelets does not depend on cytoplasmic free calcium concentration, *FEBS Lett.*, 170, 43—48, 1984.

101. **Holmsen, H., Kaplan, K. L., and Dangelmaier, C. A.**, Differential energy requirements for platelet responses. A simultaneous study of aggregation, three secretory processes, arachidonate liberation, phosphatidylinositol breakdown and phosphatidate production, *Biochem. J.*, 208, 9—18, 1982.

102. **Holmsen, H., Dangelmaier, C. A., and Rongved, S.**, Tight coupling of thrombin-induced acid hydrolase secretion and phosphatidate synthesis to receptor occupancy in human platelets, *Biochem. J.*, 222, 157—167, 1984.

103. **Billah, M. M. and Lapetina, E. G.**, Rapid decrease of phosphatidylinositol-4,5-bisphosphate in thrombin-stimulated platelets, *J. Biol. Chem.*, 257, 12705—12708, 1982.

104. **Agranoff, B. W., Murthy, P., and Seguin, E. B.**, Thrombin-induced phosphodiesteratic cleavage of phosphatidylinositol bisphosphate in human platelets, *J. Biol. Chem.*, 258, 2076—2078, 1983.

105. **Imai, A., Nakashima, S., and Nozawa, Y.**, The rapid polyphosphoinositide metabolism may be a triggering event for thrombin-mediated stimulation of human platelets, *Biochem. Biophys. Res. Commun.*, 110, 108—115, 1983.

106. **Graff, G., Nahas, N., Nikolopoulou, M., Natarajare, V., and Schmid, H.**, Possible regulation of phospholipase C activity in human platelets by phosphatidyl 4',5'-bisphosphate, *Arch. Biochem. Biophys.*, 228, 229—308, 1984.

107. **Kramer, R. M., Jakubowski, J. A., Vaillancourt, R., and Deykin, D.**, Effect of membrane cholesterol on phospholipid metabolism in thrombin-stimulated platelets, *J. Biol. Chem.*, 257, 6844—6849, 1982.

108. **Needlemen, S. W., Spector, A. A., and Hoak, J. C.**, Enrichment of human platelet phospholipids with linoleic acid diminishes thromboxane release, *Prostaglandins*, 24, 607—622, 1982.

109. **Nordvi, B. and Holmsen, H.**, Unpublished results.

UPTAKE AND RELEASE OF EICOSANOID PRECURSORS FROM PHOSPHOLIPIDS AND OTHER LIPID POOLS: LUNG

Y. S. Bakhle

INTRODUCTION

The lung has had a long-standing and highly significant involvement with the eicosanoids. Early work[1,2] showed that homogenates of lung would synthesize prostaglandins (PGs) from exogenous arachidonic acid (AA), and synthesis of PGs from endogenous AA in lung was demonstrated soon after.[3] It is now generally accepted that synthesis of eicosanoids from endogenous precursor fatty acids depends crucially upon the amount of those substrates present in the unesterified form. Because eicosanoids (or any other biologically active material) formed or released into pulmonary blood can be distributed throughout the systemic arterial circulation, the synthesis of eicosanoids in lung can have systemic effects. Thus, the supply of eicosanoid precursors in lung is important not only in the lung itself but has quasiendocrine significance. In spite of these exciting possibilities, the uptake, distribution, and release of eicosanoid precursors in lung has attracted relatively little attention when compared with similar work carried out with kidney or platelets (see previous chapter written by Holmsen).

In this short review the emphasis is on the fate of endogenous precursor acids and the much larger literature on synthesis of eicosanoids from exogenous substrates or the enzymic pathways involved subsequent to the liberation of substrate from its esterified forms will not be discussed.

UPTAKE

Uptake of arachidonate has been followed by radiochemical assay using isolated lungs from several species perfused with Krebs solution (Table 1). In most of these examples, a short infusion (3 to 4 min) of ^{14}C-AA was given and uptake calculated from the radioactivity collected in effluent. It is essential to remember that these calculations of uptake have one major source of error: oxidation of the radiolabeled AA by the oxygenases of the lung. These oxidized products are not retained by lung and thus appear in the effluent along with unchanged substrate. The amount of oxidized products will depend on the species[4] and also on the time elapsed between infusion used and assay.[5] However, because the most active oxygenase in lung, cyclooxygenase, is an intracellular enzyme, formation of cyclooxygenase products from infused substrate must be preceded by uptake into lung cells. Because of this effect, the initial uptake of AA into lung is probably higher than those values reported and may, in fact, be better estimated in the presence of cyclooxygenase inhibitors which increase uptake values.[5-8]

Uptake of AA, even allowing for errors, was extremely rapid and efficient, probably over 90% on a single pass using a Krebs solution perfusate and about 60% in the presence of albumin.[5-7] The uptake value of 76% reported by Kennedy et al.[9] was measured after a total of 45 min recirculating perfusion with ^3H-AA. The factors affecting uptake have not been systematically studied but there is agreement that in the presence of albumin, uptake is significantly reduced[5-7] and a lesser but still important reduction in uptake was noted in isolated lungs from rats made diabetic by streptozotocin.[10]

DISTRIBUTION

Distribution of radiolabel in lung after uptake of exogenous AA was studied by analysis

Table 1
UPTAKE OF RADIOLABELED ARACHIDONIC ACID BY ISOLATED PERFUSED LUNGS

Species	Concentration	Uptake	Time (min)	Ref.	Modifying factors	Ref.
By infusion						
Guinea pig	10 nM	79	45	9	Albumin	5, 6
	4 μM	80	10	6		
	18 μM	90	5	5	Cyclooxygenase inhibitors	5, 6
Rat	4 μM	90	10	7	Albumin	7
	40 μM	90	10	7	Diabetes mellitus	10
					Indomethacin	7
					Estrous cycle	12
Hamster	3 μM	80	6	82	Aspirin	82
By injection	Total amount					
Hamster	33 nmol	90	6	83	Dipyridamole	84
	50 nmol	88	6	84		
	66 nmol	75	6	83	Indomethacin	8
Rat	66 nmol	92	6	83		

Note: Uptake values (percent of administered AA) in this table have been derived by subtracting the radioactivity in the effluent measured at the times shown from the amount of radioactivity infused or injected. In this and other tables only mean values are given; the standard errors or deviations can be found in the original references.

Table 2
DISTRIBUTION OF RADIOLABELED ARACHIDONIC ACID IN LUNG

Species	Concentration	FFA[a]	NL	PL	PC	PE	PI, PS, SM	Ref.
Guinea pig	10 nM	28	6	66	52	5	4	9
	4 μM	7	17	72	22	24	25	6
	18 μM	2	20	77	36	16	24	5
Rat	4 μM	17	21	59	—	—		7
	40 μM	36	29	31	11	8	8	7
Hamster	3 μM	15	19	65	29	16	9	82
Human	40 μM	15	31	46	—	—	—	86

[a]   FFA = free fatty acid; NL = neutral lipid; PL = phospholipid; PC = phosphatidyl choline; PE = phosphatidyl ethanolamine; PI, PS, SM = phosphatidyl inositol and -serine and sphingomyelin (not separated in this analysis). Distribution is expressed as percent of the radioactivity in lung following infusion of radiolabeled arachidonate at the concentration shown.

of the label in lung homogenates into different lipid classes and is summarized in Table 2. At least two features are worth noting. First, there was no radioactivity in lung homogenates associated with cyclooxygenase products, implying that oxidation of AA led to its immediate loss from perfused lung[5-7] and that the process of homogenization (in chloroform:methanol mixtures) did not give rise to artifactual oxidation products. Second, in all examples there was rapid and extensive incorporation, within 10 min of label into phospholipid.[6,7] The label in esterified forms[5] and in the free acid fraction[11] was present as arachidonate. In rat[7] and guinea pig lung[6] distribution is only slightly affected by indomethacin but in rat lung more obviously changed over the estrous cycle.[11] Perfusion with albumin-containing solutions had the same final effect irrespective of whether the albumin was present during the initial substrate perfusion[5-7] or only added after the initial labeling had been completed.[7] Under either set of conditions, the proportion of label in the free acid pool was markedly diminished.

RELEASE

Release of arachidonate from lung has been investigated directly by measuring radioactivity released from lungs previously "loaded" with labeled AA or indirectly by measuring with radioimmunoassay or bioassay the formation of cyclooxygenase products derived from endogenous arachidonate under different conditions.

Direct Assay

There are few direct estimates of AA release from lung. In the absence of stimuli known to induce the synthesis of cyclooxygenase products, spontaneous release of label was very slow from guinea pig lung labeled with ^{14}C-AA, only about 10% of the total content being lost during a further 120 min perfusion after loading.[5] Following stimulated synthesis of cyclooxygenase products, release of label from guinea pig lung was not greater than 1% of total lung content and failed rapidly although synthesis as judged by bioassay was maintained on repeated stimulation.[5] Under another set of conditions, spontaneous release of ^{3}H from guinea pig lungs loaded with ^{3}H-AA for 45 min was followed for a further 240 min.[9] Although the efflux of ^{3}H after loading and the calculated ^{3}H uptake estimates were very similar to those reported earlier,[5] stimulated release of ^{3}H was said to be significant and related to the size of stimulus (dose of bradykinin). Since there was no clear indication of the amount of radioactivity released in absolute terms or relative to that in lung at the time of stimulation, it is not easy to compare this report[9] with that of Jose and Seale.[5]

In rat lungs loaded with ^{14}C-AA, spontaneous release was again low, about 10% of the radioactivity in lung at 10 min being lost over the next 20 min.[7] In this tissue, stimulated release of bioassayable cyclooxygenase products was much less than in guinea pig lungs[4] and very small amounts of ^{14}C were released from labeled lung by the calcium ionophore A23187.[7] The experiments of Voelkel et al.[13] have confirmed that relatively small amounts of radioactive cyclooxygenase products were released from rat isolated lungs loaded with ^{14}C-AA after stimulation by angiotensin II. The lungs contained almost 2 μCi of ^{14}C-AA and the released radioactivity increased from approximately 60 cpm/mℓ effluent before stimulation to 140 cpm/mℓ after stimulation; this stimulated release of radioactivity was inhibited by inhibition of cyclooxygenase. These results contrast with those of Kennedy et al.[9] who showed that indomethacin did not interfere with release of ^{3}H-after labeling with ^{3}H-AA.

It is therefore clear that the rapid incorporation of exogenous arachidonate into lung tissue does not lead to selective uptake into that pool of lung lipid utilized for synthesis of eicosanoids. This contrasts with the results from platelets[14,15] and is particularly disappointing in terms of providing a tracer technique for studying the metabolism of arachidonate in short-term experiments. Perhaps labeling in vivo overnight or for longer periods might improve the efficiency of this technique for following release of AA from lung lipids. However, it may also never be possible in a tissue containing over 30 different cell types[15] to attain the same efficiency of labeling as can be shown in a suspension of a single cell type like platelets (see previous chapter written by Holmsen).

Certainly, at present the most striking characteristic of AA metabolism in lung, as elu-

Table 3
UNSTIMULATED RELEASE
OF ENDOGENOUS
ARACHIDONATE IN LUNG
MEASURED AS
EICOSANOID FORMATION

In vivo	Ref.
Cat	23
Rabbit	24
Dog	18
Goat	38
Man	25
Isolated lungs	
Rat	10, 12, 23, 29, 42, 87, 88
Guinea pig	23
Lung homogenate	
Dog	43

cidated from the radioactive labeling studies, is the *resistance* of AA (once taken up) to be released either unchanged or after oxidation via cyclooxygenase or other oxygenases. This contrasts strongly with the practical experience of many workers that endogenous AA is very readily made available to the oxygenases following a variety of stimuli. There may indeed be a continuous turnover of AA in lung under normal physiological conditions.[17,18] It is relevant to note here that Blackwell et al.[19] have found a phospholipase A_2 activity in isolated lung exhibited toward substrates perfused through the pulmonary circulation.

Indirect Assay

The release of endogenous AA from its esterified forms and its consequent availability to oxygenases may be estimated by measurement of the eicosanoid products of these enzymes by bioassay or radioimmunoassay. What the indirect assay cannot yet estimate is release of AA as the free acid itself since its intrinsic biological activity is low and no relevant antisera have yet been developed. Another disadvantage of indirect assay is that AA may be oxidized to a variety of products via cyclooxygenase[20] and lipoxygenases[21] and some of these eicosanoids are themselves oxidized further in lung tissue[22] (see contributions to this volume by Willis, Smith, Kulmacz, et al.). Thus, in order to estimate the total release of AA from esterified forms, it would be necessary to measure PGs E_2, $F_{2\alpha}$, and D_2, 6-oxo-$PGF_{1\alpha}$, TXB_2, and the relevant 15-oxo derivatives together with the many hydroxy acids and leukotrienes that may be produced, apart from unchanged AA itself. In the absence of such ideal experiments, a wide variety of approximations have to be made.

Under nonstimulated conditions (Table 3), in order to estimate the physiological level of AA turnover, direct measurements of eicosanoid formation have concentrated on the assay of prostacyclin (as 6-oxo-$PGF_{1\alpha}$) in arterial blood.[23-25] There are at least two reasons for this. First, there is the hypothesis that prostacyclin could be a circulating hormone[17] because it is not inactivated in the pulmonary circulation[26] and because it is formed by the pulmonary endothelial cells. Second, in most species apart from the guinea pig, PGI_2 is a major biologically active eicosanoid detectable in perfusate leaving the lungs either isolated[4,29] or in vivo.[23,24,30,31] Evidence for the circulating hormone hypothesis[17,23-25] has now been challenged by results of many indirect[32-34] and direct[35-37] experiments showing very low, if not undetectable, amounts of 6-oxo-$PGF_{1\alpha}$ in systemic arterial blood in humans.

Table 4
STIMULATED TURNOVER OF ENDOGENOUS ARACHIDONATE IN LUNG MEASURED AS EICOSANOID FORMATION

Physical stimulus	Ref.
Hyperventilation	17, 29, 35, 89—91
Vasoconstriction	
Angiotensin II	13
Hypoxia	43, 45—48, 92
Embolization	
Particulate	49—51, 93—96
Fat emulsions	49—51
Platelets	87, 88, 94, 95
Endotoxin	55, 97—103
Anaphylaxis	3, 71, 104—110
Chemical stimulus	
Histamine	51, 63, 65—67, 111
5-Hydroxytryptamine	65, 66
Compound 48/80	112
Angiotensin II	13, 17, 90
Bradykinin	3, 9, 67, 72, 90, 113
ATP, ADP	114, 115
Calcium ionophore A23187	4, 116, 117
SRS-A, leukotrienes	5, 60, 67—69, 81, 116, 118—120
Phospholipase A_2	117, 121, 122

Nevertheless, there is evidence[18] that administration of cyclooxygenase inhibitors to conscious resting dogs modified their pulmonary vascular resistance, implying some effect of eicosanoids in normal pulmonary hemodynamics, although it could be argued that the eicosanoids concerned need not have been synthesized in lung. In another physiological situation, but this time clearly under some form of stimulus, the closing of the ductus arteriosus and the pulmonary vasodilatation following birth, cyclooxygenase inhibition also modified the normal course of events.[38] Although the exact nature of the relevant eicosanoids may still be under discussion,[39-41] an earlier event must have been release of AA from its lipid forms.

In isolated lungs of rats a basal output of cyclooxygenase products has been demonstrated. The experimental conditions vary — albumin in the perfusate and ventilation — as do the absolute amounts detected, but they all agree that in the absence of an overt stimulus, the major eicosanoid is prostacyclin. This basal output was affected by the progress of the estrous cycle[12] by gender[12,42] (being higher in females) and by streptozotocin-induced diabetes,[10] in which PGI_2 production is lower.[10] These data suggest that the supply of precursor may also be similarly affected. The synthesis of prostacyclin from endogenous precursors in dog lung homogenates[43] was increased by incubation in a hypoxic environment; there the additional problem of altered PGI_2 metabolism by the liberated PG-iso-H-dehydrogenase must make the direct extrapolation from increased PGI_2 production to increased AA precursor supply less reliable.

Increased release of eicosanoids takes place following a wide range of stimuli which have been divided roughly into two groups: physical and chemical (Table 4). This division is often arbitrary and probably unjustifiable but there are some stimuli which may be unequivocally placed in either group.

Hyperventilation is a clear example of a physical stimulus, and the formation of vasodilator eicosanoids has been proposed as a physiological means of matching ventilation with perfusion.[44] Vasoconstriction can also induce eicosanoid synthesis, particularly when caused by alveolar hypoxia in vivo, but here, too, vasodilator eicosanoids are formed perhaps in an attempt to counteract the increase in pulmonary vascular resistance.[45-48] Physical blockage of the pulmonary vasculature by inert particulate matter in isolated blood-free lungs was enough stimulus to increase eicosanoid formation.[49-51]

In relation to these comparatively simple signals, those provided by platelet embolization, endotoxin, and anaphylaxis are a complex mixture of physical and chemical stimuli. For instance, endotoxin will, apart from aggregating platelets, generate the complement fragment C5a, itself a stimulus to eicosanoid synthesis.[52] The pulmonary accumulation of leukocytes following endotoxin-induced complement activation may be another proximal event for eicosanoid formation,[53,54] since endotoxin did not stimulate eicosanoid synthesis when added to lung slices in vitro.[55]

Anaphylaxis is accompanied by extensive turnover of endogenous AA, particularly via the 5-lipoxygenase associated with leukotriene production. The bronchoconstriction induced by these agents[56-60] would be an additional stimulus for eicosanoid synthesis by airway smooth muscle.[61-64]

The chemical stimuli for eicosanoid production have been examined mostly in perfused isolated lung and here species differences are crucial. Guinea pig lungs will release many cyclooxygenase products, including TXA_2 in response to a wide variety of chemicals, endogenous and exogenous, often via receptor specific mechanisms.[63,65-67] However, in rat and particularly in human lung, release of cyclooxygenase products is much less easy to induce, i.e., greater stimulus is required as with ionophore[4] or is absent as with the leukotrienes.[4,68] It is of particular interest that two types of SRS now known to be a mixture of leukotrienes[21] formed in guinea pig or human lung, both released eicosanoids from guinea pig lung[69] but not from rat or human lung.[4] Apart from casting doubt on the relevance of experiments carried out with guinea pig lung to events in human lung, such findings suggest either a leukotriene-specific, lipase-activating, receptor in guinea pig lung, absent in other species, or else a specific lipase which requires leukotriene as some type of co-factor.

In addition, lungs other than those of guinea pig also tend to form much less TXA_2 regardless of the stimulus.[4,70,71] It is difficult to avoid the conclusion that guinea pig lung provides an unhelpful, if not actively misleading, model of AA metabolism in human lung. If an animal model must be used, rat lung would appear to be a closer paradigm.[4,7]

Inhibition of AA release from lung lipids has seldom been directly demonstrated. However, early in the studies of this topic, the inhibitory effect of mepacrine on AA release and subsequent eicosanoid synthesis was recorded, and more recently, the same drug was shown to prevent bradykinin-induced ^3H release from lungs labeled with ^3H-AA.[9] Glucocorticoids have been known for many years to alleviate the symptoms of allergic asthma[73,74] and the effects of endotoxinemia.[75-77] These effects can now be explained in terms of the steroid-induced inhibitor of phospholipase A_2, macrocortin,[78] or lipomodulin[19] (see contribution by R. Flower and F. Hirata to this Handbook). Inhibition of the phospholipase A_2 activity exhibited in guinea pig isolated lungs[19] and the eicosanoid generating ability of RCS-RF[19,80] or SRS-A[81] in the same tissue has been demonstrated with corticosteroids.

CONCLUSION

Although a relatively good understanding of the oxidative pathways of AA metabolism has emerged from work with *exogenous* substrate, the ways in which that substrate may be provided *in situ* and the influences, physiological or pathological, that play upon the uptake, distribution, and release of AA still remain largely uninvestigated. Even less is known about the fate of the other eicosanoic acids. The techniques appropriate to this study are already with us, mostly developed for experiments related to palmitate metabolism in lung; all we lack is incentive. I hope that the large gaps in our knowledge which we have tried to expose will provide such incentive.

REFERENCES

1. **Änggård, E. and Samuelsson, B.,** Biosynthesis of prostaglandins from arachidonic acid in guinea pig lung. Prostaglandins and related factors 38, *J. Biol. Chem.,* 240, 3518—3521, 1965.
2. **Nugteren, D. H. and Hazelhof, E.,** Isolation and properties of intermediates in prostaglandin biosynthesis, *Biochim. Biophys. Acta,* 326, 448—461, 1973.
3. **Piper, P. J. and Vane, J. R.,** Release of additional factors in anaphylaxis and its antagonism by anti-inflammatory drugs, *Nature (London),* 223, 29—35, 1969.
4. **Al-Ubaidi, F. and Bakhle, Y. S.,** Differences in biological activation of arachidonic acid in perfused lungs from guinea pig, rat and man, *Eur. J. Pharmacol.,* 62, 89—96, 1980.
5. **Jose, P. J. and Seale, J. P.,** Incorporation and metabolism of ^{14}C-arachidonic acid in guinea pig lungs, *Br. J. Pharmacol.,* 67, 519—526, 1979.
6. **Al-Ubaidi, F. and Bakhle, Y. S.,** The fate of exogenous arachidonic acid in guinea pig isolated lungs, *J. Physiol. (London),* 295, 445—455, 1979.
7. **Al-Ubaidi, F. and Bakhle, Y. S.,** Metabolism of ^{14}C-arachidonate in rat isolated lung, *Prostaglandins,* 19, 747—759, 1980.
8. **Uotila, P., Männistö, J., Simberg, N., and Hartiala, K.,** Indomethacin inhibits arachidonic acid metabolism via lipoxygenase and cyclooxygenase in hamster isolated lungs, *Prostaglandins Med.,* 7, 591—599, 1981.
9. **Kennedy, I., Langley, M., and Whelan, C. J.,** The release and metabolism of ^{3}H-arachidonic acid from guinea pig perfused lungs *in vitro:* a simple method for the study of the action of drugs on the release and metabolism of arachidonic acid, *J. Pharmacol. Methods,* 6, 143—151, 1981.
10. **Watts, I. A., Zakrzewski, J. T., and Bakhle, Y. S.,** Altered prostaglandin synthesis in isolated lungs of rats with streptozotocin-induced diabetes, *Thromb. Res.,* 1, 333—342, 1982.
11. **Bakhle, Y. S. and Zakrzewski, J. T.,** Unpublished experiments.
12. **Bakhle, Y. S. and Zakrzewski, J. T.,** Effects of the oestrous cycle on the metabolism of arachidonic acid in rat isolated lung, *J. Physiol. (London),* 326, 411—423, 1982.
13. **Voelkel, N. F., Gerber, J. G., McMurtry, I. F., Nies, A. S., and Reeves, J. T.,** Release of vasodilator PGI$_2$ from isolated rat lung during vasoconstriction, *Circ. Res.,* 48, 207—213, 1981.
14. **Bills, T. K., Smith, J. B., and Silver, M. J.,** Selective release of arachidonic acid from the phospholipids of human platelets in response to thrombin, *J. Clin. Invest.,* 60, 1—6, 1977.
15. **Blackwell, G. J., Duncombe, W. G., Parsons, M. F., and Vane, J. R.,** The distribution and metabolism of arachidonic acid in rabbit platelets during aggregation and its modification by drugs, *Br. J. Pharmacol.,* 59, 353—366, 1977.
16. **Sorokin, S. P.,** The cells of the lung, in *Morphology of Experimental Respiratory Carcinogenesis,* Nettesheim, P., Hanna, M. G., and Deatherage, J. W., Eds., U.S. Atomic Energy Commission, Oak Ridge, 1970, 3.
17. **Gryglewski, R. J.,** Prostacyclin as a circulatory hormone, *Biochem. Pharmacol.,* 28, 3161—3166, 1979.
18. **Walker, B. R., Voelkel, N. F., and Reeves, J. T.,** Pulmonary pressor response after prostaglandin synthesis inhibition in conscious dogs, *J. Appl. Physiol.,* 52, 705—709, 1982.
19. **Blackwell, G. J., Flower, R. J., Nijkamp, F. P., and Vane, J. R.,** Phospholipase A$_2$ activity of guinea pig isolated perfused lungs; stimulation and inhibition by anti-inflammatory steroids, *Br. J. Pharmacol.,* 62, 79—89, 1978.
20. **Kuehl, F. A., Jr., Egan, R. W., and Humes, J. L.,** Prostaglandin cyclo-oxygenase, *Prog. Lipid Res.,* 20, 97—102, 1982.
21. **Goetzl, E. J. and Scott, W. A., III, Eds.,** Proceedings of a conference on regulation of cellular activities by leukotrienes and other lipoxygenase products of arachidonic acid, *J. Allergy Clin. Immunol.,* 74(3, Part 2), 309—448, 1984.
22. **Roberts, L. J., II, Sweetman, B. J., Maas, R. L., Hubbard, W. C., and Oates, J. A.,** Clinical application of PG and TX metabolite quantification, *Prog. Lipid Res.,* 20, 117—121, 1982.

23. **Gryglewski, R. J., Korbut, R., and Ocetkiewicz, A.,** Generation of prostacyclin by lungs *in vivo* and its release into the arterial circulation, *Nature (London)*, 273, 765—767, 1978.

24. **Moncada, S., Korbut, R., Bunting, S., and Vane, J. R.,** Prostacyclin is a circulating hormone, *Nature (London)*, 273, 767—768, 1978.

25. **Hensby, C. N., Barnes, P. Y., Dollery, C. T., and Dargie, H.,** Production of 6-oxo-PGF$_{1\alpha}$ by human lung *in vivo*, *Lancet*, 2, 1162—1163, 1979.

26. **Dusting, G. J., Moncada, S., and Vane, J. R.,** Recirculation of prostacyclin (PGI$_2$) in the dog, *Br. J. Pharmacol.*, 64, 315—320, 1978.

27. **Weksler, B. B., Marcus, A. J., and Jaffe, E. A.,** Synthesis of prostaglandin I$_2$ (prostacyclin) by cultured human and bovine endothelial cells, *Proc. Natl. Acad. Sci. U.S.A.*, 74, 3922—3926, 1977.

28. **MacIntyre, D. E., Pearson, J. D., and Gordon, J. L.,** Localisation and stimulation of prostacyclin production in vascular cells, *Nature (London)*, 271, 549, 551, 1978.

29. **Korbut, R., Boyd, J., and Eling, T.,** Respiratory movements alter the generation of prostacyclin and thromboxane A$_2$ in isolated rat lungs: the influence of arachidonic acid pathway inhibitors on the ratio between pulmonary prostacyclin and thromboxane A$_2$, *Prostaglandins*, 21, 491—503, 1981.

30. **Spannhake, E. W., Hyman, A. L., and Kadowitz, P. J.,** Dependence of the airway and pulmonary vascular effects of arachidonic acid upon route and rate of administration, *J. Pharm. Exp. Ther.*, 212, 584—590, 1980.

31. **Mullane, K. M., Dusting, G. J., Salmon, J. A., Moncada, S., and Vane, J. R.,** Biotransformation and cardiovascular effects of arachidonic acid in the dog, *Eur. J. Pharmacol.*, 54, 217—228, 1979.

32. **Smith, J. B., Ogletree, M. L., Lefer, A. M., and Nicolau, K. C.,** Antibodies which antagonize the effects of prostacyclin, *Nature (London)*, 274, 64—65, 1978.

33. **Haslam, R. J. and McClenaghan, M. D.,** Measurement of circulating prostacyclin, *Nature (London)*, 292, 364—366, 1981.

34. **Steer, M. L., MacIntyre, D. E., Levine, L. E., and Salzman, E. W.,** Is prostacyclin a physiologically important circulating anti-platelet agent?, *Nature (London)*, 283, 194—195, 1980.

35. **Edlund, A., Bonfim, W., Kaijser, L., Olin, C., Patrono, C., Pinca, E., and Wennmalm, Å.,** Pulmonary formation of prostacyclin in man, *Prostaglandins*, 22, 323—332, 1981.

36. **Christ-Hazelhof, E. and Nugteren, D. H.,** Prostacyclin is not a circulating hormone, *Prostaglandins*, 22, 739—746, 1981.

37. **Blair, I. A., Barrow, S. E., Waddell, K. A., Lewis, P. J., and Dollery, C. T.,** Prostacyclin is not a circulating hormone in man, *Prostaglandins*, 23, 579—589, 1982.

38. **Leffler, C. W., Tyler, T. L., and Cassin, S.,** Effect of indomethacin on pulmonary vascular response to ventilation of fetal goats, *Am. J. Physiol.*, 234, H346—H351, 1978.

39. **Coceani, F., Olley, P. M., Bishai, I., Bodach, E., and White, E. P.,** Significance of the prostaglandin system to the control of muscle tone of the ductus arteriosus, in *Advances in Prostaglandin and Thromboxane Research*, Vol. 4, Coceani, F. and Olley, P. M., Eds., Raven Press, New York, 1978, 325—333.

40. **Coceani, F. and Olley, P. M.,** Considerations on the role of prostaglandins in the ductus arteriosus, in *Advances in Prostaglandin and Thromboxane Research*, Vol. 7, Samuelsson, B., Ramwell, P. W., and Paoletti, R., Eds., Raven Press, New York, 1980, 871—878.

41. **Clyman, R. I.,** Ductus arteriosus; developmental response to endogenous prostaglandins, oxygen and indomethacin, in *Advances in Prostaglandin and Thromboxane Research*, Vol. 7, Samuelsson, B., Ramwell, P. W., and Paoletti, R., Eds., Raven Press, New York, 1980.

42. **Maggi, F. M., Tyrrell, N., Maddox, Y., Watkins, W., Ramey, E. R., and Ramwell, P. W.,** Prostaglandin synthetase activity in vascular tissue of male and female rats, *Prostaglandins*, 19, 985—993, 1980.

43. **Hamasaki, Y., Tai, H.-H., and Said, S. I.,** Hypoxia stimulates prostacyclin generation by dog lung *in vitro*, *Prostaglandins Med.*, 8, 311—316, 1982.

44. **Smith, A. P.,** Prostaglandins and the respiratory system, in *Prostaglandins: Physiological, Pharmacological and Pathological Aspects*, Karim, S. M. M., Ed., MTP Press, Lancaster, England, 1976, 83.

45. **Said, S. I., Yoshida, T., Kitamura, S., and Vreim, C.,** Pulmonary alveolar hypoxia: release of prostaglandins and other humoral mediators, *Science*, 185, 1181—1183, 1974.

46. **Vaage, J., Bjertnaes, L., and Hauge, A.,** The pulmonary vasoconstrictor response to hypoxia; effects of inhibitors of prostaglandin biosynthesis, *Acta Physiol. Scand.*, 95, 95—101, 1975.

47. **Weir, E. K., McMurtry, I. F., Tucker, A., Reeves, J. T., and Grover, R. F.,** Prostaglandin synthetase inhibitors do not decrease hypoxic pulmonary vasoconstriction, *J. Appl. Physiol.*, 41, 714—718, 1976.

48. **Hales, C. A., Rouse, E., Buchwald, I. A., and Kazemi, H.,** Role of prostaglandins in alveolar hypoxic vasoconstriction, *Resp. Physiol.*, 29, 151—162, 1977.

49. **Lindsey, H. E. and Wyllie, J. H.,** Release of prostaglandins from embolized lungs, *Br. J. Surg.*, 57, 738—741, 1970.

50. **Piper, P. J. and Vane, J. R.,** The release of prostaglandin from lung and other tissues, *Ann. N. Y. Acad. Sci.*, 180, 363—385, 1971.

51. **Palmer, M. A., Piper, P. J., and Vane, J. R.,** Release of rabbit aorta contracting substance (RCS) and prostaglandins induced by chemical or mechanical stimulation of guinea-pig lungs, *Br. J. Pharmacol.*, 49, 226—242, 1973.

52. **Pavek, K., Piper, P. J., and Smedegård, G.,** Anaphylatoxin-induced shock and two patterns of anaphylactic shock — hemodynamics and mediators, *Acta Physiol. Scand.*, 105, 393—403, 1979.

53. **Fountain, S. W., Martin, B. A., Musclow, C. E., and Cooper, J. D.,** Pulmonary leukostasis and its relationship to pulmonary dysfunction in sheep and rabbits, *Circ. Res.*, 46, 175—180, 1980.

54. **Craddock, P. R., Hammerschmidt, D. E., White, J. G., Dalmasso, A. P., and Jacob, H. S.,** Complement (C5a)-induced granulocyte aggregation *in vitro*. A possible mechanism of complement-mediated leukostasis and leukopenia, *J. Clin. Invest.*, 60, 260—264, 1977.

55. **Feuerstein, N. C. and Ramwell, P. W.,** *In vivo* and *in vitro* effects of endotoxin on prostaglandin release from rat lung, *Br. J. Pharmacol.*, 73, 511—516, 1981.

56. **Dahlen, S. E., Hedqvist, P., Hammarström, S., and Samuelsson, B.,** Leukotrienes are potent constrictors of human bronchi, *Nature (London)*, 288, 484—486, 1980.

57. **Ghelani, A. M., Holroyde, M. C., and Sheard, P.,** Response of human isolated bronchial and lung parenchymal strips to SRS-A and other mediators of asthmatic bronchostriction, *Br. J. Pharmacol.*, 71, 107—112, 1980.

58. **Hedqvist, P., Dahlen, S. E., Gustafsson, L., Hammarström, S., and Samuelsson, B.,** Biological profile of leukotrienes C_4 and D_4, *Acta Physiol. Scand.*, 110, 331—333, 1980.

59. **Smedegård, G., Hedqvist, P., Dahlen, S.-E., Revenas, B., Hammarström, S., and Samuelsson, B.,** Leukotriene C_4 affects pulmonary and cardiovascular dynamics in monkeys, *Nature (London)*, 295, 327—329, 1982.

60. **Folco, G., Omini, C., Rossoni, G., Vigano, T., and Berti, F.,** Anticholinergic agents prevent guinea pig airway constriction induced by histamine, bradykinin and leukotriene C_4. Relationship to circulating TxA_2, *Eur. J. Pharmacol.*, 78, 159—165, 1982.

61. **Orehek, J., Douglas, J. S., Lewis, A. J., and Bouhuys, A.,** Prostaglandin regulation of airway smooth muscle tone, *Nature (London) New Biol.*, 245, 84—85, 1973.

62. **Gryglewski, R. J., Dembinska-Kiec, A., Grodzinska, L., and Panczenko, B.,** Differential generation of substances with prostaglandin-like and thromboxane-like activities by guinea pig trachea and lung strips, in *Lung Cells in Disease*, Bouhuys, A., Ed., Elsevier/North-Holland, Amsterdam, 1976, 289—307.

63. **Yen, S. S., Mathé, A. A., and Dugan, J. J.,** Release of prostaglandins from healthy and sensitized guinea pig lung and trachea by histamine, *Prostaglandins*, 11, 227—239, 1976.

64. **Burka, J. F. and Paterson, N. A. M.,** Evidence of lipoxygenase pathway involvement in allergic tracheal contraction, *Prostaglandins*, 19, 499—515, 1980.

65. **Alabaster, V. A. and Bakhle, Y. S.,** Release of smooth muscle contracting substances from isolated perfused lungs, *Eur. J. Pharmacol.*, 35, 349—360, 1976.

66. **Bakhle, Y. S. and Smith, J. W.,** Release of spasmogens from rat isolated lungs by tryptamines, *Eur. J. Pharmacol.*, 46, 31—39, 1977.

67. **Berti, F., Folco, G. C., and Omini, C.,** Pharmacological control of thromboxane A_2 in lung, *Clin. Resp. Physiol.*, 17, 509, 1981.

68. **Piper, P. J. and Samhoun, M. N.,** The mechanism of action of leukotrienes C_4 and D_4 in guinea-pig isolated perfused lung and parenchymal strips of guinea pig, rabbit and rat, *Prostaglandins*, 793—803, 1981.

69. **Sirois, P., Engineer, D. M., Piper, P. J., and Moore, E. G.,** Comparison of rat, mouse, guinea-pig and human slow reacting substance of anaphylaxis (SRS-A), *Experientia*, 35, 361—363, 1979.

70. **Alabaster, V. A.,** Metabolism of arachidonic acid and its endoperoxide (PGH_2) to myotropic products in guinea-pig and rabbit isolated lungs, *Br. J. Pharmacol.*, 69, 479—489, 1980.

71. **Schulman, E. S., Newball, H. H., Demers, L. M., Fitzpatrick, F. A., and Adkinson, N. F.,** Anaphylactic release of thromboxane A_2, prostaglandin D_2 and prostacyclin from human lung parenchyma, *Am. Rev. Resp. Dis.*, 124, 402—406, 1981.

72. **Vargaftig, B. B. and Dao-Hai, N.,** Selective inhibition by mepacrine of rabbit aorta contracting substance evoked by the administration of bradykinin, *J. Pharm. Pharmacol.*, 24, 159—161, 1972.

73. **Martin, G. L., Atkins, P. C., Dunsky, E. H., and Zweiman, B.,** Effects of theophylline, terbutaline, and prednisone on antigen-induced bronchospasm and mediator release, *J. Allergy Clin. Immunol.*, 66, 204—212, 1980.

74. **Pare, P. D. and Hogg, J. C.,** Pathophysiological basis for inhaled steroid treatment in the bronchi, in *Topical Steroid Treatment of Asthma and Rhinitis*, Mygind, N. and Clark, T. J. H., Eds., Baillerie Tindall, London, 1980, 12.

75. **Massion, W. H., Rosenbluth, B., and Kux, M.,** Protective effect of methyl prednisolone against lung complications in endotoxin shock, *South. Med. J.*, 65, 941—994, 1972.

76. **Sladen, A.,** Methyl prednisolone: pharmacologic doses in shock lung syndrome, *J. Thorac. Cardiovasc. Surg.*, 71, 800—806, 1976.

77. **Brigham, K. L., Bowers, R. E., and McKeen, C. R.,** Methyl prednisolone prevention of increased lung vascular permeability following endotoxemia in sheep, *J. Clin. Invest.,* 67, 1103—1110, 1981.

78. **Blackwell, G. J., Carnuccio, R., Di Rosa, M., Flower, R. J., Parente, L., and Persico, P.,** Macrocortin; a polypeptide causing the anti phospholipase effect of glucocorticoids, *Nature (London),* 287, 147—149, 1980.

79. **Hirata, F., Schiffmann, E., Venkatasubramanian, K., Salomon, D., and Axelrod, J.,** A phospholipase A_2 inhibitory protein in rabbit neutrophils induced by glucocorticoids, *Proc. Natl. Acad. Sci. U.S.A.,* 77, 2533—2536, 1980.

80. **Nijkamp, F. P., Flower, R. J., Moncada, S., and Vane, J. R.,** Partial purification of rabbit aorta contracting substance-releasing factor (RCS-RF) and inhibition of its activity by anti-inflammatory steroids, *Nature (London),* 263, 479—482, 1976.

81. **Engineer, D. M., Morris, H. R., Piper, P. J., and Sirois, P.,** The release of prostaglandins and thromboxanes from guinea-pig lung by slow reacting substance of anaphylaxis and its inhibition, *Br. J. Pharmacol.,* 64, 211—218, 1978.

82. **Paajanen, H., Männistö, J., and Uotila, P.,** Aspirin inhibits arachidonic acid metabolism via lipoxygenase and cyclo-oxygenase in hamster isolated lungs, *Prostaglandins,* 23, 731—741, 1982.

83. **Uotila, P., Hiltunen, A., Mäkilä, K., Männistö, J., Toivonen, H., and Hartiala, J.,** The metabolism of arachidonic acid in isolated hamster, rat and guinea-pig lungs, *Prostaglandins Med.,* 5, 149—162, 1980.

84. **Uotila, P. and Männistö, J.,** The metabolism of arachidonic acid is changed by dipyridamole in isolated hamster lungs, *Prostaglandins Med.,* 7, 19—28, 1981.

85. **Al-Ubaidi, F. and Bakhle, Y. S.,** Metabolism of vasoactive hormones in human isolated lung, *Clin. Sci.,* 58, 45, 1980.

86. **Hartiala, J., Toivonen, H., and Uotila, P.,** Demonstration of PGI_2 production by isolated perfused rat lungs with platelet aggregation test, *Prostaglandins,* 20, 127—134, 1980.

87. **Boyd, J. A. and Eling, T. E.,** Prostaglandin release and the interaction of platelets with the pulmonary vasculature of rat and guinea pig, *Thromb. Res.,* 19, 239—248, 1980.

88. **Korbut, R., Boyd, J., and Eling, T. E.,** Prostacyclin and thromboxane A_2 release in isolated rat lungs, *Prostaglandins,* 23, 67—75, 1982.

89. **Berry, E. M., Edmonds, J. F., and Wyllie, J. H.,** Release of prostaglandin F_2 and unidentified factors from ventilated lungs, *Br. J. Surg.,* 58, 189—192, 1971.

90. **Gryglewski, R. J.,** The lung as a generator of prostacyclin, in *Metabolic Activities of the Lung,* Ciba Foundation Symp. 78, Porter, R. and Whelan, J., Eds., Excerpta Medica, New York, 1980, 147—164.

91. **Splawinski, J. S. and Gryglewski, R. J.,** Release of prostacyclin by the lung, *Clin. Resp. Physiol.,* 17, 553—569, 1981.

92. **Alexander, J. M., Nyby, M. D., and Jasberg, K. A.,** Prostaglandin synthesis inhibition restores hypoxic pulmonary vasoconstriction, *J. Appl. Physiol.,* 42, 903—908, 1977.

93. **Nakano, J. and McCloy, R. B., Jr.,** Effects of indomethacin on the pulmonary vascular and airway resistance responses to pulmonary microembolization, *Proc. Soc. Exp. Biol. Med.,* 143, 218—221, 1973.

94. **Bø, G., Hognestad, J., and Vaage, J.,** The role of blood platelets in pulmonary response to microembolization with barium sulphate, *Acta Physiol. Scand.,* 90, 244—251, 1974.

95. **Vaage, J. and Piper, P. J.,** The release of prostaglandin-like substances during platelet aggregation and pulmonary microembolism, *Acta Physiol. Scand.,* 94, 8—13, 1975.

96. **Tucker, A., Weir, E. K., Reeves, J. T., and Grover, R. F.,** Pulmonary microembolism: attenuated pulmonary vasoconstriction with prostaglandin inhibitors and anti-histamines, *Prostaglandins,* 11, 31—41, 1976.

97. **Anderson, F. L., Tsagaris, T. J., Jubiz, W., and Kuida, H.,** Prostaglandin F and E levels during endotoxin-induced pulmonary hypertension in calves, *Am. J. Physiol.,* 228, 1479—1482, 1975.

98. **Fletcher, J. R. and Ramwell, P. W.,** Modification by aspirin and indomethacin of the haemodynamic and prostaglandin releasing effects of *E. coli* endotoxin in the dog, *Br. J. Pharmacol.,* 61, 175—181, 1977.

99. **Parratt, J. R. and Sturgess, R. M.,** The possible roles of histamine, 5-hydroxytryptamine and prostaglandin $F_{2\alpha}$ as mediators of the acute pulmonary effects of endotoxin, *Br. J. Pharmacol.,* 60, 209—219, 1977.

100. **Cook, J. A., Wise, W. C., and Halushka, P. V.,** Elevated thromboxane levels in rat during endotoxin shock. Protective effects of imidazole, 13-azoprostanoic acid and essential fatty acid deficiency, *J. Clin. Invest.,* 65, 227—230, 1979.

101. **Brigham, K. L. and Ogletree, M. L.,** Effects of prostaglandins and related compounds on lung vascular permeability, *Clin. Resp. Physiol.,* 17, 703—722, 1981.

102. **Demling, R. H., Smith, M., Gunther, R., Flynn, J. T., and Gee, M. H.,** Pulmonary injury and prostaglandin production during endotoxemia in conscious sheep, *Am. J. Physiol.,* 240, H348—H353, 1981.

103. **Ogletree, M. L., Oates, J. A., Brigham, K. L., and Hubbard, W. C.,** Evidence for pulmonary release of 5-hydroxy eicosatetraenoic acid (5 HETE) during endotoxemia in unanesthetized sheep, *Prostaglandins,* 23, 459—468, 1982.

104. **Liebig, R., Bernauer, W., and Peskar, B. A.**, Release of prostaglandins, a prostaglandin metabolite slow reacting substance and histamine from anaphylactic lungs and its modification by catecholamines, *Naunyn-Schmiedeberg's Arch. Pharmakol.*, 284, 279—293, 1974.

105. **Mathe, A. A. and Levine, L.**, Release of prostaglandins and metabolites from guinea-pig lung; inhibition by catecholamines, *Prostaglandins*, 4, 877—890, 1973.

106. **Dawson, W., Boot, J. R., Cockerill, A. F., Mallen, D. N. B., and Osborne, D. J.**, Release of novel prostaglandins and thromboxanes after immunological challenge of guinea-pig lung, *Nature (London)*, 262, 699—702, 1976.

107. **Benzie, R., Boot, J. R., and Dawson, W.**, A preliminary investigation of prostaglandin synthetase activity in normal sensitized and challenged sensitized guinea-pig lung, *J. Physiol. (London)*, 246, 80P—81P, 1975.

108. **Boot, J. R., Cockerill, A. F., Dawson, W., and Osborne, D. J.**, Modification of prostaglandin and thromboxane release by immunological sensitization and successive immunological challenges from guinea-pig lungs, *Int. Arch. Allergy Appl. Immunol.*, 57, 159—164, 1978.

109. **Engineer, D. M., Niederhauser, U., Piper, P. J., and Sirois, P.**, Release of mediators of anaphylaxis; inhibition of prostaglandin synthesis and the modification of release of slow reacting substance of anaphylaxis and histamine, *Br. J. Pharmacol.*, 62, 61—66, 1978.

110. **Nijkamp, F. P. and Ramakers, A. G. M.**, Prevention of anaphylactic bronchoconstriction by a lipoxygenase inhibitor, *Eur. J. Pharmacol.*, 62, 121—122, 1980.

111. **Berti, F., Folco, G. C., Nicosia, S,. Omini, C., and Parsargiklian, R.**, The role of histamine H_1 and H_2 receptors in the generation of thromboxane A_2 in perfused guinea-pig lungs, *Br. J. Pharmacol.*, 65, 629—633, 1979.

112. **Änggård, E., Bergoqvist, U., Höberg, B., Johansson, K., Thon, I. L., and Uvnäs, B.**, Biologically active principles occurring on histamine release from cat paw, guinea-pig lung and isolated rat mast cells, *Acta Physiol. Scand.*, 59, 97—110, 1963.

113. **Lefort, J. and Vargaftig, B. B.**, Role of platelets in aspirin-sensitive bronchoconstriction in the guinea-pig; interactions with salicylic acid, *Br. J. Pharmacol.*, 63, 35—42, 1978.

114. **Needleman, P., Minkes, M. S., and Douglas, J. R.**, Stimulation of prostaglandin biosynthesis by adenine nucleotides; profile of prostaglandin release by perfused organs, *Circ. Res.*, 34, 455—460, 1974.

115. **Hemker, D. P., Shebuski, R. J., and Aiken, J. W.**, Release of a prostacyclin-like substance into the circulation of dogs by intravenous adenosine-5'-diphosphate, *J. Pharm. Exp. Ther.*, 212, 246—752, 1980.

116. **Seale, J. P. and Piper, P. J.**, Stimulation of arachidonic acid metabolism by human slow reacting substances, *Eur. J. Pharmacol.*, 52, 125—128, 1978.

117. **Seeger, W., Wolf, H., Stähler, G., Neuhof, H., and Röka, L.**, Increased pulmonary vascular resistance and permeability due to arachidonate metabolism in isolated rabbit lung, *Prostaglandins*, 23, 157—173, 1982.

118. **Piper, P. J. and Samhoun, M. N.**, Stimulation of arachidonic acid metabolism and generation of thromboxane A_2 by leukotrienes B_4, C_4 and D_4 in guinea-pig lung *in vitro*, *Br. J. Pharmacol.*, 77, 267—275, 1982.

119. **Sirois, P., Borgeat, P., Jeanson, A., Roy, S., and Girard, G.**, The action of leukotriene B_4 (LTB$_4$) on the lung, *Prostaglandins Med.*, 5, 429—444, 1980.

120. **Vargaftig, B. B. and Dao-Hai, N.**, Release of vasoactive substances from guinea-pig lungs by slow reacting substance C and arachidonic acid, *Pharmacology*, 6, 99—108, 1971.

121. **Babilli, S. and Vogt, W.**, Nature of the fatty acids acting as "slow reacting substance" (SRS-C), *J. Physiol. (London)*, 177, 919, 1965.

122. **Fredholm, B. and Strandberg, K.**, Release of histamine and formation of smooth muscle stimulating principles in guinea-pig lung tissue induced by antigen and bee venom phosphatidase A, *Acta Physiol. Scand.*, 76, 446—457, 1969.

PHOSPHOLIPASES: SPECIFICITY

Carmen Vigo

Phospholipases are hydrolytic enzymes widely distributed in tissues of animals and plants.[1] The most common types of phospholipases in animal cells are the A and C type, classified according to their specificity[2] (Figure 1). Other types of mammalian phospholipases are phospholipases B and D and lysophospholipase A_1 and A_2.

PHOSPHOLIPASE A

In 1877, Bokay[3] observed that a pancreatic ferment acted upon lecithin to break it down into three components. Later, phospholipase A was found in the pancreas which can be distinguished from other pancreatic enzymes by its remarkable heat stability.[4] The presence of phospholipase A has been demonstrated in most mammalian tissues as well as in the venoms of bees and snakes.[5-14]

The pancreas of various mammals contains two kinds of phospholipase A: the heat labile phospholipase A_1 and the heat stable phospholipase A_2. Phospholipase A_1 (PLA_1) catalyzes the release of a fatty acid from the 1-position of an sn-3-phosphoglyceride such as lecithin, while phospholipase A_2 (PLA_2) catalyzes the specific hydrolysis of the fatty acid ester linkage of the 2-position.[2]

PLA_2 is the most commonly studied of the two phospholipases because in addition to its role in metabolic phospholipid turnover, the enzyme is involved in a series of vital physiological regulatory processes,[17-20] including platelet aggregation,[17] cardiac contraction and excitation,[18] prostaglandin biosynthesis,[19] and aldosterone-dependent sodium transport.[20] PLA_2 has been isolated and purified from a variety of reptilia as well as mammalian sources.[1] Structural studies of this pancreatic enzyme[21] have revealed that an active site His-48 arginine residue and Ca^{2+} are required for catalytic activity as well as Ca^{2+}.[22,23] This enzyme acts stereospecifically on all common types of 3-sn-phosphoglycerides hydrolyzing exclusively fatty acid ester bonds at the glycerol-C-2 position regardless of chain length or degree of unsaturation.[13] The sn-glycero-I-phosphoryl isomers are not at all hydrolyzed and racemic substrates are broken down by only 50%.[13] The snake venom phospholipase A slowly attacks very acidic phospholipids and becomes more active only upon addition of positively charged amphipathic molecules.[24] A reverse situation is found for the pancreatic enzymes. Human and porcine pancreatic phospholipase A attacks lecithin molecules rather slowly.[12] Addition of deoxycholate to the lecithin dispersion greatly enhances the activity of the enzyme.[13] This may result from the increased negative charge, as evidenced by the high susceptibility toward the enzyme of anionic phospholipids, such as phosphatidic acid and its monomethylester phosphatidylglycerol, cardiolipin, and, to a lesser degree, phosphatidylserine and phosphatidyl ethanolamine. With these substrates, addition of deoxycholate, though improving the quality of the dispersion, had either no stimulatory effect (phosphatidyl ethanolamine) or even inhibited the breakdown (phosphatidic acid, cardiolipin). The use of other negatively charged detergents such as sodium dodecyl sulfate (SDS) or sodium taurocholate stimulate hydrolysis of most types of lecithin but inhibit hydrolysis of lecithins containing butyric and oleic acid.[13]

Under conditions which allow a complete hydrolysis of anionic phospholipids (such as phosphatidic acid, phosphatidylglycerol, and cardiolipin) phospholipase A does not hydrolyze 2-acyl lysolecithin, glycol analogs of lecithin, phosphatidyl ethanolamine, or lecithin belonging to the β-series in which the phosphoryl nitrogenous moiety is attached at the secondary glycerol-OH group.[13] However, these lipids can be hydrolyzed by pancreatic

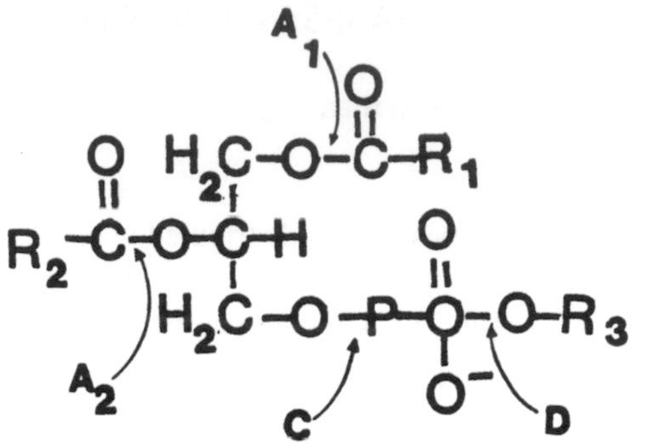

FIGURE 1.   Specificity of phospholipases.

phospholipase A using prolonged incubation periods (10- to 100-fold) or increased concentration of enzyme. Sphingomyelin fully resists enzymic breakdown.[13]

Pancreatic phospholipase A, like the snake venom enzyme, also requires at least one free phosphate ionization for its activity. Blocking this final phosphate-OH group (e.g., in phosphorus triesters) makes the compound unacceptable as a substrate for the enzyme.

In conclusion, the mammalian enzyme and the enzyme from snake venom both require the same minimum structural elements to exert their action, namely a fatty acid ester bond in a position adjacent to the alcohol-phosphate ester bond. In contrast to the snake venom enzyme, porcine pancreatic phospholipase A preferentially attacks anionic phospholipids.

PHOSPHOLIPASE C

In contrast to phospholipase A, phosphatidyl inositol-specific phospholipase C does not require Ca^{2+} for its activity. Phospholipase C has an acid optimum pH and although it acts specifically on phosphatidyl inositol, it can also act on phosphatidyl serine, phosphatidyl ethanolamine, and phosphatidyl glycerol.[25]

Phospholipase C hydrolyzes type 3 bonds releasing a diglyceride and phosphorylcholine (Figure 1). Diglyceride lipase in turn hydrolyzes the diglyceride to release free fatty acids.[25] Diglyceride lipase, which was initially found in α-toxin of *Clostridium welchii* and other bacterias, and later in most mammalian tissues, has a pH optimum of 7, and is stimulated by Ca^{2+} and reduced glutathione. In platelets, diglyceride lipase liberates 31 μmol of fatty acid per milligram of platelet particulate protein, and has sufficient activity to account for the 5 to 10 μmol of arachidonate released per 10^9 platelets upon thrombin stimulation. The lipase activity found in platelets can also hydrolyze the 1-position fatty acid.[25,26]

PHOSPHOLIPASE B AND D

Other phospholipases found in mammalian tissues are phospholipases B and D. Phospholipase B, which is secreted by the pancreas, hydrolyzes type 1 and 2 bonds while phospholipase D hydrolyzes type 4 bonds releasing phosphatidic acid and free choline. Phospholipase D can simultaneously remove the choline and add an alcohol in its place.

MECHANISMS OF INHIBITION OF PHOSPHOLIPASES

In view of the importance of phospholipases in controlling vital physiological regulatory processes, it is not surprising that in recent years they have been the subject of intensive research including purification, structure, and kinetic and modulation studies using a great

Table 1
EFFECTS OF LOCAL ANESTHETICS (5 m*M*) ON PLA₂ AND
LYSOPHOSPHOLIPASE ACTIVITIES IN RELATION TO THE
LIPID SOLUBILITIES OF THESE DRUGS[36]

Drug	Log P Octanol/water	PLA$_2$ activity in human seminal plasma[a]	Lysophospholipase activity in rat liver cytosol[a]
		% Inhibition of	
Chlorpromazine	5.3	100	100
Dibucaine	4.4	100	98
Tetracaine	3.7	86	81
Lidocaine	2.3	12	10
Cocaine	2.3	6	20
Procaine	1.9	1	3

[a] Correlation coefficients between log partition coefficients and percent inhibition statistically significant ($p < 0.001$).

variety of metals and drugs. Inhibitors of cellular phospholipases could serve to regulate the production in both normal and inflammatory processes.[27] Neutral active Ca^{2+}-dependent phospholipase A_2 accumulates in inflammatory sites and is thought to be an important mediator in inflammatory reactions. This enzyme hydrolyzes phospholipids releasing 1-acyl-lysophospholipids, which are cytotoxic and produce intense inflammation when injected intraperitoneally[28,29] and free fatty acids which are known to be membrane lytic. Some fatty acids, such as arachidonic acid, serve as precursors of prostaglandins, leukotrienes, and other eicosanoid derivatives which have a myriad of effects on membranes during inflammation.

Phospholipids (substrates for phospholipases) possess different physical properties which will determine their susceptibility toward phospholipases. These include charge, type of fatty acid, degree of unsaturation, and fatty acid chain length.[30,31] Model membranes and purified phospholipases from different sources have been extensively used to study enzyme kinetics, susceptibility toward different phospholipid subclasses, and effects of temperature, pressure, and charge of the phospholipid bilayer.[30-34] Using model membranes it is possible to elucidate the mechanisms by which drugs might affect phospholipases, either via direct interaction with the enzyme or by interaction with the bilayer.[30-36]

INHIBITION OF PHOSPHOLIPASE A₂ BY ANESTHETICS

Local and general anesthetics have been found to inhibit PLA_2, both in model systems using purified PLA_2 and in biological systems. The mechanism of inhibition of these anesthetics has also been studied.[34-37]

Anesthetics penetrate into lipid interfaces producing a positively charged interphase. Leo et al.[38] showed that the degree of penetration of different anesthetics into lecithin monolayers correlated with the degree of anesthetic potency and with the degree of inhibition of PLA_2[36] (Table 1). Several years ago, various groups demonstrated that the local anesthetics lidocaine and procaine inhibit pancreatic phospholipase A_2 in liposomal, mitochondrial, and other membrane systems.[39,40] Hendrickson and Van Dam-Mieras[41] demonstrated that these anesthetics (lidocaine and procaine) which are the least surface active local anesthetics, inhibit pancreatic PLA_2 action by interacting with the enzyme in the bulk phase. The local anesthetics with more surface activity, dibucaine and tetracaine, might also interact with the enzyme in the bulk phase, but no experimental evidence for this exists. However, since these more surface-active anesthetics penetrate readily into the substrate interface, they might inhibit the surface effects at concentrations where the bulk phase is minimal. Willman and Hendrickson,[35] using local anesthetic analogs, found that inhibition of PLA_2 by these anesthetics resulted from surface charge effects alone and not from any other effects related to structure or molecular spacing in the monolayers. Evidence for this include the findings that (1) mixed tetracaine analog, didecanoyl PC film, had the same optimum pressure as the pure lipid fiber, (2) the induction times are identical in the absence or presence of anesthetic analog (excluding the possibility of inhibition produced by different degrees of penetration of the pancreatic enzyme resulting from differences in molecular spacing,[50] and (3) inhibition by anesthetics decreases with increasing pH and decreasing surface charge.[35]

Hendrickson and Van Dam-Mieras[41] also showed that inhibition of PLA_2 activity by local anesthetics is not caused by interference with the active site but by the enzyme-anesthetic binding to the bilayer which then inhibits interaction between the positively charged enzyme and the interface characteristic of decreased penetration. For example, dibucaine and tetracaine, the most surface active anesthetics, which are soluble in the subphase, had long induction times (>10 min). These long induction times may have resulted from a repulsion between the positively charged interface, created by the anesthetic molecules that have been absorbed by the surface, and by the positively charged penetration site of the enzyme, created by the binding of the solubilized anesthetic monomers in the subphase. With the insoluble anesthetics in the subphase, penetration of the phospholipase could be detected by an immediate rise in surface pressure after enzyme injection under both the mixed and pure lipid monolayers. It is possible that positively charged failure does not necessarily inhibit penetration of the enzyme, but that inhibition results from an interaction between the positively charged anesthetic in the interface and the penetrated enzyme.

Willman and Hendrickson[35] have also shown that both the *Crotalus adamanteus* and porcine pancreatic PLA_2 enzymes are equally affected by the positive surface charge, although they have different optimum surface and cut-off pressures. According to Demel et al.,[42] the pancreatic enzyme has an optimum pressure around 10 nN/m and a cut-off pressure of approximately 18 nN/m, while *C. adamanteus* has an optimum surface pressure of 12 to .75 nN/m and cut-off pressure of 23 nN/m. These observations support the hypothesis that inhibition might be due to anesthetic-penetrated enzyme interactions, since two PLA_2 enzymes with different penetrative abilities are equally affected by the positive interface.

Studies by Kunze et al.[36] also support the hypothesis that local anesthetics inhibit PLA_2 by binding both substrate and enzyme. They suggest that the anesthetics inhibit the enzyme activity by displacing Ca^{2+} from the enzyme-Ca^{2+} complex.

Similar effects to those for local anesthetics were found for the general anesthetics ethrane, halothane, and trichloroethylene.[34] These molecules were found to inhibit PLA_2 from *Naja naja* venom by direct interaction with both the substrate and enzyme. This inhibition does not appear to be due to Ca^{2+} sequestration as it could not be reversed by a tenfold increase in Ca^{2+} concentration.

Anesthetics also inhibit PLA_2 in platelets. The local anesthetics dibucaine, tetracaine, benzocaine, QX572, and chlorpromazine and propanolol were found to be very effective inhibitors of malondialdehyde (MDA) formation in platelets induced by thrombin (Table 2), collagen, and A23187, but they were without effect on the conversion of exogenous arachidonic acid to MDA.[37] These results indicate that these drugs do not affect cyclooxygenase and/or thromboxane synthetases that are responsible for MDA formation (see contribution by Willis, in this volume of the handbook).

Local anesthetics inhibit platelet aggregation and the secretion of platelet storage granule contents elicited by ADP, thrombin, collagen, and A23187. Dibucaine and tetracaine also inhibit the rapid burst in platelet O_2 consumption brought about by platelet aggregating agents in the presence of antimycin A, which is due to oxidation of endogenous arachidonic acid.[37]

Table 2
DISTRIBUTION OF RADIOACTIVITY IN ^{14}C-ARACHIDONIC ACID-LABELED PLATELETS

	CPM (total phospholipids)	CPM (total arachidonic acid metabolites)	Total CPM
Control	17,860 (73%)	6,580 (27%)	24,440
Thrombin (THR)	10,478 (42%)	14,478 (58%)	24,835
THR + dibucaine	17,853 (78%)	5,040 (22%)	22,893
THR + chlorpromazine	20,727 (75%)	7,060 (25%)	27,803

Note: Washed platelet suspensions (1.0 mℓ containing 3 mg platelet protein) previously prelabeled with ^{14}C-arachidonic acid were incubated at 37°C with thrombin (5 U/mℓ) alone, or thrombin plus dibucaine (1.0 mM) or chlorpromazine (0.2 mM). After 5 min, platelet suspensions were extracted and chromatographed and the radioactivity in phospholipids and arachidonic acid metabolites was determined.

The two enzymes proposed to be principally involved in arachidonic acid release from platelets are PLA$_2$ and PLC. It thus seems likely that in platelets the local anesthetics inhibit both enzymes.

In seminal vesicle homogenates, tetracaine also inhibits overall endogenous production of prostaglandins, but not the conversion of added free arachidonic acid to prostaglandins. Therefore, prostaglandin synthetase activity was again unaffected.[43]

In conclusion, it appears that surface-active anesthetics inhibit phospholipases by interaction with substrate, affecting its surface charge and as a result, enzyme activity.

ANTIMALARIAL AND OTHER AMPHIPHILIC CATIONIC DRUGS: INHIBITION OF PLA AND PLC

Antimalarial drugs of the chloroquine type inhibit lipolytic processes in fat tissue in vitro[44] and phospholipase activity in various tissues.[45,47] Several of these drugs are used in the treatment of rheumatoid conditions.

PLA$_2$ activity assayed against *Escherichia coli* labeled with ^{14}C-oleate was found to be inhibited by mepacrine with an IC$_{50}$ of 33×10^{-5} M, by chloroquine (IC$_{50}$ = 6×10^{-5} M), and by primaquine (IC$_{50}$ = 66×10^{-5} M).[46] Similarly, mepacrine was found to inhibit arachidonic acid release and collagen or ADP-induced platelet aggregation by inhibiting PLA$_2$ and PLC and fibrinogen binding.[42] More than 30 cationic amphiphilic drugs, including chloroquine, 4,4'-bis (diethylaminoethoxy) α,β-diethyl diphenylethane, chlorphentermine, imipramine and chloropromazine, and 1,7-bis(p-aminophenoxy) heptane[48,52] have been reported to cause phospholipid storage in humans, laboratory animals, and cultured cells and schistosomes.[48] The cellular phospholipid content may increase by as much as two- to fourfold in some instances. Morphologically, the polar lipid accumulation appears to be confined to lysosomes which usually have a multilamellar appearance[48] and are rich in phospholipids.[49] A high percentage of these amphiphilic drugs and all the excess of phospholipid was confined to lysosomes in the liver.[50] The accumulation of phospholipid and drug, principally in lysosomes, suggests that direct inhibition of lysosomal phospholipases might be the major mechanism in drug-induced lipidosis. Chloroquine and 4,4'-bis (diethylaminoethoxy) α,β-diethyldiphenylethane are potent in vitro inhibitors of lysosomal phospholipases A and C[51] (Table 3). Three other drugs, amantadine (an anti-influenzal agent), propranolol (a β-adrenergic blocking agent), and tripelennamine (an antihistaminic), also have cationic amphiphilic structures, and inhibit lysosomal phospholipases[53] (Table 4), but they have not yet been reported to cause polar lipid storage.

Table 3
INHIBITION OF PLA AND PLC
BY CHLOROQUINE AND 4,4'-
BIS (DIETHYLAMINOETHOXY)
α,β-
DIETHYLDIPHENYLETHANE[51]

	PLA		PLC	
pH	CLQ (m*M*)	DH (m*M*)	CLQ (m*M*)	DH (m*M*)
4.4	> 10	0.23	0.33	0.20
5.4	0.3	0.04	710	6.00

Note: Concentration of chloroquine (CLQ) or
4,4'-bis (diethylaminoethoxy) α,β-di-
ethyldiphenylethane (DH) required to
produce 50% inhibition of the respective
phospholipase activities in rat liver lysosomes.[51]

Table 4
INHIBITION OF LYSOSOMAL PHOSPHOLIPASES BY
CATIONIC AMPHIPHILIC DRUGS

	IC$_{50}$ (m*M*)		
	Phospholipase		Release of ^{14}C-oleic
Drug	A	C	acid by all pathways
1,7-bis (*p*-Aminophenoxy heptane)	0.01	0.03	0.02
Chlorpromazine	0.03	0.07	0.03
Chlorphentermine	0.21	0.45	0.22
Imipramine	0.23	0.25	0.20
Propranolol	0.25	0.38	0.11
Tripelennamine	0.50	1.25	0.88
Amantadine	3.25	6.75	2.50

Note: Concentration of cationic amphiphilic drugs to produce 50% inhibition of rat
liver lysosomal phospholipases.

Despite the markedly different pharmacological actions of the amphiphilic cationic drugs,
they all have the property of inhibiting lysosomal phospholipases. One of the requirements
for the production of phospholipidosis may be the ability of a given agent to concentrate in
lysosomes. Since all these drugs are weak bases, it seems likely that they tend to accumulate
in lysosomes[50] by the mechanisms described by de Duve et al.[54]

All the inhibitors reported to date are cationic with an amino group at one pole. The
typical inhibitor molecule also has a hydrophobic portion lacking a polar functional group.
It is interesting to note that two of the most potent inhibitors of lysosomal phospholipases,
1,7-bis (*p*-aminoethoxy) heptane and 4,4'-bis (diethylaminoethoxy) α,β-diethyldiphenyl-
ethane, are bipolar, having a hydrophobic central region in the molecule with amino groups
in either end.

The foregoing studies provide firm evidence that the lysosomes are the principal site of

Table 5
**EFFECTS OF ANTIBIOTICS (10
m*M*) ON HYDROLYSIS OF EGG-
YOLK EMULSIONS BY *CR.*
ADAMANTEUS PLA[58]**

Drug	Remaining activity (%)
No addition	100
Sodium cloxacillin	95
Sodium oxacillin	80
Disodium carbenicillin	80
Minocycline hydrochloride	0
Tetracycline hydrochloride	0
Chloramphenicol	61
Dihydrostreptomycin sulfate	145
Mitomycin C	88[a]
Oleandomycin phosphate	56
Erythromycin	44

[a] This value was obtained at 7.5 m*M* drug concentration.

action of these drugs and that inhibition of lysosomal lipid catabolism results in phospholipid storage. The mechanisms of inhibition of phospholipases by some of these drugs have also been studied. Vigo et al.[34] have demonstrated that the antimalarial drugs mepacrine and chloroquine do not interact with the membrane phospholipids, but directly inhibit PLA$_2$. This is supported by studies in lysosomal phospholipases carried out by Hostetler and Matsuzawa[51,53] using cationic amphiphilic drugs, and those of Ruth et al.[55] They found that stabilization was not due to mechanical actions but rather to a direct inhibition of phospholipases. Watts and Atkins[56] observed that 1,7-bis (*p*-aminophenoxy) heptane also stabilizes rat liver lysosomes against osmotic lysis, suggesting that stabilization in this case may be due to blockade of the lytic action of these enzymes on the lipid bilayer.

INHIBITION OF PHOSPHOLIPASES BY ANTIBIOTICS

The effects of antibiotics on PLA, PLB, and PLC activities have been investigated.[57-59] Some antibiotics were found to inhibit PLC synthesis in *Bacillus cereus*,[60] to initiate phospholipase PLA$_2$ activity in human platelets in the presence of A23187, to enhance bee venom phospholipase PLA$_2$ activity by forming a complex between polymyxin B and ovolecithin liposomes,[62] and to activate phospholipase C from *Pseudomonas aeruginosa*[63] in the presence of polymyxin B. Additional experiments have been conducted to elucidate the effects of several antibiotics on the activities of PLA$_2$ from *Crotalus adamanteus*, phospholipase B from *P. notatum*, and phospholipase C from *B. cereus*[58] (Tables 5 and 6). Table 5 shows the inhibition of PLA$_2$ by some of the antibiotics that are widely used in the clinic. Two of these antibiotics, minocycline and tetracycline (10 m*M*), inhibit PLA$_2$ by 100%. This inhibition can be reversed by 10 m*M* Ca^{2+}, which suggests that they act as Ca^{2+} chelators.[64] Another antibiotic, erythromycin, has an IC$_{50}$ of 7.8 m*M*. All these antibiotics, which have different therapeutic effects, are amphiphilic drugs containing a hydrophobic region. Table 6 shows the effects of various antibiotics and copper-free bleomycin hydrochloride on PLA$_2$, PLB, and PLC activities toward mixed micelles of phosphatidylcholine and a nonionic detergent, Triton® X-100. Disodium carbenicillin (50 m*M*), erythromycin, bleomycin hy-

Table 6
EFFECTS OF ANTIBIOTICS (5 m*M*) ON HYDROLYSIS OF
EGG-YOLK PHOSPHATIDYLCHOLINE BY *CR.*
***ADAMANTEUS* PLA$_2$, *P. NOTATUM* PLB, AND *B. CEREUS* PLC**[58]

Drug	Remaining activity (%)		
	PLA$_2$ (*Cr. adamanteus*)	PLB (*P. notatum*)	PLC (*B. cereus*)
No addition	100	100	100
Chloramphenicol	84	141	77
Disodium carbenicillin	99	129	107
Erythromycin	110	94	96
Oleandomycin phosphate	81	103[a]	61[a]
Bleomycin hydrochloride (–Cu)	100	76	46
Polymyxin B sulfate	98	88	107
Chlorpromazine hydrochloride[b]	64	59	63
Dibucaine hydrochloride[b]	110	75	68

[a]   With radioactive substrate. The molecular weights of bleomycin hydrochloride and polymyxin B sulfate were calculated as 1400 and 1200, respectively.
[b]   These local anesthetics were compared with antibiotics.

drochloride (–Cu) and polymyxin B sulfate have no effect on PLA$_2$, while chloramphenicol and oleandomycin phosphate are slightly inhibitory.

With respect to PLB activity, polymyxin B sulfate (IC$_{50}$ 22.5 m*M*) is only slightly effective as compared to chlorpromazine hydrochloride (IC$_{50}$ 7.3 m*M*). Disodium carbenicillin (50 m*M*) and chloramphenicol (20 m*M*) exert 1.5- and 1.9-fold stimulation, respectively. These antibiotics have no effect on the deacylation of mono-acylglycerophospholipids by these enzymes,[57] but inhibit or stimulate the deacylation of diacylglycerophospholipids. Bleomycin hydrochloride has no effect on PLB. PLC is inhibited by chloramphenicol, oleanodomycin phosphate, and bleomycin chloride (–Cu) (Table 6). In these experiments, polymyxin B seems to activate PLC from *B. cereus*, although previously it had been reported to inhibit PLC from *Clostridium perfringens*.[57] These contradictory results might be due to the differences of the charges of the enzyme proteins of *C. perfringens*, an anionic polypeptide (pI 5.2 to 5.6), and *B. cereus*, a cationic polypeptide (pI 8 to 8.1). Chloramphenicol and chlorpromazine hydrochloride inhibit PLC with an IC$_{50}$ of 14.9 and 8.1 m*M*, respectively, while the IC$_{50}$ for bleomycin hydrochloride (–Cu) varies between 0.1 to 5 m*M*.[58] Preincubation of bleomycin hydrochloride (–Cu), 17 m*M*, with *B. cereus* PLC (16 μg protein per milliliter) for 20 min causes 100% inhibition and the presence of 1 m*M* Zn (CH$_3$COO)$_2$ abolishes this effect. In addition, Cu-free pepleomycin sulfate, one component of bleomycin, shows similar effects to bleomycin hydrochloride (–Cu), while pepleomycin sulfate –Cu complex does not inhibit PLC.[58] These results suggest that zinc, found in the metalloenzyme PLC, may participate in the inhibitory action of basic glycopeptide bleomycin hydrochloride (–Cu).

The mechanism of inhibition of phospholipases by these antibiotics is not as yet well understood, but it seems likely that the therapeutic actions of these drugs is brought about by their interactions with phospholipid membranes as suggested by Michell et al.[65] and Defrize-Quertain.[66] On the other hand, the inhibitory effect of tetracyclines and bleomycin hydrochloride (–Cu) might be due to the removal of Ca^{2+} and Zn^{2+} ions from the system, each of which is essential for the activities of PLA$_2$ and PLC, respectively.

Another group of antibiotics, the aminoglycosides, are widely used in clinical medicine

Table 7
INHIBITION OF KIDNEY
CORTEX LYSOSOMAL
PHOSPHOLIPASES BY
AMINOGLYCOSIDE
ANTIBIOTICS[59]

| Inhibitor | IC$_{50}$, mM | | |
	PLA	PLC	Total fatty acid
Amikacin	6.2	4.0	4.1
Dibekacin	3.3	2.0	3.2
Gentamicin	4.2	Ind.	3.5
Tobramycin	0.4	0.2	0.24

because of their efficacy in treatment of infections caused by Gram negative bacteria. They have also been found to inhibit phospholipases. Their use, however, is often complicated by acute nephrotoxicity involving the proximal tubular cells which may result in serious damage to the kidney. Multilamellar bodies (also called myeloid bodies or myelin figures) are a prominent early histologic feature of aminoglycoside nephrotoxicity.[67 69] These antibiotics have been shown to concentrate in the proximal tubular cells and in the kidney lysosomes,[70] and several workers have observed an increase in phospholipid content in the liver during aminoglycoside treatment.[71] These findings led to the conclusion that aminoglycosides might be inhibitors of lysosomal phospholipases.

Hostetler and Hall[59] have studied the effects of these antibiotics on lysosomal PLA and PLC in rat kidney cortex. Four of these antibiotics studied have been shown to be inhibitory (Table 7), although they are not as effective as the cationic amphiphilic drugs.[51,53] These compounds become highly concentrated in the lysosomes of kidney proximal tubular cells[60] and fibroblasts.[72] This mechanism has been postulated to be an important factor in aminoglycoside nephrotoxicity and has been termed the "lysosomal disfunction hypothesis",[71] which has been further supported using soluble lysosomal phospholipase preparations from rat kidney cortex.[59]

It is interesting to point out that streptomycin, which is not nephrotoxic, does not inhibit kidney phospholipases. In addition, Tulken and Van Hoof[73] have shown that phospholipid content of cultured fibroblasts is increased by aminoglycosides in the order amikacin (90%) > gentamycin (36%) > tobramycin (42%), which is the order found for inhibition of kidney lysosomal phospholipases in vitro.[59] There is, therefore, a close parallel between the ability to inhibit phospholipases and the degree of phospholipid storage produced in cell cultures. The molecular mechanism of aminoglycoside inhibition of lysosomal phospholipases is unknown, but according to a theory proposed by Lüllmann-Rauch and co-workers,[48,65,74,75] these drugs form complexes with phospholipids which are less susceptible to degradation by phospholipases.

INHIBITION OF PHOSPHOLIPASES BY NONSTEROIDAL ANTI-INFLAMMATORY AGENTS

Arachidonic acid mobilization and endoperoxide and prostaglandin production during inflammation is well documented.[76,78] Anti-inflammatory agents such as indomethacin and aspirin inhibit prostaglandin synthesis by all cells and tissues thus far examined.[76-79] The

Table 8

**INHIBITION BY INDOMETHACIN OF
PHOSPHOLIPID HYDROLYSIS BY VARIOUS
PLA$_2$ SOURCES[81]**

	% Inhibition with final indomethacin conc. of		
Sources of PLA$_2$	50 μM	100 μM	1000 μM
Russell viper venom	0	37	100
Bee venom	0	33	100
Pig pancreas	0	0	63
Crotalus ademanteus venom	0	0	43
PMN leukocytes	70	100	—

Note: Assays were carried out for 15 min. Assays mixtures of 0.5
mℓ contained 80 mM tris-Hcl (pH 7.5), 10 mM CaCl$_2$.

primary mechanism of action of nonsteroidal anti-inflammatory agents is widely thought to
be inhibition of cyclooxygenase;[77-80] however, in 1978, Kaplan and co-workers[81] showed
that indomethacin also inhibits PLA$_2$ activity of rabbit polymorphonuclear (PMN) leukocytes
in the presence of 10 mM Ca^{2+}. Their results showed that indomethacin at the lowest
concentration inhibitory to prostaglandin synthesis also inhibited a potent leukocyte PLA$_2$.
They also demonstrated that this inhibition is immediate and typical of a noncompetitive
inhibitor. A previous report indicated that indomethacin inhibited hydrolysis of egg lecithin
by snake venom (*Vipera ammodytes*) PLA$_2$, but the concentrations required (0.1 to 1 mM)
were 100 times higher than that needed to be anti-inflammatory.[80] Kaplan et al.[81] studied
the inhibition of PLA$_2$ from different sources by indomethacin (Table 8). Indomethacin at
concentrations which inhibit phospholipase PLA$_2$, 50 μM, in PMN leukocytes, causes no
detectable inhibition on the same enzyme extracted from Russel viper venom, bee venom,
pig pancreas, or *Crotalus adamanteus* venom. The reasons for these differences are still
unknown.

Indomethacin has also been reported to inhibit PLA$_2$ in intact platelets.[82] Purified PLA$_2$
from platelets is inhibited by indomethacin with an ID$_{50}$ of 75 μM. Aspirin, on the other
hand, had no effect on platelet PLA$_2$.

The nonsteroidal anti-inflammatory agents sodium meclofenamate and sodium flufenamate
(analogs of indomethacin) also inhibit PLA$_2$ activity from PMN leukocytes. The ID$_{50}$ is
identical for these two drugs and comparable with that for indomethacin (100 μM)[84] in the
presence of 2.5 mM Ca^{2+}. Thus, it is clear that (with the possible exception of indomethacin)
the inhibition of PLA$_2$ by these drugs is considerably less sensitive by several orders of
magnitude than the well-established inhibition of the cyclooxygenase by these agents.[77,80]

It has also been proposed that nonsteroidal anti-inflammatory agents may modulate intracellular Ca^{2+} and therefore PLA$_2$ activity, since this enzyme is critically dependent on
Ca^{2+} concentrations.[85] Franson and co-workers[84] have examined this possibility. They have
shown that inhibition of PMN leukocyte PLA$_2$ by meclofenamate is indeed dependent on
the concentration of added Ca^{2+}. Thus, with 2.5 mM Ca^{2+} the ID$_{50}$ was 0.4 mM whereas
with 0.5 mM Ca^{2+} the ID$_{50}$ was 50 mM or 10^4 times more sensitive. In similar experiments
using autoclaved *E. coli* as the membranous substrate and purified PLA$_2$ from PMN leukocytes in the presence of 0.5 mM Ca^{2+} the inhibition of this enzyme was 26% with 10
pM meclofenamate and 50% with 50 pM meclopamide. Similar results, although somewhat

less sensitive, were obtained with sodium flufenamate. Aspirin had no effect on PLA_2 activity at all concentrations tested.

The current results indicate that isolated PLA_2 is inhibited by low concentrations of nonsteroidal anti-inflammatory agents. Inhibition by these drugs at picomolar concentrations is Ca^{2+}-dependent and readily detectable with submillimolar Ca^{2+} concentrations. Macrophage cyclooxygenase is inhibited by indomethacin with an ID_{50} of 10 μM.[78] Thus, the anti-inflammatory action of these drugs may be attributed not only to inhibition of cyclooxygenase, but also to direct regulation of PLA_2. Inhibition of platelet, leukocyte, and macrophage PLA_2 activities by nonsteroidal anti-inflammatory agents may contribute to their anti-inflammatory actions.

INHIBITION OF PHOSPHOLIPASE A_2 BY ANTI-INFLAMMATORY STEROIDS

The anti-inflammatory steroids, glucocorticoids, have been successfully used in the treatment of chronic inflammatory diseases, including rheumatoid arthritis, often with rapid and dramatic results.[86-88] Their administration results in the amelioration of the signs and symptoms of chronic inflammation. It seems likely that the action of glucocorticoids is to decrease glucose utilization and to modify the metabolism of lipids and proteins by stimulation of catabolism.[89] These hormones are thought to act primarily by stimulating transcription, thus controlling the rate of synthesis of certain specific key proteins.[90] In several tissues the mechanism of steroid hormone action depends on the combination of the steroid with a cytosolic receptor protein, the translocation of this drug receptor complex to the nucleus, and the initiation of protein biosynthesis.[91-93] Danon and Assouline[94] were able to demonstrate that glucocorticoids inhibited prostaglandin biosynthesis by minced renal papilla incubated in tissue culture medium, but not in the presence of inhibitors of DNA-dependent RNA synthesis. It was later confirmed that inhibition of prostaglandin by glucocorticoids requires protein synthesis.[25,98]

From Danon's work it could not be elucidated whether the inhibition of prostaglandins by glucocorticoids was due to an inhibition of prostaglandin synthetase or decrease of precursor availability as suggested by Gryglewski[90,100a] and Hong and Levine.[100b] Flower[101] was able to show that glucocorticoids inhibit the release of prostaglandin endoperoxides and TXA_2 in the lungs of guinea pigs, without affecting the conversion of arachidonic acid to prostaglandin products.[19,102] From these experiments the authors concluded that dexamethasone induces the production of an uncharacterized "second messenger" which inhibits PLA_2.[95] Recently, these workers reported that this factor, "macrocortin", is an intracellular polypeptide whose release and synthesis is stimulated by steroids.[103] Macrocortin release from leukocytes is induced by hydrocortisone at physiological concentrations, and at a fixed concentration of hydrocortisone the amount of macrocortin released is proportional to the number of cells present. The time course of macrocortin release shows that 30 min after addition of hydrocortisone, half of the total inhibitory activity was released into the medium; by 90 min most inhibitory activity was extracellular, and by 150 min release was virtually complete.[103] This release is totally inhibited by the presence of cyclohexamide, 1 $\mu g/m\ell$. Also, partially purified macrocortin isolated from two sources (steroid-stimulated guinea pig lungs and rat peritoneal leukocytes) have a number of properties in common. Both are stable at 70°C for up to 30 min but are destroyed by boiling and have an apparent mol wt of 15,000.[103] Leukocyte-derived macrocortin inhibited phospholipase activity in the lung as estimated by a radiochemical technique; its activity was abolished by boiling.[103]

A similar PLA_2 inhibitory protein ("lipomodulin") has been isolated from glucocorticoid-treated neutrophils by Hirata and co-workers.[104] Lipomodulin formation was induced by glucocorticoids in a receptor-mediated fashion. These authors suggested that lipomodulin was involved in regulating PLA_2 and in the pathophysiology of rheumatic diseases.[105] They demonstrated that patients with systemic lupus erythematosus, rheumatoid arthritis, dermatomyositis, and periarteritis nodosum spontaneously produce antibodies against lipomodulin. These antibodies abolish lipomodulin inhibition of arachidonic acid (AA) release. In such patients, the antilipomodulin antibody increases the release of arachidonic acid and prostaglandins.

The mechanism of inhibition of PLA_2 by glucocorticoids seems, therefore, to be due to an induction of the transcription of the second messenger. Studies using purified PLA_2 and multibilayer liposomes[34] suggest that direct interaction with the A_2 is not involved.

ADDENDUM

Since this chapter was originally prepared, important new findings have occurred, both in the areas of phospholipase specificity, and in the preparation and analysis of the lipocortins. These references are given in the Addenda to the chapters by Willis and by Willis and Mead.

In addition, there has been a report of a naturally occurring marine product, "manoalide" that is an inhibitor of phosphilipase A_2 and which may have specific effects on this enzyme.[110]

REFERENCES

1. **Brocherhoff, H. and Jensen, R. G.,** *Lipolytic Enzymes,* Academic Press, New York, 1974, 194—243.
2. **Hanahan, D. J.,** in *The Enzymes,* Vol. 5, 3rd ed., Boyer, P. D., Ed., Academic Press, New York, 1971, 71—85.
3. **Bokay, A. Z.,** Uber die Verdaulichkeit des nucleins und lecicthins, *Physiol. Chem.,* 1, 157—164, 1877.
4. **Belfanti, S. and Arnaudi, C.,** A lecithinase of pancreas producing lysolecithin, *Bull. Soc. Int. Microbiol.,* 4, 399—406, 1932.
5. **Scherphof, G. L. and Van Deenen, L. L. M.,** Phospholipase A activity of rat liver mitochondria, *Biochim. Biophys. Acta,* 98, 204—206, 1965.
6. **Gallai-Hatchard, J. J. and Thompson, R. H. S.,** Phospholipase A activity of mammalian tissues, *Biochim. Biophys. Acta,* 98, 128—136, 1965.
7. **Shipolini, R. A., Callewaert, G. L., Cottrell, R. C., Dooman, S., Vernon, C. A., and Banks, B. E.,** Phospholipase A from bee venom, *Eur. J. Biochem.,* 20, 459—468, 1971.
8. **Nutter, L. J. and Privett, O. S.,** Phospholipase A properties of several snake venom preparations, *Lipids,* 1, 258—265, 1966.
9. **Rimon, A. and Shapiro, B.,** Properties and specificity of pancreatic phospholipase A_2, *Biochem. J.,* 71, 620—623, 1959.
10. **Magee, W. L., Gallai-Hatchard, J., Sanders, H., and Thompson, R. H. S.,** The purification and properties of phospholipase A from human pancrease, *Biochem. J.,* 83, 17—25, 1962.
11. **Deenen, L. L. M., van de Haas, G. H., and Heemskerk, C. H.,** Hydrolysis of synthetic mixed-acid phosphatides by phospholipase A from human pancreas, *Biochim. Biophys. Acta,* 67, 295—304, 1963.
12. **van den Bosch, H., Postema, N. M., de Haas, G. H., and van Deenen, L. L. M.,** On the positional specificity of phospholipase A from pancrease, *Biochim. Biophys. Acta,* 98, 657—659, 1965.
13. **de Haas, G. H., Postema, N. M., Nieuwenhuizen, W., and van Deenen, L. L. M.,** Purification and properties of phospholipase A from porcine pancrease, *Biochim. Biophys. Acta,* 159, 103—117, 1968.
14. **White, D. A., Pounder, D. J., and Hawthorne, J. N.,** Phospholipase A_1 activity of guinea pig pancrease, *Biochim. Biophys. Acta,* 242, 92—107, 1971.
15. **Dutilh, C. E., Van Doren, P. J., Verheul, F. E. A. M., and de Haas, G. H.,** Isolation and properties of phospholipase A_2 from ox and sheep pancrease, *Eur. J. Biochem.,* 53, 91—97, 1975.
16. **de Haas, G. H. and van Deenen, L. L. M.,** The site of action of phospholipase A on β-lecithins, *Biochim. Biophys. Acta,* 84, 469—471, 1964.
17. **Pickett, W. C., Jesse, R. L., and Cohen, P.,** Trypsin-induced phospholipase activity in human platelets, *Biochem. J.,* 160, 405—408, 1976.
18. **Giesler, G., Mentz, P., and Forster, W.,** The action of phospholipase A_2 on parameters of cardiac contraction, excitation and biosynthesis of prostaglandins, *Pharm. Res. Commun.,* 9, 117—130, 1977.

19. **Flower, R. J. and Blackwell, G. J.**, The importance of phospholipase A$_2$ in prostaglandin biosynthesis, *Biochem. Pharmacol.*, 25, 285—291, 1976.

20. **Yorio, T. and Bentley, P. L.**, Phospholipase A and the mechanism of action of aldosterone, *Nature (London)*, 271, 79—81, 1978.

21. **Dijskstra, B. W., Drenth, J., Kalk, K. H., and Van der Maelen, P. J.**, Three dimensional structure and disulfide bond connections in bovine pancreatic phospholipase A$_2$, *Biochemistry*, 13, 1466—1554, 1978.

22. **Volwerk, J. J., Pieterson, W. A., and de Haas, G. H.**, Histidine at the active site of phospholipase A$_2$, *Biochemistry*, 13, 1466—1554, 1974.

23. **Vensel, L. A. and Kantrowitz, E. R.**, An essential arginine residue in porcine phospholipase A$_2$, *J. Biol. Chem.*, 255, 7306—7310, 1980.

24. **de Haas, G. H., Bonsen, P. P. M., and van Deenen, L. L. M.**, Studies on cardiolipin 3. Structural identity of ox heart cardiolipin and synthetic diphosphatidyl glycerol, *Biochim. Biophys. Acta*, 116, 114—124, 1966.

25. **Bell, R. L., Donald, A. K., Stanford, N., and Majerus, P. W.**, Diglyceride lipase: a pathway for arachidonate release from human platelets, *Proc. Natl. Acad. Sci. U.S.A.*, 76, 3238—3241, 1979.

26. **Rittenhouse-Simmons, S.**, Production of diglyceride, from phosphatidylinositol in activated human platelets, *J. Clin. Invest.*, 63, 580—587, 1979.

27. **Thouvenot, J. P. and Douste-Blazy, L.**, Phospholipase A activity in ascitic fluids during acute pancreatites. Active phospholipasique A dans les liquides d'epanchement au cours de pancreatites aigues, *Arch. Fr. Mal. App. Dig.*, 63, 479—483, 1974.

28. **Phillips, G. B., Bachner, P., and McKay, D. C.**, Tissue effects of lysolecithin injected subcutaneously in mice, *Proc. Soc. Exp. Biol. Med.*, 119, 846—885, 1965.

29. **Munder, G. B., Modelell, M., Ferbem, E., and Fischer, H.**, in *Mononuclear Phagocytes*, Van Furth, R., Ed., Blackwell Scientific, Oxford, 1970, 445—459.

30. **Demel, R. A., Geurts Van Kessel, W. S. M., Zwaal, R. F. A., Roelofsen, B., and van Deenen, L. L. M.**, Relation between various phospholipase actions on human red cell membranes and the interfacial phospholipid pressure in monolayers, *Biochim. Biophys. Acta*, 382, 169—173, 1975.

31. **de Kruyff, B. and Demel, R. A.**, Polyene antibiotic-sterol interactions in membranes of acholeplasma laidlawaii and lecithin liposomes, *Biochim. Biophys. Acta*, 339, 57—61, 1974.

32. **Opden Kamp, J. A. F., de Gier, J., and van Deenen, L. L. M.**, Hydrolysis of phosphatidylcholine liposomes by pancreatic phospholipase A$_2$ at the transition temperature, *Biochim. Biophys. Acta*, 345, 253—257, 1979.

33. **Linden, C. D., Wright, K. L., McConnell, H. M., and Fox, C. F.**, Lateral phase separations in membrane lipids and the mechanism of sugar transport in *Escherichia coli*, *Proc. Natl. Acad. Sci. U.S.A.*, 70, 2271—2275, 1973.

34. **Vigo, C., Lewis, G. P., and Piper, P. J.**, Mechanisms of inhibition of phospholipase A$_2$, *Biochem. Pharmacol.*, 29, 623—627, 1979.

35. **Willman, C. and Hendrickson, S. H.**, Positive surface charge inhibition of phospholipase A$_2$ in mixed monolayer systems, *Arch. Biochim. Biophys.*, 191, 298—305, 1978.

36. **Kunze, H., Nahas, N., Traynor, J. R., and Wurl, M.**, Effects of local anesthetics on phospholipases, *Biochim. Biophys. Acta*, 441, 93—102, 1976.

37. **Vanerhoek, J. Y. and Feinstein, M. B.**, Local anesthetics, chloropromazine and propranolol inhibit stimulus activation of phospholipase A$_2$ in human platelets, *Mol. Pharm.*, 16, 171—180, 1978.

38. **Leo, A., Hansch, C., and Elkins, D.**, Partition coefficients and their uses, *Chem. Rev.*, 71, 525—616, 1971.

39. **Scherphof, G. L., Scarpa, A., and Van Toorenen Bergen, A.**, The effect of local anesthetics on the hydrolysis of free and membrane bound phospholipids catalyzed by various phospholipases, *Biochim. Biophys. Acta*, 270, 226—240, 1872.

40. **Waite, M. and Sisson, P.**, Effect of local anesthetics on phospholipases from mitochondria and lysosomes. A probe into the role of the calcium ion in phospholipid hydrolysis, *Biochemistry*, 11, 3098—3105, 1972.

41. **Hendrickson, H. S. and Van Dam-Mieras, M. C. E.**, Local anesthetic inhibition of pancreatic phospholipase A$_2$ action on lecithin monolayers, *J. Lipid Res.*, 17, 399—405, 1976.

42. **Demel, R. A., Geurts Van Kessel, W. S. M., Zwaal, R. F. A., Roelfsen, B., and van Deenen, L. L. M.**, Relation between various phospholipase actions on human red cells membranes and the interfacial phospholipid pressure in monolayers, *Biochim. Biophys. Acta*, 406, 97—107, 1975.

43. **Kunze, H., Bohm, E., and Vogt, W.**, Effects of local anesthetics on prostaglandin biosynthesis in vitro, *Biochim. Biophys. Acta*, 360, 260—269, 1974.

44. **Markus, H. B. and Ball, E. G.**, Inhibition of lipolytic processes in rat adipose tissue by antimalarial drugs, *Biochim. Biophys. Acta*, 187, 486—491, 1969.

45. **Blackwell, G. J., Flower, R. J., Nijkamp, F. P., and Vane, J. R.,** Phospholipase A₂ activity of guinea-pig isolated perfused lungs: stimulation and inhibition by anti-inflammatory steroids, *Br. J. Pharmacol.,* 62, 78—79, 1978.

46. **Authi, K. S. and Traynor, J. R.,** Effects of antimalarial drugs on phospholipase A₂, *Proc. Br. Pharmacol. Soc.,* 1, 469P, 1979.

47. **Winocour, P. D., Kimlough-Rathbone, R. L., and Mustard, J. R.,** The effect of phospholipase inhibitor mepacrine on platelet aggregation. The platelet release reaction and fibrinogen binding to the platelet surface, *Thromb. Haemostas.,* 45, 237—263, 1981.

48. **Lüllmann-Rauch, R.,** in *Lysosomes in Applied Biology and Therapeutics,* Vol. 6, Dingle, J. T., Jacques, P. J., and Shaw, J. H., Eds., North Holland, Amsterdam, 1979, 49—130.

49. **Tjiong, H. B., Lepthin, J., and Debuch, H.,** Lysosomal phospholipids from rat liver after treatment with different drugs, *Hoppe Seyler's Z. Physiol. Chem.,* 359, 63—69, 1978.

50. **Matsuzawa, Y. and Hostetler, K. Y.,** Studies on drug-induced lipidosis: subcellular localization of phospholipid and cholesterol in the liver of rats treated with chloroquine or 4,4'-bis(diethylaminoethoxy), *J. Lipid Res.,* 21, 202—214, 1980.

51. **Matsuzawa, Y. and Hostetler, K. Y.,** Inhibition of lysosomal phospholipase A and phospholipase C by chloroquine and 4,4'-bis(diethylaminoethoxy)α,β-diethyldiphenylethane, *J. Biol. Chem.,* 255, 5190—5194, 1980.

52. **Watts, S. D. M., Orpin, A., and MacCormic, C.,** Lysosomes and tegument pathology in the chemotherapy of schistosomiasis with 1,7,-bis(*p*-aminophenoxy)heptane, *Parasitology,* 78, 287—294, 1979.

53. **Hostetler, K. Y. and Matsuzawa, Y.,** Studies on the mechanism of drug-induced lipidosis. Cationic amphopholic drug inhibition of lysosomal phospholipase A and C, *Biochem. Pharmacol.,* 30, 1121—1126, 1981.

54. **de Duve, C., de Barsy, T., Poole, B., Trouet, A., Tulkens, P., and van Hoof, F.,** Lysomotropic agents, *Biochem. Pharmacol.,* 23, 2495—2531, 1974.

55. **Ruth, R. C., Owens, K., and Weglicki, W. B.,** Inhibition of lysosomal lipase by chlorpromazine: a possible mechanism of stabilization, *J. Pharmacol. Exp. Ther.,* 212, 361—367, 1980.

56. **Watts, S. D. M. and Atkins, A. M.,** Effects of the schistosomicide, 1,7-bis (*p*-aminophenoxy)heptane (153C52) on lysosomes and membrane stability, *Biochem. Pharmacol.,* 28, 2579—2583, 1979.

57. **Saito, K., Okada, Y., and Kawasaki, N.,** Inhibitory effect of some antibiotics on phospholipases, *J. Biochem.,* 72, 213—214, 1972.

58. **Sugatani, J., Saito, K., and Honjo, I.,** In vitro actions of some antibiotics on phospholipases, *J. Antibiotics,* 37, 734—739, 1979.

59. **Hostetler, K. Y. and Hall, L. B.,** Inhibition of kidney lysosomal phospholipases A and C by aminoglycoside antibiotics: possible mechanism of aminoglycoside toxicity, *Proc. Natl. Acad. Sci. U.S.A.,* 79, 1663—1667, 1982.

60. **Valle, K. J. and Prydz, H.,** The effect of nalidixic acid, rifampicin and chloramphenicol on the synthesis of phospholipase C in *Bacillus cereus, Acta Pathol. Microbiol. Scand. Sect. B,* 86, 25—28, 1978.

61. **Pickett, W. C., Jesse, R. L., and Cohen, P.,** Initiation of phospholipase A₂ activity in human platelets by the Ca²⁺ ionophore A23187, *Biochim. Biophys. Acta,* 486, 209—213, 1976.

62. **Mollay, C. and Krell, G.,** Enhancement of bee venom phospholipase A₂ activity by mellitin, direct lytic factor from cobra venom and polymyxin B, *FEBS Lett.,* 46, 141—144, 1974.

63. **Kusano, T., Izaki, K., and Takahashi, H.,** In vivo activation by polymyxin B of phospholipase C from *Pseudomonas aeruginosa, J. Antibiot.,* 30, 900—902, 1977.

64. **Rokos, J., Malek, P., Burger, M., Prochazka, P., and Kolc, J.,** The effect of metals on the inhibition of pancreatic lipase by chlortetracyclin, *Antibiotics Chemother.,* 9, 600—608, 1951.

65. **Michelle, R. H., Bowley, D. A. M., and Brindley, D. N.,** A possible metabolic explanation for drug-induced phospholipidosis, *J. Pharm. Pharmacol.,* 28, 331—332, 1976.

66. **Defrise-Overtain, F., Chatelain, P., and Ruyss-Chaert, J. M.,** Phospholipase inactivation induced by aminoperazine derivative: a study at the lipid-water interphase, *J. Pharm. Pharmacol.,* 30, 608—612, 1978.

67. **Kosek, J. C., Mazze, R. J., and Cousins, M. J., Nephrotoxicity** of gentamycin, *Lab. Invest.,* 36, 48—57, 1974.

68. **Houghton, D. C., Harnett, M., Campbell-Boswell, M. V., Porter, G., and Bennett, W. M.,** A light and electron microscopic analysis of gentamycin nephrotoxicity in rats, *Am. J. Pathol.,* 82, 589—612, 1976.

69. **Houghton, D. C., Campbell-Boswell, M. V., Bennett, W. M., Porter, A. J., and Brooks, R. E.,** Myeloid bodies in the renal tubules of humans: relationship of gentamycin therapy, *Clin. Nephrol.,* 10, 140—145, 1978.

70. **Morin, J. P., Viotte, G., Vandewalle, A., van Hoof, F., Tulkens, P., and Fillastre, J. P.,** Gentamicin-induced nephrotoxicity: a cell biology approach, *Kidney Int.,* 18, 583—590, 1980.

71. **Kaloyanides, G. J. and Pastoriza-Munoz, E.**, Aminoglycoside nephrotoxicity, *Kidney Int.*, 18, 571—582, 1980.

72. **Tulkens, P. and Trouet, A.**, The concept of drug-carriers in the treatment of parasitic diseases, *Biochem. Pharmacol.*, 27, 415—424, 1978.

73. **Tulkens, P. and van Hoof, F.**, Comparative toxicity of aminoglycoside antibiotics towards the lysosomes in a cell culture model, *Toxicology*, 17, 195—199, 1980.

74. **Lüllmann, H., Lüllmann-Rauch, R., and Wassermann, O.**, Lipidosis induced by amphopholic cationic drugs, *CRC Crit. Rev. Toxicol.*, 4, 185—218, 1975.

75. **Lüllmann, H., Lüllmann-Rauch, R., and Wassermann, O.**, Drug induced phospholipidosis II tissue contribution of the amphiphilic drug chlorphentermine, *Biochem. Pharmacol.*, 27, 1103—1108, 1978.

76. **Vane, J. R.**, Inhibition of prostaglandin biosynthesis as the mechanism of action of aspirin-like drugs, *Adv. Biosci.*, 9, 395—411, 1973.

77. **Willis, A. L., Davison, P., Ramwell, P. W., Brocklehurst, W. E., and Smith, B.**, in *Prostaglandins in Cellular Biology*, Ramwell, P. W. and Pharriss, B. B., Eds., Plenum Press, New York, 1972, 227—268.

78. **Humes, J. L., Bonney, R. J., Pelus, L., Dalhgreen, M. E., Sadowski, S. J., Keuhl, F. A., and Davies, P.**, Macrophages synthesize and release prostaglandins in response to inflammatory stimuli, *Nature (London)*, 269, 149—151, 1977.

79. **Smith, W. L. and Lands, W. E. M.**, Stimulation and blockade of prostaglandin biosynthesis, *J. Biol. Chem.*, 246, 6700—6702, 1972.

80. **Lands, W. E. M. and Rome, L. H.**, in *Prostaglandins: Chemical and Biological Aspects*, Karim, S. M. M., Ed., MTP, Lancaster, England, 1976, 87—137.

81. **Kaplan, L., Weiss, J., and Elsbach, P.**, Low concentrations of indomethacin inhibit phospholipase A_2 of rabbit polymorphonuclear leucocytes, *Proc. Natl. Acad. Sci. U.S.A.*, 75, 2955—2958, 1978.

82. **Deykim, D. and Russell, F.**, Indomethacin inhibits platelet phospholipase activity, *Circulation*, Suppl. 2 (Abstr.), 57—58, 1978.

83. **Hesse, R. L. and Franson, R. C.**, Modulation of purified phospholipase A_2 activity from human platelets by Ca^{2+} and indomethacin, *Biochim. Biophys. Acta*, 575, 467—470, 1979.

84. **Franson, R. C., Eisen, D., Jesse, R., and Lanni, C.**, Inhibition of highly purified phospholipase A_2 by non-steroidal anti-inflammatory agents, *Biochem. J.*, 186, 633—636, 1980.

85. **Northover, B. J.**, Effect of indomethacin and related drugs on the Ca-ion dependent secretion of lysosomal and other enzymes by neutrophil polymorphonuclear leukocytes in vitro, *Br. J. Pharmacol.*, 8, 293—296, 1977.

86. **Roth, S. H.**, *New Directions in Arthritis Therapy*, John Wright/PSG, Inc., Littleton, Mass., 1980.

87. **Hollingsworth, J. W.**, *Management of Rheumatoid Arthritis and Its Complications*, Year Book Medical Publishers, London, 1978.

88. **Zvailfer, N. J.**, Antimalarial treatment of rheumatoid arthritis, *Med. Clin. N. Am.*, 52, 759—764, 1968.

89. **Exton, J. H.**, Gluconeogenesis, *Metabolism*, 21, 945—990, 1972.

90. **Thompson, E. B. and Lippmann, M. E.**, Mechanism of action of glucocorticoids, *Metabolism*, 23, 159—202, 1974.

91. **Buller, R. E. and O'Malley, B. W.**, The biology and mechanism of steroid hormone receptor interaction with the eukaryotic nucleus, *Biochem. Pharmacol.*, 25, 1—12, 1976.

92. **Chan, L. and O'Malley, B. W.**, Mechanism of action of sex steroid hormones, *N. Engl. J. Med.*, 294, 1372—1379, 1976.

93. **Baxter, J. D. and Tomkins, G. M.**, Specific cytoplasmic glucocorticoid hormone receptors in hepatoma tissue culture cells, *Proc. Natl. Acad. Sci. U.S.A.*, 68, 932—937, 1971.

94. **Danon, A. and Assouline, G.**, Inhibition of prostaglandin biosynthesis by corticosteroids requires RNA and protein synthesis, *Nature (London)*, 273, 552—554, 1978.

95. **Flower, R. J. and Blackwell, G. J.**, Antiinflammatory steroids induce biosynthesis of a phospholipase A_2 inhibitor which prevents prostaglandin generation, *Nature (London)*, 278, 456—459, 1979.

96. **Russo-Marie, F., Paing, M., and Duval, D. J.**, Involvement of glucocorticoid receptors in steroid-induced inhibition of prostaglandin secretion, *J. Biol. Chem.*, 254, 8498—8504, 1979.

97. **Di Rosa, M. and Persico, P.**, Mechanism of inhibition of prostaglandin biosynthesis by hydrocortisone in rat leucocytes, *Br. J. Pharmacol.*, 66, 161—163, 1979.

98. **Carnucio, R., Di Rosa, M., and Persico, P.**, Hydrocortisone-induced inhibitor of prostaglandin biosynthesis in rat leucocytes, *Br. J. Pharmacol.*, 68, 14—16, 1980.

99. **Gryglewski, R. J.**, in *Prostaglandins and Thromboxanes*, Berti, F., Samuelsson, B., and Velo, G. P., Eds., Plenum Press, New York, 1977, 85.

100a. **Gryglewski, R. J., Panczenko, B., Korbut, R., Grodzinska, L., and Ocetkiewicz, A.**, Corticosteroids inhibit prostaglandin release from perfused mesenteric blood vessels of rabbits and from perfused lungs of sensitized guinea pigs, *Prostaglandins*, 10, 343—355, 1975.

100b. **Hong, L. S. C. and Levine, L.,** Inhibition of arachidonic acid release from cells as the biochemical action of antiinflammatory corticosteroids, *Proc. Natl. Acad. Sci. U.S.A.,* 73, 1730, 1976.

101. **Flower, R. J.,** Steroid anti-inflammatory drugs as inhibitors of phospholipase A_2, *Advances in Prostaglandin and Thromboxanes Research,* Galli, C. and Porcellati, G., Eds., Raven Press, New York, 1978, 105—112.

102. **Hong, L. S. C. and Levine, L.,** Inhibition of arachidonic acid release from cells as the biochemical action of antiinflammatory corticosteroids, *Proc. Natl. Acad. Sci. U.S.A.,* 73, 1730—1734, 1976.

103. **Blackwell, G. J., Carnuccio, R., Di Rosa, M., Flower, R. J., Parente, L., and Persico, P.,** Macrocortin: a polypeptide causing the anti-phospholipase effect of glucocorticoids, *Nature (London),* 287, 147—149, 1980.

104. **Hirata, F., Schiffmann, E., Venkatasubramanian, K., Salomon, D., and Axelrod, J.,** A phospholipase A_2 inhibitory protein in rabbit neutrophils induced by glucocorticoids, *Proc. Natl. Acad. Sci. U.S.A.,* 77, 2533—2536, 1980.

105. **Hirata, F., Del Carmine, R., Nelson, C. A., Axelrod, J., Schiffmann, E., Warabi, A., De Blas, A. L., Nirenberg, M., Manganiello, V., Vaughan, M., Kumagai, S., Green, I., Decker, J. L., and Steinberg, A. D.,** Presence of autiantibody of phospholipase inhibitory protein, lipomodulins, in patients with rheumatic diseases, *Proc. Natl. Acad. Sci. U.S.A.,* 78, 3190—3194, 1981.

106. **Ropes, M. W.,** *Systemic Lupus Erythematosus,* Harvard University Press, Cambridge, Mass., 1976.

107. **Bonta, J. L., Parnham, M. J., Adolfs, M. J. P., and Van Vliet, L.,** in *Perspectives in Inflammation,* Willoughby, D. A., Giroud, J. P., and Velo, G. P., Eds., University Park Press, Baltimore, 1977, 265—275.

108. **Weissmann, G. and Dingle, J. T.,** Release of lysosomal protase by ultraviolet irradiation and inhibition by hydrocortisone, *Exp. Cell Res.,* 25, 201—210, 1961.

109. **Mentz, P., Giebler, G. H., and Förster, W.,** Evidence for a direct inhibitory effect of glucocorticoids on the activity of phospholipase A_2 as a further possible mechanism of some actions of steroidal anti-inflammatory drugs, *Pharmacol. Res. Commun.,* 12, 817—827, 1980.

110. **Glaser, K. B. and Jacobs, R. S.,** Molecular Pharmacology of manoalide, *Biochem. Pharmacol.,* 35, 449—453, 1986.

Biosynthesis and Metabolism of Eicosanoids

BIOSYNTHESIS OF PROSTAGLANDINS

Richard J. Kulmacz

Arachidonic acid (AA), and several other polyunsaturated fatty acids can be converted in animal tissues to several different classes of potent biological effectors, collectively called eicosanoids. They include the classical prostaglandins, thromboxanes, prostacyclins, and leukotrienes. This chapter will be confined to the biosynthesis of the classical prostaglandins — prostaglandins G, H, D, E, and F. Biochemistry of prostacyclin (PGI_2), the leukotrienes, and eicosanoids in general is discussed in this volume by (respectively) C. Pace-Asciak, P. Borgeat, and D. L. Smith. As can be seen from Figure 1, the classical prostaglandins all derive from the parent polyunsaturated acid via an endoperoxide intermediate, prostaglandin H (PGH).

PROSTAGLANDIN H SYNTHASE

PGH synthase (EC 1.14.99.1) is the enzyme responsible for the catalysis of the initial step in the conversion of polyunsaturated fatty acids to the prostaglandins, thromboxanes, and prostacyclins: the insertion of two molecules of oxygen into the fatty acid to yield PGG (Reaction 1 in Figure 1). The enzyme has been found in a wide variety of tissues, and purified to apparent homogeneity from ovine seminal vesicles,[1-3] bovine seminal vesicles,[4] and human platelets.[5] The characteristics of PGH synthases from several sources are summarized in Table 1. They are microsomal proteins with a mol wt of about 70,000. When solubilized with detergent from ovine seminal vesicles, the synthase exists as a dimer[16,20] of two similar subunits.[18,24] Observations that antibodies raised against PGH synthase from one species are sometimes ineffective or only partially effective in recognizing the synthase from other sources[25,26] indicate that the structure of the enzymes may be more varied than suggested by the data from those preparations that have been purified to homogeneity. The pure protein has been found to possess a rather nonspecific peroxidase activity (Reaction 2 in Figure 1) which, like the cyclooxygenase activity, is dependent upon heme. The significance of the peroxidase activity is not yet clear. The K_m^{app} of the peroxidase for hydroperoxides is a function of the cosubstrate concentration.[26b]

Determination of the stoichiometry between the subunit and heme prosthetic group has been made difficult by the ability of the detergent-solubilized enzyme to bind considerable amounts of heme nonspecifically.[16] Whereas earlier reports concluded that one heme is present per subunit of the active enzyme,[8,21] the cyclooxygenase activity of the synthase has recently been shown to require only one heme per synthase dimer.[26a] This result implies that the two subunits of the synthase are not functionally identical. Binding of heme in the apoenzyme appears to induce a conformational change in the protein, as it is accompanied by a greatly decreased sensitivity to trypsin digestion.[27] Reconstitution of apoenzyme with mangano protoporphyrin IX restores cyclooxygenase, but not much peroxidase activity.[6] Two of the Mn protoporphorin IX were required per synthase dimer.[26a] Nonheme iron has been reported to be present in the enzyme,[1] but its significance is unclear, since the amount of nonheme iron appears to vary from preparation to preparation.[15]

The chemical rearrangements which occur during the cyclooxygenase reaction have been well studied.[28] Abstraction of the 13-L hydrogen appears to be the rate-determining step, and is accompanied by formation of an endoperoxide at C-9 and C-11, a hydroperoxide at C-15, and a cyclopentane ring from carbons 8 to 12.

Considerable circumstantial evidence exists to implicate a free radical in the mechanism of PGH synthase, and this topic was recently reviewed by Porter.[29] However, the epr signal

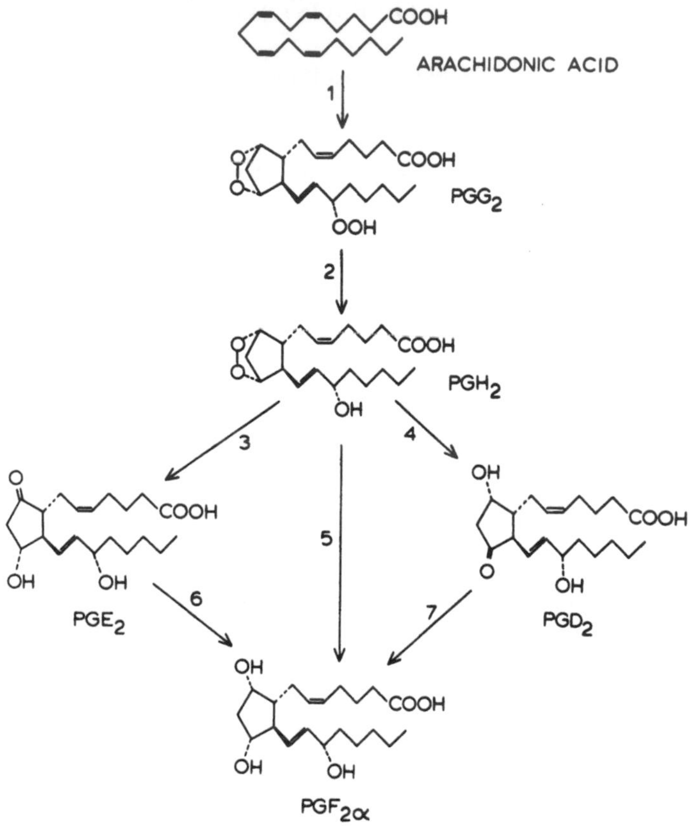

FIGURE 1.   Pathways of prostaglandin biosynthesis.

observed with the microsomal preparations of the enzyme was reported to be absent when the pure protein was examined.[22] At the present time, the classification of cyclooxygenation as a free radical reaction rests primarily on the ability of radical trapping reagents to affect the reaction kinetics, and on mimicry of the synthase by nonenzymatic model reactions.

Carbon monoxide does not inhibit the reaction,[7,13,17] although there is spectral evidence that carbon monoxide forms a complex with the holoenzyme.[17] This has been interpreted to mean that the heme in the synthase functions to activate the substrate fatty acid rather than molecular oxygen,[13] and that the synthase reaction mechanism is fundamentally different from that of other hemoprotein oxygenases, which are inhibited by carbon monoxide.[30,31]

The scheme shown in Figure 2 was proposed as a model for the catalytic mechanism of PGH synthase.[13,32] An important feature of this model is the pivotal role of lipid hydroperoxide (ROOH) as activator of the cyclooxygenase catalytic cycle.[33,34] Hydroperoxide activator is required continuously, as can be demonstrated by the ability of glutathione peroxidase to inhibit the reaction even when added after the reaction was initiated.[13] The concentration of lipid hydroperoxide required for effective activation of the cyclooxygenase (21 nM) is considerably lower than the K_m^{app} of the peroxidase for substrate (2.2 to 5.5 µM) (Table 1, References 26b and 35), allowing the cyclooxygenase to be activated at peroxide levels which are not effectively removed by the peroxidase. The process of activation of the cyclooxygenase has recently been linked with enzyme intermediates in the peroxidase catalytic cycle.[26b,26c] The cyclooxygenase activator requirement appears to be higher for synthase reconstituted with mangano protoporphyrin IX (as evidenced by a greater sensitivity to inhibition by glutathione peroxidase[101]), perhaps a consequence of its diminished[6] peroxidase activity.

Table 1
CHARACTERISTICS OF PROSTAGLANDIN H SYNTHASE

Source	Mol wt	pI	ε410 (mM/cm⁻¹)	Heme/subunit	Spec. act. (μmol fatty acid/min/mg)	K_m (substrate)	Ref.
Bovine seminal vesicles	71,000	7.0	90	1	10 at 24°	12 μM (PGG$_1$) 15 μM (15-hydroperoxy-PGE$_1$) 18 μM (15-hydroperoxy 20:3) 150 μM (H$_2$O$_2$)	4, 6—8
Ovine seminal vesicles	70,000	6.5	156	0.5	70 at 30°	5 μM (O$_2$) 2.8 μM (20:4) 20 μM (20:2) 3.5 μM (PGG$_2$) 200 μM (H$_2$O$_2$) 1.5 μM (15-hydroperoxy 20:4)	1, 9—14, 26a, 26d, 35
	126,000—151,000 (dimer) 72,000 (monomer)	7.0	61	2	18 at 25°	15 μM (20:4) 23 μM (20:3)	2, 15—19
	70,000—72,000	6.9	120	1	43 at 30°		3, 20, 21
	68,000		160		11 at 30°		22
Human platelets	79,400	7.0			1.4 at 24°		5
Rabbit kidney					0.06 at 24°	10 μM (20:4) 50 μM (O$_2$)	23

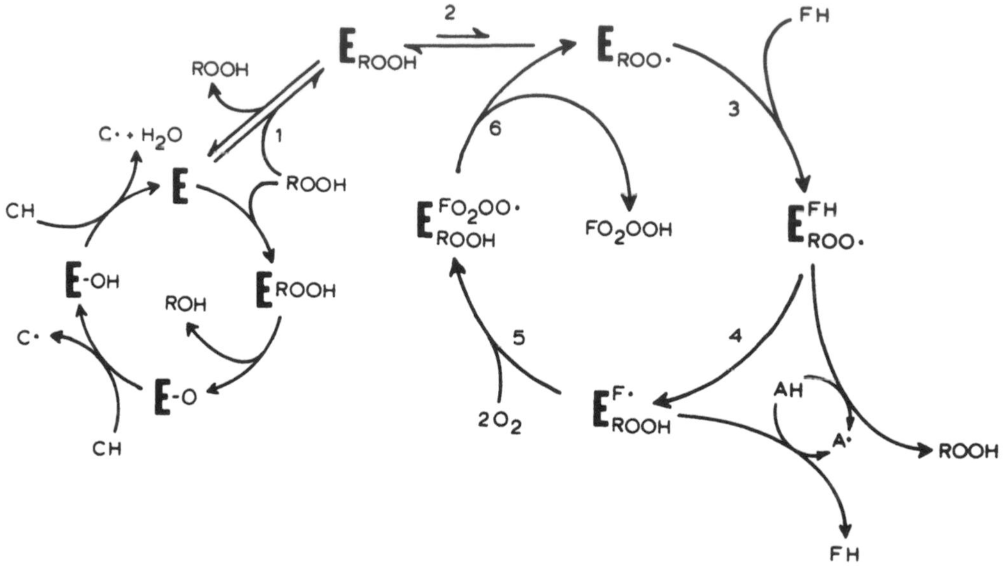

FIGURE 2. A hypothetical mechanism for prostaglandin H synthase. E represents the holoenzyme, FH the unsaturated fatty acid substrate; ROOH is lipid hydroperoxide. AH and CH are reducing agents.

The cyclooxygenase activity of PGH synthase can be affected by agents which are presumed to act as co-substrates of the peroxidase. These compounds (such as phenol, epinephrine, tryptophan, and diphenylisobenzofuran) stimulate cyclooxygenase velocity, increase the number of catalytic events per synthase molecule before self-inactivation, and are themselves oxidized.[1,4,36] In the model shown in Figure 2, this group of compounds (CH) was regarded to stimulate oxygenation by reducing the amount of enzyme in the higher oxidation states associated with peroxidase activity, and also to quench enzyme-bound free radicals (as shown for AH). The net effect of a given peroxidase co-substrate on cyclooxygenase activity can be envisioned as the balance between its ability to quench cyclooxygenase intermediates (as AH) and to reduce peroxidase intermediates (as CH).

Once the cyclooxygenase has been activated (step 2), it can activate the fatty acid substrate (step 3), which subsequently reacts with two molecules of oxygen, and rearranges to yield the free radical form of PGG. This regenerates the active form of cyclooxygenase, and diffuses from the protein as PGG (step 5).

The subcellular localization of prostaglandin H synthase has been studied in several tissues (Table 2). The enzyme has been associated with the microsomal fraction of tissue homogenates for some time. It has also been shown by immunohistochemical techniques to be concentrated around the nuclear membrane, and to be conspicuously absent from the plasma membrane of the cells examined.[39,40] The recent work of Smith and his colleagues have demonstrated that the antigenic determinants on the synthase and site of cleavage by proteinase K reside on the cytoplasmic side of the endoplasmic reticulum.[25]

PROSTAGLANDIN H-PROSTAGLANDIN E ISOMERASE

The enzymatic conversion of PGH to PGE by isomerization of the endoperoxide (Reaction 3 in Figure 1) is catalyzed by PGH-PGE isomerase (EC 5.3.99.3). A wide variety of species and tissues are able to synthesize PGE from polyunsaturated fatty acid,[41] but the relative amounts of PGH-PGE isomerase present in these tissues is difficult to ascertain since PGH synthase may be rate limiting, and since PGE can be formed nonenzymatically from PGH

Table 2
SUBCELLULAR LOCALIZATION OF PGH SYNTHASE

Tissue	Location	Method Used	Ref.
Bovine thyroid	Microsomal membrane and/or plasma membrane	Differential centrifugation	37
Rabbit renal medulla	Endoplasmic reticulum, possibly Golgi; apparatus	Differential centrifugation	38
Mouse 3T3 fibroblasts	Throughout cytoplasm and on nuclear membrane; not on plasma membrane	Peroxidase/antiperoxidase sequence with antibodies against PGH synthase	39
Mouse 3T3 fibroblasts Bovine aorta endothelium	Smooth and rough endoplasmic reticulum and nuclear membrane; not on plasma membrane	Peroxidase/antiperoxidase with antibodies against PGH synthase	40
Sheep seminal vesicles	Antigenic determinants and proteinase K-sensitive site of PGH synthase on cytoplasmic face of endoplasmic reticulum	Proteinase K digestion; monoclonal antibodies against PGH synthase	25

Table 3
CHARACTERISTICS OF PGH-PGE ISOMERASE

Source of enzyme	Specific activity	Optimal GSH conc.	pH optimum	K_m (substrate)	Ref.
Bovine seminal vesicles microsomes	1.9 μmol PGE_1 min/mg	1 mM	7—8	10 μM (PGH_1) 50 μM (PGG_1)	46
Sheep seminal vesicles microsomes	1.7 μmol PGE_2 min/mg	0.5—4 mM	5.5—7.0		43
Bovine cerebral microvessels	0.4 nmol PGE_2 min/mg	100 μM		Not saturated by 100 μM PGH_2	47
Sheep seminal vesicles	2 nmol PGE_2	<5 mM			45

rather rapidly.[42-45] The enzyme has not yet been purified to homogeneity, largely because of its extreme lability, although partial purification has been accomplished from bovine[46] and ovine seminal vesicle microsomes.[43] Curiously, Tween®-20, which was used to solubilize the bovine enzyme, was unable to solubilize the ovine enzyme.[43,46] Table 3 lists the characteristics of the PGH-PGE isomerase reported in these two studies and in two others which used PGG or PGH as the substrate (thus avoiding confusion due to the presence of PGH synthase). The enzyme requires reduced glutathione for activity, with other sulfhydryl compounds able to stabilize the enzyme but not activate it.[46,47] The role of glutathione remains unclear, since it is not consumed in the reaction with the partially purified enzyme.[46] Earlier studies with crude enzyme preparations had found that glutathione was oxidized,[48,49] perhaps due to reduction of PGG to PGH by endogenous glutathione peroxidase. Mercurials such as *p*-hydroxymercuribenzoate irreversibly inactivate the enzyme;[45-47] their action cannot subsequently be reversed by excess GSH.

Since both PGG and PGH are substrates for PGH-PGE isomerase,[45,46] PGE could conceivably be synthesized either directly from PGH or by reduction of 15-hydroperoxy PGE. The latter compound has been identified in incubations of arachidonic acid or PGG with preparations from ram seminal vesicles,[45,50] but its synthesis in vivo remains to be established. Ogino et al.[46] reported that 15-hydroperoxy PGE was not a substrate for the peroxidase activity of PGH synthase, but the reduction could be accomplished in vivo by another peroxidase, such as glutathione peroxidase.

Table 4
CHARACTERISTICS OF PGH-PGD ISOMERASE

Source of enzyme	Mol wt	Specific activity	pI	pH optimum	GSH	K_m (substrate)	Ref.
Rat brain cytosol	80,000 — 85,000 (monomeric)	1.8 μmol PGD min/mg	5.3	8.0	No effect	8 μM (PGH$_2$) 6 μM (PGG$_2$)	55, 56
Rat spleen cytosol	26,000 — 34,000 (monomeric)	0.6 μmol PGD min/mg	5.2	7-8	0.1—1 mM Required		58, 59
Sheep seminal vesicles cytosol	36,000 — 42,000				Required		42
Glutathione S-transferase (sheep lung)	47,000 (dimer) 23,500 (subunit)	1 nmol PGD min/mg	9.9		Required		53

PROSTAGLANDIN H-PROSTAGLANDIN D ISOMERASE

Prostaglandin H-prostaglandin D isomerase (E.C. 5.3.99.2) catalyzes the isomerization PGH to PGD (Reaction 4 in Figure 1). The biosynthesis of PGD attracted little interest until it was demonstrated that this prostaglandin, thought to be an intermediate in the metabolic inactivation of the endoperoxides, was a potent inhibitor of platelet aggregation.[51,52] Investigation of the biosynthesis of PGD has been complicated by the instability of PGH in aqueous buffer. Further, serum albumins and glutathiones S-transferases from several species were found to increase the amount and proportion of PGD formed during decomposition of PGH.[53-54a] Recently, Shimizu et al.[55,56] reported the purification to homogeneity of a protein from rat brain with PGH-PGD isomerase activity. This protein is probably responsible for the isomerase activity of rat brain homogenate reported earlier by Sun et al.[57] Several characteristics of this enzyme argue for the assigning it physiological importance: its K_m for PGH$_2$ is low (8 μM), it has no glutathione S-transferase activity, and it is not inhibited by 100 μM arachidonic acid, as would occur with the serum albumins.[53] Christ-Hazelhof and Nugteren[58,59] have also isolated a homogeneous protein with PGH-PGD isomerase activity, this from rat spleen. The spleen enzyme resembles that from rat brain in its cytosolic origin, isoelectric point, pH optimum, insensitivity to inhibition by arachidonic acid, and sensitivity to mercurials. The spleen enzyme, however, requires reduced glutathione, and is a much smaller protein. The characteristics reported for the isomerase from several sources are listed in Table 4, along with those of the glutathione-S-transferase isolated from sheep lung. In contrast with PGH synthase and PGH-PGE isomerase, the PGH-PGD isomerase activity has been found in the cytosol of almost all the tissues examined. The exceptions are rat skin and a rat basophilic leukemia cell line where the enzyme was found to be largely microsomal.[60,61] PGH-PGD isomerase has been found in a large variety of species and tissues (Table 5). Shimizu et al.[63] have studied the tissue distribution of the isomerase activity in the rat, and reported the highest specific activity in the brain, spinal cord, and intestines, considerably lower levels in the stomach, lung, liver, and spleen, and essentially no isomerase activity in rat heart, kidney, adrenal, or seminal vesicles. This distribution is consistent with the ability of various rat tissues to synthesize PGD from unsaturated fatty acid.[42]

CONVERSION OF PROSTAGLANDIN H TO PROSTAGLANDIN F$_{2α}$

The biological reduction of prostaglandin endoperoxides to prostaglandins of the F$_α$ series (Reaction 5, Figure 1) occur with retention of the hydrogen on C-9 of the fatty acid.[70,71] Thus, the biosynthesis of PGF$_α$ from PGH proceeds, at least in ram seminal vesicles and

Table 5
DISTRIBUTION OF PGH-PGD ISOMERASE ACTIVITY

Tissue	Ref.	Tissue	Ref.
Rat		Rabbit	
Brain	55, 57, 62, 63	Brain	63
Spleen	58, 63	Renal papillae	65
Spinal cord	63	Lung	58
Intestine	42, 63	Cow brain	63
Lung	42, 63	Pig brain	63
Stomach	42, 63	Cat brain	63
Liver	63	Mouse	
Skin	42, 60	Brain	63
Bone marrow	64	Neuroblastoma cells	63
Mast cells	61	Gerbil spleen	58
Basophilic leukemia cells	61	Monkey	
Peritoneal leukocytes	57	Lung	58
Guinea pig		Intestine	58
Brain	63	B16 malignant melanoma cells	66
Intestine	58	Human platelet	67—69

rabbit renal papillae, by a direct reduction of the endoperoxide, without initial conversion to PGE. However, no conclusive evidence has been presented that the reaction is enzymatic.[72] While there are many instances of an increase or decrease of PGF_α synthesis relative to that of other prostaglandins,[73-76] such changes may be artifacts of the extraction procedure used.[77] A pure protein has been isolated from bovine lung[77a] that catalyzes the formation of $PGF_{2\alpha}$ from either PGH_2 or PGD_2, but its substrate range suggests that it may have a physiological role other than prostaglandin biosynthesis.

PROSTAGLANDIN E 9-KETO REDUCTASE

The conversion of PGE_2 to $PGF_{2\alpha}$ (Reaction 6 in Figure 1) is catalyzed by prostaglandin 9-keto reductase. This enzyme activity has been found ubiquitously in the tissues of man,[78-81] pig,[82-84] monkey,[85,86] rabbit,[85] rat,[81] chicken,[87] frog,[88] and pigeon.[86] The enzyme activity from all sources is in the cytosolic fraction, and utilizes the hydrogen from the B-site of NADPH.[80,89]

Another enzyme activity, prostaglandin 15-hydroxy dehydrogenase, was found to be associated with the 9-keto reductase.[78,90] Subsequently, Lin and Jarabak[79] isolated two proteins from human placenta, each having both 9-keto reductase and 15-dehydrogenase activities. Chang et al.[82,83] have recently reported the purification of two proteins from pig kidney which also had both enzyme activities.

The 9-keto reductases studied have K_m's for prostaglandins which are very much higher than physiological levels (Reference 91 and Table 6). Although the K_m for the glutathione adduct of PGA is much lower[91] the physiological role of this compound is questionable. 9-Keto reductases/15-dehydrogenases have recently been shown to be able to oxidize and reduce a broad range of substrates.[80,84] Quinones have much lower K_m values and higher K_{cat} values than the prostaglandins tested, and may be the physiological substrates for these enzymes.[80,84]

PROSTAGLANDIN A AND ITS METABOLITES

The evidence for the presence of PGA in animal tissues has been reviewed recently,[95] with the conclusion that when steps are taken to avoid chemical conversion prior to assay,

Table 6
CHARACTERISTICS OF PROSTAGLANDIN E 9-KETOREDUCTASE

Source	Mol wt	15-Dehydrogenase activity	pI	k_m (substrate)	Ref.
Human erythrocyte		+		150 μM (PGE$_2$)	90
Human platelet	31,000	+			92
Human brain	31,900	Not examined	6.95, 7.85, 8.5	450 μM (PGE$_1$)	80
(and liver, kidney)	32,400		With 10 μM	17 μM (ubiquinone)	
	32,800		NADPH: 5.3, 5.6, 5.9	45 μM (menadione)	
Human placenta	31,000	+			78, 79
	31,000				
Monkey brain	33,500	Not examined	8.0		85
Pig kidney	29,500	+	4.8, 5.8	180 μM (PGE$_2$)	82 —
	29,500	+		400 μM (PGA$_2$)	84
				18 μM (PGA$_2$-GSH)	
				20 μM (menadione)	
				5 μM (9,10-phen-anthrenequinone)	
Rabbit kidney	21,800	+	5.65	320 μM (PGE$_2$)	85
	31,000	+	4.8, 6.2	200 μM (PGE$_1$)	93, 94
				13 μM (PGA$_1$-GSH)	

PGA does not occur in vivo. Thus, the PGA isomerase described in cat and rabbit sera[96,97] probably has another biological function.

REDUCTION OF PROSTAGLANDIN D

The enzymatic conversion of PGD to PGF$_\alpha$ (11-keto reductase; Reaction 7 in Figure 1) has been demonstrated with sheep blood,[98] and the pure enzyme has been isolated from bovine lung.[77a] Recent reports, however, have indicated that the K_m of the 11-ketoreductase for PGD is in the range of 0.12 to 2 mM,[77a,99,100] casting considerable doubt on the physiological importance of the enzyme.

REFERENCES

1. **Hemler, M., Lands, W. E. M,. and Smith, W. L.,** Purification of the cyclooxygenase that forms prostaglandins: demonstration of two forms of iron in the holoenzyme, *J. Biol. Chem.,* 251, 5575—5579, 1976.
2. **Van der Ouderaa, F. J., Buytenhek, M., Nugteren, D. H., and Van Dorp, D. A.,** Purification and characterisation of prostaglandin endoperoxide synthetase from sheep vesicular glands, *Biochim. Biophys. Acta,* 487, 315—331, 1977.
3. **Roth, G. J. and Siok, C. J.,** Acetylation of the NH$_2$-terminal serine of prostaglandin synthetase of aspirin, *J. Biol. Chem.,* 253, 3782—3784, 1978.
4. **Miyamoto, T., Ogino, N., Yamamoto, S., and Hayaishi, O.,** Purification of prostaglandin endoperoxide synthetase from bovine vesicular gland microsomes, *J. Biol. Chem.,* 251, 2629—2636, 1976.
5. **Ho, P. P. K., Towner, R. D., and Esterman, M. A.,** Purification and characterization of fatty acid cyclooxygenase from human platelets, *Prep. Biochem.,* 10, 597—613, 1980.
6. **Ogino, N., Ohki, S., Yamamoto, S., and Hayaishi, O.,** Prostaglandin endoperoxide synthetase from bovine vesicular gland microsomes: inactivation and activation by heme and other metalloporphyrins, *J. Biol. Chem.,* 253, 5061—5068, 1978.

7. **Ohki, S., Ogino, N., Yamamoto, S., and Hayaishi, O.,** Prostaglandin hydroperoxidase, an integral part of prostaglandin endoperoxide synthetase from bovine vesicular gland microsomes, *J. Biol. Chem.,* 254, 829—836, 1979.

8. **Ueno, R., Shimizu, T., Kondo, K., and Hayaishi, O.,** Activation mechanism of prostaglandin endoperoxide synthetase by hemoproteins, *J. Biol. Chem.,* 257, 5584—5588, 1982.

9. **Rome, L. H. and Lands, W. E. M.,** Properties of a partially-purified preparation of the prostaglandin-forming oxygenase from sheep vesicular gland, *Prostaglandins,* 10, 813—824, 1975.

10. **Hemler, M. E., Crawford, C. G., and Lands, W. E. M.,** Lipoxygenation activity of purified prostaglandin-forming cyclooxygenase, *Biochemistry,* 17, 1772—1779, 1978.

11. **Lands, W. G. M., Sauter, J., and Stone, G. W.,** Oxygen requirement for prostaglandin biosynthesis, *Prostaglandins Med.,* 1, 117—120, 1978.

12. **Hemler, M. E. and Lands, W. E. M.,** Protection of cyclooxygenase activity during heme-induced destabilization, *Arch. Biochem. Biophys.,* 201, 586—593, 1980.

13. **Hemler, M. E. and Lands, W. E. M.,** Evidence for a peroxide-initiated free radical mechanism of prostaglandin biosynthesis, *J. Biol. Chem.,* 255, 6253—6261, 1980.

14. **Kulmacz, R. J. and Lands, W. E. M.,** Characteristics of prostaglandin H synthase, *Adv. Prostaglandin Thromboxane Leukotriene Res.,* 11, 93—97, 1983.

15. **Van Dorp, D. A., Buytenhek, M., Christ-Hazelhof, E., Nugteren, D. H., and Van der Ouderaa, F. J.,** Isolation and properties of enzymes involved in prostaglandin biosynthesis, *Acta Biol. Med. Germ.,* 37, 691—699, 1978.

16. **Van der Ouderaa, F. J., Buytenhek, M., Slikkerveer, F. J., and Van Dorp, D. A.,** On the haemoprotein character of prostaglandin endoperoxide synthetase, *Biochim. Biophys. Acta,* 572, 29—42, 1979.

17. **Van der Ouderaa, F. J., Buytenhek, M., and Van Dorp, D. A.,** Characterization of prostaglandin H$_2$ synthetase, *Adv. Prostaglandin Thromboxane Res.,* 6, 139—144, 1980.

18. **Van der Ouderaa, F. J., Buytenhek, M., Nugteren, D. H., and Van Dorp, D. A.,** Acetylation of prostaglandin endoperoxide synthetase with acetylsalicylic acid, *Eur. J. Biochem.,* 109, 1—8, 1980.

19. **Van der Ouderaa, F. J. G. and Buytenhek, M.,** Purification of PGH synthase from sheep vesicular glands, *Methods Enzymol.,* 86, 60—68, 1982.

20. **Roth, G. J., Siok, C. J., and Ozols, J.,** Structural characteristics of prostaglandin synthetase from sheep vesicular gland, *J. Biol. Chem.,* 255, 1301—1304, 1980.

21. **Roth, G. J., Machuga, E. T., and Strittmatter, P.,** The heme-binding properties of prostaglandin synthetase from sheep vesicular gland, *J. Biol. Chem.,* 256, 10018—10022, 1981.

22. **Egan, R. W., Gale, P. H., Baptista, E. M., Kennicott, K. L., Vanden Heuvel, W. J. A., Walker, R. W., Fagerness, P. E., and Kuehl, F. A., Jr.,** Oxidation reactions by prostaglandin cyclooxygenase-hydroperoxidase, *J. Biol. Chem.,* 256, 7352—7361, 1981.

23. **Bhat, S. G., Yoshimoto, T., Yamamoto, S., and Hayaishi, O.,** Solubilization and partial purification of prostaglandin endoperoxide synthetase of rabbit kidney medulla, *Biochim. Biophys. Acta,* 529, 398—408, 1978.

24. **Roth, G. J., Stanford, N., Jacobs, J. W., and Majerus, P. W.,** Acetylation of prostaglandin synthetase by aspirin. Purification and properties of the acetylated protein from sheep vesicular gland, *Biochemistry,* 11, 4244—4248, 1977.

25. **DeWitt, D. L., Rollins, T. E., Day, J. S., Gauger, J. A., and Smith, W. L.,** Orientation of the active site and antigenic determinants of prostaglandin endoperoxide synthase in the endoplasmic reticulum, *J. Biol. Chem.,* 256, 10375—10382, 1981.

26. **Roth, G. J. and Machuga, E. T.,** Radioimmune assay of human platelet prostaglandin synthetase, *J. Lab. Clin. Med.,* 99, 187—196, 1982.

26a. **Kulmacz, R. J. and Lands, W. E. M.,** Prostaglandin H synthase: stoichiometry of heme cofactor, *J. Biol. Chem.,* 259, 6358—6363, 1984.

26b. **Kulmacz, R. J.,** Prostaglandin H synthase and hydroperoxides: peroxidase reaction and inactivation kinetics, *Arch. Biochem. Biophys.,* 249, 273—285, 1986.

26c. **Kulmacz, R. J., Miller, J. F., Jr., and Lands, W. E. M.,** Prostaglandin H synthase: an example of enzymic symbiosis, *Biochem. Biophys. Res. Commun.,* 130, 918—923, 1985.

26d. **Kulmacz, R. J. and Lands, W. E. M.,** Quantitative similarities in the several actions of cyanide on prostaglandin H synthase, *Prostaglandins,* 29, 175—190, 1985.

27. **Kulmacz, R. J. and Lands, W. E. M.,** Protection of prostaglandin H synthase from trypsin upon binding of heme, *Biochem. Biophys. Res. Commun.,* 104, 758—764, 1982.

28. **Samuelsson, B.,** Biosynthesis and metabolism of prostaglandins, *Prog. Biochem. Pharmacol.,* 3, 59—70, 1967.

29. **Porter, N. A.,** Prostaglandin endoperoxides, in *Free Radicals in Biology,* Vol. 4, Pryor, W. A., Ed., Academic Press, New York, 1980, 261.

30. **Yamamoto, S. and Hayaishi, O.,** Tryptophan pyrrolase of rabbit intestine: D- and L-tryptophan-cleaving enzyme or enzymes, *J. Biol. Chem.,* 242, 5260—5266, 1967.

31. **Hirata, F., Ohnishi, T., and Hayaishi, O.**, Indoleamine 2,3-dioxygenase. Characterization and properties of enzyme. O_2-complex, *J. Biol. Chem.*, 252, 4637—4642, 1977.

32. **Lands, W. E. M. and Hanel, A.**, Inhibitors and activators of prostaglandin biosynthesis, in *Comprehensive Biochemistry*, Elsevier/North-Holland, in press.

33. **Hemler, M. E., Graff, G., and Lands, W. E. M.**, Acceleration autoactivation of prostaglandin biosynthesis by PGG_2, *Biochem. Biophys. Res. Commun.*, 85, 1325—1331, 1978.

34. **Hemler, M. E., Cook, H. W., and Lands, W. E. M.**, Prostaglandin biosynthesis can be triggered by lipid peroxides, *Arch. Biochem. Biophys.*, 193, 340—345, 1979.

35. **Kulmacz, R. J. and Lands, W. E. M.**, Requirements for hydroperoxide by the cyclooxygenase and peroxidase activities of prostaglandin H synthetase, *Prostaglandins*, 25, 531—540, 1983.

36. **Marnett, L. J., Wlodawer, P., and Samuelsson, B.**, Co-oxygenation of organic substrates by the prostaglandin synthetase of sheep vesicular gland, *J. Biol. Chem.*, 250, 8510—8517, 1975.

37. **Friedman, Y., Lang, M., and Burke, G.**, Further characterization of bovine thyroid prostaglandin synthase, *Biochim. Biophys. Acta*, 397, 331—341, 1975.

38. **Bohman, S. and Larsson, C.**, Prostaglandin synthesis in membrane fractions from the rabbit renal medulla, *Acta Physiol. Scand.*, 94, 244—258, 1975.

39. **Rollins, T. E. and Smith, W. L.**, Subcellular localization of prostaglandin-forming cyclooxygenase in Swiss mouse 3T3 fibroblasts by electron microscopic immunocytochemistry, *J. Biol. Chem.*, 255, 4872—4875, 1980.

40. **Smith, W. L., Rollins, T. E., and DeWitt, D. L.**, Subcellular localization of prostaglandin forming enzymes using conventional and monoclonal antibodies, *Prog. Lipid Res.*, 20, 103—110, 1981.

41. **Christ, E. J. and Van Dorp, D. A.**, Comparative aspects of prostaglandin biosynthesis in animal tissues, *Biochim. Biophys. Acta*, 270, 537—545, 1972.

42. **Nugteren, D. H. and Hazelhof, E.**, Isolation and properties of intermediates in prostaglandin biosynthesis, *Biochim. Biophys. Acta*, 326, 448—461, 1973.

43. **Moonen, P., Buytenhek, M., and Nugteren, D. H.**, Purification of PGH-PGE isomerase from sheep vesicular glands, *Methods Enzymol.*, 86, 84—91, 1982.

44. **Hamberg, M., Svensson, J., Wakabayashi, T., and Samuelsson, B.**, Isolation and structure of two prostaglandin endoperoxides that cause platelet aggregation, *Proc. Natl. Acad. Sci. U.S.A.*, 71, 345—349, 1974.

45. **Raz, A., Schwartzman, M., and Kenig-Wakshal, R.**, Chemical and enzymatic transformations of prostaglandin endoperoxides: evidence for the predominance of the 15-hydroperoxy pathway, *Eur. J. Biochem.*, 70, 89—96, 1976.

46. **Ogino, N., Miyamoto, T., Yamamoto, S., and Hayaishi, O.**, Prostaglandin endoperoxide E isomerase from bovine vesicular gland microsomes, a glutathione-requiring enzyme, *J. Biol. Chem.*, 252, 890—895, 1977.

47. **Gerritsen, M. E., Parks, T. P., and Printz, M. P.**, Prostaglandin endoperoxide metabolism by bovine cerebral microvesseles, *Biochim. Biophys. Acta*, 619, 196—206, 1980.

48. **Nugteren, D. H., Beerthuis, R. K., and Van Dorp, D. A.**, The enzymatic conversion of all-cis 8,11,14-eicosatrienoic acid into prostaglandin E_1, *Recl. Trav. Chim. Pays-Bas*, 85, 405—419, 1966.

49. **Yoshimoto, A., Ito, H., and Tomita, K.**, Cofactor requirements of the enzyme synthesizing prostaglandin in bovine seminal vesicles, *J. Biochem.*, 68, 487—499, 1970.

50. **Samuelsson, B. and Hamberg, M.**, Role of endoperoxides in the biosynthesis and action of prostaglandins, in *Prostaglandin Synthetase Inhibitors*, Robinson, H. J. and Vane, J. R., Eds., Raven Press, New York, 1974, 107—119.

51. **Smith, J. B., Silver, M. J., Ingerman, C. M., and Kocsis, J. J.**, Prostaglandin D_2 inhibits the aggregation of human platelets, *Thromb. Res.*, 5, 291—299, 1974.

52. **Nishizawa, E. E., Miller, W. L., Gorman, R. R., Bundy, G. L., Svensson, J., and Hamberg, M.**, Prostaglandin D_2 as a potential antithrombotic agent, *Prostaglandins*, 9, 109—121, 1975.

53. **Christ-Hazelhof, E., Nugteren, D. H., and Van Dorp, D. A.**, Conversions of prostaglandin endoperoxides by glutathione-S-transferases and serum albumine, *Biochim. Biophys. Acta*, 450, 450—461, 1976.

54. **Hamberg, M. and Fredholm, B. B.**, Isomerization of prostaglandin H_2 into prostaglandin D_2 in the presence of serum albumin, *Biochim. Biophys. Acta*, 431, 189—193, 1976.

54a. **Watanabe, T., Narumiyo, S., Shimizu, T., and Hayaishi, O.**, Characterization of the biosynthetic pathway of prostaglandin D_2 in human platelet-rich plasma, *J. Biol. Chem.*, 257, 14847—14853, 1982.

55. **Shimizu, T., Yamamoto, S., and Hayaishi, O.**, Purification and properties of prostaglandin D synthetase from rat brain, *J. Biol. Chem.*, 254, 5222—5228, 1979.

56. **Shimizu, T., Yamamoto, S., and Hayaishi, O.**, Purification of PGH-PGD isomerase from rat brain, *Methods Enzymol.*, 86, 73—77, 1982.

57. **Sun, F. F., Chapman, J. P., and McGuire, J. C.**, Metabolism of prostaglandin endoperoxide in animal tissues, *Prostaglandins*, 14, 1055—1074, 1977.

58. **Christ-Hazelhof, E., and Nugteren, D. H.,** Purification and characterization of prostaglandin endoperoxide D-isomerase, a cytoplasmic, glutathione-requiring enzyme, *Biochim. Biophys. Acta,* 572, 43—51, 1979.

59. **Christ-Hazelhof, E. and Nugteren, D. H.,** Isolation of PGH-PGD isomerase from rat spleen, *Methods Enzymol.,* 86, 77—84, 1982.

60. **Kingston, W. P. and Greaves, M. W.,** Factors affecting prostaglandin synthesis by rat skin microsomes, *Prostaglandins,* 12, 51—69, 1976.

61. **Steinhoff, M. M., Lee, L. H., and Jakschik, B. A.,** Enzymatic formation of prostaglandin D$_2$ by rat basophilic leukemia cells and normal rat mast cells, *Biochim. Biophys. Acta,* 618, 28—34, 1980.

62. **Abdel-Halim, M. S., Hamberg, M., Sjoquist, B., and Anggard, E.,** Identification of prostaglandin D$_2$ as a major prostaglandin in homogenates of rat brain, *Prostaglandins,* 14, 633—643, 1977.

63. **Shimizu, T., Mizuno, N., Amano, T., and Hayaishi, O.,** Prostaglandin D$_2$, a neuromodulator, *Proc. Natl. Acad. Sci. U.S.A.,* 76, 6231—6234, 1979.

64. **Kojima, A., Shiraki, M. Takahashi, R., Orimo, H., Morita, I., and Murota, S.,** Prostaglandin D$_2$ is the major prostaglandin of arachidonic acid metabolism in rat bone marrow homogenate, *Prostaglandins,* 20, 171—176, 1980.

65. **Friesinger, G. C., Oelz, O., Sweetman, B. J., Nies, A. S., and Data, J. L.,** Prostaglandin D$_2$, another renal prostaglandin, *Prostaglandins,* 15, 969—981, 1978.

66. **Fitzpatrick, F. A. and Stringfellow, D. A.,** Prostaglandin D$_2$ formation by malignant melanoma cells correlates inversely with cellular metastatic potential, *Proc. Natl. Acad. Sci. U.S.A.,* 76, 1765—1769, 1979.

67. **Anhut, H., Peskar, B. A., Wachter, W., Grabling, B., and Peskar, B. M.,** Radioimmunological determination of prostaglandin in D$_2$ synthesis in human thrombocytes, *Experientia,* 34, 1494—1496, 1978.

68. **Ali, M., Cerskus, A. L., Zamecnik, J., and McDonald, J. W. D.,** Synthesis of prostaglandin D$_2$ and thromboxane B$_2$ by human platelets, *Thromb. Res.,* 11, 485—496, 1977.

69. **Oelz, O., Oelz, R., Knapp, H. R., Sweetman, B. J., and Oates, J. A.,** Biosynthesis of prostaglandin D$_2$. I. Formation of prostaglandin D$_2$ by human platelets, *Prostaglandins,* 13, 225—234, 1977.

70. **Hamberg, M. and Samuelsson, B.,** On the mechanism of the biosynthesis of prostaglandins E$_1$ and F$_{1\alpha}$, *J. Biol. Chem.,* 242, 5336—5343, 1967.

71. **Qureshi, Z. and Cagen, L. M.,** Prostaglandin F$_{2\alpha}$ produced by rabbit renal slices is not a metabolite of prostaglandin E$_2$, *Biochem. Biophys. Res. Commun.,* 104, 1255—1263, 1982.

72. **Wlodawer, P., Kindahl, H., and Hamberg, M.,** Biosynthesis of prostaglandin F$_{2\alpha}$ from arachidonic acid and prostaglandin endoperoxides in the uterus, *Biochim. Biophys. Acta,* 431, 603—614, 1976.

73. **Falkay, G., Herczeg, J., and Kovacs, L.,** Effect of beta-mimetic isoxsuprine on prostaglandin biosynthesis in pregnant human myometrium *in vitro, Life Sci.,* 23, 2689—2696, 1978.

74. **Moretti, R. L. and Abraham, S.,** Stimulation of microsomal prostaglandin synthesis by a vasoactive material isolated from blood plasma, *Prostaglandins,* 15, 603—622, 1978.

75. **Moretti, R. L.,** Blood borne vasoconstrictor lowers ratio of prostaglandin I$_2$ to other products formed from arachidonate, *Prostaglandins Med.,* 6, 223—232, 1981.

76. **Chan, J. A., Nagasawa, M., Takeguchi, C., and Sih, C. J.,** On agents favoring prostaglandin F formation during biosynthesis, *Biochemistry,* 14, 2987—2991, 1975.

77. **Marnett, L. J. and Bienkowski, M. J.,** Nonenzymatic reduction of prostaglandin H by lipoic acid, *Biochemistry,* 16, 4303—4307, 1977.

77a. **Watanabe, K., Yoshida, R., Shimizu, T., and Hayaishi, O.,** Enzymatic formation of prostaglandin F$_{2\alpha}$ from prostaglandin H$_2$ and D$_2$, *J. Biol. Chem.,* 260, 7035—7041, 1985.

78. **Westbrook, C., Lin, Y., and Jarabak, J.,** NADP-linked 15-hydroxyprostaglandin dehydrogenase from human placenta: partial purification and characterization of the enzyme and identification of an inhibitor in placental tissue, *Biochem. Biophys. Res. Commun.,* 76, 943—949, 1977.

79. **Lin, Y. and Jarabak, J.,** Isolating of two proteins with 9-ketoprostaglandin reductase and NADP-linked 15-hydroxyprostaglandin dehydrogenase activities, and studies on their inhibition, *Biochem. Biophys. Res. Commun.,* 81, 1227—1234, 1978.

80. **Wermuth, B.,** Purification and properties of an NADPH-dependent carbonyl reductase from human brain. Relationship to prostaglandin 9-ketoreductase and xenobiotic ketone reductase, *J. Biol. Chem.,* 256, 1206—1213, 1981.

81. **Ziboh, V. A., Lord, J. T., and Penneys, N. S.,** Alterations of prostaglandin E$_2$-9-ketoreductase activity in proliferating skin, *J. Lipid Res.,* 18, 37—43, 1977.

82. **Chang, D. G., Sun, M., and Tai, H.,** Prostaglandin 9-ketoreductase and type II 15-hydroxyprostaglandin dehydrogenase from swine kidney are alternate activities of a single enzyme protein, *Biochem. Biophys. Res. Commun.,* 99, 745—751, 1981.

83. **Chang, D. G. and Tai, H.,** Characterization of two enzyme proteins catalyzing NADP$^+$/NADPH-dependent oxidoreduction of prostaglandins at C-9 and C-15 from swine kidney, *Arch. Biochem. Biophys.,* 214, 464—474, 1982.

84. **Chang, D. G. and Tai, H.,** Prostaglandin 9-ketoreductase/type II 15-hydroxyprostaglandin dehydrogenase is not a prostaglandin specific enzyme, *Biochem. Biophys. Res. Commun.,* 101, 898—904, 1981.

85. **Stone, K. J. and Hart, M.,** Prostaglandin-E_2-9-ketoreductase in rabbit kidney, *Prostaglandins,* 10, 273—288, 1975.

86. **Lee, S. and Levine, L.,** Prostaglandin metabolism, I. Cytoplasmic reduced nicotinamide adenine dinucleotide phosphate-dependent and microsomal reduced nicotinamide adenine dinucleotide-dependent prostaglandin E 9-ketoreductase activities in monkey and pigeon tissues, *J. Biol. Chem.,* 249, 1369—1375, 1974.

87. **Hassid, A. and Levine, L.,** Multiple molecular forms of prostaglandin 15-hydroxydehydrogenase and 9-ketoreductase in chicken kidney, *Prostaglandins,* 13, 503—516, 1977.

88. **Bishai, I. and Coceani, F.,** Presence of 15-hydroyx prostaglandin dehydrogenase, prostaglandin-Δ^{13}-reductase and prostaglandin E-9-keto(α)-reductase in the frog spinal cord, *J. Neurochem.,* 26, 1167—1174, 1976.

89. **Yuan, B., Tai, C. L., and Tai, H.,** 9-Hydroyxyprostaglandin dehydrogenase from rat kidney: purification to homogeneity and partial characterization, *J. Biol. Chem.,* 255, 7439—7443, 1980.

90. **Kaplan, L., Lee, S., and Levine, L.,** Partial purification and some properties of human erythrocyte prostaglandin 9-ketoreductase and 15-hydroxyprostaglandin dehydrogenase, *Arch. Biochem. Biophys.,* 167, 287—293, 1975.

91. **Cagen, L. M. and Pisano, J. J.,** The glutathione conjugate of prostaglandin A_1 is a superior substrate for prostaglandin E 9-ketoreductase, *Fed. Proc.,* 36, 403, 1977.

92. **Watanabe, T., Shimizu, T., Narumiya, S., and Hayaishi, O.,** NADP-linked 15-hydroxyprostaglandin dehydrogenase for prostaglandin D_2 in human blood platelets, *Arch. Biochem. Biophys.,* 216, 372—379, 1982.

93. **Toft, B. S. and Hansen, H. S.,** Glutathione-prostglandin A_1 conjugate as substrate in the purification of prostaglandin 9-ketoreductase from rabbit kidney, *Prostaglandins,* 20, 735—746, 1980.

94. **Toft, B. S. and Hansen, H. S.,** Metabolism of prostaglandin E_1 and of glutathione conjugate of prostaglandin A_1 (GSH-prostaglandin A_1) by prostaglandin 9-ketoreductase from rabbit kidney, *Biochim. Biophys. Acta,* 574, 33—38, 1979.

95. **Jonsson, H. T., Jr. and Powers, R. E.,** Endogenous prostaglandins A's (PGAs): fact or artifact in biological systems, *Prog. Lipid Res.,* 20, 787—790, 1981.

96. **Jones, R. L. and Cammock, S.,** Purification, properties, and biological significance of prostaglandin A isomerase, *Adv. Biosci.,* 9, 61—70, 1973.

97. **Polet, H. and Levine, L.,** Partial purification and characterization of prostaglandin A isomerase from rabbit serum, *Arch. Biochem. Biophys.,* 168, 96—103, 1975.

98. **Hensby, C. N.,** The enzymatic conversion of prostaglandin D_2 to prostaglandin $F_{2\alpha}$, *Prostaglandins,* 8, 369—375, 1974.

99. **Reingold, D. F. and Needleman, P.,** D-prostaglandins: comparative pharmacology and metabolism by a novel pathway (prostaglandin 11-keto reductase), *Fed. Proc.,* 40, 661, 1981.

100. **Watanabe, K., Shimizu, T., and Hayaishi, O.,** Enzymatic conversion of prostaglandin D_2 to $F_{2\alpha}$ in the rat lung, *Biochem. Int.,* 2, 603—610, 1981.

101. **Kulmacz, R. J.,** Unpublished results.

LOCALIZATION OF ENZYMES RESPONSIBLE FOR PROSTAGLANDIN FORMATION

William L. Smith

ORGAN AND CELLULAR LOCALIZATION OF PROSTAGLANDIN SYNTHESIS

As with all humoral agents, two conditions must be satisfied in order for a cell to respond to a prostaglandin. First, the cell must possess the appropriate receptor-response machinery, and second, the concentration of prostaglandin must be high enough to permit occupancy of a sufficient number of receptors to evoke a response. Due to a combination of dilution and catabolism, the concentrations of prostaglandin derivatives in the circulation are too low, except in a few instances,[1,2] to elicit most biological responses.[3-10] Thus, physiologically important responses to prostaglandins are likely to occur only at or near sites of biosynthesis where prostaglandin concentrations can become elevated. As a rule, cells that synthesize a particular prostaglandin will respond to that prostaglandin.[11-17] However, there are cells that respond to but do not synthesize a particular prostaglandin; for example, platelets do not form PGI_2, but PGI_2 inhibits platelet aggregation[15,16] and the medullary thick ascending limb of Henle's loop does not synthesize PGE_2,[18] but PGE_2 inhibits chloride ion resorption in this segment of the nephron.[19] These observations lead to the working hypothesis that prostaglandins synthesized by one cell affect both the parent cell and certain receptive neighboring cells. In this way, prostaglandins serve as intercellular messengers that function to coordinate biological responses requiring the participation of more than one cell type.[20] Three examples of prostaglandins apparently functioning in this manner are in the regulation of hemostasis by TXA_2 synthesized by platelets[15-17] and by PGI_2 synthesized by the vascular endothelium[15-17] and in the regulation of tubular NaCl resorption in the kidney by PGE_2 formed by collecting tubules.[11-14,18-20]

Data on the localization of prostaglandin biosynthesis in cells of various organs are presented in Table 1. It should be remembered, however, that prostaglandin-synthesizing cells may not be the only physiologically important target cells for prostaglandins. The information in Table 1 has come from two types of studies. One approach has been to isolate specific cell types and to analyze the prostaglandins formed using radioimmunoassays (RIA), gas chromatography/mass spectrometry (GC/MS), or radio thin-layer chromatography (TLC). This approach by virtue of the assays used is the most sensitive. However, technical difficulties attendant to isolating homogeneous populations of cells make this procedure impractical in most cases. Moreover, cells cultured from primary cell sources tend to lose their ability to synthesize prostaglandins with increasing time in culture.[11] The second approach to localizing sites of prostaglandin formation is by cytochemistry, primarily immunocytochemistry of prostaglandin biosynthetic enzymes.[22] While this method can be applied to most organs, it lacks sensitivity. For example, several types of glomerular cells derived from rat kidney have been shown to form prostaglandins in culture,[23] but it has not been possible to demonstrate positive anti-PGH synthase staining in rat glomeruli.[24]

Five general statements regarding the cellular localization of prostaglandin biosynthesis can be made (Table 1): (1) virtually all organs synthesize prostaglandins; (2) in most cases, not all cells within an organ synthesize prostaglandins (e.g., kidney); (3) it is not possible to predict whether a cell can form prostaglandins on the basis of the embryonic origin (e.g., endoderm, ectoderm, mesoderm) of the cell or of the tissue type (e.g., epithelium, neural, etc.) of the cell; (4) although those cells that form prostaglandins in one species usually form prostaglandins in other species, there are some species differences (e.g., see References

Table 1
CELLULAR LOCALIZATION OF PROSTAGLANDIN BIOSYNTHESIS

Organ	Species	Major product[a,b]	Ref.
Large arteries and veins			
Homogenates of different arteries and veins	Rat, cow	PGI_2	39, 40
Homogenate, aorta	Chicken	PGE_2	41
Endothelium, aorta	Cow, pig	PGI_2	42-44
Endothelium, umbilical vein	Human	PGI_2	43,44
Smooth muscle, aorta	Rat, cow, rabbit	PGI_2	35
Endothelium and smooth muscle of intraorgan arteries (lung, intestine, stomach, brain, kidney, uterus, heart, etc.)	Cow, rabbit	PGI_2	35,36 (see below)
Formed elements of blood			
Platelets	Human, rabbit, chicken	TXA_2, PGD_2	41,45—47
Megakaryocytes	Guinea pig	TXA_2	48
Erythrocytes	Human	N. D.	—
Monocyte/macrophage-peripheral blood	Human	TXA_2, PGE_2	49
B lymphocytes	Human	TXA_2 (trace)	49
T lymphocytes	Human	N. D.	49
Peritoneal macrophages	Mouse	PGE_1, PGE_2	50
Alveolar macrophages	Rabbit	PGH	51
Bone marrow	Rat	PGD_2	52
Brain			
Brain homogenates (all regions)	Rat, guinea pig	All products	53,54
Arterioles, capillaries (cerebral cortex)	Rat, guinea pig	PGD_2	55,56
Arterioles, capillaries (cerebral cortex)	Cow	PGE_2, PGI_2	57
Arterioles, capillaries (cerebral cortex)	Cat	PGI_2	58
Nonvascular elements	Rat	N. D.	55
Bergman glia (cerebellar cortex)	Cow, pig, sheep	PGH	59
Neuroblastoma and glioma lines	Mouse, rat	PGE_2 or PGD_2	60,61
Trachea			
Tracheal strips	Guinea pig	$PGF_{2\alpha}$	62
Tracheal epithelia	Rat (9 days culture)	$PGF_{2\alpha}$	63
	Rabbit (9 days culture)	PGI_2	63
Lung			
Lung microsomes	Cat, rat, guinea pig	TXA_2, PGI_2	64—66
Type II alveolar epithelia	Fetal rat	PGE_2	67
	Adult rat	PGI_2	63
	Rabbit	N. D.	63
Clara alveolar epithelia	Rabbit	N. D.	63
Bronchial smooth muscle (Reissesen's)	Rabbit, bovine	PGI_2	35,36
Arterial smooth muscle	Rabbit, bovine	PGI_2	35,36
Arterial endothelia	Rabbit, bovine	PGI_2	35,36
Kidney			
Cortical and medullary microsomes	Rat, human	TXA_2, PGE_2 $PGF_{2\alpha}$, PGI_2	68,69
Arteries, arterioles	Cow, rabbit rat, guinea pig	PGI_2	24,36,70
Glomerular mesangium	Rat	PGE	23
	Human	N. D.	25

Table 1 (continued)
CELLULAR LOCALIZATION OF PROSTAGLANDIN BIOSYNTHESIS

Organ	Species	Major product[a,b]	Ref.
Glomerular epithelium	Rat	PGE_2, PGF_α	23
	Human	N. D.	25
Thin limb of Henle's loop	Rabbit	PGH	71,72
Cortical collecting tubule	Rabbit	PGE_2	12
Medullary and papillary collecting tubules	Rabbit, rat	PGE_2	11,73,74
Proximal, distal convoluted tubules	Rabbit	N. D.	18,75
Renomedullary interstitial cells	Rabbit	PGE_2	76
Urinary bladder			
Nonvascular smooth muscle	Rabbit, cow	PGI_2	35,36
Gastrointestinal tract			
Stomach, intestinal microsomes (muscularis and mucosa)	Dog, cow	PGI_2, TXA_2	77,78
Longitudinal smooth muscle (stomach, intestine)	Rabbit	PGI_2	35,36
Muscularis mucosa (stomach, intestine)	Rabbit	PGI_2	35,36
Epithelium (stomach, intestine)	Pig, rabbit	N. D.	36,79
Uterus			
Uterine homogenates	Rat, guinea pig, sheep	PGI_2, PGE_2, $PGF_{2\alpha}$	80
Myometrium	Rat	PGI_2	81
Stroma	Sheep, cow	PGH	82
Luminal epithelium	Sheep, cow, human	PGH	36,82,83
Grandular epithelium	Sheep, cow, human	PGH	36,82,83
Longitudinal smooth muscle	Sheep, rat, rabbit	PGI_2	35,36
Circular smooth muscle	Sheep, rat, rabbit	PGI_2	35,36
Ovary			
Corpus luteum (microsomes)	Cow	PGI_2	64
Graafian follicle (granulosa layer)	Rat, rabbit	$PGF_{2\alpha}$	84,85
Follicle (granulosa layer)	Sheep	PGE_2	86
Follicle (theca layer)	Sheep	$PGF_{2\alpha}$	86
Heart			
Coronary microsomes	Rabbit	PGI_2, PGE_2	87
Arteries and arterioles	Rabbit	PGI_2	36,88
Liver			
Liver microsomes	Rat	$PGF_{2\alpha}$, PGE_2	89
Skeletal muscle			
Muscle homogenates	Human	PGE_2	90
Skin			
Epidermis	Human	PGF_α	91
	Mouse, guinea pig	PGD_2, $PGF_{2\alpha}$	92
Dermis	Mouse, guinea pig	PGE_2	92
Adrenal gland			
Adrenal cortical tissue	Rat	PGI_2	93
Adrenal glomerulosa	Rat	(PGE_2)	94
Gall bladder, prostate, ductus deferens			
Nonvascular smooth muscle	Rabbit	PGI_2	35,36
Vesicular gland			
Vesicular gland microsomes	Sheep	PGE_2, PGI_2	95
Nonvascular smooth muscle	Rabbit	PGI_2	35,36
Pancreas			
Pancreatic tissue fragments	Rat	$PGF_{2\alpha}$	96

Table 1 (continued)
CELLULAR LOCALIZATION OF PROSTAGLANDIN BIOSYNTHESIS

Organ	Species	Major product[a,b]	Ref.
Thyroid			
Thyroid slices	Human	PGE_2, PGI_2	97
Thyroid homogenates	Mouse	PGH	98
Connective tissue			
Adherent synovial cells	Human, rheumatoid	(PGE)	99
Chondrocytes	Cow	PGE_2	100,101
Adipose tissue			
Fat cell ghosts	Rat	(PGE)	102
Spleen			
Spleen homogenates, microsomes	Rat, guinea pig, monkey	TXA_2, PGD_2	27,64,103
Some cultured cells that form prostaglandins[b]			
MDCK epithelia	Canine kidney	PGE_2, $PGF_{2\alpha}$	104,105
RPCT epithelia	Rabbit kidney	PGE_2	73
CCCT epithelia	Canine kidney	PGE_2	12
Fibroblasts	Human skin	PGI_2	106,107
Smooth muscle	Human aorta	PGI_2	106
Venous endothelia	Human umbilical cord	PGI_2	41,42,106
Neuroblastomas	Mouse	PGE_1, PGD_2	58,59
Gliomas	Rat	(PGE)	58,59
Mastocytomas	Mouse	PGE_2	108
Type II alveolar epithlia	Fetal rat lung	PGE_2	65
Glomerular epithelia	Rat kidney	PGE_2, $PGF_{2\alpha}$	23
Renomedullary interstitial cells	Rabbit kidney	PGE_2	76
Kidney cell line	Baby hamster kidney	TXA_2	107
WI-38 fibroblasts	Human lung	TXA_2	109
Some cultured cells that do not form prostaglandins			
LLC-PK$_1$	Pig kidney		104
Gliomas	Rat		59,110
Hela	Human		110
Glomerular epithelia	Human lung		25

[a] Major product is not necessarily the only product. Abbreviations: N. D., not detectable; PGH, forms prostaglandins, but major product is unknown; ( ), only product measured.

[b] A survey of tissues for TXA_2 and PGI_2 production has been reported by Sun et al.[66] Surveys of product formation by cultured cells have been reported by Ali et al.[107] and Levine and Alam.[111] A survey of tissues, organs, and cultured cells for PGD_2 form has been reported by Shimizu et al.[61] and Christ-Hazelhof and Nugteren.[27]

23 and 25); (5) vascular endothelial cells, especially those of the large vessels, all form prostaglandins (usually, but not always, PGI_2) and (6) most, if not all, vascular and non-vascular smooth muscle cells synthesize PGI_2 (e.g., see References 35 and 36).

SUBCELLULAR LOCALIZATION OF PROSTAGLANDIN BIOSYNTHETIC ENZYMES

Our understanding of the subcellular location of the proteins involved in the biosynthesis of prostaglandins is poorly developed. The goal of studies of this topic is to prepare a road map tracing the movement of arachidonic acid (AA) during its release from esterified precursors, its conversion to prostaglandins, and the subsequent exit of newly formed prostaglandins from the cell. With the exception of PGH-PGD isomerases,[26,27] other enzymes involved in prostaglandin formation are membrane bound. The PGH synthase is located on the nuclear membrane and on the endoplasmic reticulum in many prostaglandin-forming cells.[28,29] The active site and antigenic determinants of PGH synthase are located on the cytoplasmic surface of the endoplasmic reticulum.[30] One study performed by Gerrard and White[31] suggests that TXA_2 synthase is associated with the dense tubular fraction of platelets; this latter structure is the morphological equivalent of the endoplasmic reticulum of liver parenchymal cells. No studies have been reported on the location of PGH-PGE isomerase, although this enzyme is membrane bound.[32,33] PGI_2 synthase is an apparent anomaly. This latter enzyme is membrane bound[34] and appears to be associated with the inner surface of the plasma membrane and with the nuclear membrane in smooth muscle cells from rabbit urinary bladder, small intestine, vas deferens, and perhaps aorta.[35,36] Thus, PGI_2 synthesis and perhaps PGH_2 synthesis is associated with two different organelles in the same cell. The functional significance of nuclear and plasma membrane synthesis is not understood.

It is clear that biological membranes are not freely permeable to prostaglandins.[37,38] Thus, carrier mechanisms must exist to transport newly formed prostaglandins from the interior to the exterior of the cell. Virtually nothing is known regarding these carriers or potential regulation of the rates of prostaglandin exit from cells. This is an important topic which needs to be addressed particularly in the context of the vectoral transport of prostaglandins from morphologically asymmetric cells; for example, there is indirect evidence that prostaglandins may exit renal collecting tubule cells only at the basolateral surface,[13] whereas prostaglandins can apparently transverse both apical and basolateral membranes of the medullary thick limb of Henle's loop.[19]

ACKNOWLEDGMENTS

This work was supported in part by U.S.P.H.S. Grant No. AM22042 and HD10013, by a Grant-In-Aid from the Michigan Heart Association and by an Established Investigatorship from the American Heart Association.

REFERENCES

1. **Gerber, J. G., Payne, N. A., Murphy, R. C., and Nies, A. S.,** Prostacyclin produced by the pregnant uterus in the dog may act as a circulating vasodepressor substance, *J. Clin. Invest.*, 67, 632—636, 1981.
2. **Roberts, L. J., Sweetman, B. J., Lewis, R. A., Austen, K. F., and Oates, J. A.,** Increased production of prostaglandin D_2 in patients with systemic mastocytosis, *N. Engl. J. Med.*, 303, 1400—1404, 1980.
3. **Ferreira, S. H. and Vane, J. R.,** Prostaglandins: their disappearance from and release into the circulation, *Nature (London)*, 216, 868—873, 1967.
4. **Granström, E. and Samuelsson, B.,** Radioimmunoassays for prostaglandins, in *Advances in Prostaglandin Thromboxane Research*, Vol. 5, Frohlich, J. C., Ed., Raven Press, New York, 1978, 1—13.
5. **Oates, J. A., Roberts, L. J., Sweetman, B. J., Maas, R. L., Gerkins, J. F., and Taber, D. F.,** Metabolism of the prostaglandins and thromboxanes, in *Advances in Prostaglandin Thromboxane Research*, Vol. 6, Samuelsson, B., Ramwell, P., and Paoletti, R., Eds., Raven Press, New York, 1980, 35—41.
6. **Smith, J. B., Ogletree, M. L., Lefer, A. M., and Nicolaou, K. C.,** Antibiotics which antagonize the effects of prostacyclin, *Nature (London)*, 274, 64—65, 1978.
7. **Steer, M. L., MacIntyre, D. E., Levine, L., and Salzman, E. W.,** Is prostacyclin a physiologically important circulating anti-platelet agent?, *Nature (London,)* 283, 194—195, 1980.
8. **Christ-Hazelhof, E. and Nugteren, D. H.,** Prostacyclin is not a circulating hormone, *Prostaglandins*, 22, 739—746, 1981.
9. **Blair, I. A., Barrow, S. E., Waddell, K. A., Lewis, P. J., and Dollery, C. T.,** Prostacyclin is not a circulating hormone in man, *Prostaglandins*, 23, 579—589, 1982.

10. **Fitzgerald, G. A., Brash, A. R., Falardeau, P., and Oates, J. A.,** Estimated rate of prostacyclin secretion into the circulation of normal man, *J. Clin. Invest.,* 68, 1272—1275, 1981.

11. **Grenier, F. C., Rollins, T. E., and Smith, W. L.,** Kinin-induced prostaglandin synthesis by renal papillary collecting tubule cells in culture, *Am. J. Physiol.,* 241, F94—104, 1981.

12. **Garcia-Perez, A. and Smith, W. L.,** Isolation of canine collecting tubule cells using monoclonal antibodies: AVP-induced prostaglandin synthesis, *Am. J. Physiol.,* in press.

13. **Stokes, J. B. and Kokko, J. P.,** Inhibition of sodium transport by prostaglandin E_2 across the isolated perfused rabbit collecting tubule, *J. Clin. Invest.,* 59, 1099—1104, 1977.

14. **Grantham, J. J. and Orloff, J.,** Effect of prostaglandin E_1 on the permeability response of the isolated collecting tubule to vasopressin, adenosine 3',5'-monophosphate and theophylline, *J. Clin. Invest.,* 47, 1154—1161, 1968.

15. **Gorman, R. R., Fitzpatrick, F. A., and Miller, O. V.,** Reciprocal regulation of human platelet cAMP levels by thromboxane A_2 and prostacyclin, *Adv. Cyclic Nucl.,* Res. 9, 597—609, 1978.

16. **Moncada, S. and Vane, J. R.,** Arachidonic acid metabolites and the interactions between platelets and blood vessel walls, *N. Engl. J. Med.,* 300, 1142—1147, 1979.

17. **Aiken, J. W., Shebuski, R. J., Miller, O. V., and Gorman, R. R.,** Endogenous prostacyclin contributes to the efficacy of a thromboxane synthetase inhibitor for preventing coronary artery thrombosis, *J. Pharm. Exp. Ther.,* 219, 299—308, 1981.

18. **Jackson, B. A., Edwards, R. M., and Dousa, T. P.,** Vasopressin-prostaglandin interactions in isolated tubules from rat outer medulla, *J. Lab. Clin. Med.,* 96, 119—128, 1980.

19. **Stokes, J. B.,** Effect of prostaglandin E_2 on chloride transport across the thick ascending limb of Henle, *J. Clin. Invest.,* 64, 495—502, 1979.

20. **Smith, W. L.,** Renal prostaglandin biochemistry, *Mineral Electrolyte Metab.,* 6, 10, 1981.

21. **Ager, A., Gordon, J. L., Moncada, S., Pearson, J. D., Salmon, J. A., and Trevethick, M. A.,** Effects of isolation and culture on prostaglandin synthesis by porcine aortic endothelial and smooth muscle cells, *J. Cell. Physiol.,* 110, 9—16, 1982.

22. **Smith, W. L. and Rollins, T. E.,** Characteristics of rabbit anti-PGH synthase antibodies and use in immunocytochemistry, *Methods Enzymol.,* 86, 213—222, 1982.

23. **Sraer, J., Foidart, J., Chansel, D., Mahieu, P., Kouznetzova, B., and Ardaillou, R.,** Prostaglandin synthesis by mesangial and epithelial glomerular cultured cells, *FEBS Lett.,* 104, 420—424, 1979.

24. **Smith, W. L. and Bell, T. G.,** Immunohistochemical localization of the prostaglandin-forming cyclooxygenase in renal cortex, *Am. J. Physiol.,* 234, F451—457, 1976.

25. **Ardaillou, N., Sraer, J., Sraer, F. D., and Ardaillou, R.,** Prostaglandin synthesis by human isolated glomeruli and human glomerular cultured cells, in *Advances in Prostaglandin Thromboxane Research,* Paoletti, R. and Samuelsson, B., Eds., Raven Press, New York, in press.

26. **Shimizu, T., Yamamoto, S., and Hayaishi, O.,** Purification of PGH-PGD isomerase from rat brain, *Methods Enzymol.,* 86, 73—77, 1982.

27. **Christ-Hazelhof, E. and Nugteren, D. H.,** Isolation of PGH-PGD isomerase from rat spleen, *Methods Enzymol.,* 86, 77—84, 1982.

28. **Anggard, E., Bohman, S. O., Griffin, J. E., III, Larsson, C., and Maunsbach, A. B.,** Subcellular localization of the prostaglandin system in the rabbit renal papilla, *Acta Physiol. Scand.,* 84, 231—246, 1972.

29. **Rollins, T. E. and Smith, W. L.,** Subcellular localization of prostaglandin-forming cyclooxygenase in Swiss mouse 3T3 fibroblasts by electron microscopic immunocytochemistry, *J. Biol. Chem.,* 255, 4872—4875, 1980.

30. **Dewitt, D. L., Rollins, T. E., Day, J. S., Ganger, J. A., and Smith, W. L.,** Orientation of the active site and antigenic determinants of prostaglandin endoperoxide synthase in the endoplasmic reticulum, *J. Biol. Chem.,* 256, 10375, 1981.

31. **Gerrard, J. M. and White, J. G.,** Prostaglandins and thromboxanes: "middlemen" modulating platelet function in hemostasis and thrombosis, in *Progress in Hemostasis and Thrombosis,* Vol. 4, Spaeth, T. H., Ed., Greene & Stratton, New York, 1978, 87—125.

32. **Moonen, P., Buytenhek, M., and Nugteren, D. H.,** Purification of PGH-PGE isomerase from sheep vesicular glands, *Methods Enzymol.,* 86, 84—91, 1982.

33. **Ogino, N., Miyamoto, T., Yamamoto, S., and Hayaishi, O.,** Prostaglandin endoperoxide E isomerase from bovine vesicular gland microsomes, a glutathione-requiring enzyme, *J. Biol. Chem.,* 252, 890—895, 1977.

34. **Salmon, J. A., Smith, D. R., Flower, R. J., Moncada, S., and Vane, J. R.,** Further studies on the enzymatic conversion of prostaglandin endoperoxide into prostacyclin by porcine aorta microsomes, *Biochim. Biophys. Acta,* 523, 250—262, 1978.

35. **Smith, W. L., DeWitt, D. L., and Day, J. S.,** Purification, quantitation and localization of PGI_2 synthase using monoclonal antibodies, in *Advances in Prostaglandin Thromboxane Research,* Paoletti, R. and Samuelsson, B., Eds., Raven Press, New York, in press.

36. **Smith, W. L.,** Unpublished results.

37. **Bito, L. F.,** Are prostaglandins intracellular, transcellular or extracellular autocoids, *Prostaglandins,* 9, 851—855, 1975.

38. **Siegl, A. M., Smith, J. B., Silver, M. J., Nicolaou, K. C., and Ahern, D.,** Selective binding site for [³H] prostacyclin on platelets, *J. Clin. Invest.,* 63, 215—220, 1979.

39. **Skidgel, R. A. and Printz, M. P.,** PGI₂ production by rat blood vessels: diminished prostacyclin formation in veins compared to arteries, *Prostaglandins,* 16, 1—16, 1978.

40. **Gryglewski, R. J., Bunting, S., Moncada, S., Flaver, R. J., and Vane, J. R.,** Arterial walls are protected against deposition of platelet thrombi by a substance (prostaglandin X) which they make from prostaglandin endoperoxides, *Prostaglandins,* 12, 685—713, 1976.

41. **Claeys, M., Wechsung, E., Herman, A. G., and Nugteren, D. H.,** Lack of prostaglandin biosynthesis by aortic tissue of the chicken, *Prostaglandins,* 21, 739—749, 1981.

42. **Ingerman-Wojenski, C., Silver, M. J., Smith, J. B., and Macarek, E.,** Bovine endothelial cells in culture produce thromboxane as well as prostacyclin, *J. Clin. Invest.,* 67, 1292—1296, 1981.

43. **Hong, S. L.,** Effect of bradykinin and thrombin on prostacyclin synthesis in endothelial cells from calf and pig aorta and human umbilical cord vein, *Thromb. Res.,* 18, 787—795, 1980.

44. **Weksler, B. B., Marcus, A. J., and Jaffe, E. A.,** Synthesis of prostaglandin I₂ (prostacyclin) by cultured human and bovine endothelial cells, *Proc. Natl. Acad. Sci. U.S.A.,* 74, 3922—2936, 1977.

45. **VanRollins, M., Hos, S. H. K., Greenwald, J. E., Alexander, M., Dorman, N. J., Wong, L. K., and Horrocks, L. A.,** Complete separation by high performance liquid chromatography of metabolites of arachidonic acid from incubation with human and rabbit platelets, *Prostaglandins,* 20, 571—577, 1980.

46. **Harris, R. H., Fitzpatrick, T., Schmeling, J., Ryan, R., Kot, P., and Ramwell, P. W.,** Inhibition of dog platelet reactivity following 1-benzylimidazole administration, in *Advances in Prostaglandin Thromboxane Research,* Vol. 6, Samuelsson, B., Ramwell, P. W., and Paoletti, R., Eds., Raven Press, New York, 1980, 457—461.

47. **Watanabe, T., Narumiya, S., Shimizu, T., and Hayaishi, O.,** Characterization of the biosynthetic pathway for prostaglandin D₂ in human platelet-rich plasma, *J. Biol. Chem.,* in press.

48. **Miller, J. L., Stuart, M. J., and Walenga, R. W.,** Arachidonic acid metabolism in guinea pig megakaryocytes, *Biochem. Biophys. Res. Commun.,* 107, 752—759, 1982.

49. **Kennedy, M. S., Stobo, J. D., and Goldyne, M. E.,** In vitro synthesis of prostaglandins and related lipids by populations of human peripheral blood mononuclear cells, *Prostaglandins,* 20, 135—145, 1980.

50. **Bonney, R. J., Wightman, P. D., Davies, P., Sadowski, S. J., Kuehl, F. A., Jr., and Humes, J. L.,** Regulation of prostaglandin synthesis and of the selective release of lysosomal hydrolases by mouse peritoneal macrophages, *Biochem. J.,* 176, 433—442, 1978.

51. **Hsueh, W., Lamb., R., and Gonzalez-Crussi, R.,** Decreased phospholipase A₂ activity and prostaglandin biosynthesis in Bacillus Calmette-Guerin-activated alveolar macrophages, *Biochim. Biophys. Acta,* 710, 406—414, 1982.

52. **Kojima, A., Shiraki, M., Takahashi, R., Orimo, H., Morita, I., and Murota, S.,** Prostaglandin D₂ is the major prostaglandin of arachidonic acid metabolism in rat bone marrow homogenate, *Prostaglandins,* 20, 171—176, 1980.

53. **Pace-Asciak, C. and Nashat, M.,** Catabolism of prostaglandin endoperoxides into prostaglandin E₂ and F₂α by the rat brain, *J. Neurochem.,* 27, 551—556, 1976.

54. **Wolfe, L. S., Rostwarowski, K., and Marion, J.,** Endogenous formation of the prostaglandin endoperoxide metabolite thromboxane B₂ by brain tissue, *Biochem. Biophys. Res. Commun.,* 70, 907—913, 1976.

55. **Gerritsen, M. E. and Printz, M. P.,** Prostaglandin D synthase in microvessels from the rat cerebral cortex, *Prostaglandins,* 22, 553—566, 1981.

56. **Gecse, A., Ottlecz, A., Mezei, Z., Telegdy, G., Joo, F., Dux, E., and Karnushina, I.,** Prostacyclin and prostaglandin synthesis in isolated brain capillaries, *Prostaglandins,* 23, 287—297, 1982.

57. **Gerritsen, M. E., Parks, T. P., and Printz, M. P.,** Prostaglandin endoperoxide metabolism by bovine cerebral microvessels, *Biochim. Biophys. Acta,* 619, 196—206, 1980.

58. **Birkle, D. L., Wright, K. F., Ellis, C. K., and Ellis, E. F.,** Prostaglandin levels in isolated brain microvessels and in normal and norepinephrine-stimulated cat brain homogenates, *Prostaglandins,* 21, 865—877, 1981.

59. **Smith, W. L., Gutekunst, D. I., and Lyons, R. H., Jr.,** Immunocytochemical localization of the prostaglandin-forming cyclooxygenase in the cerebellar cortex, *Prostaglandins,* 19, 61—69, 1980.

60. **Hambrecht, B., Jaffe, B. M., and Philpott, G. W.,** Prostaglandin production by neuroblastoma, glioma and fibroblast cell lines; stimulation by N⁶, O² dibutyryl adenosine 3′:5′-cyclic monophosphate, *FEBS Lett.,* 36, 193—198, 1973.

61. **Shimizu, T., Mixuno, N., Amano, T., and Hayaishi, O.,** Prostaglandin D₂ a neuromodulator, *Proc. Natl. Acad. Sci. U.S.A.,* 76, 6231—6234, 1979.

62. **Burka, J. F., Ali, M., McDonald, J. W. D., and Paterson, N. A. M.,** Immunological and non-immunological synthesis and release of prostaglandins and thromboxanes from isolated guinea pig trachea, *Prostaglandins,* 22, 683—691, 1981.

63. **Eling, T.,** Personal communication.

64. **Hoyan, S. S., McNamera, D. B., Spannhake, E. W., Hyman, A. L., and Kadowitz, P. J.,** Metabolism of prostaglandin endoperoxide by microsomes from cat lung, *Prostaglandins,* 21, 531, 1981.

65. **Korbut, R., Boyd, J., and Eling, T.,** Respiratory movements alter the generation of prostacyclin and thromboxane A_2 in isolated rat lungs: the influence of arachidonic acid-pathway inhibitors on the ratio between pulmonary prostacyclin and thromboxane A_2, *Prostaglandins,* 21, 491—503, 1981.

66. **Sun, F. F., Chapman, J. P., and McGuire, J. C.,** Metabolism of prostaglandin endoperoxide in animal tissues, *Prostaglandins,* 14, 1055—1074, 1977.

67. **Taylor, L., Polgar, P., McAteer, J. A., and Douglas, W. H. J.,** Prostaglandin production by type II alveolar epithelial cells, *Biochim. Biophys. Acta,* 572, 502—509, 1979.

68. **Sraer, J., Siess, W., Moulonguet-Doleris, L., Oudinet, J., Dray, F., and Ardaillou, R.,** In vitro prostaglandin synthesis by various rat renal preparations, *Biochim. Biophys. Acta,* 710, 45—52, 1982.

69. **Hassid, A. and Dunn, M. J.,** Microsomal prostaglandin biosynthesis of human kidney, *J. Biol. Chem.,* 255, 2472—2475, 1980.

70. **McGiff, J. C. and Wong, P. Y.-K.,** Compartmentalization of prostaglandins and prostacyclin within the kidney: implications for renal function, *Fed. Proc.,* 38, 89—93, 1979.

71. **Smith, W. L., Bell, T. G., and Needleman, P.,** Increased renal tubular synthesis of prostaglandins in the rabbit kidney in response to urethral obstruction, *Prostaglandins,* 18, 269—277, 1979.

72. **Needleman, P.,** Personal communication.

73. **Bohman, S. O.,** Demonstration of prostaglandin synthesis in collecting duct cells and other cell types of the rabbit renal medulla, *Prostaglandins,* 14, 729—744, 1978.

74. **Pugliese, F., Dunn, M. J., Aikawa, M., Williams, S., Sato, M., and Hassid, A.,** Rabbit and rat renal papillary collecting tubule (RPCT) cells in culture. The interactions of arginine vasopressin (AVP), prostaglandins (PGs) and cyclic AMP (cAMP), in *Advances in Prostaglandin Thromboxane Research,* Paoletti, R. and Samuelsson, B., Eds., Raven Press, New York, in press.

75. **Currie, M. G. and Needleman, P.,** Distribution of prostaglandin synthesis along the rabbit nephron, *Fed. Proc.,* 41, 1545, 1982.

76. **Zusman, R. M. and Keiser, H. R.,** Prostaglandin E_2 biosynthesis by rabbit renomedullary interstitial culture: mechanism by agiotensin II, bradykinin, and vasopressin, *J. Biol. Chem.,* 252, 2069—2071, 1977.

77. **LeDuc, L. E. and Needleman, P.,** Regional localization of prostacyclin and thromboxane synthesis in dog stomach and intestinal tract, *J. Pharm. Exp. Ther.,* 211, 181—188, 1979.

78. **Ali, M. and McDonald, J. W. D.,** Synthesis of thromboxane B_2 and 6-keto-prostaglandin $F_{1\alpha}$ by bovine gastric mucosa and muscle microsomes, *Prostaglandins,* 20, 245—254, 1980.

79. **Bebiak, D. M., Miller, E. R., Huslig, R. L., and Smith, W. L.,** Distribution of prostaglandin-forming cyclooxygenase in the porcine stomach, *Fed. Proc.,* 38, 884, 1979.

80. **Jones, R. L., Poyser, N. L., and Wilson, N. H.,** Production of 6-oxo-prostaglandin $F_{1\alpha}$ by rat, guinea pig and sheep uteri, *In Vitro, Proc. B. P. S.,* 59, 436—437, 1977.

81. **Williams, R. I., Dembinska-Kiec, A., Zmuda, A., and Gryglewski, R. J.,** Prostacyclin formation by myometrial and decidual fractions of the pregnant rat uterus, *Prostaglandins,* 15, 343, 1978.

82. **Huslig, R. L., Fogwell, R. L., and Smith, W. L.,** The prostaglandin-forming cyclooxygenase of ovine uterus: relationship to luteal function, *Biol. Reprod.,* 21, 589—600, 1979.

83. **Rees, M. C. P., Parry, D. M., Anderson, A. B. M., and Turnbull, A. C.,** Immunohistochemical localization of cyclooxygenase in human uterus, *Prostaglandins,* 23, 207—214, 1982.

84. **Triebwasser, W. F., Clark, M. R., LeMaire, W. J., and Marsh, J. M.,** Localization and in vitro synthesis of prostaglandins in components of rabbit preovulatory graafian follicles, *Prostaglandins,* 16, 621—631, 1978.

85. **Clark, M. R.,** Personal communication.

86. **Murdoch, W. J., Dailey, R. A., and Inskeep, E. K.,** Preovulatory changes in prostaglandins E_2 and $F_{2\alpha}$ in ovine follicles, *J. Anim. Sci.,* 43, 192—205, 1981.

87. **Needleman, P., Bronson, S. D., Wyche, A., Sivakoff, M., and Nicolaou, K. C.,** Cardiac and renal prostaglandin I_2: biosynthesis and biological effects in isolated perfused rabbit tissues, *J. Clin. Invest.,* 61, 839—849, 1978.

88. **Smith, W. L. and Needleman, P.,** Immunocytochemical localization of the site of cardiac prostaglandin biosynthesis, in *Prostaglandins and the Microcirculation,* Kaley, G., Ed., Raven Press, New York, in press,

89. **Murota, S. and Morita, I.,** Prostaglandin-synthesizing system in rat liver: changes with aging and various stimuli, in *Advances in Prostaglandin Thromboxane Res.,* Vol. 8, Samuelsson, B., Ramwell, P., and Paoletti, R., Eds., Raven Press, New York, 1980, 1495—1506.

90. **Berlin, T., Cronestrand, R., Nowak, J., Sonnenfeld, T., and Wennmalm, A.,** Conversion of arachidonic acid to prostaglandins in homogenates of human skeletal muscle and kidney, *Acta Physiol. Scand.,* 106, 441—445, 1979.

91. **Hammarstrom, S., Lindgren, J. A., Marcelo, C., Duell, E. A., Anderson, T. F., and Vorhees, J. J.,** Arachidonic acid transformations in normal and psoriatic skin, *J. Invest. Dermatol.,* 73, 180—183, 1979.

92. **Ruzicka, T. and Printz, M. P.,** Arachidonic acid metabolism in guinea pig skin, *Biochim. Biophys. Acta,* 711, 391—397, 1982.

93. **Laychock, S. G. and Walker, L.,** Evidence for 6-keto-PGF$_{1\alpha}$ in adrenal cortex of the rat and effects of 6-keto-PGF$_{1\alpha}$ and PGI$_2$ on adrenal cAMP levels and steroidogenesis, *Prostaglandins,* 18, 793—811, 1979.

94. **Miller, R. T., Douglas, J. G., and Dunn, M. J.,** Dissociation of aldosterone and prostaglandin biosynthesis in the rat adrenal glomerulosa, *Prostaglandins,* 20, 449—462, 1980.

95. **Cottee, F., Flower, R. J., Moncada, S., Salmon, J. A., and Vane, J. R.,** Synthesis of 6-keto-PGF$_{1\alpha}$ by ram seminal vesicle microsomes, *Prostaglandins,* 14, 413—423, 1977.

96. **Bauduin, H., Galand, N., and Boeynaems, J. M.,** In vitro stimulation of prostaglandin synthesis in the rat pancreas by carbamylcholine, caerulein and secretin, *Prostaglandins,* 22, 35—51, 1981.

97. **Patrono, C., Rotella, C. M., Toccafondi, R. S., Aterini, S., Pinca, E., Tanini, A., and Zonefrati, R.,** Prostacyclin stimulates the adenylate cyclase system of human thyroid tissue, *Prostaglandins,* 22, 105—115, 1981.

98. **Levasseur, S., Sun, F. F., Friedman, Y., and Burke, G.,** Arachidonate metabolism in the mouse thyroid: implication of the lipoxygenase pathway in thyrotropin action, *Prostaglandins* 22, 663—673, 1981.

99. **Englis, D. J., D'Souza, S. M., Meats, J. E., Wright, J., McGuire, M. B., and Russell, R. G. G.,** Stimulation of prostaglandin biosynthesis by mononuclear cell factor added to cells from human gingiva, cartilage, synovium and endometrium, in *Advances in Prostaglandin Thromboxane Research,* Vol. 8, Samuelsson, B., Ramwell, P. W., and Paoletti, R., Eds., Raven Press, New York, 1980, 1709—1711.

100. **Copeland, M., Lippiello, L., Steensland, G., Guralnick, C., and Mankin, II. J.,** The prostaglandins of articular cartilage. I. Correlates of prostaglandin activity in a chrondrocyte culture system, *Prostaglandins,* 20, 1075—1087, 1980.

101. **Mitrovic, D., McCall, F., and Dray, F.,** The in vitro production of prostanoids by cultured bovine articular chondrocytes, *Prostaglandins,* 23, 17—28, 1982.

102. **Vigo, C., Lewis, G. P., and Piper, P. J.,** Glucocorticoids and the prostaglandin system in adipose tissue, in *Advances in Prostaglandin Thromboxane Research,* Vol. 6, Samuelsson, B., Ramwell, P. W., and Paoletti, R., Eds., Raven Press, New York, 1980, 263—265.

103. **Sors, H., Pradelles, P., Dray, F., Rigaud, M., Maclouf, J., and Bernard, P.,** Analytical methods for thromboxane B$_2$ measurement and validation of radioimmunoassay by gas liquid chromatography-mass spectrometry, *Prostaglandins,* 16, 277—290, 1978.

104. **Hassid, A.,** Transport-active renal tubular epithelial cells (MDCK and LLC-PK$_1$ in culture. Prostaglandin biosynthesis and its regulation by peptide hormones and ionophore, *Prostaglandins,* 21, 985—1001, 1981.

105. **Lewis, M. G. and Spector, A. A.,** Differences in types of prostaglandins produced by two MDCK canine kidney cell sublines, *Prostaglandins,* 21, 1025—1032, 1981.

106. **Baenziger, N. L., Becherer, P. R., and Majerus, P. W.,** Characterization of prostacyclin synthesis in cultured human arterial smooth muscle cells, venous endothelial cells and skin fibroblasts, *Cell,* 16, 967—974, 1978.

107. **Ali, A. E., Barrett, J. C., and Eling, T. E.,** Prostaglandin and thromboxane formation by fibroblasts and vascular endothelial cells, *Prostaglandins,* 20, 667—688, 1980.

108. **Koshihara, Y., Senshu, T., Kawamura, M., and Murota, S.,** Sodium n-butyrate induces prostaglandin synthetase activity in mastocytoma P-815 cells, *Biochim. Biophys. Acta,* 617, 536—539, 1980.

109. **Hopkins, N. K., Sun, F. F., and Gorman, R. R.,** Thromboxane A$_2$ biosynthesis in human lung fibroblasts WI-38, *Biochem. Biophys. Res. Commun.,* 85, 827—836, 1978.

110. **Levine, L., Hinkle, P. M., Voelkel, E. F., and Tashjian, A. H., Jr.,** Prostaglandin production by mouse fibrosarcoma cells in culture: inhibition by indomethacin and aspirin, *Biochem. Biophys. Res. Commun.,* 47, 888—896, 1972.

111. **Levine, L. and Alam, I.,** Arachidonic acid metabolism by cells in culture: analysis of culture fluids for cyclooxygenase products by radioimmunoassay before and after separation by high pressure liquid chromatography, *Prostaglandins Med.,* 3, 295—304, 1979.

THE BIOSYNTHESIS OF PROSTACYCLIN

C. Pace-Asciak

Prostacyclin (PGI_2) is a term coined by Moncada et al.[1] for a biologically active product generated from arachidonic acid (AA) and the prostaglandin endoperoxides by bovine aorta. The pathway for the formation of this product had earlier been shown to exist in the rat stomach by Pace-Asciak and Wolfe,[2] who also isolated and identified the stable end product of this pathway, 6-keto $PGF_{1\alpha}$, during incubations of AA or prostaglandin endoperoxides with homogenates of the rat stomach.[3,4] Under acidic conditions, PGI_2 is converted into 6-keto $PGF_{1\alpha}$, although it is quite stable to alkaline conditions.[5] The PGI_2 skeleton consists of a derivative of prostanoic acid containing an ether linkage between carbon atoms 6 and 9 (see Figure 1). As with other prostaglandins, the PGI_2 family is derived from certain methylene interrupted *cis* polyunsaturated fatty acids; the structural requirements are more restricted for PGI_2 synthesis in that a Δ^5 double bond is an essential requirement for 6(9) oxy ring formation (see Figure 2).

BIOSYNTHESIS

PGI_2 is formed from the prostaglandin endoperoxide PGH_2 by an enzyme which is present mostly in vascular tissue. This enzyme is present in both endothelial and smooth muscle cells.[8-10] The enzyme which converts PGH_2 to PGI_2 has been termed PGI_2 synthase[11] or 6(9)-oxy cyclase;[12] the latter terminology stresses the type of reaction that actually takes place, i.e., a cyclization between the oxygen atom at carbon atom 9 of PGH_2 with carbon atom 6 to produce a new class of prostaglandins having a 6(9)oxy ring. The proposed general mechanism for the rearrangement of PGH_2 into PGI_2 is shown in Figure 3. The importance of the Δ^5 double bond in the endoperoxide is evident since ring closure can only occur if this bond is present. In fact, PGH_1 which lacks the Δ^5 bond does not form a prostacyclin, while PGH_3 having $\Delta^{5,17}$ double bonds does form the corresponding PGI_3 (Figure 2, Reference 13).

During the process of conversion of PGH_2 into PGI_2 and 6-keto $PGF_{1\alpha}$, a hydrogen atom is lost from carbon atom 6. This was initially shown with octadeuterated [5,6,8,9,11,12,14,15]-AA and PGH_2 and analysis of 6-keto $PGF_{1\alpha}$ by mass spectrometry.[3] Subsequently, Tai et al.[14] showed that the conversion of PGH_2 into 6-keto $PGF_{1\alpha}$ could be conveniently assayed through loss of tritium from [5,6-$^3H_2$]-PGH_2 during enzymic incubations.

ENZYME PROPERTIES

The PGI_2 synthase is a membrane-bound enzyme which can be solubilized from microsomal preparations with Triton® X-100[15,16] and isolated and purified by affinity chromatography at neutral pH using an immobilized oxymethano derivative of PGH.[17] The enzyme is a hemoprotein with a subunit mol wt of approximately 50,000 daltons with a heme peak at 418 to 420 nm. The porcine synthase contains one heme per subunit and exhibits a reduced CO spectrum similar to that of cytochrome P-450 although it cannot be reduced by NADPH-cytochrome *c* reductase in the presence of NADPH.[17,18] A specific activity of 2000 μmol PGI_2/min/mg protein has been reported for the porcine synthase and 1000 μmol PGI_2/min/mg protein for a bovine synthase. The latter enzyme contains only 0.1 to 0.2 heme units per subunit.

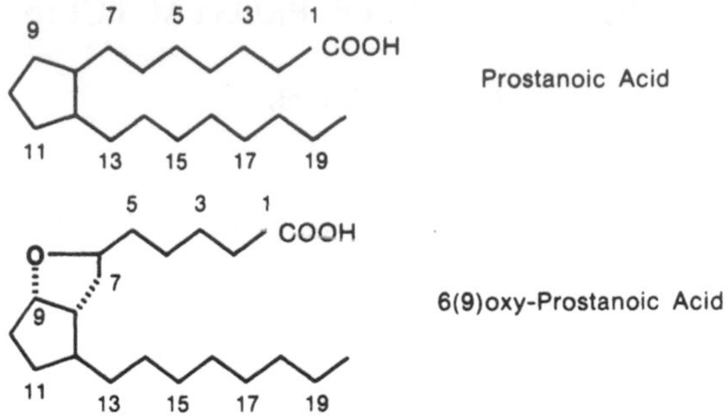

FIGURE 1.   Diagrams depicting prostanoic acid, the basic skeleton of the prostaglandins and 6(9)-oxy prostanoic acid, the basic skeleton of prostacyclin.

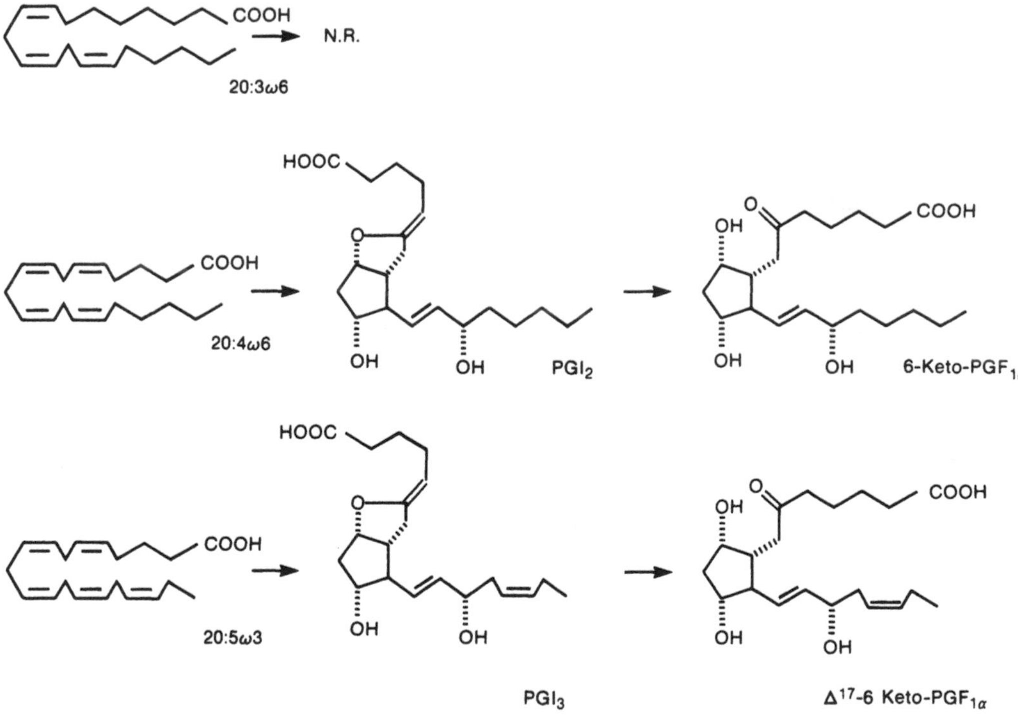

FIGURE 2.   Structures of prostacyclins and 6-keto PGF$_\alpha$ products derived from different fatty acid precursors.

CATALYTIC PROPERTIES OF PGI$_2$ SYNTHASE

Conversion of PGH$_2$ to PGI$_2$ is accompanied by inactivation of the enzyme and bleaching of the heme spectrum. Azo Analog I protects the enzyme from both inactivation and heme bleaching.[7,19] The reason for this is not known.

The enzymic reaction (K_m = 5 μM for PGH$_2$)[20] has no requirement for any co-factors, and the effect of serum albumin and other related proteins has not been investigated except that albumin has been reported to facilitate nonenzymatic conversion of PGH$_2$ into PGD$_2$.[21-23] Reduced GSH causes nonenzymatic conversion of PGH$_2$ into PGE$_2$.[24,25] Since the PGI$_2$ synthase is exceedingly sensitive to inhibition by fatty acid hydroperoxides,[26-28]

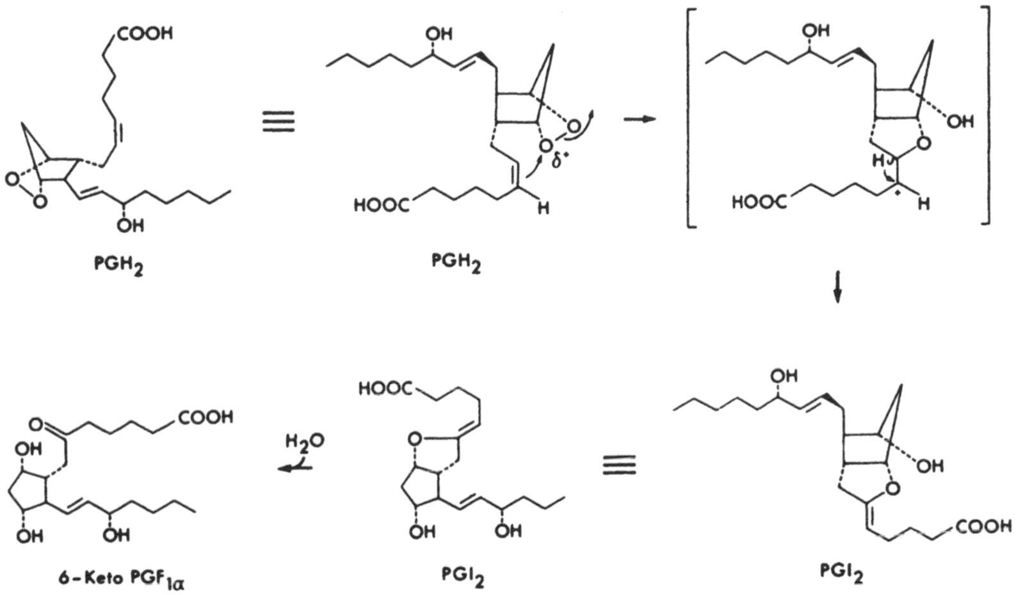

FIGURE 3. Mechanisms proposed by Just and Fried (Winter Conference on Prostaglandins, Vail, Colo., January 24—27, 1976) for the biosynthesis of prostacyclin.

antioxidants in low amounts are generally favorable toward PGI_2 synthesis by removal of these inhibitory hydroperoxides. The early observation of stimulation of PGI_2 synthesis in rat stomach homogenates by catecholamines[37] is probably related to the rapid removal of PGG_2, a hydroperoxide, converting it into PGH_2, the immediate precursor to PGI_2. Conversely, in cases of high lipid peroxidation such as vitamin E deficiency[29-31] or atherosclerosis,[32] decreased PGI_2 synthesis is evident.

TISSUE DISTRIBUTION

A variety of tissues have been shown to form and release prostacyclin detected as its hydrolysis product, 6-keto $PGF_{1\alpha}$. These include the rat stomach,[33-42] brain,[40,41] and liver,[40,42] sheep seminal vesicles,[33,35,72] bovine aorta,[44,45] and other vascular tissues from fetal and adult animals,[4,45-54] cultured vascular endothelial,[51,56] smooth muscle,[57,58] and transformed cells,[59] rat granuloma cells,[60] fibroblasts,[61] heart,[62] and lung.[63-66] The enzyme that forms PGI_2 appears to be located at least at two different sites within the same cell. Smith et al., through use of monoclonal antibodies to PGI_2 synthase and through immunofluorescent assays have shown a population of PGI_2 synthases at the inner surface of the plasma membrane and at the nuclear membrane in smooth muscle cells (see previous chapter). This suggests different roles for PGI_2 of different sites.[67]

PLASMA LEVELS

Prostacyclin is believed to be released intact (as PGI_2) with subsequent hydrolysis into 6-keto $PGF_{1\alpha}$ which is mostly resistant toward pulmonary degradation via 15-hydroxy-prostaglandin dehydrogenase.[68-70] Although intra-arterially administered PGI_2 is also mostly unaffected by this enzyme, yet 15-keto and 15-keto-13,14-dihydro metabolites of endogenous PGI_2 have been detected in the venous effluent of intact perfused organs or in plasma indicating that endogenous PGI_2 might be inactivated to some extent into these products before being released into the circulation.[71,72] Considerable lack of clarity exists regarding the origin of PGI_2 that might enter into the circulation; there is also lack of agreement regarding the amount of PGI_2 that is present in the circulation. Measurements of 6-keto $PGF_{1\alpha}$ by radioimmunoassay show a large range of values (100 pg to several nanograns per milliliter of blood).[65,71,73-84] Values of the order of 100 pg/mℓ were also initially reported by the specific method of gas chromatography/mass spectrometry (GC/MS),[52,80,82] although later studies using the highly sensitive and more specific method of negative ion chemical ionization GC/MS Blair et al.[85] reported values of plasma 6-keto $PGF_{1\alpha}$ below their limits of detection, (0.5 pg/mℓ). The latter study suggests that 6-keto $PGF_{1\alpha}$ is virtually nonexistent in the normal circulation. Using antibodies to PGI_2 which sequester PGI_2 and render it resistant to degradation, Pace-Asciak et al.[86,87] developed a method to measure the rate of entry of prostaglandins into the circulation by administering specific antibodies intravenously. Since antibody-bound PGI_2 is biologically inactive and unavailable for metabolism by enzymes in the lungs and other organs,[87,87] levels of this product increase in the circulation as a function of time after administration of antibodies. With this novel method a rate of appearance of 38 pg/min was reported for PGI_2, measured as 6-keto $PGF_{1\alpha}$ by radioimmunoassay.

These studies also showed lack of significant difference between the rate of appearance of 6-keto $PGF_{1\alpha}$ in the circulation of normotensive and hypertensive rats.[86] One limitation of these studies must be mentioned. Since it is not known to what extent PGI_2 is released under basal conditions as PGI_2 itself or as its 15-keto and 15-keto-13,14-dihydro metabolites, the above-mentioned antibody experiments offer only a rough estimate of PGI_2 generation in vivo since they do not provide a measurement of the PGI_2 metabolites as well. More information on PGI_2 production in vivo will certainly be forthcoming as improvements in the techniques for the measurements of the minute levels of these products are developed.

REFERENCES

1. **Bunting, S., Gryglewski, R., Moncada, S., and Vane, J. R.,** Arterial walls generate from PG endoperoxides a substance (PGX) which relaxes strips of coeliac arteries and inhibits platelet aggregation, *Prostaglandins*, 12, 897—913, 1976.

2. **Pace-Asciak, C. and Wolfe, L. S.,** A novel prostaglandin derivative formed from arachidonic acid by rat stomach homogenates, *Biochemistry*, 10, 3657—3664, 1971.

3. **Pace-Asciak, C.,** Isolation, structure, and biosynthesis of 6-ketoprostaglandin $F_{1\alpha}$ in the rat stomach, *J. Am. Chem. Soc.*, 98, 2348—2349, 1976.

4. **Pace-Asciak, C.,** A new prostaglandin metabolite of arachidonic acid. Formation of 6-keto $PGF_{1\alpha}$ by the rat stomach, *Experientia*, 33, 291—292, 1976.

5. **Johnson, R. A., Morton, D. R., Kinner, J. H., Gorman, R. R., McGuire, J. C., and Sun, F. F.,** The chemical structure of prostaglandin X (prostacyclin), *Prostaglandins*, 12, 915—928, 1976.

6. **Smith, W. L., DeWitt, D. L., and Allen, M. L.,** Bimodal distribution of prostaglandin I_2 synthase antigen in smooth muscle, *J. Biol. Chem.*, 258, 5922—5926, 1983.

7. **Smith, W. L., DeWitt, D. L., and Day, J. S.,** Purification, quantitation, and localization of PGI_2 synthase using monoclonal antibodies, in *Advances in Prostaglandin, Thromboxane and Leukotriene Research*. Vol. 2, Samuelsson, B., Paoletti, R., and Ramwell, P., Eds., Raven Press, New York, 1983, 87—92.

8. **Moncada, S., Herman, A. G., Higgs, E. A., and Vane, J. R.,** Differential formation of prostacyclin (PGX or PGI_2) by layers of the arterial wall: an explanation for the anti-thrombotic properties of vascular endothelium, *Thromb. Res.*, 2, 323—344, 1977.

9. **Christofinis, G. J., Moncada, S., Bunting, S., and Vane, J. R.,** Prostaglandin release by rabbit aorta and human umbilical vein endothelial cells after prolonged subculture, in *Prostacyclin*, Vane, J. R., and Bergstrom, S., Eds., Raven Press, New York, 1979, 77—84.

10. **Tansik, R. L., Namm, D. H., and White, H. L.,** Synthesis of 6-keto PGF$_{1\alpha}$ by cultured aortic smooth muscle cells and stimulation of its formation in a coupled system with platelet lysates, *Prostaglandins,* 15, 399—408, 1978.

11. **Johnson, R. A., Morton, D. R., Kinner, J. H., Gorman, R. R., McGuire, J. C., Sun, F. F., Whittaker, N., Bunting, S., Salmon, J., Moncada, S., and Vane, J. R.,** The chemical structure of prostaglandin X (prostacyclin), *Prostaglandins,* 12, 915—928, 1976.

12. **Pace-Asciak, C. and Nashat, M.,** Mechanistic studies on the biosynthesis of 6-ketoprostaglandin F$_{1\alpha}$, *Biochim. Biophys. Acta,* 487, 495—507, 1977.

13. **Needleman, P., Raz, A., Minkes, M. S., Ferrendelli, J. A., and Sprecher, H.,** Triene prostaglandins, prostacyclin and thromboxane biosynthesis and unique biological properties, *Proc. Natl. Acad. Sci. U.S.A.,* 76, 944—948, 1979.

14. **Tai, H-H., Hsu, C-T., Tai, C. L. and Sih, C. J.,** Synthesis of tritium labeled arachidonic acid and its use in development of a sensitive assay for prostacyclin synthetase, *Biochemistry,* 19, 1989—1993, 1980.

15. **Wlodawer, P. and Hammerström, S.,** Conversions of prostaglandin endoperoxides by prostacyclin synthase from pig aorta, *Prostaglandins,* 19, 969—976, 1980.

16. **Watanabe, K., Yamamoto, S., and Hayaishi, O.,** Reactions of prostaglandin endoperoxides with prostaglandin I synthetase solubilized from rabbit aorta microsomes, *Biochem. Biophys. Res. Commun.,* 87, 192—199, 1979.

17. **Ullrich, V. and Haurand, M.,** Thromboxane synthase as a cytochrome P-450 enzyme, in *Advances in Prostaglandin, Thromboxane and Leukotriene Research,* Samuelsson B., Paoletti, R., and Ramwell, P., Eds., Vol. 11, Raven Press, New York, 1983, 105—110.

18. **Ullrich, V., Castle, L., and Weber, P.,** Spectral evidence for the cytochrome P450 nature of prostacyclin synthetase, *Biochem. Pharmacol.,* 30, 2033—2036, 1981.

19. **Smith, W. L., DeWitt, D. L., and Day, J. S.,** Purification, quantitation, and localization of PGI$_2$ synthase using monoclonal antibodies, in *Advances in Prostaglandin, Thromboxane and Leukotriene Research,* Samuelsson, B., Paoletti, R., and Ramwell, P., Eds., Vol. 2, Raven Press, New York, 1983, 87—92.

20. **Salmon, J. A., Smith, D. R., Flower, R. J., Moncada, S., and Vane, J. R.,** Further studies on the enzymatic conversion of PG enderoperoxide into prostacyclin in porcine aorta microsomes, *Biochim. Biophys. Acta,* 523, 250—262, 1978.

21. **Shimizu, T., Yamamoto, S., and Hayaishi, O.,** Purification and properties of prostaglandin D synthetase from rat brain, *J. Biol. Chem.,* 254, 5222—5228, 1979.

22. **Christ-Hazelhof, E., Nugteren, D. H., and Van Dorp, D. A.,** Conversions of prostaglandin endoperoxides by glutathione-S-transferases and serum albumins. *Biochim. Biophys. Acta,* 450, 450—461, 1976.

23. **Hamberg, M. and Fredholm, B. B.,** Isomerization of prostaglandin H$_2$ into prostaglandin D$_2$ in the presence of serum albumin, *Biochim. Biophys. Acta,* 431, 189—193, 1976.

24. **Ogino, N., Miyamoto, T., Yamamoto, S., and Hayaishi, O.,** Prostaglandin endoperoxide E isomerase from bovine vesicular gland microsomes. A glutathione-requiring enzyme, *J. Biol. Chem.,* 252, 890—895, 1977.

25. **Moonen, P., Buytenhek, M., and Nugteren, D. H.,** Purification of PGH-PGE isomerase from sheep vesicular glands, *Methods Enzymol.,* 86, 84—91, 1982.

26. **Bunting, S., Gryglewski, R., Moncada, S., and Vane, J. R.,** Arterial walls generate from PG endoperoxides a substance (PGX) which relaxes strips of coeliac arteries and inhibits platelet aggregation, *Prostaglandins,* 12, 897—913, 1976.

27. **Salmon, J. A., Smith, D. R., Flower, R. J., Moncada, S., and Vane, J. R.,** Further studies on the enzymatic conversion of PG enderoperoxide into prostacyclin by porcine aorta microsomes, *Biochim. Biophys. Acta,* 523, 250—262, 1978.

28. **Ham, E. A., Egan, R. W., Soderman, D. D., Gale, P. H., and Kuehl, F. A., Jr.,** Peroxidase-dependent deactivation of prostacyclin synthetase, *J. Biol. Chem.,* 254, 2191—2194, 1979.

29. **Carpenter, M.,** Antioxidant effects on the prostaglandin endoperoxide synthase product profile, *Fed. Proc.,* 40, 189—194, 1981.

30. **Okuma, M., Takayama, H., and Uchino, H.,** Generation of prostacyclin-like substance and lipid peroxidation in vitamin E-deficient rats, *Prostaglandins,* 19, 527—536, 1981.

31. **Chan, A. C. and Leith, M. K.,** Decreased prostacyclin synthesis in vitamin E-deficient rabbit aorta, *Am. J. Clin. Nutr.,* 34, 2341—2347, 1981.

32. **Gryglewski, R. J., Dembinska-Kiec, A., Zmuda, A., and Gryglewska, T.,** Prostacyclin and thromboxane A$_2$ biosynthesis capacities of heart, arteries and platelets at various stages of experimental atherosclerosis in rabbits, *Atherosclerosis,* 31, 385—394, 1978.

33. **Pace-Asciak, C. and Wolfe, L. S.,** A prostanoic acid derivative formed in the enzymatic conversion of tritiated arachidonic acid into prostaglandins by rat stomach homogenates, *Chem. Commun.,* 1234—1235, 1970.

34. **Pace-Asciak, C. and Wolfe, L. S.,** Polyhydroxylated by-products of the enzymatic conversion of tritiated arachidonic acid into prostaglandins by sheep seminal vesicles, *Chem. Commun.,* 1235—1236, 1970.

35. **Pace-Asciak, C. and Wolfe, L. S.,** A novel prostaglandin derivative formed from arachidonic acid by rat stomach homogenates, *Biochemistry,* 10, 3657—3664, 1971.
36. **Pace-Asciak, C.,** Polyhydroxy cyclic ethers formed from tritiated arachidonic acid by acetone powders of sheep seminal vesicles, *Biochemistry,* 10, 3664—3669, 1971.
37. **Pace-Asciak, C.,** Prostaglandin synthetase activity in the rat stomach fundus. Activation by L-norepinephrine and related compounds, *Biochim. Biophys. Acta,* 280, 161—171, 1972.
38. **Pace-Asciak, C.,** Catecholamine induced increase in prostaglandin E biosynthesis in homogenates of the rat stomach fundus, *Adv. Biosci.,* 9, 29—33, 1973.
39. **Pace-Asciak, C. and Nashat, M.,** Transformation of prostaglandin H2 into 6(9)-oxy-11,15-dihydroxyprosta-7, 13-dienoic acid by the rat stomach fundus, *Biochim. Biophys. Acta,* 424, 323—325, 1976.
40. **Pace-Asciak, C.,** Presented at the Winter Conference on Prostaglandins, Vail, Colo., January 24-27, 1976.
41. **Pace-Asciak, C. and Nashat, M.,** The 6(9)-oxy pathway in prostaglandin biosynthesis — isolation of an intermediate, 6-keto-$PGF_{1\alpha}$ and its occurrence in several tissues, *Proc. Can. Fed. Biol. Soc.,* 19, 268, 1976.
42. **Pace-Asciak, C. and Nashat, M.,** Catabolism of prostaglandin endoperoxides into prostaglandin E_2 and $F_{2\alpha}$ by the rat brain, *J. Neurochem.,* 27, 551—556, 1976.
43. **Pace-Asciak, C. and Rangaraj, G.,** The 6-ketoprostaglandin $F_{1\alpha}$ pathway in the lamb ductus arteriosis, *Biochim. Biophys. Acta,* 486, 583—585, 1977.
44. **Cottee, F., Flower, R. J., Moncada, S., Salmon, J. A., and Vane, J. R.,** Synthesis of 6-keto-$PGF_{1\alpha}$ by ram seminal vesicle microsomes, *Prostaglandins,* 14, 413—423, 1977.
45. **Moncada, S., Gryglewski, R. J., Bunting, S., and Vane, J. R.,** An enzyme isolated from arteries transforms prostaglandin eneroperoxides to an unstable substance that inhibits platelet aggregation, *Nature (London),* 263, 663—665, 1976.
46. **Gryglewski, R. J., Bunting, S., Moncada, S., Flower, R. J., and Vane, J. R.,** Arterial walls are protected against deposition of platelet thrombi by a substance (PGX) which they make from prostaglandin endoperoxides, *Prostaglandins,* 12, 685—714, 1976.
47. **Powell, W. and Solomon, S.,** Formation of 6-oxoprostaglandin $F_{1\alpha}$ by arteries of the fetal calf, *Biochem. Biophys. Res. Commun.,* 75, 815—822, 1977.
48. **Sun, F. F. and McGuire, J. C.,** Metabolism of PGs and PG endoperoxides in rabbit tissues during pregnancy: differences in enzyme activities between mother and fetus, in *Advances in Prostaglandin Thromboxane Research,* Vol. 4, Coceani, F. and Olley, P. M., Eds., Raven Press, 1978, 75—85.
49. **Terragno, N. A., Terragno, A. McGiff, J. C., and Rodriguez, D. J.,** Synthesis of prostaglandins by the ductus arteriosis of the bovine fetus, *Prostaglandins,* 14, 721—727, 1977.
50. **Terragno, N. A., Terragno, A., and McGiff, J. C.,** Prostacyclin (PGI_2) in production by renal blood vessels: relationship to an endogenous prostaglandin synthesis inhibitor (EPSI), *Clin. Res.,* 26, 545A, 1978.
51. **Skidgel, R. A. and Printz, M. P.,** PGI_2 production by rat blood vessels: diminished prostacyclin formation in veins compared to arteries, *Prostaglandins,* 16, 1—16, 1978.
52. **Wolfe, L. S., Rostworoski, K., and Manku, M.,** Measurement of prostaglandin synthesis and release from rat aortic tissue and from the perfused mesenteric artery by gas chromatography mass spectrometric methods, in *Prostacyclin,* Vane, J. R. and Bergström, S., Eds., Raven Press, New York, 1979, 113—118.
53. **Villa, S. and de Gaetano, G.,** Prostacyclin-like activity in rat vascular tissues: fast, long-lasting inhibition by treatment with lysine acetylsalicylate, *Prostaglandins,* 14, 1117—1124, 1977.
54. **MacIntyre, D. E., Pearson, J. D., and Gordon, J. L.,** Localization and stimulation of prostacyclin production in vascular cells, *Nature (London),* 271, 549—551, 1978.
55. **Weksler, B. B., Marcus, A. J., and Jaffe, E. A.,** Synthesis of prostaglandin I_2 (prostacyclin) by cultured human and bovine endothelial cells, *Proc. Natl. Acad. Sci. U.S.A.,* 74, 3922—3926, 1977.
56. **Christofinis, G. J., Moncada, S., Bunting, S., and Vane, J. R.,** Prostacyclin release by rabbit aorta and human umbilical vein endothelial cells after prolonged subculture, in *Prostacyclin,* Vane, J. R. and Bergström, S., Eds., Raven Press, New York, 1979, 77—84.
57. **Tansik, R. L., Namm, D. H., and White, H. L.,** Synthesis of Prostaglandin 6-keto-F_a by cultured smooth muscle cells and stimulation of its formation in a coupled system with platelet lysates, *Prostaglandins,* 15, 399—408, 1978.
58. **Fitzpatrick, F. A., Stringfellow, D. A., Maclouf, J., and Rigaud, M.,** Analytical methods that reveal prostacyclin synthesis by cell cultures — glass capillary gas chromatography with electron capture detection, in *Prostacyclin,* Vane, J. R. and Bergström, S., Eds., Raven Press, New York, 1979, 55—64.
59. **Chang, W.-C., Murota, S., Matsuko, M., and Tsurufugi, S.,** A new prostaglandin transformed from arachidonic acid in carrageenin-induced granuloma, *Biochem. Biophys. Res. Commun.,* 72, 1259—1264, 1976.
60. **Baenziger, N. L., Dillander, M. J., and Majerus, P. W.,** Cultured human skin fibroblasts and arterial cells produce labile platelet-inhibitory prostaglandin, *Biochem. Biophys. Res. Commun.,* 78, 294—301, 1977.

61. **Raz, A., Isakson, P. C., Minkes, M. S., and Needleman P.**, Characterization of a novel metabolic pathway of arachidonate in coronary arteries which generates a potent endogenous coronary vasodilator, *J. Biol. Chem.*, 252, 1123—1126, 1977.

62. **Dawson, W., Boot, J., Cockerill, A., Mallen, D., and Osborne, D.**, Release of novel prostaglandins and thromboxanes after immunological challenge of guinea pig lung, *Nature (London)*, 262, 699—702, 1976.

63. **Gryglewski, R. J., Korbut, R., and Ocetkiewicz, A.**, Generation of prostacyclin by lungs in vivo and its release into the arterial circulation, *Nature (London)*, 273, 765—767, 1978.

64. **Hensby, C. N., Barnes, P. J., Dollery, C. T., and Dargie, H.**, Production of 6-oxo-PGF$_{1\alpha}$ by human lung in vivo, *Lancet*, 2, 1162—1163, 1979.

65. **Edlund, A., Bomfin, W., Kaijser, L., Olin, C., Patrono, C., Pinca, E., and Wennmalm, A.**, Pulmonary formation of prostacyclin in man, *Prostaglandins*, 22, 323—332, 1981.

66. **Smith, W. L., DeWitt, D. L., and Allen, M. L.**, Bimodal distribution of the prostaglandin I$_2$ synthase antigen in smooth muscle cells, *J. Biol. Chem.*, 258, 5922—5926, 1983.

67. **Dusting, G. J., Moncada, S., and Vane, J. R.**, Recirculation of prostacyclin (PGI$_2$) in the dog, *Br. J. Pharmacol.*, 64, 315—320, 1978.

68. **Armstrong, J. M., Lattimer, N., Moncada, S., and Vane, J. R.**, Comparison of the vasodepressor effects of prostacyclin and 6-oxo-prostaglandin F$_{1\alpha}$ with those of prostaglandin E$_2$ in rats and rabbits, *Br. J. Pharmacol.*, 62, 125—130, 1978.

69. **Pace-Asciak, C. R., Carrara, M. C., and Nicolaou, K. C.**, Prostaglandin I$_2$ has more potent hypotensive properties than prostaglandin E$_2$ in the normal and spontaneously hypertensive rat, *Prostaglandins*, 15, 999—1004, 1978.

70. **Machleidt, C., Foerstermann, U., Anhut, H., and Hertting, G.**, Formation and elimination of prostacyclin metabolites in the cat in vivo as determined by radioimmunoassay of unextracted plasma, *Eur. J. Pharmacol.*, 74, 19—26, 1981.

71. **Fitzgerald, G. A., Brash, A. R., Falardeau, P., and Oates, J. A.**, Estimated rate of prostacyclin secretion into the circulation of normal man, *J. Clin. Invest.*, 68, 1272—1276, 1981.

72. **Haslam, R. J. and McClenaghan, M. D.**, Measurement of circulating prostacyclin, *Nature (London)*, 292, 364—366, 1981.

73. **Demers, L. M., Harrison, T. S., Halbert, D. R., and Santen, R. J.**, Effect of prolonged exercise on plasma prostaglandin levels, *Prostaglandins Med.*, 6, 413—418, 1981.

74. **Bult, H., Beetens, J., and Herman, A. G.**, Blood levels of 6-oxo-prostaglandin F$_{1\alpha}$ during endotoxin-induced hypotension in rabbits, *Eur. J. Pharmacol.*, 63, 47—56, 1980.

75. **MacKenzie, I. Z., MacLean, D. A., and Mitchell, M. D.**, Prostaglandins in the human fetal circulation in mid-trimester and term pregnancy, *Prostaglandins*, 20, 649—654, 1980.

76. **Mitchell, M. D.**, A sensitive radioimmunoassay for 6-keto-prostaglandin F$_{1\alpha}$: preliminary observations on circulating concentrations, *Prostaglandins Med.*, 1, 13—21, 1978.

77. **Morris, H. G., Sherman, N. A., and Shepperdson, F. T.**, Variables associated with radioimmunoassay of prostaglandins in plasma, *Prostaglandins*, 21, 771—788, 1981.

78. **Ylikorkala, O. and Viinikka, L.**, Maternal plasma levels of 6-keto-prostaglandin F$_{1\alpha}$ during pregnancy and puerperium, *Prostaglandins Med.*, 7, 95—99, 1981.

79. **Hensby, C. N., Fitzgerald, G. A., Friedman, L. A., Lewis, P. J., and Dollery, C. T.**, Measurement of 6-oxo-PGF$_{1\alpha}$ in human plasma using gas chromatography-mass spectrometry, *Prostaglandins*, 18, 731—736, 1979.

80. **Roy, A. C. and Karim, S. M. M.**, Hourly and day to day variation in circulating levels of 6-ketoprostaglandin F$_{1\alpha}$ and thromboxane B$_2$, *IRCS Med. Sci.*, 9, 616—617, 1981.

81. **Dollery, C. T., Friedman, L. A., Hensby, C. N., Kohner, E., Lewis, P. J., Porta, M., and Webster, J.**, Circulating prostacyclin may be reduced in diabetes, *Lancet*, 2, 1365—Letter, 1979.

82. **Mitchell, M. D., Anderson, A. B. M., Brunt, J. D., Clover, L., Ellwood, D. A., Robinson, J. S., and Thurnbull, A. C.**, Concentrations of 6-oxo-prostaglandin F$_{1\alpha}$ in the maternal and fetal plasma of sheep during spontaneous and induced parturition, *J. Endocrinol.*, 83, 141—148, 1979.

83. **Blair, I. A., Barrow, S. E., Waddell, K. A., Lewis, P. J., and Dollery, C. T.**, Prostacyclin is not a circulating hormone in man, *Prostaglandins*, 23, 579—589, 1982.

84. **Pace-Asciak, C. R., Carrara, M. C., and Levine, L.**, Antibodies to 5,6-dihydroprostaglandin I$_2$ trap endogenously produced prostaglandin I$_2$ in the rat circulation, *Biochim. Biophys. Acta*, 620, 186—192, 1980.

85. **Pace-Asciak, C. R., Carrara, M. C., Levine, L., and Nicolaou, K. C.**, PGI$_2$-specific antibodies administered *in vivo* suggest against a role for endogenous PGI$_2$ as a circulating vasodepressor hormone in the normotensive and spontaneously hypertensive rat, *Prostaglandin*, 20, 1053—1060, 1980.

86. **Pace-Asciak, C., Rosenthal, A., and Domazet, Z.**, Comparison between the *in vivo* rate of metabolism of prostaglandin I$_2$ and its blood pressure lowering response after intravenous administration in the rat, *Biochim. Biophys. Acta*, 574, 182—186, 1979.

BIOCHEMISTRY OF THE LEUKOTRIENES

Pierre Borgeat

Leukotrienes (LTs), prostaglandins, and thromboxanes are products of the oxidative metabolism of arachidonic acid (AA) in mammalian cells. LTs are linear unsaturated C20 fatty acids carrying polar substituants (one or more oxygenated groups and a peptidic group), in contrast with eicosanoids derived from the cyclooxygenases which are cyclized oxygenated derivatives of C20-unsaturated fatty acids. LTs are further characterized by the presence of three conjugated double bonds (conjugated triene) which account for the typical UV absorption pattern of this novel family of compounds. The discovery of the leukotriene pathway was closely associated to biochemical and pharmacological studies on the "slow reacting substance of anaphylaxis" (SRS-A) and the reader should refer to reviews for detailed reports on the structure elucidation of SRS-A[1,2] and the LTs,[3-6] and to other chapters dealing with the biological and pharmacological properties of these putative mediators of inflammation and allergy.

STRUCTURE AND BIOSYNTHESIS

The 5-lipoxygenase catalyzes the dioxygenation of AA at C-5 and initiates the synthesis of LTs; this enzyme and the LTs were first reported in 1976 and 1979, respectively, following studies on the metabolism of AA in rabbit peritoneal and human blood polymorphonuclear leukocytes (PMNL), where 5-HETE and LTB$_4$ were found to be the major metabolites of the fatty acid.[7-11] Since then, considerable progress has been made in the field; indeed, an impressive number of novel structures has been described and their mechanism of formation has been elucidated in most cases. Table 1 shows the structures of the metabolites of AA derived from the 5-lipoxygenase and summarizes the pertinent biochemical data.

Recently, another family of LTs derived from the 15-lipoxygenase has been reported (see Table 2 for structures and biochemical data).[12-15] The biosynthesis of these compounds involves the formation of an unstable (14S,15S)- 14(15)-oxido-(Z,Z,E,E)-5,8,10,12-eicosatetraenoic acid from a 15-HPETE precursor,[12-14,16] in analogy with the mechanism of formation of LTA$_4$ from 5-HPETE. It is noteworthy that the products of the nonenzymatic hydrolysis of the 14,15-epoxy acid have been isolated in incubation mixtures of human or porcine leukocytes with AA or 15-HPETE, whereas compounds corresponding to LTB$_4$, LTC$_4$, LTD$_4$, or LTE$_4$ (formed enzymatically) have not been detected;[12-14,16] however, the direct addition of the synthetic epoxy acid to rat basophilic leukemia cell (RBL-1) suspensions led to the formation of the glutathione conjugate, i.e., the 14-S-glutathionyl-15-hydroxy-5,8,10,12-eicosatetraenoic acid.[16] The 15-lipoxygenase derived metabolites of AA were found to be formed in larger amounts in eosinophils than in other classes of human leukocytes.[15]

OTHER TRANSFORMATIONS OF THE 5-LIPOXYGENASE

It was shown in early studies on the metabolism of C20 unsaturated fatty acids in PMNL that these cells catalyze the formation of 8(S)-8-hydroxy-(E,Z,Z)-9,11,14-eicosatrienoic acid from exogenous all-*cis*-8,11,14-eicosatrienoic acid (C20:3 ω6);[7] under the same conditions, the corresponding 8-hydroxy-eicosatetraenoic acid was not produced from AA, but rather 5-HETE.[7] More recently, the 8-hydroxy-9-S-glutathionyl-10,12,14-eicosatrienoic acid was detected in the incubation mixture of mouse mastocytoma cells with the ionophore A23187 and C20:3 ω6, indicating the formation of 8-hydroperoxy-9,11,14-eicosatrienoic acid and the further metabolism of this hydroperoxy acid into the allylic epoxy acid 8(9)-oxido-

Table 1
METABOLISM OF C20:4ω6 THROUGH THE 5-LIPOXYGENASE PATHWAY

Structure	Nomenclature (1), trivial name (2), abbreviation (3), formula (mol wt) (4)	Comments
	(1) all-*cis*-5,8,11,14-eicosatetraenoic acid (2) Arachidonic acid (3) C20:4ω6 (4) $C_{20}H_{32}O_2$ (304)	C20:4ω6 is a major constituent of cellular lipids and is present mainly as esters (phospholipids, glycerides, and cholesterol esters. Its metabolism by lipoxygenases or the cyclooxygenase first involves the action of various lipases which release the free acid into the cellular milieu.[29-31] Although arachidonic acid is not the exclusive substrate for leukotriene synthesis (see text), it was found to be the most important precursor in the systems so far studied.
	(1) (5S)-5-Hydroperoxy-(E,Z,Z,Z)-6,8,11,14-eicosatetraenoic acid (3) 5-HPETE (4) $C_{20}H_{32}O_4$ (336)	5-HPETE is the product of the action of the 5-lipoxygenase on C20:4ω6.[7] The *trans-cis* conjugated diene α to a hydroperoxy group is a structural characteristic of lipoxygenase products.[7,23] The 5-HPETE is difficult to isolate from biological material in view of its rapid metabolism into 5-HETE or LTA₄ (see below).[11] The 5-lipoxygenase has been purified from human leukocytes. It requires calcium, ATP, and other factors for maximal activity.[9,65-68]
	(1) (5S)-5-Hydroxy-(E,Z,Z,Z)-6,8,11,14-eicosatetraenoic acid (3) 5-HETE (4) $C_{20}H_{32}O_3$ (320)	5-HETE is likely the product of the action of a peroxidase on 5-HPETE. The enzyme(s) responsible for this particular reaction has not been characterized. 5-HETE is a major metabolite of AA in rabbit peritoneal neutrophils[7] and in ionophore A23187 stimulated human peripheral PMNL (see Table 4).[9] This compound was found to be rapidly incorporated (acylation) in cellular lipids following its synthesis in PMNL and macrophages.[69-71]

LTA$_4$ is derived from 5-HPETE. The reaction is enzymatic (enzyme recently characterized)[67] and involves the stereospecific removal of a proton at C-10 of 5-HPETE and loss of water.[11,72] As expected for an allylic epoxide, LTA$_4$ is highly unstable; in aqueous buffer at pH 7.4 and 25°C, the time for 50% decomposition is less than 10 sec; it instantaneously decomposes at acid pH. LTA$_4$ is stabilized under alkaline conditions[11] and upon binding to serum albumin.[73] LTA$_4$ undergoes facile nucleophilic substitution (hydrolysis products are described below).[11,74] The biosynthesis of LTA$_4$ in rabbit peritoneal PMNL incubated with C20:4ω6 is maximal within 90 sec and is followed by rapid enzymatic and nonenzymatic conversions into various more polar compounds (see below).[11] The formation of LTA$_4$ in biological systems can be demonstrated by trapping with methanol and measurement of the stable methanolysis products [epimers at C-12, (5S,12SR)-5-hydroxy-12-methoxy-(E,E,Z)-6,8,10,14-eicosatetraenoic acids].[11,75] LTA$_4$ could be extracted from biological media[55] and measured directly by HPLC.[58,59]

The Δ6-*trans*-LTB$_4$ and Δ6-*trans*-12-epi-LTB$_4$, are the products of the nonenzymatic hydrolysis of LTA$_4$, which occurs spontaneously in aqueous medium.[11,73] These compounds are formed upon incubation of PMNL suspensions with C20:4ω6 and the ionophore A23187 and are considered as by-products in LTA$_4$ biosynthesis. Their occurrence as a pair of epimers (C-12) suggested the involvement of a carbonium ion in the mechanism of hydrolysis of LTA$_4$[11] Labeling experiments with isotopes (H$_2$ ^{18}O and ^{18}O$_2$) indicated that hydroxyl groups at C-12 were derived from water and that the hydroxyl at C-5 originated from the 5(6)-epoxy ring of LTA$_4$.[11] The two stereoisomers of LTB$_4$ are also formed during the peroxidase catalyzed oxidation of LTC$_4$ in PMNL.[49,51]

(1) (5S, 6S)-5(6)-Oxido-(E,E,Z,Z)-7,9,11,14-eicosatetraenoic acid
(2) Leukotriene A$_4$
(3) LTA$_4$
(4) C$_{20}$H$_{30}$O$_3$ (318)

(1) (5S,12R)-5,12-Dihydroxy-(E,E,Z)-6,8,10,14-eicosatetraenoic acid
(2) Δ6-*trans*-Leukotriene B$_4$
(3) Δ6-*trans*-LTB$_4$
(4) C$_{20}$H$_{32}$O$_4$ (336)

(1) (5S,12S)-5,12-Dihydroxy-(E,E,Z)-6,8,10,14-eicosatetraenoic acid
(2) Δ6-*trans*-12-epi-Leukotriene B$_4$
(3) Δ6-*trans*-12-epi-LTB$_4$
(4) C$_{20}$H$_{32}$O$_4$ (336)

Table 1 (continued)

METABOLISM OF C20:4ω6 THROUGH THE 5-LIPOXYGENASE PATHWAY

Structure	Nomenclature (1), trivial name (2), abbreviation (3), formula (mol wt) (4)	Comments
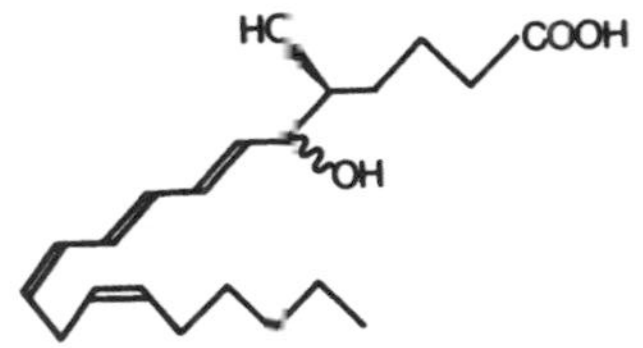	(1) 5,6-Dihydroxy-7,9,11,14-eicosatetraenoic acid (4) $C_{20}H_{32}O_4$ (336)	These dihydroxy acids are products of the nonenzymatic hydrolysis of LTA_4 and their mechanism of formation likely involves a carbonium ion.[11] These compounds are produced during incubation of PMNL suspensions with the ionophore A23187 and C20:4ω6, and also appear to be by-products in LTA_4 biosynthesis. The exact stereochemistry of the two compounds has not been determined; however, their mechanism of formation strongly suggest that the stereochemistry depicted here [(5S,6RS)-5,6,-dihydroxy-(E,E,Z,Z)-7,9,11,14-eicosatetraenoic acids], is correct.
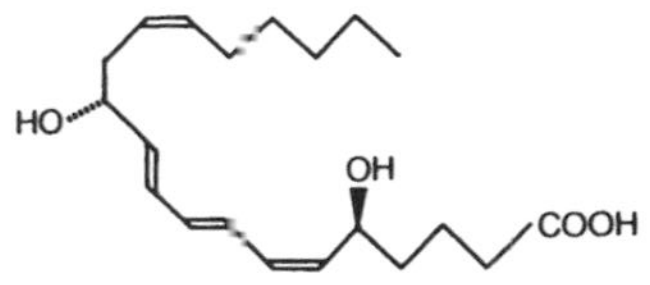	(1) (5S,12R)-5,12-Dihydroxy-(Z,E,E,Z)-6,8,10,14-eicosatetraenoic acid (2) Leukotriene B_4 (3) LTB_4 (4) $C_{20}H_{32}O_4$ (336)	LTB_4 is the product of the enzymatic hydrolysis of LTA_4.[11] It is formed in PMNL incubated with C20:4ω6 and the ionophore A23187 (Table 4). Hydroxyl groups (at C-5 and C-12) in LTB_4 are derived respectively from the epoxide ring of LTA_4 and from water. The direct addition of synthetic LTA_4 to suspensions of human leukocytes clearly demonstrated that LTA_4 is an intermediate in LTB_4 synthesis.[25,76]
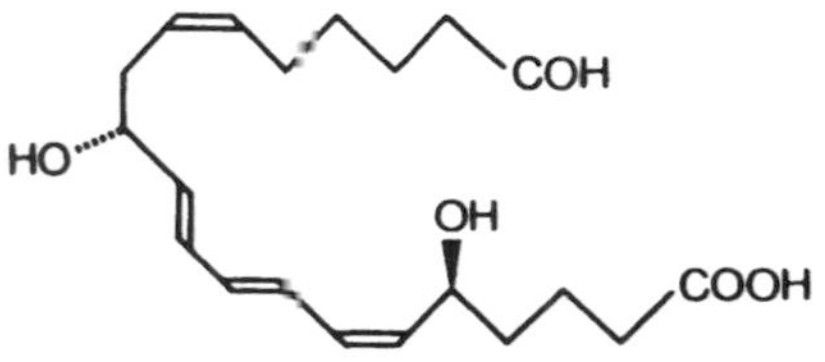	(1) (5S,12R)5,12,20-Trihydroxy-(Z,E,E,Z)-6,8,10,14-eicosatetraenoic acid (2) ω-Hydroxy-Leukotriene B_4 (3) ω-OH-LTB_4 (4) $C_{20}H_{32}O_5$ (352)	LTB_4 undergoes ω-oxidation when incubated with human blood PMNL.[40-42] The enzymatic system involved has not been characterized. The first step of oxidation leads to ω-OH-LTB_4, and a second step results in the formation of ω-carboxy-leukotriene B_4 (ω-COOH-LTB_4). The Δ^6-*trans*- LTB_4, Δ^6-*trans*-12-epi-LTB_4, as well as the 5S,12S-DiHETE[26] (see Table 3) also undergo ω-oxidation in PMNL suspensions, but at a slower rate than LTB_4,[145] indicating that these reactions show some specificity for LTB_4.

LTC$_4$ is the product of the conjugation of glutathione and LTA$_4$.[52,77] The glutathione transferase involved in this reaction has not yet been characterized. LTC$_4$ is formed in human PMNL stimulated with the ionophore A23187 (Table 4) or incubated directly with the precursor LTA$_4$;[75] inhibition of glutathione synthesis or trapping of the tripeptide led to decreased LTC$_4$ synthesis in rat basophilic leukemia cells and macrophages.[79-81]

LTD$_4$ is formed upon hydrolysis of the peptide chain of LTC$_4$ and loss of the γ-glutamyl residue.[53,82,83] The peptidase involved has not yet been characterized. Radiolabeled 5-HPETE was incorporated in rat basophilic leukemia cell SRS (LTD$_4$) in support of the proposed mechanism of formation of LTD$_4$.[84] Among other sources, LTD$_4$ is released from human and guinea pig anaphylactic lungs (Table 4). The occurrence of LTC$_4$ or LTD$_4$ in biological material likely depends on the presence or absence of the peptidase that catalyzes the transformation of LTC$_4$ into LTD$_4$. The porcine kidney γ-glutamyl transpeptidase was shown to catalyze this reaction.[53,85] Other studies on the enzymes involved in LTD$_4$ synthesis were reported.[86-88]

LTE$_4$ is the product of the action of a peptidase on LTD$_4$, resulting in the loss of the glycyl residue.[89] Such a peptidase activity is present in human plasma.[43]

The Δ^{11}-*trans* isomer of LTC$_4$ as well as Δ^{11}-*trans*-LTD$_4$ and Δ^{11}-*trans*-LTE$_4$ have been isolated from biological sources (Table 4).[82,89-91] It was proposed that the isomerization of the Δ^{11} double bond occurs upon reversible addition of thiyl radicals at C-12.[92] That the formation of Δ^{11}-*trans* isomers of LTC$_4$, D$_4$, and E$_4$ also occurs during extraction and purification procedures should, however, not be excluded.

(1) (5*S*,6*R*)-5-Hydroxy-6-*S*-glutathionyl-(E,E,Z,Z)-7,9,11,14-eicosatetraenoic acid
(2) Leukotriene C$_4$
(3) LTC$_4$
(4) C$_{30}$H$_{47}$O$_9$N$_3$S$_1$ (625)

(1) (5*S*,6*R*)-5-Hydroxy-6-*S*-cysteinylglycyl-(E,E,Z,Z)-7,9,11,14-eicosatetraenoic acid
(2) Leukotriene D$_4$
(3) LTD$_4$
(4) C$_{25}$H$_{40}$O$_6$N$_2$S$_1$ (496)

(1) (5*S*,6*R*)-5-Hydroxy-6-*S*-cysteinyl-(E,E,Z,Z)-7,9,11,14-eicosatetraenoic acid
(2) Leukotriene E$_4$
(3) LTE$_4$
(4) C$_{23}$H$_{37}$O$_5$N$_1$S$_1$ (439)

(1) (5*S*,6*R*)-5-Hydroxy-6-*S*-glutathionyl-(E,E,E,Z)-7,9,11,14-eicosatetraenoic acid
(2) Δ^{11}-*trans*-Leukotriene C$_4$
(3) Δ^{11}-*trans*-LTC$_4$
(4) C$_{30}$H$_{47}$O$_9$N$_3$S$_1$ (625)

Table 2

METABOLISM OF C20:4ω6 THROUGH THE 15-LIPOXYGENASE PATHWAY

Structure	Nomenclature (1), trivial name (2), abbreviation (3), formula (mol wt) (4)	Comments
	(1) (15S)-15-Hydroperoxy-(Z,Z,Z,E)-5,8,11,13-eicosatetraenoic acid (3) 15-HPETE (4) $C_{20}H_{32}O_4$ (336)	The 15-HPETE is the product of the action of the 15-lipoxygenase on C20:4ω6. It is formed in PMNL (and other cells and tissues) incubated with C20:4ω6 and it is rapidly reduced into 15-HETE,[9,13-15,93-98] or metabolized into the allylic epoxide described below.[12,14] Two 15-lipoxygenases were recently purified from rabbit reticulocytes[98,99] and peritoneal PMNL.[97] The partially purified enzymes showed a high degree of specificity for the C-15 position of C20:4ω6. It is noteworthy that contrary to the 5-lipoxygenase, the 15-lipoxygenase of human PMNL does not require activation.
	(1) (15S)-15-Hydroxy-(Z,Z,Z,E)-5,8,11,13-eicosatetraenoic acid (3) 15-HETE (4) $C_{20}H_{32}O_3$ (320)	The 15-HETE is the product of the reduction of 15-HPETE by a peroxidase (enzyme(s) not characterized).[98]
	(1) (14S,15S)-14(15)-Oxido-(Z,Z,E,E)-5,8,10,12-eicosatetraenoic acid (4) $C_{20}H_{30}O_3$ (318)	This epoxy acid is formed in PMNL from C20:4ω6 or 15-HPETE;[12-14,16] its mechanism of formation is believed to be similar to that of LTA₄. The enzymes involved appear to be the 15- and 12-lipoxygenases.[144] The addition of the synthetic 14,15-epoxy acid to suspensions of rat basophilic leukemia cells resulted in the formation of a glutathione conjugate (at C-14).[16]
	(1) (8S,15S)-8,15-Dihydroxy(Z,E,E,E)-5,9,11,13-eicosatetraenoic acid and (8R,15S)-8,15-Dihydroxy-(Z,E,E,E)-5,9,11,13-Eicosatetraenoic acid (4) $C_{20}H_{32}O_4$ (336)	The two dihydroxy acids (epimers at C-8) are products of the hydrolysis (nonenzymatic) of the allylic epoxide described earlier, a reaction that occurs spontaneously in aqueous media. The mechanism of formation of the dihydroxy acids appeared to be similar to that of Δ⁶-*trans*-LTB₄ and Δ⁶-*trans*-12-epi-LTB₄ from LTA₄. These dihydroxy acids are formed upon incubation of PMNL with C20:4ω6 or 15-HPETE.[12-14]

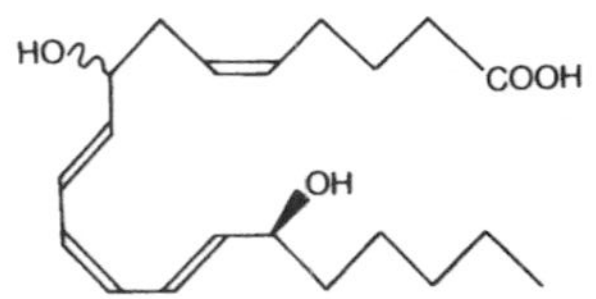

These dihydroxy acids (epimers at C-8) differ from the compounds described above in the geometry of the Δ^{11} double bond. The geometry of the conjugated triene (*trans-cis-trans*) and the incorporation of molecular oxygen into both hydroxyl groups (at C-8 and C-15) suggested a double dioxygenation mechanism in the formation of the two compounds. However, the lack of stereochemical purity at C-8 argues against this hypothesis and the mechanism of formation of the pair of dihydroxy acids remains unclear. The two compounds are formed upon incubation of PMNL with C20:4ω6 or 15-HPETE.[14]

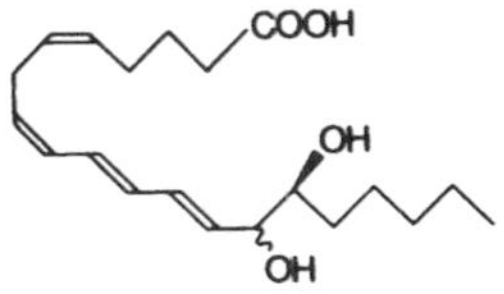

Isomeric 14,15-dihydroxy acids (four different compounds) are formed upon incubation of PMNL with C20:4ω6 or 15-HPETE. Two of these are products of the hydrolysis (nonenzymatic) of the allylic epoxide shown earlier and likely correspond to the compounds depicted here, i.e., (14RS,15S)-dihydroxy-(Z,Z,E,E)-5,8,10,12-eicosatetraenoic acids; their mechanism of formation appear to be similar to that of the 5,6-dihydroxy acids shown in Table 1.[12-14] The 14R,15S-dihydroxy acid can also be formed through a double lipoxygenation mechanism.[144]

(1) (8S,15S)-8,15-Dihydroxy-(Z,E,Z,E)-5,9,11,13-eicosatetraenoic acid and (8R,15S)-8,15-Dihydroxy-(Z,E,Z,E)-5,9,11,13-eicosatetraenoic acid

(4) $C_{20}H_{32}O_4$ (336)

(1) 14,15-Dihydroxy-5,8,10,12-eicosatetraenoic acids

(4) $C_{20}H_{32}O_4$ (336)

10,12,14-eicosatrienoic acid.[17] The involvement of the 5-lipoxygenase and of the enzymes catalyzing the synthesis of LTA_4 and LTC_4 (rather than specific enzymes) in these transformations of C20:3 ω6 in PMNL and RBL-1 cells is not excluded since the existence of specific enzymes has not been demonstrated.

All-*cis*-5,8,11-eicosatrienoic acid (C20:3ω9) and all-*cis*-5,8,11,14,17-eicosapentaenoic acid (C20:5ω3) were shown to be metabolized through the 5-lipoxygenase pathway, the former leading to the formation of 5-hydroxy-6-*S*-glutathionyl-7,9,11-eicosatrienoic acid, and the latter, to the formation of LTB_5 and LTC_5 (i.e., to compounds corresponding to LTB_4 and LTC_4 with an additional double bond at Δ^{17}) in RBL-1 cells[18] and mouse mastocytoma cells.[19-21] The 5-hydroxy-6-*S*-cysteinylglycyl and 5-hydroxy-6-*S*-cysteinyl derivatives of C20:5ω3 (LTD_5 and LTE_5) and C20:3ω9 were also formed in RBL-1 cells incubated with the fatty acids and the ionophore A23187.[18] In a separate study, series of synthetic polyunsaturated fatty acids were incubated with RBL-1 cells;[22] it was shown that the position of a pair of double bonds at Δ^5 and Δ^8 was critical for the reaction with the 5-lipoxygenase (reaction of fatty acids with double bonds at Δ^6 and Δ^9 was slower) and that chain length was of secondary importance. Interestingly, two hydroxylated derivatives of AA with double bonds at Δ^5 and Δ^8 were found to react with the 5-lipoxygenase; the addition of (12*S*)-12-hydroxy-(Z,Z,E,Z)-5,8,10,14-eicosatetraenoic acid (12-HETE) (a major metabolite of AA in human platelets)[23] or of 15-HETE to human PMNL suspensions led to further oxygenation of the hydroxy acids at C-5 and to the formation of 5*S*,12*S*-DiHETE[24-26] and 5*S*,15*S*-DiHETE (see Table 3 for biochemical data and structures).[27,28]

CONTROL OF LEUKOTRIENE BIOSYNTHESIS

The concept that the formation of the various AA metabolites is dependent on substrate availability is well accepted,[29] and the addition of AA to cells and tissues in vitro usually results in the rapid synthesis of prostaglandins thromboxanes, and hydroxy acids. The concentration of free AA in cells under normal conditions is low as compared to the total amount of the fatty acid present in the form of esters. The formation of cyclooxygenase and of lipoxygenase products is thus initially controlled by the activity of the various lipases which make AA available to metabolizing enzymes.[30,31] The effects of the divalent cation ionophore A23187, which unspecifically stimulates the release of AA (likely through stimulation of Ca^{2+}-dependent phospholipase A_2) and the synthesis of cyclooxygenase and lipoxygenase products in a variety of systems, support this concept.[32] The synthesis of 5-HETE and LTs also depends on substrate availability; addition of AA to suspensions of rabbit peritoneal PMNL (glycogen-induced) results in the formation of 5-HETE and LTB_4.[7,8] However, in other systems, such as in human blood PMNL, monocytes, the C-5 lipoxygenase will not readily transform exogenous AA (or only to a small extent), whereas incubation of these cells in the presence of the ionophore A23187 leads to synthesis of substantial amounts of 5-HETE and LTs from endogenous AA.[9,33] These data indicate that the ionophore not only causes the release of AA but also activates[9] the C-5 lipoxygenase involved in the further transformation of the fatty acid.[9] Thus, the 5-lipoxygenase, unlike other known dioxygenases, must be activated to produce LTs (at least in some systems). Previous studies on the biosynthesis of SRS-A clearly indicated that the release of the substance by lungs and leukocytes was increased upon immunologic stimulation.[34-37] This was recently confirmed when lungs of sensitized guinea pigs were found to release LTD_4 in vitro upon stimulation with the specific antigen; similarly, macrophages release LTC_4 after antigen-antibody (Ag-Ab) challenge. It thus appears that in cells carrying IgE (immunoglobulin E) receptor, 5-lipoxygenase product synthesis is activated upon receptor stimulation. Besides the ionophore A23187 and antigen-challenge, several inflammatory stimuli, namely phagocytosis, the anaphylatoxin C5a, the synthetic chemotactic peptide formyl-methionyl-leucyl-phenylala-

Table 3
DOUBLE DIOXYGENATION OF C20:4ω6 BY LIPOXYGENASES

Structure	Nomenclature (1), trivial name (2), abbreviation (3), formula (mol wt) (4)	Comments
	(1) (5S,12S)-5,12-Dihydroxy(E,Z,E,Z)-6,8,10,14-eicosatetraenoic acid (3) 5S,12S-DiHETE (4) $C_{20}H_{32}O_4$ (336)	The 5S,12S-DiHETE is formed by successive dioxygenations of C20:4ω6 by the 12- and the 5-lipoxygenase (followed by reduction). Thus, the hydroxyl groups in the 5S,12S-DiHETE are both derived from molecular oxygen. The compound is formed in systems where both lipoxygenases are active such as in suspensions of human PMNL and platelets or in porcine blood PMNL incubated in the presence of the ionophore A23187 and C20:4ω6. The synthesis of the 5S,12S-DiHETE can proceed either from 5-HETE or 12-HETE.[24-26,71]
	(1) (5S,15S)-5,15-Dihydroxy-(E,Z,Z,E)-6,8,11,13-eicosatetraenoic acid (3) 5S,15S-DiHETE (4) $C_{20}H_{32}O_4$ (336)	The 5S,15S-DiHETE is formed by successive dioxygenation of C20:4ω6 by the 15- and 5-lipoxygenases (and reduction of the hydroperoxy groups to hydroxy groups). The compound is formed in biological systems where the two lipoxygenases are active, such as in human PMNL (neutrophils or eosinophils) incubated in the presence of the ionophore A23187 and C20:4ω6.[27,28,72]

nine, the platelet activating factor, and platelet-derived 12-HPETE, induce the synthesis of LTs. The 15-lipoxygenase product 15-HETE is either a potent inhibitor or stimulator of the 5-lipoxygenase in PMNL and a mast/basophil cell line, respectively.[38,39] Table 4 summarizes the various LT sources reported to date.

METABOLISM OF LEUKOTRIENES

The present knowledge on the metabolism of LTs is yet very fragmentary. LTB_4 undergoes two successive steps of ω-oxidation in human blood leukocytes, leading to ω-hydroxy-LTB_4 and ω-carboxy-LTB_4 (Table 1), the latter retaining little of the biological activity of the parent compound.[40-42]

Recent studies indicated two pathways for the inactivation or degradation of LTC_4, one involving hydrolysis of the glutationyl group by peptidases, and the other, a peroxidase-mediated oxidation of the sulfur atom. Human plasma was shown to rapidly convert LTC_4 and LTD_4 into LTE_4.[43] In vivo experiments have shown that tritium labeled 14,15-dihydro-LTC_4 (also named LTC_3) is rapidly transformed into 14,15-dihydro-LTE_4 (LTE_3) upon intravenous injection to monkeys and rapidly disappears from the circulation.[44] Guinea pig lung homogenates transform 14,15-dihydro-LTC_4 into 14,15-dihydro-LTD_4 (LTD_3), whereas liver and kidney homogenates were inactive. Interestingly, the same homogenates transform 14,15-dihydro-LTD_4 into 14,15-dihydro-LTE_4 and 14,15-dihydro-LTC_4.[45] An enzyme preparation from porcine kidney transforms LTD_4 and LTD_5 into LTE_4 and LTE_5, respectively.[46] The intravenous injection of 14,15-dihydro-LTC_4 in the mouse also results in the formation of 14,15-dihydro-LTD_4 or 14,15-dihydro-LTE_4 in various organs.[47] Uptake and metabolism of 14,15-dihydro-LTC_4 by isolated rat organs and cells was also studied and the formation of products resulting from loss of amino acid(s) from the peptidic group was again observed.[48]

LTC_4 was also shown to undergo oxidative metabolism and inactivation by intact eosinophils or eosinophil peroxidase in a reaction requiring H_2O_2 and halides ions.[49-51] Detailed analysis of the products formed indicated that LTC_4 was successively transformed into diastereoisomeric sulfoxides of LTC_4 and into Δ^6-trans-LTB_4 and Δ^6-trans-12-epi-LTB_4 (practically devoid of biological activity) in a reaction probably involving an S-chlorosulfonium ion as the reactive intermediate.[51] The oxidative metabolism of LTC_4 by peroxidase may constitute a more efficient inactivation pathway in comparison with the peptidase catalyzed degradation which leads to LTE_4, a compound that retains significant biological activity.

ANALYSIS OF LEUKOTRIENES AND OTHER LIPOXYGENASE PRODUCTS

Except for the highly polar peptidolipids LTC_4, D_4, and E_4, and the unstable epoxy acid LTA_4, lipoxygenase products can be extracted from biological media using classical organic solvent extraction procedures.[10] LTC_4, D_4, and E_4 are extracted by adsorption on hydrophobic polymers (XAD resins).[52,53] A method for the extraction of HETEs and prostaglandins using octadecylsilyl silica was also reported.[54] A procedure for the extraction of LTA_4 as the methyl ester has also been reported.[55]

LTs and several other lipoxygenases products contain conjugated double bonds and show elevated extinction coefficients ($\epsilon \sim 30,000$ at 230 to 235 nm for conjugated dienes and $\epsilon \sim 45,000$ at 270 to 280 nm for conjugated trienes).[7,10,56] This physical property of several lipoxygenase products accounts for the rapid development of high performance liquid chromatography (HPLC) for the analysis of these compounds. Silica gel HPLC was used for analysis of HETEs, DiHETEs, LTB_4, and its metabolites[10,25,28,33] but is not suitable for analysis of the more polar LTC_4, D_4, and E_4; reversed phase HPLC (octadecyl silica) is widely used for purification and analysis of hydroxy acids[9] and the peptidolipids LTC_4, D_4,

Table 4
SOURCES AND STIMULI FOR THE SYNTHESIS OF 5-LIPOXYGENASE PRODUCTS

Compounds	Cells and tissues	Stimuli	Ref.
5-HETE	Rabbit peritoneal PMNL	C20:4ω6	7,8,10
		Cytochalasin-B and platelet activating factor	100
	Human blood PMNL	Ionophore A23187	9,33,70,101,102,103
	Human lymphocytes	C20:4ω6 and phytohemagglutinin	104
		C20:4ω6 or ionophore A23187 or concanavalin A	105
	Human promyelocytic leukemia cells	Formyl-Met-Leu-Phe	106
	Synovial fluid	Rheumatic diseases	107
	Guinea pig peritoneal PMNL	Formyl-Met-Leu-Phe	108
		Ionophore A23187	108,109
	Rat peritoneal macrophages	Ionophore A23187	110
	Rat basophilic leukemia cell homogenate	C20:4ω6	65,66
	Rat basophilic leukemia cells	Ionophore A23187	111
	Mouse peritoneal macrophages	C20:4ω6	94
	Mouse mastocytoma cell homogenate	C20:4ω6	112
	Murine mast/basophil cell line	15-HETE and C20:4ω6	39
	Porcine blood leukocytes	Ionophore A23187	24
	Sheep lung	Endotoxemia	64
	Pleural exudate	Carrageenan	113
LTB₄	Rabbit peritoneal PMNL	C20:4ω6	8,10
		Cytochalasin-B and platelet-activating factor	100
	Rabbit alveolar macrophages	Zymosan	114
	Human blood PMNL	Ionophore A23187	9,33,42,102,115
		C20:4ω6 and 12-HPETE	33
		Zymosan	116
	Human PMNL	Platelet activating factor	117
	Human lymphocytes	C20:4ω6, ionophore A23187, or concanavalin A	105
	Human eosinophils	Ionophore A23187	15
	Synovial fluid	Rheumatic diseases	107,118
	Human promyelocytic leukemia cells	Formyl-Met-Leu-Phe	106
	Rat basophilic leukemia cell	Ionophore A23187	111
	Rat basophilic leukemia cell homogenate	C20:4ω6	65,66
	Rat peritoneal PMNL	Ionophore A23187	119
	Rat peritoneal macrophages	Ionophore A23187	110
	Mouse peritoneal macrophages	Zymosan	120
	Mouse mastocytoma cells homogenate	C20:4ω6	112
	Murine mast/basophil cell line	15-HETE and C20:4ω6	39
	Porcine blood leukocytes	Ionophore A23187	14, 24
	Horse eosinophils	Ionophore A23187	121
	Pleural exudate	Carrageenan	113
	Guinea pig peritoneal PMNL	Ionophore A23187	1090
ω-Hydroxy-LTB₄	Human leukocytes	Ionophore A23187	40
		Formyl-Met-Leu-Phe	41

Table 4 (continued)
SOURCES AND STIMULI FOR THE SYNTHESIS OF 5-LIPOXYGENASE PRODUCTS

Compounds	Cells and tissues	Stimuli	Ref.
ω-Carboxy-LTB$_4$	Human leukocytes	Ionophore A23187	40
		Formyl-Met-Leu-Phe	41
LTC$_4$	Murine mastocytoma cells	Ionophore A23187	52,77,122,123
	Rat basophil leukemia cells	Ionophore A23187	89,91
	Rat basophil leukemia cell homogenate	C20:4ω6	84
	Human PMNL	Ionophore A23187	124,125
	Human monocytes	Cancer extract-antitumor Ab	126
	Human lung	Ag-Ab	127
	Bone marrow-derived mast cells	Ionophore A23187	128
	Rat peritoneal cavity	Ag-Ab	127
	Rat alveolar macrophages	Ag-Ab	129
	Rat peritoneal mononuclear cells	Ionophore A23187	130
	Mouse pulmonary macrophages	Zymosan	131
	Mouse peritoneal macrophages	Zymosan, bacteria	120,132—136,137
		Ag-Ab	131
	Dog spleen	C20:4ω6	138
	Horse eosinophils	Ionophore A23187	121
	Human eosinophils	Ionophore A23187	58
Δ^{11}-*trans*-LTC$_4$	Murine mastocytoma cells	Ionophore A23187	52,92
	Human PMNL	Ionophore A23187	124
LTD$_4$	Rat basophil leukemia cells	Ionophore A23187	53,82,89,91
	Rat basophil leukemia cell homogenate	Ionophore A23187	78
	Rat peritoneal cavity	Ag-Ab	127
	Mouse peritoneal macrophages	Zymosan, bacteria	134,136
	Guinea pig lung	Ag-Ab	83,139,140
		Anaphylatoxin (C5a-des-arg)	141
	Human lung	Ag-Ab	127
	Cat paws	Compound 48/80	142
	Horse eosinophils	Ionophore A23187	121
LTE$_4$	Rat basophil leukemia cells	Ionophore A23187	18,89,91
	Rat peritoneal cavity	Ag-Ab	142
	Cat paws	Compound 48/80	143
Δ^{11}-*trans*-LTE$_4$	Rat basophil leukemia cells	Ionophore A23187	92

and E$_4$.[52,53,57,58] Detection limits are in the range of 1 to 10 ng. HPLC methods for the analysis of LTA$_4$ were also reported.[55,59] More recently, radioimmunological methods for the analysis of LTC$_4$ and LTB$_4$ have been reported;[60-62] a GC/MS assay of 5-HETE was also described.[63,64]

REFERENCES

1. **Brocklehurst, W. E.,** Pharmacological mediators of hypersensitivity reactions, in *Clinical Aspects of Immunology,* Vol. 6, Gell, P. F. H. and Coombs, R. R. A., Eds., 611—632, Blackwell Scientific, Oxford, 1968, 611—632.
2. **Orange, R. P., Murphy, R. C., Karnovsky, M. L., and Austen, K. F.,** The physicochemical characteristics and purification of slow-reacting substance of anaphylaxis, *J. Immunol.,* 110, 760—770, 1973.
3. **Samuelsson, B., Hammarström, S., and Borgeat, P.,** Pathways of arachidonic acid metabolism, in *Advances in Inflammation Research,* Vol. 1, Weissman, G., Samuelsson, B., Paoletti, R., Eds., Raven Press, New York, 1979, 405—412.
4. **Samuelsson, B.,** The leukotrienes: a new group of biologically active compounds including SRS-A, *Trends Pharmacol. Sci.,* 1, 227—230, 1980.
5. **Sirois, P. and Borgeat, P.,** From slow reacting substance of anaphylaxis (SRS-A) to leukotriene D_4 (LTD$_4$), *Int. J. Immunopharmacol.,* 2, 281—293, 1980.
6. **Borgeat, P. and Sirois, P.,** The leukotrienes: a major step in the understanding of immediate hypersensitivity reactions, *J. Med. Chem.,* 24, 121—126, 1981.
7. **Borgeat, P., Hamberg, M., and Samuelsson, B.,** Transformation of arachidonic acid and homo-γ-linolenic acid in rabbit polymorphonuclear leukocytes, *J. Biol. Chem.,* 252, 8772, 7816—7820, 1977.
8. **Borgeat, P. and Samuelsson, B.,** Transformation of arachidonic acid by rabbit polymorphonuclear leukocytes: formation of a novel dihydroxyeicosatetraenoic acid, *J. Biol. Chem.,* 254, 2643—2646, 1979.
9. **Borgeat, P. and Samuelsson, B.,** Metabolism of arachidonic acid in polymorphonuclear leukocytes. Effects of the ionophore A23187, *Proc. Natl. Acad. Sci. U.S.A.,* 76, 2148—2152, 1979.
10. **Borgeat, P. and Samuelsson, B.,** Metabolism of arachidonic acid in polymorphonuclear leukocytes: structural analysis of novel hydroxylated compounds, *J. Biol. Chem.,* 254, 7865—7869, 1979.
11. **Borgeat, P. and Samuelsson, B.,** Metabolism of arachidonic acid in polymorphonuclear leukocytes: unstable intermediate in formation of dihydroxy acids, *Proc. Natl. Acad. Sci. U.S.A.,* 76, 3213—3217, 1979.
12. **Lundberg, U., Radmark, O., Malmsten, C., and Samuelsson, B.,** Transformation of 15-hydroperoxy-5,9,11,13-eicosatetraenoic acid into novel leukotrienes, *FEBS Lett.,* 126, 127—132, 1981.
13. **Jubiz, W., Radmark, O., Lindgren, J. A., Malmsten, C., and Samuelsson, B.,** Novel leukotrienes: products formed by initial oxygenation of arachidonic acid at C-15, *Biochem. Biophys. Res. Commun.,* 99, 976—986, 1981.
14. **Maas, R. L., Brash, A. R., Oates, J. A.,** A second pathway of leukotriene biosynthesis in porcine leukocytes, *Proc. Natl. Acad. Sci. U.S.A.,* 78, 5523—5527, 1981.
15. **Turk, J., Maas, R. L., Brash, A. R., Jackson, Roberts, L., II, and Oates, J. A.,** Arachidonic acid 15-lipoxygenase products from human eosinophils, *J. Biol. Chem.,* 257, 7068—7076, 1982.
16. **Sok, D. E., Han, C. O., Shieh, W. R., Zhou, B. N., and Sih, C. J.,** Enzymatic formation of 14,15-leukotriene A and C(14)-sulfur-linked peptides, *Biochem. Biophys. Res. Commun.,* 104, 1363—1370, 1982.
17. **Hammarström, S.,** Conversion of dihomo-γ-linolenic acid to an isomer of leukotriene C_3, oxygenated at C-8, *J. Biol. Chem.,* 256, 7712—7714, 1981.
18. **Orning, L., Bernström, K., and Hammarström, S.,** Formation of leukotrienes E_3, E_4 and E_5 in rat basophilic leukemia cells, *Eur. J. Biochem.,* 120, 41—45, 1981.
19. **Hammarström, S.,** Conversion of 5,8,11-eicosatrienoic acid to leukotrienes C_3 and D_3, *J. Biol. Chem.,* 256, 2275—2279, 1981.
20. **Murphy, R. C., Pickett, W. C., Culp, B. R., and Lands, W. E. M.,** Tetraene and pentaene leukotrienes: selective production from murine mastocytoma cells after dietary manipulation, *Prostaglandins,* 22, 613—623, 1981.
21. **Hammarström, S.,** Conversion of ^{14}C-labeled eicosapentaenoic acid (n-3) to leukotriene C_5, *Biochem. Biophys. Acta,* 663, 575—577, 1981.
22. **Jakschik, B. A., Sams, A. R., Sprecher, H., and Needleman, P.,** Fatty acids structural requirements for leukotriene biosynthesis, *Prostaglandins,* 20, 401—410, 1980.
23. **Hamberg, M. and Samuelsson, B.,** Prostaglandin endoperoxides. Novel transformation of arachidonic acid in human platelets, *Proc. Natl. Acad. Sci. U.S.A.,* 71, 3400—3404, 1974.
24. **Borgeat, P., Picard, S., Vallerand, P., and Sirois, P.,** Transformation of arachidonic acid in leukocytes. Isolation and structural analysis of a novel dihydroxy derivative, *Prostaglandins Med.,* 6, 557—570, 1981.
25. **Borgeat, P., Fruteau de Laclos, B., Picard, S., Drapeau, J., Vallerand, P., and Corey, E. J.,** Studies on the mechanism of formation of the 5S, 12S-dihydroxy-6,8,10,14(E,Z,E,Z)-eicosatetraenoic acid in leukocytes, *Prostaglandins,* 23, 713—724, 1982.
26. **Lindgren, J. A., Hansson, G., and Samuelsson, B.,** Formation of novel hydroxylated eicosatetraenoic acids in preparations of human polymorphonuclear leukocytes, *FEBS Lett.,* 128, 329—335, 1981.

27. **Maas, R. L., Turk, J., Oates, J. A., and Brash, A. R.,** Formation of a novel dihydroxy acid from arachidonic acid by lipoxygenase-catalyzed double oxygenation in rat mononuclear cells and human leukocytes, *J. Biol. Chem.,* 257, 7056—7067, 1982.

28. **Borgeat, P., Picard, S., Drapeau, J., and Vallerand, P.,** Metabolism of arachidonic acid in leukocytes: isolation of a 5,15-dihydroxyeicosatetraenoic acid, *Lipids,* 17, 676—681, 1982.

29. **Ramwell, P. W., Leovey, E. M. K., and Sintetos, A. L.,** Regulation of the arachidonic acid cascade, *Biol. Reprod.,* 16, 70—87, 1977.

30. **Irvine, R. F.,** How is the level of free arachidonic acid controlled in mammalian cells?, *Biochem. J.,* 204, 3—16, 1982.

31. **Lapetina, E. G.,** Regulation of arachidonic acid production: role of phospholipases C and A_2, *Trends Pharmacol. Sci.,* 115—118, 1982.

32. **Knapp, H. R., Oelz, O., Roberts, J., Sweetman, B. J., Oates, J. A., and Reed, P. W.,** Ionophores stimulate prostaglandin and thromboxane biosynthesis, *Proc. Natl. Acad. Sci. U.S.A.,* 74, 4251—4255, 1977.

33. **Maclouf, J., Fruteau de Laclos, B., and Borgeat, P.,** Stimulation of leukotriene biosynthesis in human blood leukocytes by platelet-derived 12-hydroperoxy-eicosatetraenoic acid, *Proc. Natl. Acad. Sci. U.S.A.,* 79, 6042—6046, 1982.

34. **Burka, J. F. and Eyre, P.,** The immunological release of slow-reacting substance of anaphylaxis from bovine lung, *Can. J. Physiol. Pharmacol.,* 52, 1201—1204, 1974.

35. **Ishizaka, T., Ishizaka, K., Orange, R. P., and Austen, K. F.,** Pharmacologic inhibition of the antigen-induced release of histamine and slow reacting substance of anaphylaxis (SRS-A) from monkey lung tissues mediated by human Ig E, *J. Immunol.,* 106, 1267—1273, 1971.

36. **Liebig, R., Bernauer, W., and Peskar, B. A.,** Release of prostaglandins, a prostaglandin metabolite, slow-reacting substance and histamine from anaphylactic lungs, and its modification by catecholamines, *Naunyn-Schmiedeberg's Arch. Exp. Pathol. Pharmacol.,* 284, 279—293, 1974.

37. **Morris, H. R., Piper, P. J., Taylor, G. W., and Tippins, A. B.,** Comparative studies on immunologically and non-immunologically produced slow reacting substances from man, guinea-pig and rats, *Br. J. Pharmacol.,* 67, 179—184, 1979.

38. **Vanderhoek, J. Y., Bryant, R. W., and Bailey, J. M.,** Inhibition of leukotriene biosynthesis by the leukocyte product 15-hydroxy-5,8,11,13-eicosatetraenoic acid, *J. Biol. Chem.,* 255, 10064—10066, 1980.

39. **Vanderhoek, J. Y., Tare, N. S., Bailey, J. M., Goldstein, A. L., and Pluznik, D. H.,** New role for 15-hydroxyeicosatetraenoic acid. Activator of leukotriene biosynthesis in PT-18 mast/basophil cells, *J. Biol. Chem.,* 257, 12191—12195, 1982.

40. **Hansson, G., Lindgren, J. A., Dahlen, S. E., Hedqvist, P., and Samuelsson, B.,** Identification and bilogical activity of novel ω-oxidized metabolites of leukotriene B_4 from human leukocytes, *FEBS Lett.,* 130, 107—112, 1981.

41. **Salmon, J. A., Simmons, P. M., and Palmer, R. M. J.,** Synthesis and metabolism of leukotriene B_4 in human neutrophils measured by specific radioimmunoassay, *FEBS Lett.,* 146, 18—22, 1982.

42. **Jubiz, W., Radmark, O., Malmsten, C., Hansson, G., Lindgren, J. A., Palmblad, J., Uden, A. M., and Samuelsson, B.,** A novel leukotriene produced by stimulation of leukocytes with formylmethionylleucylphenylalanine, *J. Biol. Chem.,* 257, 6106—6110, 1982.

43. **Parker, C. W., Koch, D., Huber, M. M., and Falkenhein, S. F.,** Formation of the cysteinyl form of slow reacting substance (leukotriene E_4) in human plasma, *Biochem. Biophys. Res. Commun.,* 97, 1038—1046, 1980.

44. **Hammarström, S., Bernström, K., Orning, L., Dahlen, S. E., Hedqvist, P., Smedegard, G., and Revenäs, B.,** Rapid in vivo metabolism of leukotriene C_3 in the monkey, *Macaca irus, Biochem. Biophys. Res. Commun.,* 101, 1109—1115, 1981.

45. **Hammarström, S.,** Metabolism of leukotriene C_3 in the guinea pig. Identification of metabolites formed by lung, liver and kidney, *J. Biol. Chem.,* 256, 9573—9578, 1981.

46. **Bernström, K. and Hammarström, S.,** Metabolism of leukotriene D by porcine kidney, *J. Biol. Chem.,* 256, 9579—9582, 1981.

47. **Appelgren, L. E. and Hammarström, S.,** Distribution and metabolism of ^{3}H-labeled leukotriene C_3 in the mouse, *J. Biol. Chem.,* 257, 531—535, 1982.

48. **Ormstad, K., Uehara, N., Orrenius, S., Orning, L., and Hammarström, S.,** Uptake and metabolism of leukotriene C_3 by isolated rat organs and cells, *Biochem. Biophys. Res. Commun.,* 104, 1434—1440, 1982.

49. **Goetzl, E. J.,** The conversion of leukotriene C_4 to isomers of leukotreiene B_4 by human eosinophil peroxidase, *Biochem. Biophys. Res. Commun.,* 106, 270—275, 1982.

50. **Henderson, W. R., Jörg, A., and Klebanoff, S. J.,** Eosinophil peroxidase-mediated inactivation of leukotrienes B_4, C_4, and D_4, *J. Immunol.,* 128, 2609—2613, 1982.

51. **Lee, C. W., Lewis, R. A., Corey, E. J., Barton, A., Oh, H., Tauber, A. I., and Austen, K. F.,** Oxidative inactivation of leukotriene C₄ by stimulated human polymorphonuclear leukocytes, *Proc. Natl. Acad. Sci. U.S.A.,* 79, 4166—4170, 1982.

52. **Murphy, R. C., Hammarström, S., and Samuelsson, B.,** Leukotriene C: a slow-reacting substance from murine mastocytoma cells, *Proc. Natl. Acad. Sci. U.S.A.,* 76, 4275—4279, 1980.

53. **Orning, L., Hammarström, S., and Samuelsson, B.,** Leukotriene D: a slow reacting substance from rat basophilic leukemia cells, *Proc. Natl. Acad. Sci. U.S.A.,* 77, 2014—2017, 1980.

54. **Powell, W. S.,** Rapid extraction of oxygenated metabolites of arachidonic acid from biological samples using octadecylsilyl silica, *Prostaglandins,* 20, 947—957, 1980.

55. **Radmark, O., Malmsten, C., Samuelsson, B., Goto, G., Marfat, A., and Corey, E. J.,** Leukotriene A: isolation from human polymorphonuclear leukocytes, *J. Biol. Chem.,* 255, 11828—11831, 1980.

56. **Corey, E. J., Clark, D. A., Goto, G., Marfat, A., Mioskowski, C., Samuelsson, B., and Hammarström, S.,** Stereospecific total synthesis of a "slow reacting substance" of anaphylaxis, Leukotriene C-1, *J. Am. Chem. Soc.,* 102, 1436—1439, 1980.

57. **Mathews, W. R., Rokach, J., and Murphy, R. C.,** Analysis of leukotrienes by high-pressure liquid chromatography, *Anal. Biochem.,* 118, 96—101, 1981.

58. **Borgeat, P., Fruteau de Laclos, B., Rabinovitch, H., Picard, S., Braquet, P., Hébert, J., and Laviolette, M.,** Generation and structures of the lipoxygenase products. Eosinophil-rich human polymorphonuclear leukocyte preparations characteristically release leukotriene C₄ on ionophore A23187 challenge, *J. Allergy Clin. Immunol.,* 74, 310—315, 1984.

59. **Wynalda, M. A., Morton, D. R., Kelly, R. C., and Fitzpatrick, F. A.,** Liquid chromatographic determination of intact leukotriene A₄, *Anal. Chem.,* 54, 1079—1082, 1982.

60. **Levine, L., Morgan, R. A., Lewis, R. A., Austen, K. F., Clark, D. A., Marfat, A., and Corey, E. J.,** Radioimmunoassay of the leukotrienes of slow reacting substance of anaphylaxis, *Proc. Natl. Acad. Sci. U.S.A.,* 78, 7692—7696, 1981.

61. **Young, R. N., Kakushima, M., and Rokach, J.,** Studies on the preparation of conjugates of leukotriene C₄ with proteins for development of an immunoassay for SRS-A (1), *Prostaglandins,* 23, 603—613, 1982.

62. **Salmon, J. A., Simmons, P. M., and Palmer, R. M. J.,** A radioimmunoassay for leukotriene B₄, *Prostaglandins,* 24, 225—235, 1982.

63. **Hubbard, W. C., Phillips, M. A., and Taber, D. F.,** Selective synthesis of octadeuterated (±)-5-HETE for use in GC-MS quantitation of 5-HETE, *Prostaglandins,* 23, 61—65, 1982.

64. **Ogletree, M. L., Oates, J. A., Brigham, K. L., and Hubbard, W. C.,** Evidence for pulmonary release of 5-hydroxyeicosatetraenoic acid (5-HETE) during endotoxemia in unanesthetized sheep, *Prostaglandins,* 23, 459—468, 1982.

65. **Jakschik, B. A., Sun, F. F., Lee, L. H., and Steinhoff, M. M.,** Calcium stimulation of a novel lipoxygenase, *Biochem. Biophys. Res. Commun.,* 95, 103—110, 1980.

66. **Jakschik, B. A. and Lee, L. H.,** Enzymatic assembly of slow reacting substance, *Nature (London),* 287, 51—52, 1980.

67. **Rouzer, C. A., Matsumoto, T., and Samuelsson, B.,** Single protein from human leukocytes possesses 5-lipoxygenase and leukotriene A₄ synthase activities, *Proc. Natl. Acad. Sci. U.S.A.,* 83, 857—861, 1986.

68. **Parker, C. W. and Aykent, S.,** Calcium stimulation of the 5-lipoxygenase from RBL-1 cells, *Biochem. Biophys. Res. Commun.,* 109, 1011—1016, 1981.

69. **Stenson, W. F. and Parker, C. W.,** 12-L-Hydroxy-5,8,10,14-eicosatetraenoic acid, a chemotactic fatty acid, is incorporated into neutrophil phospholipids and triglyceride, *Prostaglandins,* 18, 285—292, 1979.

70. **Stenson, W. F. and Parker, C. W.,** Metabolism of arachidonic acid in ionophore-stimulated neutrophilis. Esterification of hydroxylated metabolite into phospholipids, *J. Clin. Invest.,* 64, 1457—1465, 1979.

71. **Pawlowski, N. A., Scott, W. A., Andreach, M., and Cohn, Z. A.,** Uptake and metabolism of mono-hydroxy-eicosatetraenoic acids by macrophages, *J. Exp. Med.,* 155, 1653—1664, 1982.

72. **Maas, R. L., Ingram, C. D., Taber, D. F., Oates, J. A., and Brash, A. R.,** Stereospecific removal of the D_R hydrogen atom at the 10-carbon of arachidonic acid in the biosynthesis of leukotriene A₄ by human leukocytes, *J. Biol. Chem.,* 257, 13515—13519, 1982.

73. **Fitzpatrick, F. A., Morton, D. R., and Wynalda, M. A.,** Albumin stabilizes leukotriene A₄, *J. Biol. Chem.,* 257, 4680—4683, 1982.

74. **Corey, E. J., Arai, Y., and Mioskowski, C.,** Total synthesis of (±)5,6-oxido-7,9-trans-11,14-cis-eicosapentaenoic acid, a possible precursor of SRS-A, *J. Am. Chem. Soc.,* 101, 6748—6749, 1979.

75. **Hammarström, S. and Samuelsson, B.,** Detection of leukotriene A₄ as an intermediate in the biosynthesis of leukotrienes C₄ and D₄, *FEBS Lett.,* 122, 83—86, 1980.

76. **Radmark, O., Malmsten, C., Samuelsson, B., Clark, D. A., Goto, G., Marfat, A., and Corey, E. J.,** Leukotriene A: stereochemistry and enzymatic conversion to leukotriene B, *Biochem. Biophys. Res. Commun.,* 92, 954—961, 1980.

77. **Hammarström, S., Murphy, R. C., Samuelsson, B., Clark, D. A., Mioskowski, C., and Corey, E. J.,** Structure of leukotriene C identification of the amino acid part, *Biochem. Biophys. Res. Commun.,* 91, 1266—1272, 1979.

78. **Jakschik, B. A., Harper, T., and Murphy, R. C.,** Leukotriene C_4 and D_4 formation by particulate enzymes, *J. Biol. Chem.,* 257, 5346—5349, 1982.

79. **Parker, C. W., Fischman, C. M., and Wedner, H. J.,** Relationship of biosynthesis of slow reacting substance to intracellular glutathione concentrations, *Proc. Natl. Acad. Sci. U.S.A.,* 77, 6870—6873, 1980.

80. **Rouzer, C. A., Scott, W. A., Griffith, O. W., Hamill, A. L., and Cohn, Z. A.,** Depletion of glutathione selectively inhibits synthesis of leukotriene C by macrophages, *Proc. Natl. Acad. Sci. U.S.A.,* 78, 2532—2536, 1981.

81. **Rouzer, C. A., Scott, W. A., Griffith, O. W., Hamill, A. L., and Cohn, Z. A.,** Arachidonic acid metabolism in gluthione-deficient macrophages, *Proc. Natl. Acad. Sci. U.S.A.,* 79, 1621—1625, 1982.

82. **Morris, H. R., Taylor, G. W., Piper, P. J., Samhoun, M. N., and Tippins, J. R.,** Slow reacting substances (SRSs): the structure identification of SRSs from rat basophil leukemia (RBL-1) cells, *Prostaglandins,* 19, 185—201, 1980.

83. **Morris, H. R., Taylor, G. W., Piper, P. J., and Tippins, J. R.,** Structure of slow-reacting substance of anaphylaxis from guinea-pig lung, *Nature (London),* 285, 104—107, 1980.

84. **Parker, C. W., Koch, D., Huber, M. M., and Falkenhein, S. F.,** Incorporation of radiolabel from [1-^{14}C]5-hydroperoxy-eicosatetraenoic acid into slow reacting substance, *Biochem. Biophys. Res. Commun.,* 94, 1037—1043, 1980.

85. **Orning, L. and Hammarström, S.,** Kinetics of the conversion of leukotriene C by λ-glutamyl transpeptidase, *Biochem. Biophys. Res. Commun.,* 106, 1304—1309, 1982.

86. **Anderson, M. E., Allison, R. D., and Meister, A.,** Interconversion of leukotrienes catalyzed by purified λ-glutamyl transpeptides: concomitant formation of leukotriene D_4 and λ-glutamyl amino acids, *Proc. Natl. Acad. Sci. U.S.A.,* 79, 1088—1091, 1982.

87. **Morris, H. R., Taylor, G. W., Jones, C. M., Piper, P. J., Samhoun, M. N., and Tippins, J. R.,** Slow reacting substances (leukotrienes): enzymes involved in their biosynthesis, *Proc. Natl. Acad. Sci. U.S.A.,* 79, 4838—4842, 1982.

88. **Sok, D. E., Pai, J. K., Atrache, V., Kang, Y. C., and Sih, C. J.,** Enzymatic inactivation of RSR-Cys-Gly (Leukotriene D), *Biochem. Biophys. Res. Commun.,* 101, 222—229, 1981.

89. **Parker, C. W., Falkenhein, S. F., and Huber, M. M.,** Sequential conversion of the glutathionyl side chain of slow reacting substance (SRS) to cysteinyl-glycine and cysteine in rat basophilic leukemia cells stimulated with A-23187, *Prostaglandins,* 20, 863—886, 1980.

90. **Clark, D. A., Goto, G., Marfat, A., Corey, E. J., Hammarström, S., and Samuelsson, B.,** 11-trans-Leukotriene C: a naturally occurring slow reacting substance, *Biochem. Biophys. Res. Commun.,* 94, 1133—1139, 1980.

91. **Sok, D. E., Pai, J. K., Atrache, V., and Sih, C. J.,** Characterization of slow reacting substances (SRSs) of rat basophilic leukemia (RBL-1) cells: effect of cysteine on SRS profile, *Proc. Natl. Acad. Sci. U.S.A.,* 77, 6481—6485, 1980.

92. **Atrache, V., Sok, D. E., Pai, J. K., and Sih, C. J.,** Formation of 11-trans- slow reacting substances, *Proc. Natl. Acad. Sci. U.S.A.,* 78, 1523—1526, 1981.

93. **Hamberg, M., Hedqvist, P., and Radegran, K.,** Identification of 15-hydroxy-5,8,11,13-eicosatetraenoic acid (15-HETE) as a major metabolite of arachidonic acid in human lung, *Acta Physiol. Scand.,* 110, 219—221, 1980.

94. **Rabinovitch, H., Durand, J., Gualde, N., and Rigaud, M.,** Metabolism of polyunsaturated fatty acids by mouse peritoneal macrophages: the lipoxygenase metabolic pathway, *Agents Actions,* 11, 580—583, 1981.

95. **Claeys, M., Coene, M. C., Herman, A. G., Van Der Planken, M., Jouvenaz, G. H., and Nugteren, D. H.,** 15-Hydroxy-5,8,11,13-eicosatetraenoic acid (15-HETE) formation by rabbit peritoneal tissue, *Agents Actions,* 11, 589—591, 1981.

96. **Claeys, M., Coene, M. C., Herman, A. G., Van Der Planken, M., Jouvenaz, G. H., and Nugteren, D. H.,** 15-Hydroxy-5,8,11,13-eicoasatetraenoic acid (15-HETE) formation by rabbit peritoneal tissue, *Arch. Int. Pharmacodyn.,* 250, 305—308, 1981.

97. **Narumiya, S., Salmon, J. A., Cottee, F. H., Weatherley, B. C., and Flower, R. J.,** Arachidonic acid 15-lipoxygenase from rabbit peritoneal polymorphonuclear leukocytes, *J. Biol. Chem.,* 256, 9583—9592, 1981.

98. **Bryant, R. W., Bailey, J. M., Schewe, T., and Rapoport, S. M.,** Positional specificity of a reticulocyte lipoxygenase. Conversion of arachidonic acid to 15S-hydroperoxy-eicosatetraenoic acid, *J. Biol. Chem.,* 257, 6050—6055, 1982.

99. **Rapoport, S. M., Schewe, T., Weisner, R., Halangk, W., Ludwig, P., Janicke-Hohne, M., Tannert, C., Hiebsch, C., and Klatt, D.,** The lipoxygenase of reticulocytes. Purification, characterization and biological dynamics of the lipoxygenase; its identity with the respiratory inhibitors of the reticulocytes, *Eur. J. Biochem.,* 96, 545—561, 1979.

100. **Chilton, F. H., O'Flaherty, J. T., Walsh, C. E., Thomas, M. J., Wykle, R. L., DeChatelet, L. R., and Waite, B. M.,** Stimulation of the lipoxygenase pathway in polymorphonuclear luekocytes by 1-0-alkyl-2-0-acetyl-SN-glycero-3-phosphocholine, *J. Biol. Chem.,* 257, 5402—5407, 1982.

101. **Goetzl, E. J. and Sun, F. F.,** Generation of unique mono-hydroxyeicosatetraenoic acids from arachidonic acid by human neutrophils, *J. Exp. Med.,* 150, 406—411, 1979.

102. **Walsh, C. E., Waite, B. N., Thomas, M. J., and DeChatelet, L. R.,** Release and metabolism of arachidonic acid in human neutrophils, *J. Biol. Chem.,* 256, 7228—7234, 1981.

103. **Walsh, C. E., DeChatelet, L. R., Thomas, M. J., O'Flaherty, J. T., and Waite, M.,** Effect of phagocytosis and ionophores on release and metabolism of arachidonic acid from human neutrophils, *Lipids,* 16, 120—124, 1981.

104. **Parker, C. W., Stenson, W. F., Huber, M. G., and Kelly, J. P.,** Formation of thromboxane B_2 and hydroxyarachidonic acids in purified human lymphocytes in the presence and absence of PHA, *J. Immunol.,* 122, 1572—1577, 1979.

105. **Geotzl, E. J.,** Selective feedback inhibition of the 5-lipoxygenation of arachidonic acid in human T-lymphocytes, *Biochem. Biophys. Res. Commun.,* 101, 344—350, 1981.

106. **Bonser, R. W., Siegel, M. I., McConnell, R. T., and Cuatrecas, P.,** Chemotactic peptide stimulated endogenous arachidonic acid metabolism in HL60 granulocytes, *Biochem. Biophys. Res. Commun.,* 102, 1269—1275, 1981.

107. **Klickstein, L. B., Shapleigh, C., and Goetzl, E. J.,** Lipoxygenation of arachidonic acid as a source of polymorphonuclear leukocyte chemotactic factors in synovial fluid and tissue in rheumatoid arthritis and spondyloarthritis, *J. Clin. Invest.,* 66, 1166—1170, 1980.

108. **Bokoch, G. M. and Reed, P. W.,** Stimulation of arachidonic acid metabolism in the polymorphonuclear leukocyte by an N-formylated peptide. Comparison with ionophore A23187, *J. Biol. Chem.,* 255, 10223—10226, 1980.

109. **Bokoch, G. M. and Reed, P. W.,** Evidence for inhibition of leukotriene A_4 synthesis by 5,8,11,14-eicosatetraynoic acid in guinea pig polymorphonuclear luekocytes, *J. Biol. Chem.,* 256, 4156—4159, 1981.

110. **Doig, M. V. and Ford-Hutchinson, A. W.,** The production and characterization of products of the lipoxygenase enzyme system released by rat peritoneal macrophages, *Prostaglandins,* 20, 1007—1019, 1980.

111. **Ford-Hutchinson, A. W., Piper, P. J., and Samhoun, M. N.,** Generation of leukotriene B_4, its all-*trans* isomers and 5-hydroxyeicosatetraenoic acid by rat basophilic leukaemia cells, *Br. J. Pharmacol.,* 76, 215—220, 1982.

112. **Koshihara, Y., Mizumura, M., and Murota, S.,** Enhancement of 5-lipoxygenase activity in mastocytoma P-815 cells by *n*-butyrate treatment, *J. Biol. Chem.,* 257, 7302—7305, 1982.

113. **Siegel, M. I., McConnell, R. T., Bonser, R. W., and Cuatrecasas, P.,** The production of 5-HETE and leukotriene B in rat neutrophils from carageenan pleural exudates, *Prostaglandins,* 21, 123—132, 1981.

114. **Hsueh, W. and Sun, F. F.,** Leukotriene B_4 biosynthesis by alveolar macrophages, *Biochem. Biophys. Res. Commun.,* 106, 1085—1091, 1982.

115. **Ford-Hutchinson, A. W., Bray, M. A., Doig, M. V., Shipley, M. E., and Smith, M. J. H.,** Leukotriene B, a potent chemokinetic and aggregating substance released from polymorphonuclear leukocytes, *Nature (London),* 286, 264—265, 1980.

116. **Claesson, H. E., Lundberg, U., and Malmsten, C.,** Serum-coated zymosan stimulates the synthesis of leukotriene B_4 in human polymorphonuclear leukocytes. Inhibition by cyclic AMP, *Biochem. Biophys. Res. Commun.,* 99, 1230—1237, 1981.

117. **Lin, A. H., Morton, D. R., and Gorman, R. R.,** Acetyl glyceryl ether phosphorylcholine stimulates leukotriene B_4 synthesis in human polymorphonuclear leukocytes, *J. Clin. Invest.,* 70, 1058—1065, 1982.

118. **Davidson, E. M., Rae, S. A., and Smith, M. J. H.,** Leukotriene B_4 in synovial fluid, *J. Pharm. Pharmacol.,* 34, 410, 1982.

119. **Ford-Hutchinson, A. W., Bray, M. A., Cunningham, F. M., Davidson, E. M., and Smith, M. J. H.,** Isomers of leukotriene B_4 possess different biological potencies, *Prostaglandins,* 21, 143—152, 1980.

120. **Humes, J. L., Sadowski, S., Galavage, M., Goldenberg, M., Subers, E., Bonney, R. J., and Kuehl, F. A., Jr.,** Evidence for two sources of arachidonic acid for oxidative metabolism by mouse peritoneal macrophages, *J. Biol. Chem.,* 257, 1591—1594, 1982.

121. **Jörg, A., Henderson, W. R., Murphy, R. C., and Klebanoff, S. J.,** *J. Exp. Med.,* 155, 390—402, 1982.

122. **Hammarström, S., Samuelsson, B., Clark, D. A., Goto, G., Marfat, A., Mioskowski, C., and Corey, E. J.,** Stereochemistry of leukotriene C-1, *Biochem. Biophys. Res. Commun.,* 92, 946—953, 1980.

123. **Samuelsson, B., Borgeat, P., Hammarström, S., and Murphy, R. C.,** Leukotrienes: a new group of biologically active compounds, in *Advances in Prostaglandin and Thromboxane Research,* Vol. 6, Samuelsson, B., Ramwell, P., and Paoletti, R., Eds., Raven Press, New York, 1979, 1—18.

124. **Hansson, G., and Radmark, O.,** Leukotriene C_4: isolation from human polymorphonuclear leukocytes, *FEBS Lett.,* 122, 87—90, 1980.

125. **Aehringhaus, U., Wöbling, R. H., König, W., Patrono, C., Peskar, B. M., and Peskar, B. A.,** Release of leukotriene C_4 from human polymorphonuclear leukocytes as determined by radioimmunoassay, *FEBS Lett.,* 146, 111—114, 1982.

126. **Thomson, D. M. P., Phelan, K., Morton, D. G., and Bach, M. K.,** Armed human monocytes challenged with a sensitizing cancer extract release substances pharmacologically similar to leukotrienes, *Int. J. Cancer,* 30, 299—306, 1982.

127. **Lewis, R. A., Austen, K. F., Drazen, J. M., Clark, D. A., Marfat, A., and Corey, E. J.,** Slow reacting substances of anaphylaxis: identification of leukotrienes C-1 and D from human and rat sources, *Proc. Natl. Acad. Sci. U.S.A.,* 77, 3710—3714, 1980.

128. **Razin, E., Mencia-Huerta, J. M., Lewis, R. A., Corey, E. J., and Austen, D. F.,** Generation of leukotriene C_4 from a subclass of mast cells differentiated *in vitro* from mouse bone marrow, *Proc. Natl. Acad. Sci. U.S.A.,* 79, 4665—4667, 1982.

129. **Rankin, J. A., Hitchcock, M., Merrill, W., Bach, M. K., Brashler, J. R., and Askenase, P. W.,** IgE-dependent release of leukotriene C_4 from alveolar macrophages, *Nature (London),* 297, 329—330, 1982.

130. **Bach, M. K., Brashler, J. R., Hammarström, S., and Samuelsson, B.,** Identification of leukotriene C-1 as a major component of slow-reacting substance from rat mononuclear cells, *J. Immunol.,* 125, 115—117, 1980.

131. **Rouzer, C. A., Scott, W. A., Hamill, A. L., and Cohn, Z. A.,** Synthesis of leukotriene C and other arachidonic acid metabolites by mouse pulmonary macrophages, *J. Exp. Med.,* 155, 720—733, 1982.

132. **Rouzer, C. A., Scott, W. A., Cohn, Z. A., Blackburn, P., and Manning, J. M.,** Mouse peritoneal macrophages release leukotriene C in response to a phagocytic stimulus, *Proc. Natl. Acad. Sci. U.S.A.,* 77, 4928—4932, 1980.

133. **Humphrey, H. P., Coote, J., Butchers, P. R., Wheeldon, A., Vardey, C. J., and Skidmore, I. F.,** The release of prostaglandin and slow-reacting substance from mouse macrophages, *Agents Actions,* 11, 577—578, 1981.

134. **Roubin, R., and Benveniste, J.,** Release of leukotrienes C (LTC) and D (LTD) from inflammatory macrophages during phagocytosis of zymosan and bacteria, *Agents Actions,* 11, 578—579, 1982.

135. **Scott, W. A., Pawlowski, N. A., Murray, H. W., Andreach, M., Zrike, J., and Cohn, Z. A.,** Regulation of arachidonic acid metabolism by macrophage activation, *J. Exp. Med.,* 155, 1148—1160, 1982.

136. **Roubin, R., Mencia-Huerta, J. M., and Benveniste, J.,** Release of platelet-activating factor (PAF-acether) and leukotrienes C and D from inflammatory macrophages, *Eur. J. Immunol.,* 12, 141—146, 1982.

137. **Rouzer, C. A., Scott, W. A., Hamill, A. L., and Cohn, Z. A.,** Dynamics of leukotriene C production by macrophages, *J. Exp. Med.,* 152, 1136—11247, 1980.

138. **Malik, K. U. and Wong, P. Y. K.,** Leukotriene C_4: the major lipoxygenase metabolite of arachidonic acid in dog spleen, *Biochem. Biophys. Res. Commun.,* 103, 511—520, 1981.

139. **Baker, S. R., Boot, J. R., and Osborne D. J.,** A comparative study of the physical properties and enzymatic reactions of slow reacting substance of anaphylaxis and some leukotriene D_4 isomers, *Prostaglandins,* 23, 569—577, 1982.

140. **Morris, H. R., Taylor, G. W., Rokach, J., Girard, Y., Piper, P. J., Tippins, J. R., and Samhoun, M. N.,** Slow reacting substance of anaphylaxis, SRS-A: assignment of the stereochemistry, *Prostaglandins,* 20, 601—607, 1980.

141. **Stimler, N. P., Bach, M. K., Bloor, C. M., and Hugli, T. E.,** Release of leukotrienes from guinea pig lung stimulated by C5adesArg anaphylatoxin, *J. Immunol.,* 128, 2247—2252, 1982.

142. **Lewis, R. A., Drazen, J. M., Austen, K. F., Clark, D. A., and Corey, E. J.,** Identification of the C(6)-S-conjugate of leukotriene A with cysteine as a naturally occurring slow reacting substance of anaphylaxis (SRS-A). Importance of the 11-cis-geometry for biological activity, *Biochem. Biophys. Res. Commun.,* 96, 271—277, 1980.

143. **Houglum, J., Pai, J. K., Atrache, V., Sok, D. E., and Sih, C. J.,** Identification of the slow reacting substances from cat paws, *Proc. Natl. Acad. Sci. U.S.A.,* 77, 5688—5692, 1980.

144. **Maas, R. L. and Brash, A. R.,** Evidence for a lipoxygenase mechanism in the biosynthesis of epoxide and dihydroxy leukotrienes from 15(S)-hydroperoxyicosatetraenoic acid by human platelets and porcine leukocytes, *Proc. Natl. Acad. Sci. U.S.A.,* 80, 2884—2888, 1983.

145. **Borgeat, P.,** Unpublished information.

EICOSANOID TRANSPORT SYSTEMS:
MECHANISMS, PHYSIOLOGICAL ROLES, AND INHIBITORS

Laszlo Z. Bito

It has been demonstrated during the past 15 years that some tissues actively transport prostaglandins (PGs) and related autacoids and that these transport systems play a key role in the pharmacokinetics of eicosanoids. Special aspects of the PG transport system have been reviewed previously by Bito and associates,[1-5] and a general review of PG transport processes and their inhibition is in preparation. The following is intended to serve as a compendium of the relevant literature and to provide an overview of the accumulated findings.

SURVEY OF VERTEBRATE TISSUES FOR APPARENT PG TRANSPORT CAPACITY

In vitro studies have established that tissues or slices of tissues that have an organic acid transport system accumulate such organic acids from incubation media against concentration gradients.[6-11] Incubation with radioactively labeled substrates, followed by determination of tissue radioactivity, has been used to demonstrate accumulation of minute quantities of such substrates.[8,11] Thus, a tissue-to-medium distribution ratio (T/M) of ^{14}C or ^{3}H greater than unity after incubation of tissues in media containing ^{14}C- or ^{3}H-labeled PGs for periods sufficient to approach steady state can be taken as a *preliminary* indication that a given tissue has an active transport system for PGs and related eicosanoids.

The T/M ratios for the accumulation of PGs by a variety of mammalian (Table 1) and other vertebrate (Table 2) organs and tissues show that the choroid plexus, anterior uvea, kidney cortex, and liver of vertebrates can accumulate PGs against an apparent concentration gradient, suggesting that these tissues have the capacity for active transport of PGs. In rabbits, but not in cats, the major tissues of the female reproductive tract (uterus, cervix, and vagina) also show concentrative PG accumulation. Pieces of lung tissue also yield T/M ratios greater than unity when special precautions are taken to minimize the amount of retained air, as was done in experiments using rhesus monkeys, but not in earlier experiments using rabbits, rats, or cats.[4,16]

Some authors, working with organic acids other than PGs, have routinely presented accumulation ratios corrected for solid content or extracellular volume; i.e., activity or concentration per unit volume of tissue water (Tw) or intracellular water (Iw) divided by activity per unit volume of incubation medium. Such expressions clearly increase the apparent accumulation ratio, but may well overstate the concentration gradient between the incubation medium and any tissue compartment. In rabbit ciliary processes, for example, the Iw/M ratios suggest a 46-fold concentration gradient between the intra- and extracellular compartments (Table 3). However, in tissues that are covered by tight-junctional epithelia, such as the ciliary processes or choroid plexus, the PG concentration in sequestered extracellular compartments can also be maintained well above that of the medium as a result of active transepithelial transport. Thus, the implicit assumption that the PG accumulation is limited to intracellular compartments is unjustified. Although autoradiographic studies do indeed suggest that PGs accumulate in the stroma of the ciliary processes, the techniques used preclude estimation of the existing concentration gradient.[18,19] In the absence of quantitative information on the distribution of the accumulated activity, corrections that assume localization of the accumulated substance in specific tissue compartments cannot be justified.

Table 1
ACCUMULATION (T/M)ᵃ OF PGE₁ AND PGF₂α BY TISSUES OF FIVE MAMMALIAN SPECIES

	Species										
Organ system or tissue	Rhesus[4,b]		Rabbit[4,12-15]		Rat[16]		Woodchuck[17]	Cat[16]	
	E_1	$F_{2\alpha}$	E_1	$F_{2\alpha}$	E_1	$F_{2\alpha}$	E_1	E_1	$F_{2\alpha}$
CNS and eye									
Brain			0.7 ± 0.05 (6)	0.6 ± 0.03 (4)	0.6 ± 0.04 (6)	0.6 ± 0.03 (6)	0.9 ± 0.07 (6)	0.9 ± 0.05 (9)	0.9 ± 0.08 (4)
Choroid plexus		21.4	4.9 ± 0.95 (8)	10.5 ± 0.60 (130)			4.3 ± 0.69 (4)	3.5 ± 0.45 (6)	
Anterior uvea		8.6 ± 1.16 (4)	3.5 ± 0.50 (8)	5.3 ± 0.71 (215)			3.5 ± 0.42 (6)	1.7 ± 0.14 (10)	1.4 ± 0.17 (6)
Cornea			1.1 ± 0.05 (8)						
Gastrointestinal									
Intestine			0.5 ± 0.02 (8)		0.8 ± 0.05 (9)	0.9 ± 0.05 (9)			
Liver	5.4 ± 0.44 (4)	1.4 ± 0.10 (17)	1.4 ± 0.18 (11)		3.6 ± 0.25 (14)	5.2 ± 0.45 (9)	1.5 ± 0.07 (9)	1.5 ± 0.16 (8)	
Gallbladder			0.9 ± 0.05 (6)						
Urinary									
Kidney									
Cortex	2.7 ± 0.36 (4)	5.9 ± 0.4 (28)	1.8 ± 0.09 (8)	4.3 ± 0.29 (204)	3.1 ± 0.24 (15)	3.0 ± 0.25 (9)	1.7 ± 0.30 (8)		
Medulla			0.8 ± 0.03 (6)				0.89	0.9 ± 0.09 (6)	
Female reproductive									
Ovary			0.8 ± 0.04 (6)		1.0 ± 0.14 (5)			0.6 ± 0.04 (6)	0.4 ± 0.04 (3)

Uterus	1.2 ± 0.6 (4)	1.5 ± 0.30 (5)	2.6 ± 0.38 (8)	5.1 ± 0.53 (16)	1.3 ± 0.06 (17)	1.2 ± 0.03 (6)	1.3 ± 0.15 (6)	0.8 ± 0.03 (12)	0.8 ± 0.07 (4)
Cervix		0.9 ± 0.05 (7)	8.0 ± 1.14 (16)	8.8 ± 1.34 (36)	1.0 ± 0.02 (5)	1.6 ± 0.36 (5)		0.8 ± 0.12 (3)	
Vagina	0.8 ± 0.05 (4)	0.9 ± 0.05 (12)	2.9 ± 0.58 (7)	3.5 ± 0.14 (134)	1.1 ± 0.04 (4)	1.0 ± 0.07 (8)		0.7 ± 0.04 (6)	
Other									
Adipose									
White fat			0.04 ± 0.08 (8)		0.4 ± 0.03 (15)	0.2 ± 0.03 (6)	0.4 ± 0.03 (6)	0.7 ± 0.05 (10)	0.3 ± 0.03 (10)
Brown fat							1.1 ± 0.05 (6)		
Aorta	0.8 ± 0.03 (14)		1.4 ± 0.14 (6)	0.9 ± 0.05 (5)			0.8 ± 0.05 (6)	1.9 ± 0.11 (9)	0.9 ± 0.06 (4)
Diaphragm			1.0 ± 0.05 (6)	1.1 ± 0.06 (6)					
Heart			1.1 ± 0.07 (7)	1.0 ± 0.06 (6)					
Lung	1.7 ± 0.02 (4)	2.2 ± 0.2 (6)	1.2 ± 0.13 (8)	1.4 ± 0.09 (15)	1.7 ± 0.15 (9)	1.3 ± 0.19 (6)		0.9 ± 0.05 (6)	0.6 ± 0.03 (6)
Spleen	0.9 ± 0.06 (4)	0.9 ± 0.09 (16)	0.3 ± 0.01 (8)	0.7 ± 0.03 (6)	0.4 ± 0.03 (6)	0.5 ± 0.03 (6)		0.5 ± 0.07 (12)	0.4 ± 0.06 (4)
Subcutaneous connective			0.9 ± 0.03 (6)		0.7 ± 0.04 (6)			0.9 ± 0.02 (6)	0.7 ± 0.03 (3)

Note: Values represent the mean ± SEM(n).

[a] Tissue to medium ratio: (cpm/mg tissue)/(cpm/$\mu\ell$ medium) after 15 to 60 min of incubation in tissue culture media containing ^3H- or ^{14}C-labeled PGE_1, or PGE_2, or $PGF_{2\alpha}$.

[b] Unpublished data.

Table 2
ACCUMULATION (T/M) OF PGE$_1$, PGF$_{2\alpha}$, OR PGA$_1$ BY ORGANS OR TISSUES OF FOUR NONMAMMALIAN VERTEBRATES (VALUES REPRESENT MEAN ± S.E.M. (n))

							Species	
	Turkey[4]		Dogfish[16]				Scup[16]	Sea bass[16]
Organ or tissue	E$_1$	F$_{2\alpha}$	E$_1$	A$_1$	F$_{2\alpha}$		F$_{2\alpha}$	F$_{2\alpha}$
CNS and eye								
Brain	0.9 ± 0.5 (4)	0.8 ± 0.05 (4)						
Choroid plexus and tela choroidea			7.7 ± 0.71 (6)	21.7 ± 3.00 (6)	7.1 ± 0.55 (22)		7.6 ± 0.72 (2)	16.4 ± 1.28 (2)
Anterior uvea	2.3 ± 0.25 (14)	1.8 ± 0.21 (14)						
Iris					1.9 ± 0.16 (23)		2.7 ± 0.17 (3)	2.1 ± 0.11 (3)
Cornea	1.38							
Pecten	1.8 ± 0.15 (15)	1.5 ± 0.13 (15)						
Liver	2.5 ± 0.35 (9)	1.4 ± 0.10 (9)			1.6 ± 0.19 (6)		1.3 ± 0.10 (3)	0.9 ± 0.04 (3)
Gut or Stomach					0.3 ± 0.06 (6)		0.5 ± 0.04 (3)	0.6 ± 0.03 (3)
Gallbladder	1.2 ± 0.07 (10)	1.1 ± 0.03 (10)						
Kidney cortex	2.3 ± 0.20 (10)	1.7 ± 0.08 (10)	3.2 ± 0.78 (3)	3.7 ± 0.01 (3)	4.3 ± 0.27 (6)		4.1 ± 0.24 (3)	2.6 ± 0.23 (3)
Adipose White fat	0.3 ± 0.02 (9)	0.2 ± 0.02 (9)						
Heart	1.7 ± 0.07 (9)	1.5 ± 0.09 (9)						

Lung	1.2 ± 0.07 (12)	1.0 ± 0.01 (12)					
Spleen	0.8 ± 0.03 (8)	0.7 ± 0.04 (8)	0.8 ± 0.05 (6)	0.5 ± 0.05 (3)	0.7 ± 0.08 (3)		
Gill			0.3 ± 0.01 (3)	0.4 ± 0.06 (3)	0.2 ± 0.04 (6)	0.3 ± 0.01 (3)	0.3 ± 0.03 (3)

Table 3
**THE APPARENT ^{3}H-PGE$_1$ ACCUMULATION BY
RABBIT TISSUES EXPRESSED ON THE BASIS
OF ^{3}H ACTIVITY OF TOTAL TISSUE WATER
(Tw)^a AND INTRACELLULAR WATER (Iw)^a
RELATIVE TO THE ^{3}H ACTIVITY IN THE
INCUBATION MEDIUM (VALUES REPRESENT
MEAN ± S.E.M.)^{12}**

Tissues	(n)	T/M^b	Tw/M	Iw/M
Brain	(6)	0.5 ± 0.04	0.6 ± 0.04	0.4 ± 0.05
Ciliary body^c	(6)	7.0 ± 1.01	8.1 ± 1.14	46.6 ± 7.22
Liver	(9)	1.6 ± 0.07	2.2 ± 0.09	3.3 ± 0.34
Kidney cortex	(6)	2.4 ± 0.14	3.0 ± 0.17	4.4 ± 0.37
Cervix	(10)	6.3 ± 1.29	7.6 ± 1.56	15.4 ± 3.50
Vagina	(8)	2.7 ± 0.36	3.3 ± 0.44	6.2 ± 1.06
Spleen	(4)	0.3 ± 0.01	0.4 ± 0.02	0.4 ± 0.01
Subcutaneous connective	(6)	0.7 ± 0.03	0.8 ± 0.02	0.3 ± 0.08

^{a} T = ^{3}H activity per unit weight of tissue; Tw = ^{3}H activity per unit volume of total tissue water; Iw = ^{3}H activity in intracellular water.
^{b} M = ^{3}H activity per unit volume of incubation medium.
^{c} The bulk of the iris was removed from the anterior uvea.

CONCENTRATIVE ACCUMULATION OF PGs BY TISSUES OF MARINE INVERTEBRATES

Several invertebrate tissues incubated in media containing radioactively labeled PGs under appropriate conditions yielded T/M ratios well above unity (Table 4). Concentrative uptake of PGs by some of these tissues, such as the accumulation of PGF$_{2\alpha}$ by the gills of bay scallops, exhibited such characteristics of organic acid transport processes as temperature dependence and saturability.[16,20] However, the very large accumulation of PGA$_1$ by several invertebrate tissues had characteristics that were inconsistent with such transport mechanisms. The accumulation of PGA$_1$ by scallop gills, for example, was blocked by ethacrynic acid but not by probenecid,[20] and the accumulated PGA$_1$ was not readily elutable.[16] This suggests that at least some invertebrate tissues may incorporate PGs into structural components and/ or may possess pathways of PG metabolism that do not exist or that play only minor roles in mammalian tissues. A better understanding of such pathways will be required before the distribution and physiological significance of PG transport capacity in these invertebrates can be elucidated.

THE EXCLUSION OF PGS FROM THE INTRACELLULAR VOLUME OF SOME MAMMALIAN CELLS AND TISSUES

Under the same conditions that produce concentrative PG accumulation (T/M >1) in several vertebrate tissues, other tissues, including the brain, spleen, white fat, and subcu-

Table 4
ACCUMULATION (T/M) OF PGE$_1$, PGA$_1$, OR PGF$_{2\alpha}$ BY TISSUES OR ORGANS OF FOUR INVERTEBRATE SPECIES (VALUES REPRESENT MEAN ± S.E.M. (n))[16]

	Species					
	Sea scallop			Clam	Spider crab	Squid
	E$_1$	A$_1$	F$_{2\alpha}$	F$_{2\alpha}$	F$_{2\alpha}$	F$_{2\alpha}$
Tissue or organ						
Iris						2.3 ± 0.39 (2)
Gill	53.8 ± 3.9 (4)	212.5 ± 34.0 (4)	14.1 ± 0.06 (4)	2.4 ± 0.09 (3)	0.9 ± 0.16 (6)	0.6 ± 0.06 (6)
Liver or hepatic caeca	1.8 ± 0.34 (4)	3.5 ± 0.41 (4)	1.3 ± 0.08 (6)	1.9 ± 0.33 (3)	2.3 ± 0.27 (6)	
Kidney				2.7 ± 0.29 (4)		
Gut or stomach			1.4 ± 0.04 (6)	2.3 ± 0.13 (3)	2.2 ± 0.30 (6)	
Intestine	5.9 ± 1.92 (4)	6.5 ± 0.46 (4)	1.3 ± 0.04 (2)			
Mantle with eyes	5.7 ± 0.59 (4)	13.2 ± 2.86 (4)	2.4 ± 0.06 (2)			

taneous connective tissues of all species studied yielded T/M ratios less than unity (Tables 1 and 2). Since these tissues do not have subcompartments that are covered by tight-junctional epithelia, the concentration of solutes in their extracellular fluid volume must, under steady-state conditions, equal that of the medium. Based on this assumption, the calculated maximum PG concentration in the total intracellular water volume of slices of rabbit brain, spleen, or subcutaneous connective tissue remained well below that of the medium (Table 3), indicating that the cellular membranes in these tissues are not readily permeable to PGs.

More detailed studies demonstrated that rabbit erythrocytes are impermeable to E and F PGs, since PG entry into these cells could not be demonstrated even when they were suspended in ^{3}H-PG-containing media for up to 24 hr.[21] Rabbit erythrocytes were similarly shown to exclude thromboxane B$_2$, PGI$_2$, and 6-keto PGF$_{1\alpha}$ from their intracellular volume.[22] These findings contrast sharply with the long-held assumption that because PGs can be extracted into organic solvents, they must pass through cell membranes without appreciable restriction. Since the cell membranes of all tissues have similar properties, the impermeability of rabbit erythrocytes to PGs implies that the effective passage of these autacoids into the cytoplasm of cells or across tight-junctional epithelia requires special mechanisms, generally referred to as carrier-mediated, facilitated, or active transport processes.

THE REQUIREMENT FOR FACILITATED OR ACTIVE PG TRANSPORT SYSTEMS IN SOME VERTEBRATE TISSUES

Biologically active eicosanoids are found in the extracellular compartments of the body and, in most in vitro systems, endogenously produced PGs are found in higher concentrations in the incubation medium than within the incubated cells or tissues.[23] These observations can be best explained on the basis of mediation of the synthesis of PGs by membrane-bound enzyme systems that are oriented toward the extracellular space.[2] Because receptors that mediate the effects of PGs are located on the exterior surfaces of cells, the action of PGs does not necessarily require transmembrane transport. However, 15-hydroxy PG dehydrogenase and other enzymes responsible for the inactivation and metabolism of PGs are intracellular.[24] Thus, in all tissues that can effectively metabolize PGs, at least a subgroup of cells must have transmembrane PG transport function.[2]

Furthermore, several organ systems, such as the eye,[25] brain,[26] thymus,[27] and parts of the testes,[28] have tight-junctional endothelial or epithelial barriers that are interposed between the circulating blood and the parenchyma of these organs. It has recently been shown that the mucosa of the rabbit vagina is impermeable to sucrose, indicating that this organ also has a tight-junctional epithelium.[29] The passage of PGs or their initial metabolites across such barrier systems must require carrier-mediated, facilitated, or active transport processes. Thus, PGs can be generally regarded as extracellular autacoids that are transported into cells or across cellular membranes to be eliminated, inactivated, metabolized, or excreted.[2] In fact, the tissues that have been shown to accumulate PGs concentratively (Tables 1 and 2) correspond to the tissues that, based on these considerations, can be expected to require active PG transport processes.

In contrast, the in vitro concentrative accumulation of PGs by vertebrate tissues, under the conditions of the experiments described above, could not be correlated with the known distribution of PG metabolizing enzymes, which may be regarded as potential binding sites, and could not be accounted for by any other aspects of PG pharmacokinetics, such as binding to receptors.[12,30] However, concentrative PG accumulation cannot, by itself, be regarded as evidence for PG transport. In order to establish the existence of PG transport, a transmembrane flux that is saturable and is subject to specific inhibition must be demonstrated in at least one organ system. Furthermore, experimental evidence must be presented that indicates that concentrative PG accumulation by each tissue in question indeed represents a transport mechanism.

EXPERIMENTAL DEMONSTRATION OF ACTIVE TRANSMEMBRANE PG TRANSPORT

Of the tissues shown to accumulate PGs concentratively (Tables 1, 2, and 4), the rabbit vagina, which in young animals is a tube of thin translucent membrane, was judged best suited for demonstrating the existence of transmembrane PG transport. PGs were shown to be transported unidirectionally from the mucosal to the serosal side of a bladder-like preparation of this organ even against several-fold concentration gradients.[29] This transmembrane transport of ^{3}H-PGF$_{2\alpha}$ was inhibited by probenecid, bromcresol green, or by a large excess of PGF$_{2\beta}$, as well as by lowered temperature or by iodoacetic acid under anaerobic conditions. Furthermore, PGF$_{2\alpha}$ placed into the lumen and later recovered from the serosal side was identified as a biologically active PG and could again be transported across this tissue when placed into the lumen of another preparation, demonstrating that PGF$_{2\alpha}$ was not converted into an impermeable molecular species during its first transport across this tissue.[29] Such experiments demonstrate the existence of a physiological mechanism for the active transmembrane transport of PGs and substantiate the assumption that in vitro accumulation of PGs does indeed reflect transmembrane PG transport capacity at least in some tissues.

DEMONSTRATION OF PG TRANSPORT ACTIVITY IN OTHER TISSUES AND ORGAN SYSTEMS

Table 5 presents different types of experimental evidence for the existence of facilitated, carrier-mediated, or active PG transport processes in most of the mammalian tissues and organ systems included in Table 1, which have been shown to actively accumulate PGs. Table 5 also summarizes the apparent physiological significance of the transport system of

Table 5

VERTEBRATE TISSUES AND ORGAN SYSTEMS WHICH HAVE AN APPARENT OR DEMONSTRATED CAPACITY TO TRANSPORT PGs OR OTHER EICOSANOIDS AND THE DEMONSTRATED (!) OR ASSUMED (?) ROLE OF THESE TRANSPORT SYSTEMS (NUMBERS IN PARENTHESES INDICATE REFERENCES)

Tissue or organ	Nature of evidence			Physiological role of the PG transport system	Consequence of the inhibition or blockade of PG transport
	In vitro inhibition of PG accumulation	Perfused organ or transmembrane transport studes	In vivo evidence of PG transport		
Choroid plexus	(12,14,17,30,31)		(4,32—38)	Removal of PGs from brain (!)	Accumulation of PGs in CSF (!); enhancement of PG effects on brain, including induction of epileptiform activity (!)
Eye					
Anterior uvea	(12,14,15,30,31,39)		(4,40)	Removal of PGs from mammalian eye (!)	Enhancement of adverse PG effects on retina (!); may contribute to recurrent nature of some forms of uveitis (?)
Retinal choroid	(41)				
Pecten	(4)		(42,43)	Removal of PGs from avian eye (?)	
Lung	(17)	(44—51)	(3)	Facilitation of pulmonary PG metabolism (!)	Inhibition of pulmonary PG metabolism (!)
Kidney cortex	(12,14,17,30,31)	(52,53)	(3,54,55)	Facilitation of renal PG metabolism Renal excretion of PGs and their metabolites (!) Delivery of PGs from one region of kidney to other renal sites (?)	Limits PG excretion to filtered fraction (!) and blocks PG metabolism (!); leads to renal excretion of active PGs (!)
Vagina	(12)	(29)	(56)	Effective absorption of PGs from vaginal lumen (!); Removal of PGs from semen (?) Delivery of PGs from vaginal lumen to ovaries (?)	Decreases the rate of PG absorption from vaginal lumen (!)

Table 5 (continued)

VERTEBRATE TISSUES AND ORGAN SYSTEMS WHICH HAVE AN APPARENT OR DEMONSTRATED CAPACITY TO TRANSPORT PGs OR OTHER EICOSANOIDS AND THE DEMONSTRATED (!) OR ASSUMED (?) ROLE OF THESE TRANSPORT SYSTEMS (NUMBERS IN PARENTHESES INDICATE REFERENCES)

| Tissue or organ | Nature of evidence | | | Physiological role of the PG transport system | Consequence of the inhibition or blockade of PG transport |
	In vitro inhibition of PG accumulation	Perfused organ or transmembrane transport studies	In vivo evidence of PG transport		
Uterus	(12,17)			Removal of PGs and their metabolites from the fetus (?) Delivery of PGs from the fetus to maternal circulation (?)	Accumulation of PGs and their metabolites in fetus (?)
Liver	(12,17)	(57)	(58)	Delivery of PG metabolites to the biliary system (?)	Inhibition of the metabolism of PGs in the liver (?)

these tissues, as deduced from in vitro and/or in vivo studies, and lists the known or presumed effects of the inhibition of this transport system. Such evidence has established the existence of PG transport by the choroid plexus, anterior uvea, retinal choroid, lung, and kidney cortex (Table 5). However, similar evidence to support the existence of a PG transport function has not yet been presented for several organs, such as the liver, that appear (on the basis of in vitro accumulation studies) to have a PG transport capacity (Tables 1, 2, and 4) or that can be assumed, on the basis of physiological and/or morphological considerations, to require such a transport function.

OTHER VERTEBRATE TISSUES THAT MAY HAVE PG TRANSPORT CAPACITY

The liver plays an important role in PG metabolism.[57-59] It can be assumed that delivery of eicosanoids to intracellular enzymes in this organ, as in the kidney and lungs, requires an initial step of carrier-mediated transport. Furthermore, biliary excretion of PGs and initial PG metabolites can also be expected to require such transport processes. Liver slices of all mammals (Table 1) and most other vertebrates and invertebrates studied so far (Tables 2 and 4) have shown concentrative PG accumulation. This can be taken only as a suggestion, not as proof of the existence of active PG transport in this tissue. Although perfusion studies similar to those performed on the lung and kidney should be done on the liver, preliminary experiments with radioactively labeled PGs were discouraged by the wide range of PG breakdown products formed in the this organ.

The pecten of the avian eye also accumulates PGs against a concentration gradient (Table 2). Although no other transport function has yet been demonstrated in this organ, the apparent PG transport function of the pecten deserves further attention, since this organ may play an important role in maintaining the chemical environment of the avascular avian retina.[60,61]

It should be noted that most tissues are composed of many cell types; if PG transport occurs in only a small subcompartment of the tissue, while other compartments exclude PGs, the T/M ratio for the whole tissue may not always exceed unity. Thus, tissues that yield T/M ratios for PG accumulation close to unity, as do brown fat, the aorta, and the heart (Tables 1 and 2), may have subcompartments that have PG transport capacity. In fact, all tissues that can effectively metabolize PGs can be expected to contain a subcomponent of cells with plasma membranes that have PG transport systems. In addition, the effective delivery of PGs to their target cells may also require transmembrane transport in some tissues. Thus, for example, the accessibility of PGs from the blood to the musuclar coats of some blood vessels may require a transendothelial transport process.

Although the uterus, cervix, and vagina of rabbits yielded T/M ratios for PGE_1 and $PGF_{2\alpha}$ ranging from 2.6 to 8.8 (Tables 1 and 3), and although active transmembrane PG transport by the rabbit vagina has been demonstrated,[12] the T/M ratios for these tissues were close to unity in rhesus monkeys, rats, and woodchucks, and consistently fell below unity in the cat. It should be noted, however, that the female reproductive organs appear to contain greater muscle and/or connective tissue mass in the latter four species than in young rabbits. Thus, the ratio of the transporting cells (presumably epithelial) to nontransporting cell types may be lower in the vagina of other species than in rabbits. The possibility that the female reproductive tract in all or most mammals has PG transport capacity deserves further attention, since this capacity in these tissues might have important implications with regard to both the normal physiological role and the therapeutic use of eicosanoids in reproductive processes.

Finally, there are several mammalian tissues of great physiological significance, most notably the glands of the endocrine system, that have not yet been studied even preliminarily for PG transport capacity. The thymus, for example, is quite likely to have a PG transport system, since this organ has been shown to have a blood-tissue barrier system,[27] which presumably has some properties similar to those of the blood-brain and blood-ocular barriers.

THE SCOPE AND SPECIFICITY OF THE PG TRANSPORT SYSTEM

Although the concentrative accumulation, transmembrane transport, or excretion of a substance is the best evidence of its active transport, some eicosanoids cannot be easily assayed in the small quantities typically used in such transport studies or are not commercially available in radioactively labeled form. It has been well established, however, that substances transported by a particular carrier system compete for those carriers and, when present in an excess, decrease the transmembrane flux of all other substances using that transport system.[9,62] Thus, the demonstration that an appropriate concentration of an eicosanoid inhibits the accumulation or transmembrane flux of a radioactively labeled PG that is a known substrate for the PG transport system provides evidence that the eicosanoid in question is also a substrate for the same transport system.

A list of eicosanoids that have been shown to be substrates for the PG transport system by one or more of the above-mentioned criteria is presented in Table 6, together with references to the supporting evidence. All eicosanoids studied to date, including several leukotrienes as well as analogs, metabolites, and isomers of naturally occurring eicosanoids, have been found to be substrates for the PG transport system. Because the PG transport system is a subclass of the organic acid transport system,[15,30,53] it is reasonable to assume, unless it is proven otherwise for some members of this as yet only partially described group of substances, that all eicosanoids that are organic acids are substrates for this transport system.

It should be noted, however, that the transport systems of some tissues may show considerable specificity and may, in fact, exclude some eicosanoids. PGA_1 and PGI_2, for example, were shown to be less efficiently removed from the pulmonary circulation (and hence are less efficiently inactivated) than E or F PGs.[45,47,65] Whether this specificity is oriented toward the free form of such eicosanoids or toward their ligands or adducts remains to be established. It is not clear whether either PGA_1 or PGI_2 is present within the pulmonary tissue in a free, unbound, and chemically unaltered state. In fact, PGA_1 forms sulfhydryl adducts very readily in some mammalian tissues[21,66,67] and apparently even more extensively in most marine invertebrates.[16,20]

THE TRANSPORT OF OTHER EICOSANOIDS

Most studies on the PG transport system were completed before several important arachidonic acid derivatives, such as the thromboxanes and leukotrienes, were discovered and became available for research. Also, many of these more recently identified members of the arachidonic acid cascade have very short half-lives in aqueous media. Therefore, their transport cannot be studied by any of the methods described. The transmembrane transport of molecular species such as thromboxane A_2, which are transformed spontaneously with a half-time of seconds to less biologically active forms, may be a moot point because such compounds could not accumulate in extracellular compartments and because their transformation is not dependent on intracellular enzymes. However, the more stable products of such spontaneous transformation, such as thromboxane B_2 (TXB_2), can be expected to require transmembrane transport processes similar to those required by PGs.

Recent experiments show that TXB_2 and other eicosanoids that do not have a pentane ring, as well as compounds such as PGI_2, which have ring structures different from that of other PGs, are substrates for the PG transport system (Table 6). This demonstrates that a PG-like pentane ring is not a requirement for an eicosanoid to be a substrate for this transport system. In fact, we can expect all eicosanoids, including leukotrienes, to be substrates for this system as long as they are free acids. Thus, the PG transport system can be more appropriately referred to as the eicosanoid transport system.

Table 6
EICOSANOIDS WHOSE ACTIVE OR FACILITATED TRANSPORT HAS BEEN DEMONSTRATED OR IS EVIDENT FROM THE STUDIES REFERENCED IN PARENTHESES

Eicosanoid	Nature of evidence			
	Conc. accumulation of eicosanoid by tissue pieces or slices	Inhibition of conc. PG accumulation based on in vitro studies	Transmembrane transport or perfused organ studies	In vivo studies and observations
PGA_1	(12,16,17)		(45,49,53)	
PGE_1	(4,12,16,17)	(30)	(45,47—49,53,57,59)	(3,4,33—35,40,42,43)
PGE_2	(15)	(15)	(49,58)	(4,33,36,43,54,55)
$PGF_{1\alpha}$	(12,16,17)		(49,57)	
$PGF_{2\alpha}$	(12—17,30)	(14,15,30)	(29,31,44—50,52,53,58)	(3,4,33—35,37,42,54,56)
$PGF_{2\beta}$	(14)	(14,30)		(56)
PGI_2	(63)			
TXB_2	(63)			
15-keto-$PGF_{2\alpha}$		(30)		
15-keto-$PGF_{2\beta}$		(30)		
6-keto-$PGF_{1\alpha}$	(63)			
16,16-Dimethyl PGE_2			(51)	
(15S)-15-Methyl-PGE_2			(51)	
(15S)-15-Methyl-$PGF_{2\alpha}$		(30)	(49)	
7-oxa-13-Prostynoic acid		(30)		
Leukotriene C_4	(64)			

Table 7

THE INHIBITORY EFFECTS OF SOME
EICOSANOIDS ON ^{3}H-PGF$_{2\alpha}$
ACCUMULATION BY THE CHOROID
PLEXUS, ANTERIOR UVEA, AND KIDNEY
CORTEX[30]

	Percent inhibition at a 10^{-4} M concentration		
	Choroid plexus	Anterior uvea	Kidney cortex
PGF$_{2\alpha}$	84	91	79
PGA$_1$	92	91	76
PGE$_1$	87	92	69
PGF$_{2\beta}$	76	80	70
15-keto-PGF$_{2\alpha}$	51	79	57
15-keto-PGF$_{2\beta}$	73	67	48
(15S)-15-Methyl-PGF$_{2\alpha}$	82	74	58
7-oxa-13-Prostynoic acid	82	97	74

INHIBITORS OF THE EICOSANOID TRANSPORT SYSTEM AND THEIR EFFECTS ON THE PHARMACOKINETICS OF EICOSANOIDS

Most compounds that have been shown to inhibit the transport of eicosanoids are listed in Tables 7, 8, and 9. Table 7 presents the eicosanoids shown to inhibit the concentrative accumulation of ^{3}H-labeled PGF$_{2\alpha}$ by the anterior uvea, choroid plexus, and kidney cortex. Because these eicosanoids are themselves substrates for the PG transport system, it can be expected that this inhibition is due to competition for transport sites. Because many of these eicosanoids are biologically active, the possibility cannot be ruled out that at least part of this inhibition is due to the effect of these eicosanoids on metabolic activity, sodium transport, or other cellular functions required for effective PG transport. However, PGF$_{2\beta}$, a relatively inactive isomer of PGF$_{2\alpha}$, has been shown to have an inhibitory effect on PGF$_{2\alpha}$ transport comparable to that of biologically active PGs.[30,32] Thus, all eicosanoids and their derivatives or metabolites, provided they are organic acids but irrespective of whether they have other biological activities, must be assumed to affect the transport of all other eicosanoids. This implies, for example, that during hemorrhagic shock increased levels of circulating PG metabolites can be expected to interfere with the pulmonary uptake and inactivation of biologically active PGs, as well as with their general pharmacokinetics and renal excretion. Such inhibition of PG transport at these and other sites is likely to extend the biological half-lives of circulating biologically active eicosanoids, and to modify the rates of removal of eicosanoids from the extracellular fluids of the eye, brain, and other organs that are separated from the circulation by a blood-tissue barrier system.

Because most nonsteroidal anti-inflammatory drugs are organic acids, it is not surprising that they are highly effective inhibitors of the PG transport system (Table 8). This implies that some of the observed experimental, clinical, or side effects of these drugs may be due not only to their ability to inhibit PG synthesis, but also to their capacity to modify the pharmacokinetics of eicosanoids. In addition, it should be noted that when present in concentrations below those required to inhibit PG transport, some of these nonsteroidal anti-inflammatory agents stimulate PG accumulation.[14] Although direct stimulation of transport cannot be ruled out, this potentiation may simply be caused by inhibition of endogenous

Table 8
THE INHIBITORY EFFECTS OF NONSTEROIDAL ANTI-INFLAMMATORY AGENTS AND OTHER DRUGS ON THE PG TRANSPORT FUNCTION OF THE CHOROID PLEXUS, ANTERIOR UVEA, AND KIDNEY CORTEX[14]

	Concentration (μM) required for 50% inhibition of PG transport		
	Choroid plexus	Anterior uvea	Kidney cortex
Cyclooxygenase inhibitors			
Indomethacin	10	8	12
Phenylbutazone	70	18	42
Oxyphenbutazone	38	6	74
Ibuprofen	112	11	27
Pirprofen	80	21	41
D-Naproxen	36	24	23
l-Naproxen	96	28	30
Aspirin	>1000	1582	>1000
l-methyl-5-methoxy-indoleacetic acid	698	298	171
Other drugs			
Dexamethasone phosphate	1189	451	873
Benzylpenicillin Na	342	122	463
Papaverine hydrochloride	14	38	80
Fursemide	80	52	80
Probenecid	106	13	42
Bromcresol green	4	0.1	3
Iodipamide[29]	10	10	100

Table 9
DRUGS AND OTHER COMPOUNDS THAT SHOW NO INHIBITORY EFFECT OR ONLY A WEAK INHIBITORY EFFECT ON PG TRANSPORT

	Percent inhibition at 10^{-3} M concentration[14]		
	Choroid plexus	Anterior uvea	Kidney cortex
Paracetamol	1	7	0
Caffeine	27	39	63
Diphenhydramine	38	37	40
Gentisic acid	0	4	0
Phenelzine	23	27	14
β-Resorcyclic acid	30	43	39
Na salicylate	0	31	18
Octanoic acid	22[a]	0[a]	0[a]
NaI	0[a]	0[a]	0[a]

[a]   Percent inhibition at 10^{-4} M concentration.[30]

PG synthesis and hence a consequent decrease in competition between the exogenous tracer and endogenously released PGs.

Other drugs that have been found to inhibit PG transport, including probenecid and some other classical inhibitors of the organic acid transport system, are also listed in Table 8. Table 9 presents a list of substances found to have no inhibitory effect or only a partial inhibitory effect on PG accumulation even at the relatively high concentration of 10^{-3} M.

THE KINETICS OF THE PG TRANSPORT SYSTEM AND ITS ENERGY AND IONIC DEPENDENCE

Detailed studies on the initial accumulation of $PGF_{2\alpha}$ by the isolated anterior uvea of rabbit eyes have demonstrated that this process conforms to Michaelis-Menten kinetics, exhibiting an apparent K_t of 34 mM and a J_{max} of 13.6 nmol/g of tissue per minute. The $PGF_{2\alpha}$ transport rate was inhibited by the absence of glucose or oxygen, cold (0°C), 0.1 mM iodoacetic acid, and by 10^{-5} M ouabain. Significant reduction of $PGF_{2\alpha}$ accumulation was also observed when the sodium in the incubation medium was replaced by lithium or mannitol.[15] The eicosanoid transport system of the choroid plexus was found to have similar kinetic parameters and dependence on metabolic energy and sodium.[63] These studies indicate that the PG transport system represents a saturable, energy- and/or sodium-dependent mechanism.

THE RELATIONSHIP BETWEEN THE EICOSANOID TRANSPORT SYSTEM AND OTHER TRANSPORT SYSTEMS

Probenecid and bromcresol green, two classical inhibitors of the organic acid transport system, are effective inhibitors of the concentrative accumulation and transmembrane transport of eicosanoids in all systems tested thus far (Table 10). In comparative studies, these two agents were more effective inhibitors of eicosanoid transport than *p*-aminohippuric acid.[15,30,53] Furthermore, sodium iodide had no inhibitory effect on this transport system even at a concentration of 10^{-4} M (Table 9).[30] These findings suggest that the eicosanoid transport system is a subclass of the organic acid transport system, and that the iodide transport system of these tissues does not play a primary role in the eicosanoid transport system. Using Barany's classification of the organic acid transport system,[10] the PG transport system can therefore be regarded as a subclass of the kidney-like system. The fact that PG transport is also inhibited by iodipamide (Table 8) suggests, however, that PGs may also use a subcomponent of the liver-like system[68,69] for their transport.

Saturation and competitive inhibition studies indicate that PGE_2 and $PGF_{2\alpha}$ share the same carriers within the organic acid transport system, whereas PAH interacts with, but is not primarily transported by, the PG system.[15,53,63]

THE EFFECTS OF INFLAMMATORY PROCESSES ON PG TRANSPORT

The foregoing discussion clearly indicates that local or systemic accumulation of PGs or PG metabolites, as may occur during hemorrhagic shock, renal failure, or inflammation, can be expected to decrease the efficacy of PG transport as a result of saturation or competitive inhibition. Thus, such conditions must decrease the rate of inactivation, metabolism, and excretion of PGs and other eicosanoids. In addition, it has been shown that intraocular inflammation of short duration temporarily depresses the PG transport capacity of the anterior uvea, whereas more prolonged or more severe inflammation has long-lasting and possibly permanent effects on anterior uveal transport functions.[39,70,71] Because it is likely that inflammatory processes in other tissues have similar effects on their PG transport capacity,

Table 10
TISSUES, ORGANS, AND SYSTEMS IN WHICH PROBENECID AND/OR BROMCRESOL GREEN WERE SHOWN, DIRECTLY OR BY INFERENCE, TO INHIBIT THE CONCENTRATIVE ACCUMULATION OR ACTIVE TRANSPORT OF AN EICOSANOID

Observation	Tissue or organ system	Species	Eicosanoid	Ref.
Inhibition of in vitro concentrative PG accumulation by tissues or tissue slices	Choroid plexus, anterior uvea, kidney cortex	Rabbit	$PGF_{2\alpha}$, PGE_1, PGE_2	13—15,30
Inhibition of PG clearance from ventriculo-cisternal perfusates	CNS	Rabbit	$PGF_{2\alpha}$	32
Enhancement of adverse effect of exogenous PGs	CNS Eye	Rabbit	$PGF_{2\alpha}$, PGE_1, PGE_2	3,4,33—36,42,43
Increased PG levels in CSF	CNS	Human Rabbit	$PGF_{2\alpha}$, PGF, TXB_2	37,38
Inhibition of PG uptake from in vitro perfusates	Kidney Lung	Rabbit Rat Dog	$PGF_{2\alpha}$, A_1, E_1, E_2, M-PGE_2ME^a	3,44—46,49—53
Inhibition of excretion of eicosanoids	Kidney	Rabbit Chicken Rat Dog	PGF_1, $F_{1\alpha}$, $PGF_{2\alpha}$-metabolite[b]	3,52—55
Inhibition of transmembrane PG transport in vitro	Vagina	Rabbit	$PGF_{2\alpha}$	29

[a] (15*S*)-15-methyl PGE_2 methyl ester.
[b] 13,14-dihydro, 15-keto-$PGF_{2\alpha}$.

the possibility must be investigated that inflammatory and releated pathological processes in the lungs, liver, and kidney as well as the BBB can modify the PG transport processes of these organs and hence the pharmacokinetics of PGs and related autacoids.

CONSEQUENCES OF THE INHIBITION OF THE PG TRANSPORT SYSTEM OF PERIPHERAL ORGAN SYSTEMS

The demonstrated or predictable effects of the inhibition of the PG transport system in different tissues and organs are listed in Table 5. A review of that table clearly suggests that general inhibition of this transport system can directly or indirectly affect all organ systems of the body. Because the brain and intraocular tissues do not have adequate capacity to inactivate or metabolize PGs, the carrier-mediated removal of PGs across the blood-brain and blood-ocular barriers represents the major and possibly the only mechanism for terminating the action of locally produced PGs.[4] PG transport can, therefore, be expected to play a very important role in the normal physiological functions of these organs. Indeed, it has already been shown that inhibition of PG transport greatly enhances the effects of exogenous PGs on the retina and the brain.[33-35,42,43] It has been suggested that at least some forms of epileptogenic foci may be associated with focal inhibition of PG transport across the blood-brain barrier, and with the ensuing local accumulation of PGs and possibly other organic acids that use the same transport system.[33,35] It has also been suggested that a damaged ocular PG transport function may be the cause of or may contribute to cystoid macular edema[72,73] and the recurrent nature of some forms of uveitis.[39]

It has been demonstrated that, at least in the rabbit, transmembrane PG transport is responsible for the effective removal of PGs from the vaginal lumen into the circulation.[56] However, the biological significance of this transport system and the consequences of its inhibition are yet to be elucidated. Because PGs remain biologically active during their transport across the vaginal wall, the possibility that the function of this transport is related to the delivery of PGs from the vaginal lumen to some remote site of PG action, such as the ovaries, must be considered.[12,56]

Although it has been suggested that uterine PG transport may be required for the effective removal of PGs from the fetus, preliminary studies on implanted and nonimplanted regions of the rabbit uterus have been inconclusive.[12] The possible dependence of PG passage across the placenta on facilitated or active transport should be further investigated, since such dependence would imply that a variety of drugs known to be effective inhibitors of PG transport could affect the removal of eicosanoids and other organic acids from the fetal circulation.

SUMMARY AND CONCLUSIONS

Because the precursors of eicosanoids and the enzymes required for their synthesis are contained within cell membranes, endogenously produced eicosanoids in most, and possibly all tissues can be released directly into extracellular compartments from the plasmalemma or from internal membrane systems connected to the plasmalemma. In fact, most cells contain cytoplasmic enzymes that can readily metabolize PGs and most other eicosanoids. Thus, PGs released into the cytoplasm are unlikely to escape metabolism, and hence are not expected to have a biological effect. On the other hand, the effective inactivation and metabolism of PGs and most other eicosanoids present in extracellular compartments must depend on their delivery to intracellular enzymes. The fact that the basic cell membrane is an effective barrier to the free diffusion of PGs and other eicosanoids implies that delivery of such eicosanoids from extracellular to cytoplasmic sites requires facilitated or active transport processes. Furthermore, eicosanoids released into the extracellular fluid compartments of the eye and the brain must be removed by facilitated or active transport processes across the blood-brain and blood-ocular barriers.[1,74] On the basis of these considerations, a hypothesis has been proposed that regards PGs as extracellular autacoids that depend primarily on transmembrane transport processes for their elimination, inactivation, metabolism, and excretion.[2]

Evidence obtained over the past decade has established the existence of active PG transport processes and demonstrated the key role this transport system plays in the removal of these autacoids from the cerebrospinal and intraocular fluids, in the pulmonary and renal metabolism of PGs, and in the renal excretion of PGs. In addition, it has been shown that the absorption of PGs from the vaginal lumen of at least one species, the rabbit, is mediated by PG transport processes. Although there is some evidence that several other tissues, specifically the liver, brown fat, and the rabbit cervix and uterus, have effective PG transport systems, the role of PG transport in these systems remains to be elucidated.

Substantial evidence indicates that not only classical PGs, but all eicosanoids that are organic acids are substrates for the PG transport system. Thus, we must expect the PG transport systems to have critical importance in the pharmacokinetics and pharmacodynamics of all eicosanoids. We also expect the inhibition of eicosanoid transport to have direct and/ or indirect effects on all organ systems of the body. In fact, the manipulation of this transport system must be regarded as a potential method for modifying the distribution and pharmacokinetics of endogenously produced, as well as therapeutically administered eicosanoids.

On the other hand, modification of the distribution and elimination of eicosanoids must be considered in interpreting the effects and side effects of some drugs, such as the nonsteroidal anti-inflammatory agents, that are known to or, on the basis of their chemical structure, can be expected to affect the eicosanoid transport system.

ACKNOWLEDGMENTS

I wish to thank my collaborators who are listed as co-authors on my publications in the reference list. I also wish to thank Dr. Olivia C. Miranda for her help with the preparation of this manuscript. The work reviewed here that originated from my laboratory was supported by research grant EY 00333 from the National Eye Institute, N.I.H., U.S. Public Health Service.

REFERENCES

1. **Bito, L. Z.**, Absorptive transport of prostaglandins from intraocular fluids to blood: a review of recent findings, *Exp. Eye Res.*, 16, 299—309, 1973.
2. **Bito, L. Z.**, Are prostaglandins intracellular, transcellular or extracellular autocoids?, *Prostaglandins*, 9, 851—855, 1975.
3. **Bito, L. Z., Wallenstein, M., and Baroody, R.**, The role of transport processes in the distribution and disposition of prostaglandins, in *Prostaglandins and Thromboxane Research*, Vol. 1, Samuelsson, B. and Paoletti, R., Eds., Raven Press, New York, 1976, 297—303.
4. **Bito, L. Z. and Wallenstein, M. C.**, Transport of prostaglandins across the blood-brain and blood-aqueous barriers and the physiological significance of these absorptive transport processes, in *The Ocular and Cerebrospinal Fluids*, Bito, L. Z., Davson, H., and Fenstermacher, J. D., Eds., Academic Press, New York, 1977, 229—243.
5. **Bito, L. Z. and Merritt, S. Q.**, Therapeutic implications of facilitated PG transport, *Prostaglandins Ther.*, (Upjohn Co.), 4, 3, 1978.
6. **Forster, R. P.**, Use of thin kidney slices and isolated renal tubules for direct study of cellular transport kinetics, *Science*, 108, 65—67, 1948.
7. **Cross, R. J. and Taggart, J. V.**, Renal tubule transport: accumulation of para-aminohippurate by rabbit kidney slices, *Am. J. Physiol.*, 161, 181—190, 1950.
8. **Becker, B.**, The transport of organic anions by the rabbit eye. I. In vitro iodopyracet (diodrast) accumulation by ciliary body-iris preparations, *Am. J. Ophthalmol.*, 50, 862—867, 1960.
9. **Barany, E. H.**, Characterization of simple and composite uptake systems in cells and tissues by competition experiments, *Acta Physiol. Scand.*, 83, 220—234, 1971.
10. **Barany, E. H.**, Inhibition by hippurate and probenecid of in vitro uptake of iodipamide and *o*-iodohippurate. A composite uptake system for iodipamide in choroid plexus, kidney cortex and anterior uvea of several species, *Acta Physiol. Scand.*, 86, 12—27, 1972.
11. **Barany, E. H.**, In vitro uptake of bile acids by choroid plexus, kidney cortex and anterior uvea. I. The iodipamide-sensitive system in the rabbit, *Acta Physiol. Scand.*, 93, 250, 1975.
12. **Bito, L. Z.**, Accumulation and apparent active transport of prostaglandins by some rabbit tissues *in vitro*, *J. Physiol. (London)*, 221, 371—387, 1972.
13. **Bito, L. Z. and Baroody, R. A.**, Concentrative accumulation of ^{3}H-prostaglandins by some rabbit tissues *in vitro*: the chemical nature of the accumulated ^{3}H-labelled substances, *Prostaglandins*, 7, 131—140, 1974.
14. **Bito, L. Z. and Salvador, E. V.**, Effects of anti-inflammatory agents and some other drugs on prostaglandin biotransport, *J. Pharmacol. Exp. Ther.*, 198, 481—488, 1976.
15. **DiBenedetto, F. E. and Bito, L. Z.**, The kinetics and energy dependence of prostaglandin transport processes. I. In vitro studies on the rate of PGF$_{2\alpha}$ accumulation by the rabbit anterior uvea, *Exp. Eye Res.*, 30, 175—182, 1980.
16. **Bito, L. Z.**, Comparative study of concentrative prostaglandin accumulation by various tissues of mammals and marine vertebrates and invertebrates, *Comp. Biochem. Physiol.*, 43A, 65—82, 1972.

17. **Salvador, E. V., Roberts, J. C., and Bito, L. Z.,** Concentrative accumulation of prostaglandins *in vitro* at 7° and 37°C by tissues of normothermic and hibernating woodchucks (*Marmota monax*), *Comp. Biochem. Physiol.,* 60A, 173—176, 1978.
18. **Ehinger, B.,** Localization of the uptake of prostaglandin E₁ in the eye, *Exp. Eye Res.,* 17, 43—47, 1973.
19. **Bhattacherjee, P.,** Autoradiographic localization of intravitreally- or intracamerally-injected [³H]prostaglandins, *Exp. Eye Res.,* 18, 181—188, 1974.
20. **Bito, L. Z., DiBenedetto, F. E., DeRousseau, C. J., and Bito, J. W.,** Unpublished data.
21. **Bito, L. Z. and Baroody, R. A.,** Impermeability of rabbit erythrocytes to prostaglandins, *Am. J. Physiol.,* 229, 1580—1584, 1975.
22. **Baroody, R. A. and Bito, L. Z.,** The impermeability of the oasic cell membrane to thromboxane-B₂, prostacyclin and 6-keto-PGF₁, *Prostaglandins,* 21, 133—142, 1981.
23. **Crowshaw, K.,** The incorporation of [1-¹⁴C] arachidonic acid into the lipids of rabbit renal slices and conversion to prostaglandins E₂ and F₂ₐ, *Prostaglandins,* 3, 607—620, 1973.
24. **Anggard, E., Larsson, C., and Samuelsson, B.,** The distribution of 15-hydroxy prostaglandin dehydrogenase and prostaglandin-13-reductase in tissues of the swine, *Acta Physiol. Scand.,* 81, 396—404, 1971.
25. **Raviola, G.,** The structural basis of the blood-ocular barriers, in *The Ocular and Cerebrospinal Fluids,* Bito, L. Z., Davson, H., and Fenstermacher, J. D., Eds., Academic Press, New York, 1977, 27—63.
26. **Brightman, M. W.,** Morphology of blood-brain interfaces, *Exp. Eye Res. (Suppl.),* 25, 1—25, 1977.
27. **Raviola, E. and Karnovsky, M. J.,** Evidence for a blood-thymus barrier using electron-opaque tracers, *J. Exp. Med.,* 136, 466—498, 1972.
28. **Dym, M. and Fawcett, D. W.,** The blood-testis barrier in the rat and the physiological compartmentation of the seminiferous epithelium, *Biol. Reprod.,* 3, 308—326, 1971.
29. **Bito, L. Z.,** Saturable, energy-dependent, transmembrane transport of prostaglandins against concentration gradients, *Nature (London),* 256, 134—136, 1975.
30. **Bito, L. Z., Davson, H., and Salvador, E. V.,** Inhibition of *in vitro* concentrative prostaglandin accumulation by prostaglandins, prostaglandin analogues and by some inhibitors of organic anion transport, *J. Physiol. (London),* 256, 257—271, 1976.
31. **Bito, L. Z.,** Prostaglandin biotransport and its inhibition by some nonsteroidal anti-inflammatory agents, in *Second Congress of the Hungarian Pharmacological Society, Symposium on Prostaglandins,* Knoll, J. and Kelemen, K., Eds., Akademiaia Kiado, Budapest, 1976, 172.
32. **Bito, L. Z., Davson, H., and Hollingsworth, J. R.,** Facilitated transport of prostaglandins across the blood-cerebrospinal fluid and blood-brain barriers, *J. Physiol. (London),* 256, 273—285, 1976.
33. **Wallenstein, M. C. and Bito, L. Z.,** Prostaglandin E₁-induced alterations in visually evoked response and production of epileptiform activity, *Neuropharmacology,* 16, 687—694, 1977.
34. **Wallenstein, M. C. and Bito, L. Z.,** Hyperthermic effects of supracortically applied prostaglandins after systemic pretreatment with inhibitors of prostaglandin transport and synthesis, *J. Pharmacol. Exp. Ther.,* 204, 454—460, 1978.
35. **Wallenstein, M. C. and Bito, L. Z.,** Prostaglandin E₁-induced latent epileptogenic foci, *Electroenceph. Clin. Neurophysiol. J.,* 46, 106—109, 1979.
36. **Crawford, I. L., Kennedy, J. I., Lipton, J. M., and Ojeda, S. R.,** Effects of central administration of probenecid on fevers produced by leukocytic pyrogen and PGE₂ in the rabbit, *J. Physiol. (London),* 287, 519—533, 1979.
37. **Gross, H. A., Dunner, D. L., Lafleur, D., Meltzer, H. L., and Fieve, R. R.,** Prostaglandin F in patients with primary affective disorder, *Biol. Psychiatr.,* 12, 347—357, 1977.
38. **Spagnuolo, C., Petroni, A., Blasevich, M., and Galli, C.,** Differential effects of probenecid on the levels of endogenous PGF₂ₐ and TxB₂ in brain cortex, *Prostaglandins,* 18, 311—315, 1979.
39. **Bito, L. Z.,** The effects of experimental uveitis on anterior uveal prostaglandin transport and aqueous humor composition, *Invest. Ophthalmol.,* 13, 959—966, 1974.
40. **Bito, L. Z. and Salvador, E. V.,** Intraocular fluid dynamics. III. The site and mechanism of prostaglandin transfer across the blood intraocular fluid barriers, *Exp. Eye Res.,* 14, 233—241, 1972.
41. **DiBenedetto, F. E. and Bito, L. Z.,** Unpublished data, 1979.
42. **Wallenstein, M. C. and Bito, L. Z.,** The effects of intravitreally injected prostaglandin E₁ on retinal function and their enhancement by a prostaglandin-transport inhibitor, *Invest. Ophthal. Visual Sci.,* 17, 795—799, 1978.
43. **Siminoff, R. and Bito, L. Z.,** The effects of prostaglandins and arachidonic acid on the electroretinogram: evidence for functional cyclooxygenase activity in the retina, *Curr. Eye Res.,* 1, 635—642, 1982.
44. **Bito, L. Z. and Baroody, R. A.,** Inhibition of pulmonary prostaglandin metabolism by inhibitors of prostaglandin biotransport (probenecid and bromcresol green), *Prostaglandins,* 10, 633—639, 1975.
45. **Bito, L. Z., Baroody, R. A., and Reitz, M. E.,** Dependence of pulmonary prostaglandin metabolism on carrier-mediated transport processes, *Am. J. Physiol.,* 232, E382—E387, 1977.
46. **Wicks, T. C., Ramwell, P. W., Kot, P. A., and Rose, J. C.,** Inhibition of prostaglandin-induced pulmonary vasoconstriction by organic acid transport inhibitors (40120), *Proc. Soc. Exp. Biol. Med.,* 157, 677—680, 1978.

47. **Anderson, M. W. and Eling, T. E.**, PG removal and metabolism by isolated perfused rat lung, *Prostaglandins*, 11, 645—677, 1976.
48. **Eling, T. E. and Anderson, M. W.**, Studies on the biosynthesis, metabolism and transport of prostaglandins by the lung, *Agents Actions*, 6, 543—546, 1976.
49. **Eling, T. E., Hawkins, H. J., and Anderson, M. W.**, Structural requirements for, and the effects of chemicals on, the rat pulmonary inactivation of prostaglandins, *Prostaglandins*, 14(1), 51—60, 1977.
50. **Hawkins, H. J., Smith, J. B., Nicolaou, K. C., and Eling, T. E.**, Studies of the mechanisms involved in the fate of prostacyclin (PGI$_2$) and 6-keto-PGF$_{1\alpha}$ in the pulmonary circulation, *Prostaglandins*, 16, 871—884, 1978.
51. **Bakhle, Y. S., Jancar, S., and Whittle, B. J. R.**, Uptake and inactivation of prostaglandin E$_2$ methyl analogues in the rat pulmonary circulation, *Br. J. Pharmacol.*, 62, 275—280, 1978.
52. **Bito, L. Z.**, Inhibition of renal prostaglandin metabolism and excretion by probenecid, bromcresol green and indomethacin, *Prostaglandins*, 12, 639—646, 1976.
53. **Bito, L. Z. and Baroody, R. A.**, A comparison of renal prostaglandin and p-aminohippuric acid transport processes, *Am. J. Physiol.*, 234, F80—F88, 1978.
54. **Rennick, B. R.**, Renal tubular transport of prostaglandins: inhibition by probenecid and indomethacin, *Am. J. Physiol.*, 233, F133—F137, 1977.
55. **Rosenblatt, S. G., Patak, R. V., and Lifschitz, M. D.**, Organic acid secretory pathway and urinary excretion of prostaglandin E in the dog, *Am. J. Physiol.*, 235, F473—F479, 1978.
56. **Bito, L. Z. and Spellane, P. J.**, Saturable, "carrier-mediated", absorption of prostaglandin F$_{2\alpha}$ from the *in vivo* rabbit vagina and its inhibition by prostaglandin F$_{2\alpha}$, *Prostaglandins*, 8, 345—352, 1974.
57. **Dawson, W., Jessup, S. J., McDonald-Gibson, W., Ramwell, P. W., and Shaw, J. E.**, Prostaglandin uptake and metabolism by the perfused rat liver, *Br. J. Pharmacol.*, 39, 585—598, 1970.
58. **Anderson, F. L., Jubiz, W., and Tsagaris, T. J.**, Degradation of prostaglandin E$_2$ and F$_{2\alpha}$ by the canine liver, *Am. J. Physiol.*, 231(2), 426—429, 1976.
59. **Osborne, D. J. and Boot, J. R.**, PGE$_1$ metabolism by the perfused rat liver, *Prostaglandins*, 17(6), 863—872, 1979.
60. **Bito, L. Z. and DeRousseau, C. J.**, Transport functions of the blood-retinal barrier system and the micro-environment of the retina, in *The Blood-Retinal Barriers*, Vol. 32, Cuhna-Vaz, J. G., Ed., Plenum Press, New York, 1980, 133—163.
61. **Bito, L. Z., DiBenedetto, F. E., and Stetz, D.**, Homeostasis of the retinal micro-environment. I. Magnesium, potassium and calcium distributions in the avian eye, *Exp. Eye Res.*, 34, 229—237, 1982.
62. **Stein, W. D.**, *The Movement of Molecules Across Cell Membranes*, New York, Academic Press, 1967.
63. **DiBenedetto, F. E. and Bito, L. Z.**, The transport of prostaglandins and other eicosanoids by the choroid plexus: its characterization and physiological significance, *J. Neurochem.*, 46, 1725—1731, 1986.
64. **Spector, R. and Goetzl, E. J.**, Leukotriene C$_4$ transport by the choroid plexus in vitro, *Science*, 228, 325, 1985.
65. **Dusting, G. J., Moncada, S., and Vane, J. R.**, Recirculation of prostacyclin (PGI$_2$) in the dog, *Br. J. Pharmacol.*, 64, 315—320, 1978.
66. **Smith, J. B., Silver, M. J., Ingerman, C. M., and Kocsis, J. J.**, Uptake and inactivation of A-type prostaglandins by human red cells, *Prostaglandins*, 9, 135—145, 1975.
67. **Ham, E. A., Oien, H. G., Ulm, E. H., and Kuehl, F. A., Jr.**, The reaction of PGA$_1$ with sulfhydryl groups; a component in the binding of A-type prostaglandins to proteins, *Prostaglandins*, 10, 217—229, 1975.
68. **Barany, E. H.**, The liver-like anion transport system in rabbit kidney, uvea and choroid plexus. I. Selectivity of some inhibitors, direction of transport, possible physiological substrates, *Acta Physiol. Scand.*, 88, 412—429, 1973.
69. **Barany, E. H.**, The liver-like anion transport system in rabbit kidney, uvea and choroid plexus. II. Efficiency of acidic drugs and other anions as inhibitors, *Acta Physiol. Scand.*, 88, 491—504, 1973.
70. **Bito, L. Z.**, Inhibition of uveal prostaglandin transport in experimental uveitis, in *Prostaglandins and Cyclic AMP: Biological Actions and Clinical Applications*, Kahn, R. and Land, W., Eds., Academic Press, London, 1973, 213—214.
71. **Bengtsson, E. and Ehinger, B.**, The effect of experimental uveitis on the uptake of prostaglandin E$_1$ in the rabbit iris-ciliary body, *Acta Ophthal.*, 55, 688—695, 1977.
72. **Tennant, J. L.**, Is cystoid macular edema reversible by oral indocin, yes or no?, in *Current Concepts in Cataract Surgery*, Emery, J. M. and Paton, D., Eds., C. V. Mosby, St. Louis, 1976, 310.
73. **Miyake, K.**, Prevention of cystoid macular edema after lens extraction by topical indomethacin II. A control study in bilateral extractions, *Jpn. J. Ophthalmol.*, 22, 80—94, 1978.
74. **Bito, L. Z.**, Absorptive transport of prostaglandins and other eicosanoids across the blood-brain barrier system and its physiological significance, in *The Blood-Brain Barrier in Health and Disease*, Suckling, A. J., Rumsby, M. G., and Bradbury, M. W. B., Eds., Ellis Horwood, Chichester, England, 1986, 109—121.

COMPARATIVE METABOLISM AND FATE OF THE EICOSANOIDS

L. Jackson Roberts, II

In the normal physiologic situation, the biological actions of arachidonic acid (AA) metabolites are limited to the local site of biosynthesis as opposed to exerting systemic effects. Therefore, prostaglandins and other AA metabolites are generally considered to be local rather than circulating hormones. A major reason that AA metabolites released at a local site of biosynthesis do not exert systemic effects under normal circumstances is because these compounds are rapidly and efficiently metabolized to biologically inactive metabolites. Thus, the biologically active unmetabolized compounds are prevented from reaching the systemic circulation in sufficient concentration to exert effects at sites distant to the origin of biosynthesis. In addition, some arachidonic metabolites are chemically very unstable, in particular thromboxane A_2 (TXA_2) and prostacyclin (PGI_2), so that spontaneous nonenzymatic chemical degradation to biologically inactive compounds may be an additional mechanism which limits the biological actions of these compounds to the local site of formation. Such general considerations are reviewed in the first chapter by A. L. Willis.

Information regarding the specific structures of the metabolic products of prostaglandins (PG) and thromboxanes (TX) has substantial practical value. As mentioned above, because of very rapid and efficient metabolism, only very small concentrations of unmetabolized PG and TX are present in the circulation. Therefore, it is extremely difficult to measure the miniscule concentrations of unmetabolized PG and TX in the circulation as an index of endogenous production of these compounds. However, circulating concentrations of metabolites of PG and TX are present in greater amounts than the unmetabolized compounds and quantification of metabolites, therefore, methodologically is a more accurate means to assess endogenous production of PG and TX. In addition, formed elements of blood produce substantial quantities of PG and TX and substantial quantities of these compounds are formed and released by platelets and leukocytes during blood sampling and plasma isolation. As a result, when unmetabolized PG and TX are quantified in plasma, the levels measured almost entirely reflect artifactual production of these compounds by formed elements of blood during sample processing rather than the true endogenous circulating concentrations. Therefore, it has been well established that quantification of circulating metabolites of PG and TX represents a much more reliable means to assess endogenous release of PG and TX than does quantification of circulating unmetabolized PG and TX.[1-3] In addition, because of the extremely low circulating concentrations of unmetabolized PG and TX, only miniscule quantities of these compounds are filtered and excreted from the circulation into the urine. However, PG and TX metabolites are primarily excreted by the kidney and thus are present in measurable concentrations in the urine. Although measurable quantities of unmetabolized PGs and TX are present in urine, their origin derives almost entirely from renal biosynthesis so that they are excreted directly into the urine, thereby escaping metabolic degradation at sites such as the lung and liver.[4,5] Therefore, although quantification of unmetabolized urinary PG and TX provides a reasonable index of renal PG and TX production, it does not provide an accurate total index of endogenous PG and TX biosynthesis by the entire body or reflect synthesis of these compounds by organs other than the kidney. In contrast, however, quantification of urinary metabolites of PG and TX has proven to be a very accurate and reliable means to assess endogenous PG and TX production.[1-3,6]

The metabolic fate of the cyclooxygenase products PGE_2, $PGF_{2\alpha}$, PGD_2, PGI_2, and TXB_2 have been extensively investigated both in experimental animals and humans.[7-26] Only very limited information has been obtained regarding the metabolic fate of the lipoxygenase products of AA metabolism.[27-30] Therefore, the primary focus of the following discussion will be directed toward outlining current knowledge regarding the metabolism of PG and TX. It is not possible for this discussion to detail all of the specifics of the metabolism of PG and TX. There are a number of common features shared in the metabolism of all of the PG and TX, although specific differences exist. The general scheme of pathways of metabolic transformation of PG and TX, therefore, will be outlined, and where significant and important deviations from this general scheme exist for specific compounds, these will be discussed.

The enzymatic metabolic degradation of PG and TX occurs in multiple organs of the body including the liver,[31-40] lung,[33,35,37-46] kidney,[35,37-39,47-53] blood vessels,[54] platelets,[55] and erythrocytes.[56] In most instances the liver and lung are probably quantitatively the most important sites of metabolism of the PG and TX.[40] An important initial pathway of metabolic transformation of many of these compounds involves the 15-hydroxyprostaglandin dehydrogenase and Δ^{13} reductase enzymes which convert PG to 15-keto-13,14-dihydro metabolites.[57,58] The 15-hydroxyprostaglandin dehydrogenase is a cytosolic enzyme with widespread distribution in the body including the lung, liver, kidney, and other organs.[37] Different types of this enzyme exist which require either NAD^+ or $NADP^+$ as a cofactor.[39] This is an extremely important and efficient pathway of metabolism of some compounds, in particular PGE_2 and $PGF_{2\alpha}$. An important site of conversion of PGE_2 and $PGF_{2\alpha}$ to their respective 15-keto-13,14-dihydro metabolites is the lung. Greater than 90% of PGE_2 and $PGF_{2\alpha}$ entering the pulmonary circulation are converted in one pass through the lung to 15-keto-13,14-dihydro-PGE_2 and 15-keto-13,14-dihydro PGF_2, respectively.[40] 15-Keto-13,14-dihydro metabolites of PGE_2 and $PGF_{2\alpha}$ are essentially devoid of biological activity[58] and the extraordinary efficiency of this pathway of metabolism in the lung is a primary mechanism which prevents unmetabolized PGE_2 and $PGF_{2\alpha}$ from reaching the systemic circulation in concentrations sufficient to exert biological effects. In order for pulmonary metabolism of PG and TX to occur, these compounds must be initially taken up by the lung from the pulmonary circulation. This uptake process in the lung, at least for some of the PG, appears to occur by an active process.[59,60] For example, PGI_2 is a good substrate for the 15-hydroxyprostaglandin dehydrogenase and is readily metabolized by homogenates of lung in vitro.[44,45] However, the majority of PGI_2 entering the pulmonary circulation escapes metabolic inactivation by the lung.[44,46,61] This has been interpreted as indicative of a limited capacity of the lung to extract PGI_2. Thus, the lung is not an important site of metabolic inactivation of PGI_2. Both the liver and kidney, however, have been shown to readily metabolize PGI_2 by pathways including the 15-hydroxyprostaglandin dehydrogenase;[34,51] perhaps net cellular uptake of PGI_2 is greater in these organs. Metabolic transformation involving the 15-hydroxyprostaglandin dehydrogenase is of relatively minor importance in the metabolism of PGD_2 and TXB_2.[18,22-26] PGD_2 has been shown to be a poor substrate for the 15-hydroxyprostaglandin dehydrogenase[62] and incubations in vitro of TXB_2 with this enzyme obtained from guinea pig liver does not result in the formation of detectable amounts of 15-keto-13,14-dehydro-TXB_2.[62a] In addition, 15-keto-13,14-dihydro thromboxane ring metabolites have not been identified in urine following intravenous infusion of TXB_2 into monkey and man.[22-26] However, formation of 15-keto-13,14-dihydro-TXB_2 has been shown to occur in antigen-challenged sensitized guinea pig lungs and the formation of this metabolite increases with successive antigen challenges.[63,64] Although it is not entirely clear, these studies at least raise the question of whether TXA_2 itself may be metabolized directly by the 15-hydroxyprostaglandin dehydrogenase. These studies have, however, conclusively demonstrated an interesting phenomenon of immunologic modulation of thromboxane metabolism.

Because of the efficient metabolism of PGE_2 and $PGF_{2\alpha}$ by the lung to their respective 15-keto-13,14-dihydro metabolites, 15-keto-13,14-dihydro-PGE_2 and 15-keto-13,14-dihydro-$PGF_{2\alpha}$ are major circulating metabolites of PGE_2 and $PGF_{2\alpha}$, respectively.[8,11] Quantification of these metabolites in plasma, therefore, can serve as a useful index of endogenous production of PGE_2 and $PGF_{2\alpha}$.[2,65] However, these metabolites have a very short $t^{1/2}$ in the circulation of approximately 8 min.[8] Because of the rapid formation but short $t^{1}/_{2}$ of these metabolites, the quantification of these plasma metabolites is most useful in detecting release of PGE_2 and $PGF_{2\alpha}$ which occurs suddenly over a brief period of time. Recently, tetranor-15-keto-13,14-dihydro-$PGF_{2\alpha}$ has been shown to be a major circulating metabolite of $PGF_{2\alpha}$ during prolonged continuous infusion of $PGF_{2\alpha}$ in rats.[66] Although this metabolite of $PGF_{2\alpha}$ is not formed as rapidly as 15-keto-13,14-dihydro-$PGF_{2\alpha}$, it disappears from the circulation at a slower rate than does 15-keto-13,14-dihydro-$PGF_{2\alpha}$. Because of the accumulation of this metabolite in the circulation, the quantification of this compound may provide a more consistent and accurate index of endogenous production of $PGF_{2\alpha}$. Whether accumulation of this or similar metabolites of $PGF_{2\alpha}$ also occurs in man, however, remains to be investigated. Following intravenous infusion of PGI_2 in humans, the major product which appears in the circulation is 6-keto-$PGF_{1\alpha}$.[67] Also identified in approximately one fifth abundance relative to 6-keto-$PGF_{1\alpha}$ were the PGI_2 metabolites 2,3-dinor-6,15-diketo-13,14-dihydro-$PGF_{1\alpha}$ and 2,3-dinor-6,15-diketo-13,14-dihydro-20-carboxyl-$PGF_{1\alpha}$. It is unclear whether the 6-keto-$PGF_{1\alpha}$, which appears in the circulation during intravenous infusion of PGI_2, represents in vivo hydrolysis of circulating PGI_2 to 6-keto-$PGF_{1\alpha}$ or ex vivo hydrolysis of PGI_2 during plasma processing. Regardless, however, quantification of 6-keto-$PGF_{1\alpha}$ in plasma does represent a means to assess endogenous production of PGI_2 in man. However, the circulating concentrations of 6-keto-$PGF_{1\alpha}$ in normal individuals has been found to be extremely low, less than 3 pg/mℓ, requiring sophisticated, sensitive methodology for accurate quantification.[68] Thus, quantification of urinary metabolites of PGI_2 in most instances probably represents a simpler approach for assessing endogenous production of PGI_2.

Although the urinary metabolites formed following intravenous infusions of TXB_2 and PGD_2 have been studied, the spectrum of metabolites of these compounds that appear predominately in the circulation has not been investigated.

Major pathways of additional metabolic transformation of PG and TX involve processes of β- and ω-oxidation. β-Oxidation of these compounds results primarily in the loss of either 2 or 4 carbons from the upper side chain forming either 2,3-dinor or 2,3,4,5-tetranor metabolites. β-Oxidation can also occur on the lower side chain of these compounds, usually resulting in loss of only two carbons, but this is quantitatively a relatively less important pathway of metabolic transformation. Differences exist between PG and TX regarding the extent to which these compounds are β- and ω-oxidized in vivo. Whereas the major urinary metabolites of PGE_2 and $PGF_{2\alpha}$ are 2,3,4,5-tetranor compounds, the major urinary metabolites of TXB_2, PGD_2, and PGI_2 are 2,3-dinor compounds.[8,12,13,18,22-26] In addition, ω-oxidation is a major pathway of metabolism of $PGF_{2\alpha}$ and PGE_2 but a relatively less important pathway of metabolism of TXB_2, PGD_2, and PGI_2. Other relatively more minor pathways or prostaglandin metabolism include C-19 hydroxylation and ω-hydroxylation at C-20. ω-Hydroxylation occurs in the microsomal fractions from liver, lung, and kidney cortex and requires NADPH.[33,69-73] The enzymes responsible for C-20 and C-19 hydroxylation appear to be separate distinct enzymes which have different properties.[33]

Although specific deviations for particular PG and TX have been discussed, in summary, major pathways of metabolic transformation of PG and TX involve dehydrogenation of the C-15 alcohol group by the 15-hydroxyprostaglandin dehydrogenase, reduction of the Δ^{13} double bond, β-oxidation, and ω-oxidation. The general scheme of these metabolic transformations of $PGF_{2\alpha}$ is illustrated in Figure 1. The major pathways of metabolism of PGE_2 are the same as those depicted for $PGF_{2\alpha}$ in Figure 1. As previously discussed, for PGE_2 and $PGF_{2\alpha}$, the initial products formed from 15-hydroxyprostaglandin dehydrogenase and Δ^{13} reductase enzymes, 15-keto-13,14-dihydro-PGE_2 and 15-keto-13,14-dihydro-$PGF_{2\alpha}$, respectively, are major circulating metabolites of PGE_2 and $PGF_{2\alpha}$ in plasma. These metabolites are then subsequently transformed by processes of β- and ω-oxidation and the end products are primarily excreted by the kidney. The final product of these pathways of metabolism

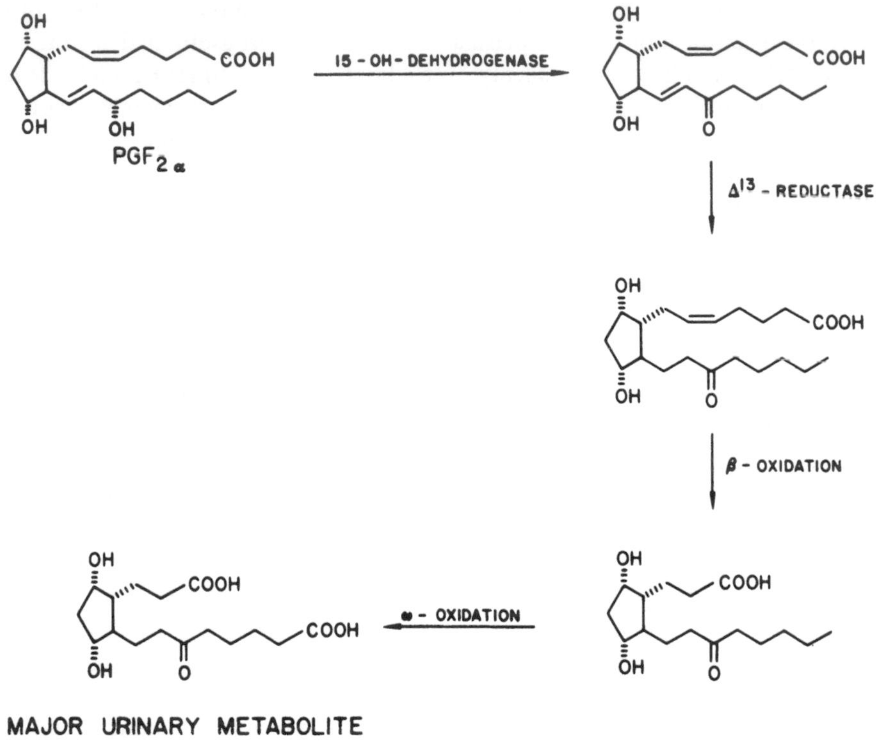

MAJOR URINARY METABOLITE

FIGURE 1.   Outline of the major pathways of metabolism of PGF$_{2\alpha}$. The primary metabolic transformations of PGE$_2$ are similar.

for PGF$_{2\alpha}$ depicted in Figure 1 does, in fact, appear in the urine as the most abundant urinary metabolite.[12] The analogous PGE-ring metabolite is also the major urinary metabolite of PGE$_2$.[8] PGI$_2$ and TXB$_2$, however, are not as extensively β- and ω-oxidized as PGE$_2$ and PGF$_{2\alpha}$ and the major metabolites of these compounds in urine are 2,3-dinor-6-keto-PGF$_{1\alpha}$ and 2,3-dinor-TXB$_2$, respectively.[19-25]

There are important unique pathways of metabolism of TXB$_2$ and PGD$_2$ which deserve to be mentioned specifically. Although the major urinary metabolite of TXB$_2$ is 2,3-dinor-TXB$_2$, a very prominent pathway of metabolism of TXB$_2$ involves dehydrogenation of the hemiacetal hydroxyl group at C-11 of TXB$_2$.[23,26] The initial product of this pathway of metabolism, 11-dehydro-TXB$_2$, is the second most abundant urinary metabolite of TXB$_2$. These metabolic pathways of transformation of TXB$_2$ are outlined in Figure 2. 11-Dehydro-TXB$_2$ is then extensively metabolized by processes of β- and ω-oxidation. The nature of the 11-hydroxythromboxane dehydrogenase enzyme is not known except that this enzymatic activity has been found to be present in the 100,000 × g supernatant of guinea pig liver and utilizes NAD$^+$ as a cofactor.[62a] The identification of 11-dehydro-TXB$_2$ metabolites necessitated the establishment of a new nomenclature for the thromboxanes.[23,74]

The recent investigation of the metabolic fate of PGD$_2$ in the nonhuman primate revealed a unique major pathway of metabolism of this prostaglandin involving a 11-keto-reductase enzyme.[18] This enzymatic activity converts the PGD-ring to a PGF-ring by reducing the C-11 keto group of PGD$_2$ to a hydroxyl group. 11-Keto-reductase activity was first described to be present in the cellular fraction of sheep blood.[75] The major urinary metabolite of PGD$_2$ in the monkey was found to be 2,3-dinor-PGF$_{2\alpha}$ and in total, PGF-ring metabolites of PGD$_2$ were present in an approximately twofold greater abundance in the urine than were metabolites which retained the original PGD-ring. Following intravenous infusion of radiolabeled

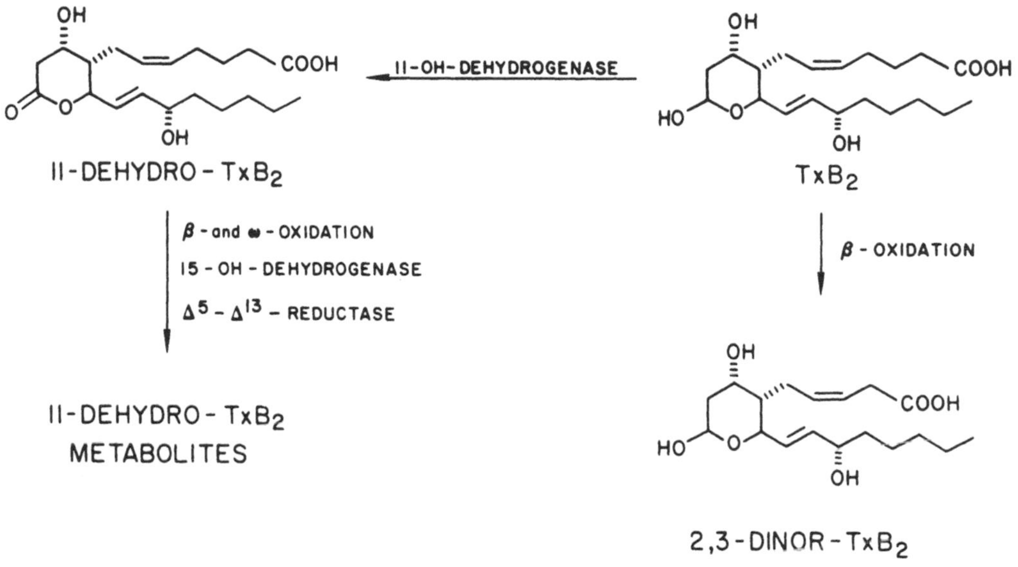

FIGURE 2. Outline of the major pathways of metabolism of TXB$_2$.

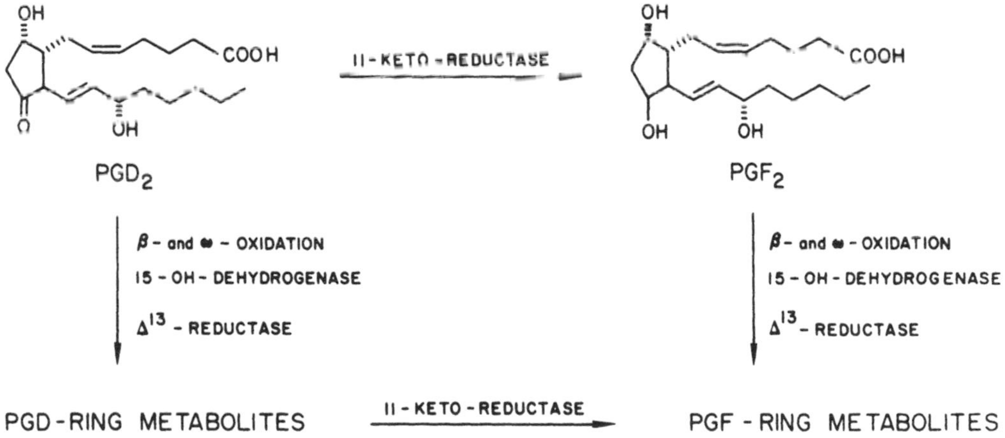

FIGURE 3. Outline of the major pathways of metabolism of PGD$_2$.

PGD$_2$, radiolabeled PGF$_{2\alpha}$ itself also appeared in urine. The metabolic fate of PGD$_2$ is summarized in Figure 3. The discovery that PGD$_2$ is metabolically transformed in substantial part to PGF-ring metabolites and PGF$_{2\alpha}$ itself in vivo raises two important questions that remain to be answered. The first is to what extent does PGF$_{2\alpha}$ play a role in the known biological effects of PGD$_2$. The second question of importance is whether previous studies in which elevated levels of PGF-ring metabolites have been found, such as following antigen challenge in human asthmatics,[76] anaphylaxis,[77] pregnancy,[78] and during flushing associated with medullary carcinoma of the thyroid,[79] actually result from increased production of PGD$_2$ rather than PGF$_{2\alpha}$. Until such studies can be repeated with simultaneous measurements of both a PGF- and a PGD-ring metabolite, these questions remain unanswered. The study described earlier of PGD$_2$ metabolism was conducted in a nonhuman primate. However, there is very suggestive evidence that conversion of PGD$_2$ to PGF-ring metabolites also occurs to a major extent in man. In a patient with marked overproduction of PGD$_2$ associated with systemic mastocytosis,[80] in addition to a series of PGD-ring urinary metabolites, markedly increased urinary excretion of six PGF-ring metabolites was also found.[62a] In addition, an approximately 50-fold increase in the circulating plasma level of 15-keto-13,14-dihydro-$PGF_{2\alpha}$ has been measured in a patient with mastocytosis during a severe episode of flushing.[81] These observations, however, are not conclusive as studies involving infusion of radiolabeled PGD_2 in man have not been performed. Thus, it cannot be definitely concluded that the increased levels of PGF-ring metabolites in these patients originate from PGD_2 rather than $PGF_{2\alpha}$ overproduction as well.

Similar to the 11-keto-reductase, 9-keto-reductase activity has been identified in several tissues which converts PGE_2 to $PGF_{2\alpha}$.[38,47] The reverse reaction, 9-hydroxydehydrogenase activity, has also been described.[34,48,55,82] Although the interconversion of PGE_2 and $PGF_{2\alpha}$ may be of biological importance locally at various sites in the body, it does not play an important role in the overall metabolism of $PGF_{2\alpha}$ and PGE_2 in man since only a relatively small quantity of a radiolabeled PGF-ring metabolite has been found in human urine following the administration of radiolabeled PGE_2[83] and no PGE-ring metabolites were identified following the administration of radiolabeled $PGF_{2\alpha}$ to humans.[10,12,13]

Recent reports have also described the ability of various cells and organs to convert PGI_2 to 6-keto-PGE_1.[34,55] This finding seemed to be of potential significance in that 6-keto-PGE_1 was shown to inhibit platelet aggregation, although some debate existed regarding its potency in this regard.[84,85] Because 6-keto-PGE_1 is a relatively stable compound in contrast to PGI_2, it was questioned whether conversion of PGI_2 to 6-keto-PGE_1 was an important mechanism by which the antiplatelet effects resulting from PGI_2 production might persist for a longer period of time than would the direct antiplatelet effect of PGI_2 itself. However, a recent study employing a mass spectrometric assay for 6-keto-PGE_1 failed to detect any increase in plasma levels of 6-keto-PGE_1 during intravenous infusion of PGI_2 in normal human volunteers.[86] These studies employed intravenous infusions of exogenous PGI_2. Cognizant of this fact, however, unless there are significant differences in the metabolism of exogenous vs. endogenously released PGI_2, it appears that in man, PGI_2 is not converted to an appreciable extent to 6-keto-PGE_1 and 6-keto-PGE_1 is unlikely to be a mediator of the biological effects of PGI_2. Such considerations are extremely important, since 6-keto-PGE_1 causes constriction of human coronary arteries, a possibly dangerous consequence in metabolism of PGI_2 infused in clinical studies (see first chapter by Willis).

It is also of importance to briefly discuss spontaneous nonenzymatic chemical degradation of PG and TX as an additional mechanism of biological inactivation of these compounds. The potential importance of nonenzymatic chemical degradation is probably limited to the labile compounds TXA_2 and PGI_2. TXA_2 is an extremely unstable compound which spontaneously degrades in aqueous solutions to TXB_2 with a $t^1/_2$ of approximately 32 sec.[87] PGI_2 is more stable than TXA_2, requiring a period of minutes to hydrolyze to 6-keto-$PGF_{1\alpha}$, and plasma albumin has also been shown to stabilize PGI_2.[88] The importance of spontaneous chemical degradation of TXA_2 and PGI_2 as a determinant of the duration of the biological activity in vivo of these compounds is not entirely clear. Since intravenous infusion of PGI_2 results in increased urinary excretion of 6-keto-$PGF_{1\alpha}$,[89] this indicates either that PGI_2 is in part directly excreted intact by the kidney or that 6-keto-$PGF_{1\alpha}$ is excreted following degradation in the circulation of PGI_2 to 6-keto-$PGF_{1\alpha}$. Although intravenous infusions of PGI_2 and 6-keto-$PGF_{1\alpha}$ yield the same urinary metabolites, the relative abundance of the urinary metabolites formed differ for the two compounds infused.[20] This suggests that PGI_2, at least in part, may be enzymatically degraded itself prior to hydrolysis to 6-keto-$PGF_{1\alpha}$. Other evidence in support of direct metabolic transformation of PGI_2 is the finding that PGI_2 is a better substrate than 6-keto-$PGF_{1\alpha}$ for the 15-hydroxyprostaglandin dehydrogenase. An outline of the metabolic fate of PGI_2 is depicted in Figure 4. As mentioned previously, there are some data which possibly suggest that TXA_2 can also be metabolized directly prior to

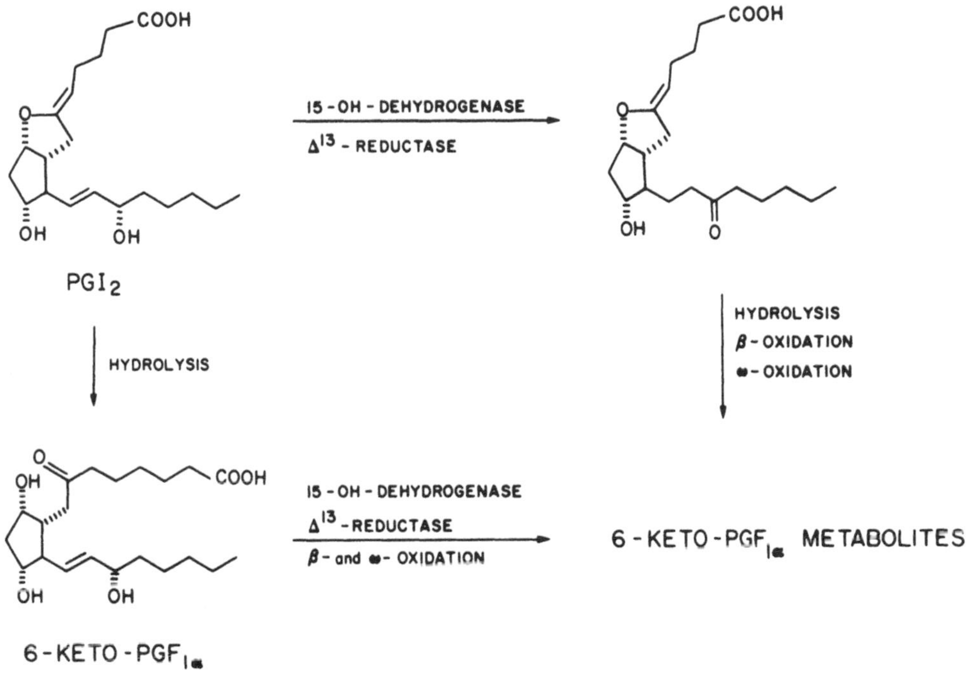

FIGURE 4. Outline of the major pathways of metabolism of PGI₂.

hydrolysis to TXB$_2$, although there is considerably more doubt about this than the possibility of direct metabolic transformation of PGI$_2$.

In addition to the possibility that endogenously formed TXA$_2$ itself may be enzymatically metabolized directly prior to hydrolysis to TXB$_2$, very complex interactions of TXA$_2$ with plasma proteins have also been described which may potentially affect the duration of biological activity of TXA$_2$ in vivo. Some data suggest that TXA$_2$ may be covalently bound to plasma albumin, which would result in biological inactivation of TXA$_2$.[90] However, there is also some evidence suggesting that noncovalent binding of TXA$_2$ to serum albumin and possibly other plasma proteins may prolong the biological activity of TXA$_2$.[91-93] When all of these factors are considered (spontaneous chemical degradation, enzymatic metabolism, and plasma protein binding effects), the potential determinants of the biological activity of endogenously formed TXA$_2$ can potentially interact in a very complex way, which are illustrated in Figure 5.

In summary, knowledge of the fate and metabolism of PG, TX, and lipoxygenase products is extremely important in understanding the biological effects of these compounds in vivo, and the quantification of metabolites of these compounds is the most accurate and reliable means by which to quantitatively assess the endogenous production of these compounds. Considerable important information regarding the metabolism of some of these compounds, in particular the lipoxygenase products, has not yet been obtained. However, continued investigations in this area will undoubtedly expand our understanding and ability to interpret the biological effects of these compounds in vivo and provide the background biochemical information necessary to better assess and define the physiologic and pathophysiologic involvement of the compounds in various physiologic and pathologic processes in man.

ADDENDUM

Since the original writing of this chapter, important new findings have been made related

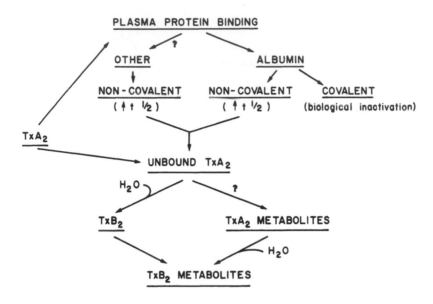

FIGURE 5.   Proposed biological fate and determinants of the duration of biological activity of endogenously formed TXA$_2$.

to the metabolism of PGD$_2$. As a consequence of these recent findings, some of the statements made above regarding the metabolism of PGD$_2$ require clarification and correction as below.

It was stated above that a major pathway of metabolism of PGD$_2$ was via a 11-ketoreductase pathway to PGF$_2$ and PGF$_2$ metabolites. Although it was originally thought that both the C-9 and C-11 hydroxyl groups of the PGF$_2$ compounds were both oriented alpha in relation to the cyclopentane ring, identical to that of PGF$_{2\alpha}$, this has subsequently been found not to be the case. Studies of the metabolic fate of PGD$_2$ in humans revealed that almost all of the PGF-ring metabolites of PGD$_2$ tested did not form a derivative with butylboronic acid, indicating that the C-9 and C-11 hydroxyls are *trans*.[94,95] Subsequent in vitro studies established that human 11-ketoreductase stereospecifically transforms PGD$_2$ to $9_\alpha,11_\beta$-PGF$_2$ rather than to PGF$_{2\alpha}$,($9_\alpha,11_\alpha$-PGF$_2$).[96,97] Of considerable interest and importance is that $9_\alpha,11_\beta$-PGF$_2$ has been shown to be a biologically active metabolite of PGD$_2$ and is present in human plasma and urine.[96,98-100] A more recent discovery of great interest is that PGD$_2$ has been found to be a very labile compound in vivo undergoing extensive isomerization presumably involving multiple chiral centers and double bond location and geometry. Subsequent reduction of these PGD$_2$ isomers by 11-ketoreductase then yields more stable PGF$_2$ compounds. At least 16 different isomeric forms of PGF$_2$ have been shown to be present in both human plasma and urine.[101] Future studies aimed at elucidating the precise structures of these PGF$_2$ compounds and their spectrum of biological actions will contribute importantly to our understanding of the biological consequences of PGD$_2$ release in vivo. These metabolic findings also bring into question the reliability of methods for measuring PGF$_{2\alpha}$ and its metabolites in biological fluids since both immunoassays and physical methods of analysis such as GC/MS may simultaneously measure PGF$_{2\alpha}$ or its metabolites and at least in part some of the PGF$_2$ metabolites arising from PGD$_2$.[102]

REFERENCES

1. **Samuelsson, B.**, Quantitative aspects of prostaglandin synthesis in man, *Adv. Biosci.*, 9, 7—14, 1973.

2. **Samuelsson, B. and Green, K.**, Endogenous levels of 15-keto-dihydro-prostaglandins in human plasma, *Biochem. Med.*, 11, 298—303, 1974.

3. **Samuelsson, B., Granstrom, E., Green, K., Hamberg, M., and Hammarstrom, S.**, Prostaglandins, *Ann. Rev. Biochem.*, 44, 669—695, 1975.

4. **Frolich, J. C., Wilson, T. W., Sweetman, B. J., Smigel, M., Nies, A. S., Carr, K., Watson, J. T., and Oates, J. A.**, Urinary prostaglandins: identification and origin, *J. Clin. Invest.*, 55, 763—770, 1975.

5. **Williams, W. M., Frolich, J. C., Nies, A. S., and Oates, J. A.**, Urinary prostaglandins: site of entry into tubular fluid, *Kidney Int.*, 11, 256—260, 1977.

6. **Roberts, L. J., II, Sweetman, B. J., Maas, R. L., Hubbard, W. C., and Oates, J. A.**, Clinical application of PG and TX metabolite quantification, *Prog. Lipid Res.*, 20, 117—121, 1981.

7. **Green, K.**, Metabolism of prostaglandin E_2 in the rat, *Biochemistry*, 10, 1072—1086, 1971.

8. **Hamberg, M. and Samuelsson, B.**, On the metabolism of prostaglandins E_1 and E_2 in man, *J. Biol. Chem.*, 246, 6713—6721, 1972.

9. **Hamberg, M. and Samuelsson, B.**, On the metabolism of prostaglandins E_1 and E_2 in the guinea pig, *J. Biol. Chem.*, 247, 3495—3502, 1972.

10. **Granstrom, E.**, On the metabolism of prostaglandin $F_{2\alpha}$ in female subjects. Structure of two C_{14} metabolites, *Eur. J. Biochem.*, 25, 581—589, 1972.

11. **Granstrom, E.**, On the metabolism of prostaglandin $F_{2\alpha}$ in female subjects. Structures of two metabolites in blood, *Eur. J. Biochem.*, 27, 462—469, 1972.

12. **Granstrom, E. and Samuelsson, B.**, On the metabolism of prostaglandin $F_{2\alpha}$ in female subjects, *J. Biol. Chem.*, 246, 5254—5263, 1971.

13. **Granstrom, E. and Samuelsson, B.**, On the metabolism of prostaglandin $F_{2\alpha}$ in female subjects. II. Structure of six metabolites, *J. Biol. Chem.*, 246, 7470—7485, 1971.

14. **Green, K.**, The metabolism of prostaglandin $F_{2\alpha}$ in the rat, *Biochim. Biophys. Acta*, 231, 419—444, 1971.

15. **Sun, F. F.**, Metabolism of prostaglandin $F_{2\alpha}$ in the rat, *Biochim. Biophys. Acta*, 348, 249—262, 1974.

16. **Sun, F. F.**, Metabolism of prostaglandin $F_{2\alpha}$ in rhesus monkeys, *Biochim. Biophys. Acta*, 369, 95—110, 1974.

17. **Svanborg, K. and Bygdeman, M.**, Metabolism of prostaglandin $F_{2\alpha}$ in the rabbit, *Eur. J. Biochem.*, 28, 127—135, 1972.

18. **Ellis, C. K., Smigel, M. D., Oates, J. A., Oelz, O., and Sweetman, B. J.**, Metabolism of prostaglandin D_2 in the monkey, *J. Biol. Chem.*, 254, 4152—4163, 1979.

19. **Rosenkranz, B., Fischer, C., Reimann, I., Weimer, K. E., Beck, G., and Frolich, J. C.**, Identification of the major metabolite of prostacyclin and 6-keto-prostaglandin in $F_{1\alpha}$ in man, *Biochim. Biophys. Acta*, 619, 207—213, 1980.

20. **Rosenkranz, B., Fischer, C., Weimer, K. E., and Frolich, J. C.**, Metabolism of prostacyclin and 6-keto-prostaglandin $F_{1\alpha}$ in man, *J. Biol. Chem.*, 255, 10194—10198, 1980.

21. **Sun, F. F. and Taylor, B. M.**, Metabolism of prostacyclin in rat, *Biochemistry*, 17, 4096—4101, 1978.

22. **Roberts, L. J., II, Sweetman, B. J., Morgan, J. L., Payne, N. A., and Oates, J. A.**, Identification of the major urinary metabolite of thromboxane B_2 in the monkey, *Prostaglandins*, 13, 631—647, 1977.

23. **Roberts, L. J., II, Sweetman, B. J., and Oates, J. A.**, Metabolism of thromboxane B_2 in the monkey, *J. Biol. Chem.*, 253, 5305—5318, 1978.

24. **Kindahl, H.**, Metabolism of thromboxane B_2 in the cynomolgus monkey, *Prostaglandins*, 13, 619—629, 1977.

25. **Roberts, L. J., II, Sweetman, B. J., Payne, N. A., and Oates, J. A.**, Metabolism of thromboxane B_2 in man. Identification of the major urinary metabolite, *J. Biol. Chem.*, 252, 7415—7417, 1977.

26. **Roberts, L. J., II, Sweetman, B. J., and Oates, J. A.**, Metabolism of thromboxane B_2 in man. Identification of twenty urinary metabolites, *J. Biol. Chem.*, 256, 8384—8393, 1981.

27. **Hammarstrom, S.**, Metabolism of leukotriene C_3 in the guinea pig, *J. Biol. Chem.*, 256, 9573—9578, 1981.

28. **Bernstrom, K. and Hammarstrom, S.**, Metabolism of leukotriene D by porcine kidney, *J. Biol. Chem.*, 256, 9579—9582, 1981.

29. **Hammarstrom, S.**, Metabolism of leukotriene C_3, in *Advances in Prostaglandin, Thromboxane, and Leukotriene Research*, Vol. 9, Samuelsson, B. and Paoletti, R., Eds., Raven Press, New York, 1982, 83—101.

30. **Parker, C. W., Del Koch, M. M. H., and Falkenheim, S.**, Formation of the cysteinyl form of slow reacting substance (leukotriene E_4) in human plasma, *Biochem. Biophys. Res. Commun.*, 97, 1038—1046, 1980.

31. **Hamberg, M.**, Metabolism of prostaglandins in rat liver mitochondria, *Eur. J. Biochem.*, 6, 135—146, 1968.

32. **Hamberg, M. and Israelsson, U.**, Metabolism of prostaglandin E_2 in guinea pig liver, *J. Biol. Chem.*, 245, 5107—5114, 1970.

33. **Powell, W. S.**, ω-Oxidation of prostaglandins by lung and liver microsomes: changes in enzyme activity induced by pregnancy, pseudopregnancy, and progesterone treatment, *J. Biol. Chem.*, 253, 6711—6718, 1978.

34. **Wong, P. Y.-K., Malik, K. U., Desiderio, D. M., McGiff, J. C., and Sun, F. F.**, Hepatic metabolism of prostacyclin (PGI_2) in the rabbit: formation of a potent novel inhibitor of platelet aggregation, *Biochem. Biophys. Res. Commun.*, 93, 486—494, 1980.

35. **Powell, W. S.**, Distribution of prostaglandin ω-hydroxylases in different tissues, *Prostaglandins*, 19, 701—710, 1980.

36. **Taylor, B. M. and Sun, F. F.**, Tissue distribution and biliary excretion of prostacyclin metabolites in the rat, *J. Pharmacol. Exp. Ther.*, 214, 24—30, 1980.

37. **Anggard, E., Larsson, C., and Samuelsson, B.**, The distribution of 15-hydroxy prostaglandin dehydrogenase and prostaglandin-Δ^{13}-reductase in tissues of the swine, *Acta Physiol. Scand.*, 81, 396—404, 1971.

38. **Lee, S.-C. and Levine, L.**, Prostaglandin metabolism. I. Cytoplasmic reduced nicotinamide adenine dinucleotide phosphate-dependent and microsomal reduced nicotinamide adenine dinucleotide-dependent prostaglandin E 9-ketoreductase activities in monkey and pigeon tissues, *J. Biol. Chem.*, 249, 1369—1375, 1974.

39. **Lee, S.-C. and Levine, L.**, Prostaglandin metabolism. II. Identification of two 15-hydroxyprostaglandin dehydrogenase types, *J. Biol. Chem.*, 250, 548—552, 1975.

40. **Ferreira, S. H. and Vane, J. R.**, Prostaglandins: their disappearance from and release into the circulation, *Nature (London)*, 216, 868—873, 1967.

41. **McGiff, J. C., Terragno, N. A., Strand, J. C., Lee, J. B., and Lonigro, A. J.**, Selective passage of prostaglandins across the lung, *Nature (London)*, 223, 742—745, 1969.

42. **Sun, F. F. and Armour, S. B.**, Prostaglandin 15-hydroxy dehydrogenase and Δ^{13} reductase levels in the lungs of maternal, fetal, and neonatal rabbits, *Prostaglandins*, 7, 327—338, 1974.

43. **Jose, P., Niedernauser, V., Piper, P. J., Robinson, C., and Smith, A. P.**, Degradation of prostaglandin $F_{2\alpha}$ in the human pulmonary circulation, *Thorax*, 31, 713—719, 1976.

44. **Wong, P. Y.-K., McGiff, J. C., Sun, F. F., and Malik, K. V.**, Pulmonary metabolism of prostacyclin (PGI_2) in the rabbit, *Biochem. Biophys. Res. Commun.*, 83, 731—738, 1978.

45. **McGuire, J. C. and Sun, F. F.**, Metabolism of prostacyclin. Oxidation by rhesus monkey lung 15-hydroxyl prostaglandin dehydrogenase, *Arch. Biochem. Biophys.*, 189, 92—96, 1978.

46. **Waldman, H. M., Kot, A. P. A., Rose, J. C., and Ramwell, P. W.**, Effect of lung transit on systemic depressor responses to arachidonic acid and prostacyclin in dogs, *J. Pharmacol. Exp. Ther.*, 204, 289—293, 1978.

47. **Lee, S.-C., Pong, S.-S., Katzen, D., Wu, K.-Y., and Levine, L.**, Distribution of prostaglandin E 9-ketoreductase and types I and II 15-hydroxyprostaglandin dehydrogenase in swine kidney medulla and cortex, *Biochemistry*, 14, 142—145, 1975.

48. **Pace-Asciak, C.**, Prostaglandin 9-hydroxydehydrogenase activity in the adult rat kidney, *J. Biol. Chem.*, 250, 2789—2794, 1975.

49. **Katzen, D. R., Pong. S.-S., and Levine, L.**, Distribution of prostaglandin E 9-ketoreductase and NAD^+-dependent and $NADP^+$-dependent 15-hydroxyprostaglandin dehydrogenase in the renal cortex and medulla of various species, *Res. Commun. Chem. Pathol. and Pharmacol.*, 12, 781—787, 1975.

50. **Larsson, C. and Anggard, E.**, Regional differences in the formation and metabolism of prostaglandins in the rabbit kidney, *Eur. J. Pharmacol.*, 21, 30—36, 1973.

51. **Wong, P. Y.-K., McGiff, J. C., Cagen, L., Malik, K. U., and Sun, F. F.**, Metabolism of prostacyclin the rabbit kidney, *J. Biol. Chem.*, 254, 12—14, 1979.

52. **Granstrom, E.**, Metabolism of prostaglandin $F_{2\alpha}$ in swine kidney, *Biochim. Biophys. Acta*, 239, 120—125, 1971.

53. **Oates, J. A., Sweetman, B. J., Green, K., and Samuelsson, B.**, Identification and assay of tetranor-prostaglandin E_1 in human urine, *Anal. Biochem.*, 74, 546—559, 1976.

54. **Wong, P. Y.-K., Sun, F. F., and McGiff, J. C.**, Metabolism of prostacyclin in blood vessels, *J. Biol. Chem.*, 253, 5555—5557, 1978.

55. **Wong, P. Y.-K., Lee, W. H., Chao, P. H.-W., Reiss, R. F., and McGiff, J. C.**, Metabolism of prostacyclin by 9-hydroxyprostaglandin dehydrogenase in human platelets. Formation of a potent inhibitor of platelet aggregation and enzyme purification, *J. Biol. Chem.*, 255, 9021—9024, 1980.

56. **Kaplan, L., Lee, S.-C., and Levine, L.**, Partial purification and same properties of human erythrocyte prostaglandin 9-ketoreductase and 15-hydroxyprostaglandin dehydrogenase, *Arch. Biochem. Biophys.*, 167, 287—293, 1975.

57. **Anggard, E. and Samuelsson, B.**, Metabolism of prostaglandin E_1 in guinea pig lung: the structure of two metabolites, *J. Biol. Chem.*, 239, 4097—4102, 1964.

58. **Samuelsson, B., Granstrom, E., Green, K., and Hamberg, M.,** Metabolism of prostaglandins, *Ann. N.Y. Acad. Sci.,* 180, 138—163, 1971.

59. **Anderson, M. W. and Eling, T. E.,** Prostaglandin removal and metabolism by isolated perfused rat lung, *Prostaglandins,* 11, 645—677, 1976.

60. **Bito, L. Z. and Baroody, R. A.,** Inhibition of pulmonary prostaglandin metabolism by inhibitors of prostaglandin biotransport (probenecid and bromocresol green), *Prostaglandins,* 10, 633—639, 1975.

61. **Gerkens, J. F., Friesinger, G. C., Branch, R. A., Shand, D. G., and Gerber, J. G.,** A comparison of the pulmonary, renal, and hepatic extractions of PGI_2 and PGE_2-PGI_2 a potential circulating hormone, *Life Sci.,* 22, 1837—1842, 1978.

62. **Sun, F. F., Armour, S. B., Bockstanz, V. R., and McGuire, J. C.,** Studies on 15-hydroxyprostaglandin dehydrogenase from monkey lung, in *Advances in Prostaglandin and Thromboxane Research,* Vol. 1, Samuelsson, B. and Paoletti, R., Eds., Raven Press, New York, 1976, 163—169.

62a. **Roberts, L. J., II,** Unpublished data.

63. **Dawson, W., Boot, J. R., Cockerill, A. F., Mallen, D. N. B., and Osborne, D. J.,** Release of novel prostaglandins and thromboxanes after immunological challenge of guinea pig lung, *Nature (London),* 699—702, 1976.

64. **Boot, J. R., Cockerill, A. F., Dawson, W., Mallen, D. N. B., and Osborne, D. J.,** Modification of prostaglandin and thromboxane release by immunological sensitization and successive immunological challenges from guinea pig lung, *Int. Arch. Allergy Appl. Immunol.,* 57, 159—164, 1978.

65. **Hubbard, W. C. and Watson, J. T.,** Determination of 15-keto-13,14-dihydro metabolites of PGE_2 and $PGF_{2\alpha}$ in plasma using high performance liquid chromatography and gas chromatography-mass spectrometry, *Prostaglandins,* 12, 21—35, 1976.

66. **Edwards, N. S. and Pace-Asciak, C. R.,** Identification of a "urinary-type" metabolite of prostaglandin $F_{2\alpha}$ in the rat circulation, *Biochim. Biophys. Acta,* 711, 369—371, 1982.

67. **Rosenkranz, B., Fischer, C., and Frolich, J. C.,** Prostacyclin metabolites in human plasma, *Clin. Pharmacol. Ther.,* 29, 420—424, 1981.

68. **Blair, I. A., Barrow, S. E., Waddell, K. A., Lewis, P. J., and Dollery, C. T.,** Prostacyclin is not a circulating hormone in man, *Prostaglandins,* 23, 579—589, 1982.

69. **Israelsson, U. M., Hamberg, M., and Samuelsson, B.,** Biosynthesis of 19-hydroxyprostaglandin A_1, *Eur. J. Biochem.,* 11, 390—394, 1969.

70. **Kupfer, D., Navarro, J., and Piccolo, D. E.,** Hydroxylation of prostaglandins A_1 and E_1 by liver microsomal monooxygenase: characteristics of the enzyme system in the guinea pig, *J. Biol. Chem.,* 253, 2804—2811, 1978.

71. **Powell, W. S.,** Metabolism of prostaglandins, prostaglandin analogs, and thromboxane B_2 by lung and liver microsomes from pregnant rabbits, *Biochim. Biophys. Acta,* 575, 335—349, 1979.

72. **Powell, W. S. and Solomon, S.,** Formation of 20-hydroxyprostaglandins by lungs of pregnant rabbits, *J. Biol. Chem.,* 253, 4609—4616, 1978.

73. **Navarro, J., Piccolo, D. E., and Kupfer, D.,** Hydroxylation of prostaglandin E_1 by kidney cortex microsomal monooxygenase in the guinea pig, *Arch. Biochem. Biophys.,* 191, 125—133, 1978.

74. **Samuelsson, B., Hamberg, M., Roberts, L. J., II, Oates, J. A., and Nelson, N. A.,** Nomenclature for thromboxanes, *Prostaglandins,* 16, 857—860, 1978.

75. **Hensby, C. N.,** The enzymatic conversion of prostaglandin D_2 to prostaglandin $F_{2\alpha}$, *Prostaglandins,* 8, 369—375, 1974.

76. **Green, K., Hedqvist, P., and Ivanborg, N.,** Increased plasma levels of 15-keto-13,14-dihydro-prostaglandin $F_{2\alpha}$ after allergen-provoked asthma in man, *Lancet,* 2, 1419—1421, 1974.

77. **Strandberg, K. and Hamberg, M.,** Increased excretion of $5_\alpha,7_\alpha$-dihydroxy-11-keto-tetranor-prostanoic acid in anaphylaxis in the guinea pig, *Prostaglandins,* 6, 159—170, 1974.

78. **Hamberg, M.,** Quantitative studies on prostaglandin synthesis in man. III. Excretion of the major urinary metabolite of prostaglandins $F_{1\alpha}$ and $F_{2\alpha}$ during pregnancy, *Life Sci.,* 14, 247—252, 1974.

79. **Roberts, L. J., II, Hubbard, W. C., Bloomgarden, Z. T., Bertagna, X. Y., McKenna, T. J., Rabinowitz, D., and Oates, J. A.,** Prostaglandins: role in the humoral manifestations of medullary carcinoma of the thyroid and inhibition by somatostatin, *Trans. Assoc. Am. Phys.,* 92, 286—291, 1979.

80. **Roberts, L. J., II, Sweetman, B. J., Lewis, R. A., Austen, K. F., and Oates, J. A.,** Increased production of prostaglandin D_2 in patients with systemic mastocytosis, *N. Engl. J. Med.,* 303, 1400—1404, 1980.

81. **Roberts, L. J., II, Turk, J. W., and Oates, J. A.,** Shock syndrome associated with mastocytosis: pharmacologic reversal of the acute episode and therapeutic prevention of recurrent attacks, *Adv. Shock Res.,* 8, 145—152, 1982.

82. **Pace-Asciak, C.,** Activity profiles of prostaglandin 15 and 9-hydroxydehydrogenase and 13-reductase in the developing rat kidney, *J. Biol. Chem.,* 250, 2795—2800, 1975.

83. **Hamberg, M. and Wilson, M.,** Structures of new metabolites of prostaglandin E_2 in man, *Adv. Biosci.,* 9, 39—48, 1973.

84. **Wong, P. Y.-K., McGiff, J. C., Sun, F. F., and Lee, W. H.,** 6-Keto-prostaglandin E$_1$ inhibits the aggregation of human platelets, *Eur. J. Pharmacol.,* 60, 245—248, 1979.

85. **Miller, O. V., Aiken, J. W., Shebuski, R. J., and Gorman, R. R.,** 6-Keto-prostaglandin E$_1$ is not equipotent to prostacyclin (PGI$_2$) as an antiaggregatory agent, *Prostaglandins,* 20, 391—400, 1980.

86. **Jackson, E. K., Goodman, R. P., FitzGerald, G. A., Oates, J. A., and Branch, R. A.,** Assessment of the extent to which exogenous prostaglandin I$_2$ is converted to 6-keto-prostaglandin E$_1$ in human subjects, *J. Pharmacol. Exp. Ther.,* 221, 183—187, 1982.

87. **Hamberg, M., Svensson, J., and Samuelsson, B.,** Thromboxanes: a new group of biologically active compounds derived from prostaglandin endoperoxides, *Proc. Natl. Acad. Sci. U.S.A.,* 72, 2994—2998, 1975.

88. **Wynalda, M. A. and Fitzpatrick, F. A.,** Albumins stabilize prostaglandin I$_2$, *Prostaglandins,* 20, 853—861, 1980.

89. **Sun, F. F., Taylor, D. M., McGuire, G. C., Wong, P. Y.-K., Malik, K. V., and McGiff, J. C.,** Metabolic disposition of prostacyclin, in *Prostacyclin,* Vane, J. R. and Bergstrom, S., Eds., Raven Press, New York, 1979, 119.

90. **Maclouf, J., Kindahl, H., Granstrom, E., and Samuelsson, B.,** Thromboxane A$_2$ and prostaglandin H$_2$ form covalently linked derivatives with human serum albumin, in *Advances in Prostaglandin and Thromboxane Research,* Vol. 6, Samuelsson, B., Ramwell, P. W., and Paoletti, R., Eds., Raven Press, New York, 1980, 283—286.

91. **Folco, G., Granstrom, E., and Kindahl, H.,** Albumin stabilizes thromboxane A$_2$, *FEBS Lett.,* 82, 321—324, 1977.

92. **Lagarde, M., Velardo, B., Blanc, M., and Dechavanne, M.,** Fatty acids bound to serum albumin decrease the half-life of thromboxane A$_2$, *Prostaglandins,* 20, 275—283, 1980.

93. **Smith, J. B., Ingerman, C., and Silver, M. J.,** Persistence of thromboxane A$_2$-like material and platelet release-inducing activity in plasma, *J. Clin. Invest.,* 58, 1119—1122, 1976.

94. **Roberts, L. J., II and Sweetman, B. J.,** Metabolic fate of endogenously synthesized prostaglandin D$_2$ in a human female with mastocytosis, *Prostaglandins,* 30, 383—400, 1985.

95. **Liston, T. E. and Roberts, L. J., II,** Metabolic fate of radiolabelled prostaglandin D$_2$ in a normal human male volunteer, *J. Biol. Chem.,* 260, 13172—13180, 1985.

96. **Liston, T. E. and Roberts, L. J., II,** Transformation of prostaglandin D$_2$ to 9$_\alpha$,11$_\beta$,15(S)-Trihydroxy-5(Z),13(E)-dien-l-oic acid: A unique biologically active prostaglandin produced enzymatically *in vivo* in humans, *Proc. Natl. Acad. Sci. U.S.A.,* 82, 6030—6034, 1985.

97. **Watanabe, K., Iguchi, Y., Iguchi, S., Arai, Y., Hayaishi, O., and Roberts, L. J., II,** Stereospecific conversion of prostaglandin D$_2$ to (5Z,13E)-(15S)-9$_\alpha$,11$_\beta$-trihydroxyprosta-5,13-dien-i-oic acid (9$_\alpha$,11$_\beta$-prostaglandin F$_2$) and of prostaglandin H$_2$ to prostaglandin F$_{2\alpha}$ by bovine lung prostaglandin F synthase, *Proc. Natl. Acad. Sci. U.S.A.,* 83, 1583—1587, 1986.

98. **Pugliese, G., Spokas, E. G., Marcinkrewicz, E., and Wang, P.Y.-K.,** Hepatic transformation of prostaglandin D$_\alpha$ to a new prostanoid, 9$_\alpha$,11$_\beta$-PGF$_2$, that inhibits platelet aggregation and constricts blood vessels, *J. Biol. Chem.,* 260, 14621—14625, 1985.

99. **Roberts, L. J., II, Seibert, K., Liston, T. E., Tantengco, M. V., and Robertson, R. M.,** PGD$_2$ is transformed by human coronary arteries to 9$_\alpha$,11$_\beta$-PGF$_2$ which contracts human coronary artery rings, *Adv. Prostaglandin, Thromboxane, and Leukotriene Research,* in press.

100. **Seibert, K., Sheller, J. R., and Roberts, L. J., II,** 9$_\alpha$,11$_\beta$-Prostaglandin F$_2$: formation and metabolism by human lung and contractile effects on human bronchial smooth muscle, *Proc. Natl. Acad. Sci. U.S.A.,* in press.

101. **Wendelborn, D. F., Seibert, K., and Roberts, L. J., II,** Isomeric PGF$_2$ compounds: a new family of eicosanoids produced *in vivo,* *Adv. Prostaglandin, Thromboxane, and Leukotriene Research,* in press.

102. **Roberts, L. J., II,** Recent metabolic findings bring into question the reliability of existing methods for quantification of PGF$_{2\alpha}$ and its metabolites in human biological fluids, *Adv. Prostaglandin, Thromboxane, and Leukotriene Research,* in press.

EICOSANOID METABOLIZING ENZYMES AND METABOLITES

Donald L. Smith, K. John Stone, and Anthony L. Willis

The ubiquity of 15-hydroxyprostaglandin dehydrogenase (Table 1), which metabolizes the primary prostaglandins to 15-keto-derivatives, suggests that prostaglandins have a purely local function and act at or near their site of synthesis. Early findings which suggested that prostaglandin metabolites are devoid of significant biological activity have supported this view.[80]

More recently, the above concept has been modified with the recognition that while PGE and PGF are almost completely inactivated by a single passage through the pulmonary circulation, PGA and PGC emerge relatively unscathed.[77,89,90] These hypotensive prostaglandins may, therefore, act as circulatory hormones and have effects on blood vessels and organs remote from their site of synthesis.[91] The availability of pure prostaglandin metabolites has facilitated detailed studies of their biological properties, and it is now realized that certain metabolites may be more potent or have a different spectrum of biological activity from their primary prostaglandin substrates, e.g., dihydro-PGE_2 has about half the activity of PGE_2 on uterine smooth muscle,[88] but is virtually inactive against other tissues. Pulmonary metabolism of PGE_2 to dihydro-PGE_2 might, therefore, be construed as a mechanism for generating a selective agent which stimulates the uterus. On the other hand, keto-$PGF_{2\alpha}$ is twice as potent as $PGF_{2\alpha}$ in producing contractions of bronchial musculature[84] so that pulmonary metabolism of $PGF_{2\alpha}$ could be regarded as a biological amplification mechanism. The ability of some metabolites to sensitize smooth muscle to stimulation by other agonists[87] also suggests that prostaglandin metabolism may be involved in producing hyperalgesia in inflamed tissue.[92] Such hyperalgesia, produced by small amounts of PGE, becomes maximal only after prolonged infusion[93] or after injected PGE_2 has been eliminated from the tissue.[94]

Metabolites of PGD_2 recently identified include both F-ring metabolites (some of them identical to metabolites of $PGF_{2\alpha}$) and D-ring metabolites. The F-ring metabolites presumably result from either prior conversion, via a 11-keto reductase,[85,126,133] of PGD_2 to PGF_2 or of PGD-ring metabolites to PGF-ring metabolites. Metabolism of PGI_2 and TXB_2 appears to involve many of the same pathways as metabolism of other prostaglandins (hydrolysis, β- and ω-oxidation, 15-hydroxydehydrogenase, Δ^{13}-reductase, and in the case of thromboxane, Δ^5-reductase). In addition, 11-hydroxydehydrogenase[126] is an important enzyme apparently unique to TXB_2, forming metabolites with a keto group at C-11 (11-dehydro-TXB_2 and related metabolites). Table 1 (Prostaglandin Metabolizing Enzymes) and Table 2 (Prostaglandin Metabolites) have been updated from an earlier CRC volume[265] to include metabolites of PGD_2, PGI_2 and TXB_2, and recently discovered metabolizing enzymes. Also newly included is Table 3, which describes more recently discovered lipoxygenase products, including leukotrienes. More detailed descriptions of the biosynthetic transformations to form arachidonate lipoxygenase metabolites (including source, stimuli, enzyme mechanisms, and regulation) are reviewed in the chapter by Borgeat in this volume, in other chapters in this series, and elsewhere.[259-262]

ADDENDUM

Since the original writing, studies by Roberts have shown that most of the F-ring metabolites of PGD_2 have a β-hydroxyl group at C-11, and that 11-keto reductase converts PGD_2 to 9α,11β-PGF_2, an active metabolite that is further metabolized to other compounds (see addendum to the chapter by Roberts in this volume). Many of these products are shown in Table 2.

Table 1
PROSTAGLANDIN METABOLIZING ENZYMES

Reference[a]	Enzyme	Action	Location	Mol wt	Cofactor	Km (μM)	Inhibitors	Comment
(a)	Prostaglandin-9-keto-reductase (PKR)	Reduction of keto to α-hydroxyl						PGE_1, PGF_2 are substrates
			Pigeon, monkey tissues[1]		NADPH			Reversible, cytoplasmic
			Monkey liver[1]		NADH			Reversible, microsomal
			Monkey brain[2]		NADH			Not separated from PGDH
			Guinea pig liver[3]					Cytoplasmic, dihydroPGE₂ and dihydroketo-PGE_2 are substrates
			Sheep blood[4]					Cellular fraction, not reversible, NADH and NADPH are not required; PGD_2 is a substrate[8,5]
			Rat tissues[5]		NADH		Iodoacetamide, Ellman's reagent, p-mercuribenzoate	
			Chicken heart[6,7]	45,000—50,000	NADPH		p-Mercuribenzoate, Ellman's reagent	Purified 2,600-fold; PGA_1 is a substrate, PGD_2 is not / Purified 8-13-fold
			Swine kidney cortex, medulla[1,8]					
			Human erythrocytes[9]		NADPH	$90(E_2)$	NADP⁺	Cytoplasmic, not separated from PGDH; not inhibited by indomethacin
			Human kidney cortex[10]		NADPH		Indomethacin, ethacrynic acid[10]	
			Rabbit kidney cortex[11]	21,800	NADPH	$320(E_2)$	Indomethacin,[10] non-thiazide diuretics[10]	Cytoplasmic, isoelectric point 5.65; PGA_2, not a substrate
			Frog spinal cord[12] / Human skin[6]		NADPH			Enzyme activity elevated in psoriasis[6]
(b)	Prostaglandin-9-hydroxydehydrogenase	Oxidation of α-hydroxyl to keto	Rat kidney[13-16]		NAD⁺		Indomethacin	Only uses dihydroketo-$PGF_{2\alpha}$ as substrate
(c)	Prostaglandin E-dehydrase	Dehydration of PGE to PGA	Serum from several species[17] / Human serum[17]				N-Ethylmaleimide, iodoacetate	PGE_1 used as substrate

	Enzyme	Reaction	Source	MW	Cofactor	K_m (nM)	Inhibitors	Remarks
(d)	Prostaglandin A-isomerase	Isomerization of PGA to PGC	Serum from several species[17,18]				N-Ethylmaleimide, Ellman's reagent	PGA_1 and PGA_2 used as substrates
			Rabbit serum[19]	110,000			N-Ethylmaleimide	Purified 120-fold, not inhibited by other sulphydryl blockers
			Human serum[17]			50(A_1)		Purified 50-fold
			Plasma from several species[20]				N-Ethylmaleimide, iodoacetate	19-Hydroxy-PGA_1, 15-epi-PGA_2 are substrates
			Cat plasma[20]				Dihydro-PGA_1	Purified 48-fold
(e)	Prostaglandin C-isomerase	Isomerization of PGC to PGB	Serum from several species[17]					PGC_1 used as substrate
			Human serum[17]				N-Ethylmaleimide, iodoacetate	
			Crystallized human and rabbit serum albumin[17]					
(f)	Prostaglandin-15-hydroxydehydrogenase (PGDH)	Oxidation of α-hydroxyl to keto	Tissues from several species[2,21-25]		NAD+			Specific for 15(S)-hydroxyl PGE and PGF substrates
			Human placenta[26]	42,000	NAD+	$2.9(E_1)$, $4(E_2)$, $57(F_{1α})$, $26(F_{2α})$	Ellman's reagent, phenylmercury acetate	Purified 900-fold, no subunits
			Human placenta[27]					Purified 3,100-fold
			Human placenta[28,29]		NAD+	$27(E_1)$, $10(E_2)$, $32(A_2)$, $33(A_1)$, $59(F_{2α})$	Furosemide, ethacrynic acid, mersalyl	Purified 107-fold isoenzymic
			Chicken heart[2,6]	60,000–70,000	NAD+		NADH, thyroid hormones	Not inhibited by NADPH PGA_1 is a substrate; PGA_2, PGB_1, PGB_2 are poor substrates
			Cat myocardium[30]		NAD+		p-Mercuribenzoate, CaCl_2	
			Swine lung[31-33]		NAD+	$77(E_1)$, $12(E_2)$, $13(E_1)$, $25(F_{1α})$, $31(F_{2α})$		Dinor-PGF_{1α} and tetra-nor compounds are not substrates
			Pig lung[34]				Polyphloretin phosphate	
			Guinea pig lung[35]				Polyphloretin phosphate	
			Bovine lung[36]	40,000	NAD+	$3.4(E_1)$	Indomethacin, aspirin, non-thiazide diuretics[104]	Purified sevenfold, PGB is not a substrate; inhibition by diuretics is non-competitive[104]
			Frog spinal cord[12]		NAD+			

Table 1 (continued)
PROSTAGLANDIN METABOLIZING ENZYMES

Reference	Enzyme	Action	Location	Mol wt	Cofactor	Km (μM)	Inhibitors	Comment
	Prostaglandin-15-hydroxydehydrogenase (PGDH) (continued)		Guinea pig lung				U46619, U44069, Pinane TXA$_2$, and EPO11[239]	PGD$_2$ and PGB$_2$ are not substrates
			Monkey lung[3,7]					
			Rat kidney[10,14]		NAD$^+$, NADP$^+$		Furosemide	Purified 33-fold; activated twofold in vitro by psychotropic drugs of the chlorpromazine and imipramine type[106]
			Swine kidney cortex[8,38]		NAD$^+$		Indomethacin, ethacrynic acid[10] Thyroid hormones	Purified 30-fold
			Swine kidney medulla[8] Rabbit kidney[10,39]		NADP$^+$ NAD$^+$, NADP$^+$			
			Rabbit kidney papilla[104]		NAD	15(A$_1$); 6(A$_2$)	PGB$_1$; nonthiazide diuretics.	Specific for PGA series; not inhibited by NADH or NADPH; inhibition by diuretics and PGB$_1$ is competitive. Specific location in renal papilla is unusual.
			Rabbit colon[240]		NAD$^+$		Sulfasalazine, homosulfasalazine, dihomosulfasalazine, carbenoxolone, diphloretin phosphate	Sulfasalazine and its homo and dihomo analogs also inhibited bovine lung and placental PGDH[240]
			Monkey brain[1,2]		NADP$^+$		NADPH	Not inhibited by NADH, thyroid hormones, associated with PKR
			Human erythrocytes[9]		NADP$^+$	35(E$_2$)	NADPH	Purified 110-fold, not inhibited by thyroid hormones or indomethacin, associated with PKR
			Monkey brain[2]		NADP$^+$			Purified 73-fold

	Enzyme	Reaction	Source	Cofactor	Inhibitor/Substrate	MW	Comments
(g)	Prostaglandin-Δ^{13}-reductase	Reduction of *trans* double bond	Swine tissues[21,24], Guinea pig lung[23], Frog spinal cord[12], Chicken heart[6], Human placenta[101]	NADH			Keto-PGE, keto-PGF$_\alpha$ substrates
(h)	Prostaglandin-15-ketodehydrogenase	Reduction of keto to α-hydroxyl	Guinea pig liver[3], Swine kidney[24]	NADPH, NADH	*p*-Mercuribenzoate, 7(keto-PGE$_2$)	70,000–80,000	No PGDH activity; Cytoplasmic enzyme
(i)	"β-Oxidase"	Loss of C_2 β-oxidation of α-chain	Rat liver[40], Rat lung, kidney[41]	Carnitine			PGE, PGF, PGA-derivatives substrates; Mitochondrial enzyme; Mitochondrial enzyme
(j)	"ψ-Hydroxylation"	Introduction of β-hydroxyl[b] to C_{19}	Guinea pig liver[42]	NADPH			PGA, substrate
(k)	"ω-Hydroxylation"	Introduction of hydroxyl to C_{20}	Guinea pig liver[42]	NADPH			Microsomal enzyme, possibly P-450[105]
(l)	Dinorprostaglandin-Δ^{13}-reductase	Reduction of *cis* double bond					Microsomal enzyme, possibly P-450[105]
(m)	"ω-Hydroxyprostaglandin dehydrogenase	Oxidation of terminal hydroxyl to carboxyl					Formation of metabolites 32 and 36 (Table 2) suggests that this enzyme utilizes dinor-prostaglandins
(n)	"β-Oxidase"	β-Oxidation of ω-chain					Must occur before β-oxidation of the ω-chain; does not metabolize PG$_3$ substrates
(o)		δ-Lactonization					Does not metabolize PG$_3$ substrates
(p)		γ-Lactonization					Might occur non-enzymatically during metabolite isolation
(q)	Prostaglandin E-ketoreductase	Reduction of keto to β-hydroxyl					Might occur non-enzymatically during metabolite isolation; Guinea pig kidney extracts are incapable of reducing the 9-keto function of PGE$_2$[102]

Table 1 (continued)
PROSTAGLANDIN METABOLIZING ENZYMES

Reference	Enzyme	Action	Location	Mol wt	Cofactor	Km (µM)	Inhibitors	Comment
								Possibly dihydroketo-PGE$_2$ is used as a substrate. The oxidation of dinor-PGE$_{1\beta}$ (metabolite 33, Table 5) indicates that "β-oxidase" will accept substrates with a 9β-hydroxyl
(r)	"Prostaglandin-8-epimerase"	Isomerization of α-chain to β-chain						See metabolites 1 and 26 (Table 2)
(s)		Loss of α-Hydroxyl						Presumably occurs via dehydration, see metabolite 41 (Table 2)
(t)	11-Hydroxy thromboxane dehydrogenase	Oxidation of α-hydroxyl to keto	Guinea pig liver[126]		NAD$^+$			100,000 × g supernatant
(u)	11-ketoprostaglandin reductase	Reduction of keto to hydroxyl	Sheep blood[85]					Cellular fraction

a Reference letters refer to target sites – see Figure 1.

b The geometry of this hydroxyl group has been assumed to be the same as that determined for the 19-hydroxyl group of 19-hydroxy-PGB isolated from human seminal fluid.

Table 2
PROSTAGLANDIN METABOLITES

Structure	Chemical name	Molecular formula	Mol wt	Substrate	Source	Biological action	Presumed metabolic route from primary prostaglandin[a]	Comment
1	11,15-Dihydroxy-8-iso-9-ketoprosta-5,13-dienoic acid (8-iso-PGE₂)	$C_{20}H_{32}O_5$	352	PGE_2	Guinea pig liver[b,43]		(r)	8-iso-PGE₁ has 8.5% of E₁-vasodilator activity in dog.[81] unusual metabolic route
2	11-Hydroxy-9,15-diketoprost-13-enoic acid (keto-PGE₁)	$C_{20}H_{32}O_5$	352	PGE_1	Swine tissues[b,30,32,33,44] Rat testicle[b,22] Rabbit kidney cortex[39] Bovine lung[b,36] Human placenta[b,28]	16% of PGE₁-activity on isolated rabbit duodenum, 8% PGE₁-activity on guinea pig ileum; inactive on blood pressure of rabbit and guinea pig.[80] has 2% of PGE₁-vasodilator activity in dog.[81] inactive on ADP-induced aggregation of platelets from man or rabbit.[108]	(f)	
3	11-Hydroxy-9,15-diketo-prosta-5,13-dienoic acid (keto-PGE₂)	$C_{20}H_{30}O_5$	350	PGE_2	Swine lung[b,33] Human placenta[b,28] Cat heart[b,3] Monkey brain[b,3] Chicken heart[b,3] Rabbit, placenta[b,139] Rabbit lung[b,158] Rabbit kidney[b,164]	1% of PGE₂-activity on intestinal smooth muscle.[87] 5—20% of PGE₂-activity on respiratory smooth muscle;[87] 10% of PGE₂-activity on rat stomach strip, colon, and chick rectum, 100—180% on guinea pig trachea. inactive on rat uterus, guinea pig ileum, rabbit aorta, and pulmonary artery.[88] potentiates the response of respiratory smooth muscle to histamine and acetylcholine;[87] inactive on human and rabbit ADP-induced platelet aggregation.[108]	(f)	Keto-PGE₂ from rabbit plasma and kidney appears to be formed directly from arachidonate.[139,164] This may occur through a Ca⁺⁺-dependent, non-PG-dehydrogenase pathway[159]

Table 2 (continued)
PROSTAGLANDIN METABOLITES

	Structure	Chemical name	Molecular formula	Mol wt	Substrate	Source	Biological action	Presumed metabolic route from primary prostaglandin[a]	Comment
4		11-Hydroxy-9,15-diketoprosta-5,13,17-trienoic acid (keto-PGE3)	$C_{20}H_{28}O_5$	348	PGE3	Swine lung[b,33] Cat heart[b,30] Human placenta[b,28]		(f)	
5		11-Hydroxy-9,15-diketo-prostanoic acid (dihydroketo-PGE1)	$C_{20}H_{34}O_5$	354	PGE1, PGE1 (i.v.)	Rabbit kidney[b,39] Guinea pig lung[b,45] Swine tissues[b,33,44] Rat plasma[46] Pregnant rabbit lung[b,152]	12% of PGE1-activity on rabbit duodenum; inactive on guinea pig uterus and ileum and on B.P. of rabbit and guinea pig;[20] inactive on human and rabbit ADP-induced platelet aggregation.[108]	(f,g)	
6		11-Hydroxy-9,15-diketoprost-5-enoic acid (dihydroketo-PGE2)	$C_{20}H_{30}O_5$	352	PGE2, PGE2 (i.v.), Keto-PGE2, Dihydroketo-PGF2α	Guinea pig lung,[b] liver[b,3,43,47] Human blood[48] Chicken heart[b,6] Rat kidney[b,13,15] Rabbit lung[b,152,158] Human placenta (maternal side)[b,159,160]	Inactive on intestinal smooth muscle;[47] 1% of PGE2-activity on rat stomach strip, colon, and chick rectum, 100% in guinea pig trachea, inactive on rat uterus, guinea pig ileum, rabbit aorta, and pulmonary artery.[48] potentiates the response of respiratory smooth muscle to histamine and acetylcholine;[47] inactive on human and rabbit ADP-induced platelet aggregation.[108]	(f,g)	Main metabolite
7		11-Hydroxy-9,15-diketo-prosta-5,17-dienoic acid (dihydroketo-PGE3)	$C_{20}H_{30}O_5$	350	PGE3	Guinea pig lung[b,49]		(f,g)	
8		11,15-Dihydroxy-9-keto-prostanoic acid (dihydro-PGE1)	$C_{20}H_{36}O_5$	356	PGE1, PGE1 (i.v.)	Guinea pig lung[b,45] Swine tissues[b,44] Rat plasma[46]	14—35% of PGE1-activity on isolated rabbit duodenum, guinea pig uterus, and ileum; lowers blood pressure in rabbit and guinea pig;[30] 20% PGE1-vasodilator activity in dog.[61] Inhibits ADP-induced platelet aggregation in rabbit and human	(f,g,h)	

No.	Name	Formula	MW	PG	Source	Activity	Ref.	Notes
9	11,15-Dihydroxy-9-ketoprost-5-enoic acid (dihydro-PGE₂)	C₂₀H₃₄O₅	354	PGE₂	Guinea pig lung, liver[b,3,43,47]	platelet-rich plasma; has more than twice the activity of PGE₁ on human platelets[108] 1% of PGE₂-activity on intestinal smooth muscle;[47] 5—20% of PGE₂ activity on respiratory smooth muscle;[47] 10% of PGE₂-activity on guinea pig ileum, 20% on rat stomach strip, colon, and chick rectum, 30—50% on rat uterus, 100% on guinea pig trachea and inactive on rabbit aorta and pulmonary artery.[88] potentiates the response of respiratory smooth muscle to histamine and acetylcholine;[87] potentiates ADP-induced platelet aggregation in human, but not rabbit platelet-rich plasma[108]	(f,g,h)	
10	11,15-Dihydroxy-9-keto-prosta-5,17-dienoic acid (dihydro-PGE₃)	C₂₀H₃₂O₅	352	PGE₃	Guinea pig lung[b,49]		(f,g,h)	
11	11,15,19-Trihydroxy-9-ketoprost-13-enoic acid (19-hydroxy-PGE₁)	C₂₀H₃₂O₆	368	PGE₁[c]	Human semen[50]		(j)	Endogenous; occurs in semen in larger amounts than 19-OH-PGA and 19-OH-PGB
					Rabbit liver[b,163,165]			In rabbit liver, may be formed by P-450 system
12	11,15,19-Trihydroxy-9-ketoprosta-5,13-dienoic acid (19-hydroxy-PGE₂)	C₂₀H₃₀O₆	366	PGE₂[c]	Human semen[59]		(j)	Endogenous; occurs in semen in larger amounts than 19-OH-PGA and 19-OH-PGB
					Rabbit liver[b,163] Human seminal vesicles[b,167]			In rabbit liver, may be formed by P-450 system

Table 2 (continued)
PROSTAGLANDIN METABOLITES

Structure		Chemical name	Molecular formula	Mol wt	Substrate	Source	Biological action	Presumed metabolic route from primary prostaglandin[a]	Comment
	13	9,13-Dihydroxy-7-ketodinor prost-11-enoic acid (dinor-PGE₁)	$C_{18}H_{30}O_5$	326	PGE₁, PGE₂	Rat liver[b,51,53]		(i)	
	14	9-Hydroxy-7,13-diketodinor prostanoic acid	$C_{18}H_{30}O_5$	326	Dihydroketo-PGE₁	Rat liver[b,51]		(f,g,i)	
	15	9-Hydroxy-7,13-diketodinor prost-3-enoic acid	$C_{18}H_{28}O_5$	324	PGE₂ (i.v.)	Human urine[52]		(f,g,i)	
	16	9-Hydroxy-7-13-diketodinor prost-3-ene-1,18-dioic acid	$C_{18}H_{26}O_7$	354	PGE₂ (i.v.)	Human urine[52]		(f,g,i,k,m)	
	17	9,13-Dihydroxy-7-keto-prostanoic acid (dihydrodinor-PGE₁)	$C_{18}H_{32}O_5$	328	Dihydro-PGE₁	Rat liver[b,51]		(f,g,h,i)	
	18	7,11-Dihydroxy-5-keto-tetranorprost-9-enoic acid (tetranor-PGE)	$C_{16}H_{26}O_5$	298	PGE₂ (i.v.), PGE₂, PGE₁	Rat urine[53] Rat liver[b,53]		(i,l,i)	

No.	Structure	Name	Formula	MW	Precursor	Source	Activity	Refs.	Comments
19		7-Hydroxy-5,11-diketo-tetranorprostanoic acid	$C_{16}H_{26}O_5$	298	Dihydroketo-PGE₁, PGE₂ (i.v.)	Rat liver[b,51] Rat urine[53] Human urine[52,154]		(f,g,i,i) (f,g,i,i)	
20		7,16-Dihydroxy-5,11-diketotetranorprostanoic acid	$C_{16}H_{26}O_6$	314	Metabolite 19 (i.v.) PGE₂ (i.v.)	Rat urine[53] Rat urine[53]		(f,g,i,l,i,k)	
21		7-Hydroxy-5,11-diketo-tetranorprosta-1,16-dioic acid	$C_{16}H_{24}O_7$	328	PGE₂ (i.v.) PGE₁ (i.v.)	Rat and human urine[53,154] Human urine[48,154]		(f,g,i,i,k,m) (f,g,i,i,k,m)	Major metabolite Major metabolite
22		7,11-Dihydroxy-5-keto-tetranorprostanoic acid (dihydrotetranor-PGE)	$C_{16}H_{28}O_5$	300	Dihydro-PGE₁	Rat liver[b,51]		(f,g,h,i,i)	
23		7,11-Dihydroxy-5-keto-tetranorprosta-1,16-dioic acid	$C_{16}H_{26}O_7$	330	PGE₂ (i.v.)	Human urine[52]		(f,g,h,i,l,i,k,m)	
24		9,11,15-Trihydroxyprost-13-enoic acid (PGF₁α)	$C_{20}H_{36}O_5$	356	PGE₁	Sheep blood cells[b,4]	See Table 3	(a)	
25		9,11,15-Trihydroxyprosta-5,13-dienoic acid (PGF₂α)	$C_{20}H_{34}O_5$	354	PGE₂ and PGD₂ PGE₂	Sheep blood cells[b,4,45] Human erythrocyte[b,9] Rabbit kidney[b,11,162] Guinea pig liver[b,43]	See Table 3 Potent bronchoconstrictor[82]	(u)	First demonstration of PGE → PGF

Table 2 (continued)
PROSTAGLANDIN METABOLITES

	Structure	Chemical name	Molecular formula	Mol wt	Substrate	Source	Biological action	Presumed metabolic route from primary prostaglandin[a]	Comment
26		9,11,15-Trihydroxy-8-isoprosta-5,13-dienoic acid (8-iso-PGF$_{2\alpha}$)	C$_{20}$H$_{34}$O$_5$	354	PGE$_2$	Rat tissues[b,5]; Monkey urine[133]; Guinea pig liver[b,3,43]		(r)	Tentative, unusual metabolic route
27		9,11-Dihydroxy-15-keto-prost-13-enoic acid (keto-PGF$_{1\alpha}$)	C$_{20}$H$_{34}$O$_5$	354	PGF$_{1\alpha}$	Rat stomach[b,55]; Swine lung[b,33]; Cat heart[b,30]		(f)	
28		9,11-Dihydroxy-15-keto-prosta-5,13-dienoic acid (keto-PGF$_{2\alpha}$)	C$_{20}$H$_{32}$O$_5$	352	PGF$_{2\alpha}$ (i.v.); PGF$_{2\alpha}$, PGE$_2$	Guinea pig urine[56]; Human erythrocyte[b,9]; Swine lung[b,33]; Cat heart[b,30]; Monkey brain[b,2]; Chicken heart[b,2]; Rabbit kidney[b,162]	Less than 3% of PGF$_{2\alpha}$-bronchoconstrictor activity in vitro;[42] more active than PGF$_{2\alpha}$ on guinea pig ileum and trachea, gerbil and rat colon, rat stomach strip and uterus, human bronchial muscle[44]	(f)	
29		9,11-Dihydroxy-15-keto-prostanoic acid (dihydroketo-PGF$_{1\alpha}$)	C$_{20}$H$_{36}$O$_5$	356	PGF$_{1\alpha}$	Rat stomach[b,55]; Rat blood[258]		(f,g)	
30		9,11-Dihydroxy-15-ketoprost-5-enoic acid (dihydroketo-PGF$_{2\alpha}$)	C$_{20}$H$_{34}$O$_5$	354	PGE$_2$; PGF$_{2\alpha}$ (i.v.); PGE$_2$ (i.v.); keto-PGF$_{2\alpha}$	Guinea pig liver[b,3,43]; Swine kidney[b,34]; Guinea pig lung[b,23]; Human blood[57,138]; Chicken heart[b,6]; Monkey urine[145]; Rat urine[144]; Lungs of pregnant rabbit[b,152]; Rat blood[258]; Rabbit kidney[b,162]	Little spasmogenic activity on most tissues;[3] less than 3% of PGF$_{2\alpha}$-bronchoconstrictor activity[57]	(f,g)	Major metabolite

No.	Structure	Name	Formula	MW	Precursor (administration)	Source	Activity	Ref.	Metabolite note
31	(chemical structure)	9,11,15-Trihydroxyprost-5-enoic acid (dihydro-PGF2α)	$C_{20}H_{36}O_5$	356	PGE2 PGF2α PGF2α (i.v.)	Guinea pig liver[b,3,43] Swine kidney[b,24] Human blood[57] Monkey urine[145] Rat urine[144]	Twice as potent as PGF2α as a bronchoconstrictor[82]	(f,g,h)	Minor metabolite
32	(chemical structure)	7,9,13-Trihydroxydinorprost 11-enoic acid (dinor-PGF1α)	$C_{18}H_{32}O_5$	328	PGF1α PGF1α (s.c.) PGF2α (i.v.)	Rat liver[b,51] Rat urine[58] Rat urine[59,60]		(i)	Major metabolite
33	(chemical structure)	7β,9α,13-Trihydroxydinor-prost-11-enoic acid (dinor-PGF1β)	$C_{18}H_{32}O_5$	328	PGF1β	Rat liver[b,51]		(i,l)	Unusual PGF1α metabolite
34	(chemical structure)	7,9-Dihydroxy-13-ketodinor-prost-3-enoic acid	$C_{18}H_{30}O_5$	326	PGF2α (i.v.) and PGD2 (i.v.)	Monkey urine[61,133,145] Human urine[134]		(f,g,i) or (f,g,i,u)	Not a natural metabolite of PGF1α
35	(chemical structure)	7,9,13-Trihydroxydinorprost-3-enoic acid (dihydrodinor-PGF2α)	$C_{18}H_{32}O_5$	328	PGF2α (i.v.) and PGD2 (i.v.)	Monkey urine[61,133]		(f,g,h,i) or (f,g,h,i,u)	Stereochemistry at C-9, C-10 not established (from PGD2)
36	(chemical structure)	7,9,13-Trihydroxydinor-prosta-11,15-dienoic acid	$C_{18}H_{30}O_5$	326	PGF3α (i.v.)	Rat urine[62]		(i,l)	
37	(chemical structure)	7,9,18-Trihydroxy-13-ketodinorprost-3-enoic acid	$C_{18}H_{30}O_6$	342	PGF2α (i.v.) and PGD2 (i.v.)	Human urine[63] Monkey urine[61,133] Human plasma[138]		(f,g,i,k) or (f,g,i,k,u)	Stereochemistry at C-9, C-10 not established (from PGD2)
38	(chemical structure)	7,9-Dihydroxy-13-ketodinor-prost-3-ene-1,18-dioic acid	$C_{18}H_{28}O_7$	356	PGF2α (i.v.) and PGD2 (i.v.)	Human urine[63] Monkey urine[61,133,145] Human plasma[138]		(f,g,i,k,m) or (f,g,i,k,m,u)	Stereochemistry at C-9, C-10 not established (from PGD2)

Table 2 (continued)
PROSTAGLANDIN METABOLITES

Structure	Chemical name	Molecular formula	Mol wt	Substrate	Source	Biological action	Presumed metabolic route from primary prostaglandin[a]	Comment
39	7,9-Dihydroxy-13-keto(dinor, ω-dinor)prost-3-ene-1,16-dioic acid	$C_{16}H_{24}O_7$	328	PGF$_{2\alpha}$(i.v.) and PGD$_2$(i.v.)	Human urine[63,134,145] Monkey urine[61,133]		(f,g,i,k,m) or (f,g,i,k,m,n,u)	Stereochemistry at C-9, C-10 not established (from PGD$_2$)
40	7,9,13-Trihydroxy-(dinor, ω-dinor)prost-3-ene-1,16-dioic acid γ-lactone	$C_{16}H_{24}O_6$	312	PGF$_{2\alpha}$ (i.v.) and PGD$_2$ (i.v.)	Human urine[63,134] Human plasma[138] Monkey urine[145]		(f,g,h,i,k,m,n,p)	Stereochemistry at C-9, C-10 not established (from PGD$_2$)
41	7,9-Dihydroxy(dinor, ω-tetranor)-prost-3-ene-1,14-dioic acid	$C_{14}H_{22}O_6$	286	PGF$_{2\alpha}$ (i.v.)	Human urine[64,65] Monkey urine[61,145] Human plasma[138]		(f,g,h,i,k m,s,n)	Unusual metabolite
42	5,7,11-Trihydroxytetranor-prost-9-enoic acid (tetranor-PGF$_{1\alpha}$)	$C_{16}H_{28}O_5$	300	PGF$_{2\alpha}$ (i.v.)	Rat urine[59,60] Rabbit urine[66]		(i,l,i)	
43	5,7,11-Trihydroxytetranor-prosta-9,13-dienoic acid (tetranor-PGF$_{3\alpha}$)	$C_{16}H_{26}O_5$	298	PGF$_{3\alpha}$ (i.v.)	Rat urine[62]		(i,l,i)	Major metabolite
44	5,7,11-Trihydroxytetranor-prosta-9,13-dienoic acid-5 lactone	$C_{16}H_{24}O_4$	280	PGF$_{3\alpha}$ (i.v.)	Rat urine[62]		(i,l,i,o)	

No.	Structure	Name	Formula	M.W.	Precursor	Source	Ref.	Notes
45		5,7,11,15-Tetrahydroxy-tetranorprost-9-enoic acid (15-hydroxytetranor-PGF$_{1\alpha}$)	C$_{16}$H$_{28}$O$_6$	316	PGF$_{2\alpha}$ (i.v.)	Rat urine[59,60]	(i,l,i,j)	
46		5,7,11,15-Tetrahydroxy-tetranorprosta-9,13-dienoic acid (15-hydroxytetranor-PGF$_{3\alpha}$)	C$_{16}$H$_{26}$O$_6$	314	PGF$_{3\alpha}$ (i.v.)	Rat urine[62]	(i,l,i,j)	Tentative
47		5,7,11,16-Tetrahydroxy-tetranorprost-9-enoic acid (16-hydroxytetranor-PGF$_{1\alpha}$)	C$_{16}$H$_{28}$O$_6$	316	PGF$_{2\alpha}$ (i.v.)	Rat urine[29,60]	(i,l,i,k)	
48		5,7,11,16-Tetrahydroxy-tetranorprosta-9,13-dienoic acid (16-hydroxytetranor-PGF$_{3\alpha}$)	C$_{16}$H$_{26}$O$_6$	314	PGF$_{3\alpha}$ (i.v.)	Rat urine[62]	(i,l,i,k)	Tentative
49		5,7,11-Trihydroxytetranor-prosta-1,16-dioic acid (16-carboxytetranor-PGF$_{1\alpha}$)	C$_{16}$H$_{26}$O$_7$	330	PGF$_{2\alpha}$ (i.v.)	Man,[63] rabbit,[56,66] monkey,[61] and rat[67] urine	(i,l,i,k,m)	
50		5,7-Dihydroxy-11-keto-tetranorprostanoic acid	C$_{16}$H$_{26}$O$_5$	300	PGF$_{2\alpha}$ (i.v.) and PGD$_2$ (i.v.)	Human plasma[138] Rat urine[59,69,144] Guinea pig urine[56] Rabbit urine[66] Monkey urine[62,133,145] Rat blood[218]	(f,g,i,l,j) or (f,g,i,l,i,u)	Stereochemistry at C-5 or C-7 not established (from PGD$_2$)
51		5,7-Dihydroxy-11-keto-tetranorprostanoic acid δ-lactone	C$_{16}$H$_{24}$O$_4$	282	PGF$_{2\alpha}$ (i.v.)	Guinea pig urine[56]	(f,g,i,l,i,o)	Tentative

Table 2 (continued)
PROSTAGLANDIN METABOLITES

Structure		Chemical name	Molecular formula	Mol wt	Substrate	Source	Biological action	Presumed metabolic route from primary prostaglandin[a]	Comment
	52	5β,7α-Dihydroxy-11-keto-tetranorprostanoic acid	$C_{16}H_{28}O_5$	300	PGE$_1$ (i.v.), PGE$_2$ (i.v.), PGE$_2$ (i.v.), PGE$_2$	Guinea pig urine[68], Guinea pig urine[69], Rat urine[53], Guinea pig liver[h,70]		(f,g,q,i,i,i)	Major metabolite, Major metabolite
	53	5,7-Dihydroxy-11-keto-tetranorprost-13-enoic acid	$C_{16}H_{26}O_5$	298	PGE$_{3\alpha}$ (i.v.)	Rat urine[62]		(f,g,i,i,i)	
	54	5,7,15-Trihydroxy-11-keto-tetranorprostanoic acid	$C_{16}H_{28}O_6$	316	PGF$_{2\alpha}$ (i.v.)	Monkey and rat urine[61,144,145]		(f,g,i,i,i,j)	
	55	5,7,16-Trihydroxy-11-ketotetranorprostanoic acid	$C_{16}H_{28}O_6$	316	PGF$_{2\alpha}$ (i.v.) and PGD$_2$ (i.v.)	Rat urine[59,60,144], Human urine[63], Monkey urine[61,133,145], Rat blood[258]		(f,g,i,i,k) or (f,g,i,i,k,u)	Stereochemistry at C-5 or C-7 not established (from PGD$_2$)
	56	5,7,16-Trihydroxy-11-keto-tetranorprostanoic acid δ-lactone	$C_{16}H_{26}O_5$	298	PGF$_{2\alpha}$ (i.v.)	Human urine[63]		(f,g,i,i,k,o)	
	57	5,7-Dihydroxy-11-keto-tetranorprosta-1,16-dioic acid	$C_{16}H_{26}O_7$	330	PGF$_{2\alpha}$ (i.v.), PGD$_2$ (i.v.), PGE$_1$ (i.v.)	Human plasma[158], Human urine[71,72,134,154], Monkey urine[61,133,145], Rat urine[61,144], Rabbit urine[66], Rat blood[258]		(f,g,i,i,k,m) or (f,g,i,i,k,m,u)	Major later appearing metabolite of PGF$_{2\alpha}$, Major metabolite of PGF$_{2\alpha}$

No.	Structure	Name	Formula	M.W.	Precursor	Found in	Ref.	Notes
58		5,7-Dihydroxy-11-keto-tetranorprosta-1,16-dioic acid δ-lactone	$C_{16}H_{22}O_6$	312	$PGF_{2\alpha}$(i.v.), PGE_1(i.v.)	Human urine[63,71,72,134,146,154] Rat urine[60]	(f,g,i,i,i,k m,o)	Major metabolite of $PGF_{2\alpha}$ Stereochemistry at C-7, C-8 not established in human (from PGD_2)
59		5,7,11-Trihydroxytetranor-prostanoic acid (dihydrotetranor-$PGF_{2\alpha}$)	$C_{16}H_{30}O_5$	302	$PGF_{2\alpha}$(i.v.)	Rabbit urine[66] Monkey urine[61] Rat urine[145]	(f,g,h,i,i,i)	
60		5,7,11,15-Tetrahydroxy-tetranorprostanoic acid	$C_{16}H_{30}O_6$	318	$PGF_{2\alpha}$(i.v.)	Rat urine[59,60]	(f,g,h,i,i,i,j)	
61		5,7,11,15-Tetrahydroxy-tetranorprost-13-enoic acid	$C_{16}H_{28}O_6$	316	$PGF_{3\alpha}$(i.v.)	Rat urine[62]	(f,g,h,i,i,i,j)	Tentative
62		5,7,11,16-Tetrahydroxy-tetranorprostanoic acid	$C_{16}H_{30}O_6$	318	$PGF_{2\alpha}$(i.v.)	Rat urine[59,60]	(f,g,h,i,i,i,k)	
63		5,7,11,16-Tetrahydroxy-tetranorprost-13-enoic acid	$C_{16}H_{28}O_6$	316	$PGF_{3\alpha}$(i.v.)	Rat urine[62]	(f,g,h,i,i,i,k)	Tentative
64		5,7,11-Trihydroxytetranor-prosta-1,16-dioic acid	$C_{16}H_{28}O_7$	332	$PGF_{2\alpha}$(i.v.)	Human plasma[138] Rabbit urine[66] Human urine[43] Monkey urine[145]	(f,g,h,i,i,i,k,m)	

Table 2 (continued)
PROSTAGLANDIN METABOLITES

Structure		Chemical name	Molecular formula	Mol wt	Substrate	Source	Biological action	Presumed metabolic route from primary prostaglandin[a]	Comment
	65	5,7,11-Trihydroxytetranor-prosta-1,16-dioic acid δ-lactone	$C_{16}H_{26}O_6$	314	PGF$_{2\alpha}$ (i.v.) and PGD$_2$ (i.v.)	Human urine[63,134]		(f,g,h,i,l,i,k,m,o) or (f,g,h,i,k,m,o,l,j)	Stereochemistry at C-7, C-8 not established (from PGD$_2$)
	66	5,7-Dihydroxy-11-keto (tetranor-ω-dinor)prosta-1,14-dioic acid	$C_{14}H_{22}O_7$	302	PGF$_{2\alpha}$ (i.p. or i.v.)	Human urine[64,73] Rat urine[144] Monkey urine[145] Rat blood[258]		(f,g,i,k,m,n,l,j) or (f,g,i,l,i,k,m,n)	
	67	5,7-Dihydroxy-11-keto (tetranor-ω-dinor)prosta-1,14-dioic acid δ-lactone	$C_{14}H_{20}O_6$	284	PGF$_{2\alpha}$ (i.p. or i.v.) PGF$_{2\alpha}$ (i.v.)	Human urine[64,73] Monkey urine[61]		(f,g,i,k,m,n,l,i,o) or (f,g,i,l,i,k,m,n,o)	
	68	5,7,11-Trihydroxy(tetranor, ω-dinor)prosta-1,14-dioic acid	$C_{14}H_{22}O_7$	304	PGF$_{2\alpha}$ (i.p. or i.v.)	Human urine[64,73]		(f,g,h,i,l,i,k,m,n)	
	69	5,7,11-Trihydroxy(tetranor, ω-dinor)prosta-1,14-dioic acid δ-lactone	$C_{14}H_{22}O_6$	286	PGF$_{2\alpha}$ (i.p. or i.v.)	Human urine[64,73]		(f,g,h,i,l,i,k,m,n,o)	
	70	15-Hydroxy-9-ketoprosta-10,13-dienoic acid (PGA$_1$)	$C_{20}H_{32}O_4$	336	PGF$_1$	Serum of several species including human[b,17]	32% of PGE$_1$-vasodilator activity in dog[61]	(c)	Little inactivation on passage through lung

No.	Name	Formula	MW	Designation	Source	Activity	Ref.	Comments
71	9,15-Diketoprost-10-enoic acid (dihydroketo-PGA$_1$)	C$_{20}$H$_{32}$O$_4$	336	PGA$_1$	Rabbit renal cortex[b,74]	Inactive on rat blood pressure[74]	(c,f,g)	
72	15-Hydroxy-9-ketoprost-10-enoic acid (dihydro-PGA$_1$)	C$_{20}$H$_{34}$O$_4$	338	PGA$_1$	Rabbit renal cortex[b,74]	As active as PGA$_1$ in lowering rat blood pressure[74]	(c,f,g,h)	
73	15,19-Dihydroxy-9-ketoprosta-10,13-dienoic acid (19-hydroxy-PGA$_1$)	C$_{20}$H$_{32}$O$_5$	352	PGA$_1$	Human and guinea pig liver[2,43] Rabbit liver[b,163] Human seminal plasma[75]	Contracts rat fundus. lowers cat blood pressure	(c,f,g,h,j)	Microsomal metabolite; may be formed by P-450 system[105,163] Some endogenous material may be artifact[50]
74	15,19-Dihydroxy-9-ketoprosta-5,10,13-trienoic acid (19-hydroxy-PGA$_2$)	C$_{20}$H$_{30}$O$_5$	350	PGA$_2$	Guinea pig liver[b,42] Human seminal plasma[75]		(e,j)	May be formed by P-450 system[105]; is found endogenously but may be artifact[50]
75	15,20-Dihydroxy-9-ketoprosta-10,13-dienoic acid (20-hydroxy-PGA$_1$)	C$_{20}$H$_{32}$O$_5$	352	PGA$_1$	Guinea pig liver[b,42] Rabbit liver[b,163]		(c,k)	May be formed by P-450 system[105]
76	13-Hydroxy-7-ketodinorprosta-8,11-dienoic acid (dinor-PGA$_1$)	C$_{18}$H$_{28}$O$_4$	308	PGA$_1$	Rat liver[b,51]		(c,j)	
77	11-Hydroxy-5-ketotetranorprosta-6,9-dienoic acid (tetranor-PGA)	C$_{16}$H$_{24}$O$_4$	280	PGA$_1$ Tetranor-PGE (i.v.)	Rat liver[b,51] Rat urine[53]		(c,j,i)	
78	5,11-Diketotetranorprosta-6,9-diene-1,16-dioic acid	C$_{16}$H$_{20}$O$_6$	308	PGE$_2$ (i.v.)	Human urine[52]		(c,f,j,i,j,k m)	

Table 2 (continued)
PROSTAGLANDIN METABOLITES

Structure		Chemical name	Molecular formula	Mol wt	Substrate	Source	Biological action	Presumed metabolic route from primary prostaglandin[a]	Comment
	79	15-Hydroxy-9-ketoprosta-11,13-dienoic acid (PGC$_1$)	C$_{20}$H$_{32}$O$_4$	336	PGE$_1$	Serum of several species including human[b,17] Plasma of several species[b,76,77]	Depressor effect on cat and dog blood pressure;[77,78] more potent vasodilator than PGA$_1$[77]	(c,d)	Not inactivated on passage through the lung
	80	15-Hydroxy-9-ketoprosta-5,11,13-trienoic acid (PGC$_2$)	C$_{20}$H$_{30}$O$_4$	334	PGE$_2$	Serum of cat[78]	Depressor effect on cat and dog blood pressure[78]	(c,d)	
	81	15-Hydroxy-9-ketoprosta-8(12),13-dienoic acid (PGB$_1$)	C$_{20}$H$_{32}$O$_4$	336	PGE$_1$ PGA$_1$ PGA$_1$ PGA$_1$	Serum of several species including human[b,17,18] Cat plasma[79] Plasma of several species[76] Rabbit renal cortex[b,74]	5% of PGA$_1$-depressor activity;[78] no vasodilator activity in dog[81]	(c,d,e)	
	82	15,19-Dihydroxy-9-ketoprosta-8(12),13-dienoic acid (19-hydroxy-PGB$_1$)	C$_{20}$H$_{32}$O$_5$	352	PGE^c	Human seminal plasma[75]		(c,j,d,e)	Endogenous, may be artifact[50]
	83	15,19-Dihydroxy-9-ketoprosta-5,8(12),13-trienoic acid (19-hydroxy-PGB$_2$)	C$_{20}$H$_{30}$O$_5$	350	PGE^c	Human seminal plasma[75]		(c,j,d,e)	Endogenous, may be artifact[50]
	84	13-Hydroxy-7-ketodinor-prosta-6(10),11-dienoic acid (dinor-PGB$_1$)	C$_{18}$H$_{22}$O$_4$	308	PGB$_1$	Rat liver[b,51]		(i)	

No.	Structure	Name	Formula	MW	Administered	Source	Ref.	Comments
85		11-Hydroxy-5-ketotetranor-prosta-4(8),9-dienoic acid (tetranor-PGB)	$C_{16}H_{24}O_4$	280	PGE₂ (i.v.)	Rat urine[53]	(c,d,e,i,l,i)	
86		11,15-Dihydroxy-5-keto-tetranorprosta-4(8),9-dienoic acid (15-hydroxytetranor-PGB)	$C_{16}H_{24}O_5$	296	PGE₂ (i.v.) Tetranor-PGB (i.v.) Tetranor-PGE (i.v.)	Rat urine[53] Rat urine[53] Rat urine[53]	(c,d,e,i,l,i,j) or (i,l,i,c,d,e,j)	
87		11,16-Dihydroxy-5-keto-tetranorprosta-4(8),9-dienoic acid (16-hydroxytetranor-PGB)	$C_{16}H_{24}O_5$	296	PGE₂ (i.v.) Tetranor-PGB (i.v.) Tetranor-PGE (i.v.)	Rat urine[53] Rat urine[53] Rat urine[53]	(c,d,e,i,l,i,k) or (i,l,i,c,d,e,k)	
88		11-Hydroxy-5-ketotetranor-prosta-4(8),9-diene-1,16-dioic acid (16-carboxytetranor-PGB)	$C_{16}H_{22}O_6$	310	PGE₂ (i.v.) Tetranor-PGB (i.v.) Tetranor-PGE (i.v.)	Rat urine[53] Rat urine[53] Rat urine[53]	(c,d,e,i,l,i,k,m) or (i,l,i,c,d,e,k,m)	
89		9α,11,15(S)-Trihydroxy-2,3-dinorthromba-5-*cis*, 13-*trans*-dienoic acid (2,3-dinor-TXB₂)	$C_{18}H_{30}O_6$	342	TXB₂ (i.v.)	Human urine[112,113,241] Rabbit urine[242] Monkey urine[113,114,115] Rabbit plasma[242]	(i)	Major metabolite (except rabbits) Major metabolite
90		9α,15(S)-Dihydroxy-11-keto-thromba-5-*cis*, 13-*trans*-dienoic acid (11-dehydro-TXB₂)	$C_{20}H_{32}O_6$	368	TXB₂ (i.v.)	Human urine[169,241] Human plasma[158,241] Monkey urine[216] Rabbit urine and plasma[242]	(t)	A major product (man and rabbit)
91		8-(1,3-Dihydroxypropyl)-9,12(S)-dihydroxy-5-*cis*, 10-*trans*-heptadecadienoic acid	$C_{20}H_{36}O_6$	372	TXB₂ (i.v.)	Human urine[199] Rabbit urine and plasma[242]	(*)	*C-11, C-12 Dihydroxylation mechanism unknown.
92		9α,11,15(S)-Trihydroxy-2,3,4,5-tetranor-thromb-13-*trans*-enoic acid (tetranor-TXB₂)	$C_{16}H_{28}O_6$	316	TXB₂ (i.v.)	Human urine[109] Rat urine[11] Rabbit urine[242]	(i,l,i)	Major metabolite

Table 2 (continued)
PROSTAGLANDIN METABOLITES

	Structure	Chemical name	Molecular formula	Mol wt	Substrate	Source	Biological action	Presumed metabolic route from primary prostaglandin[a]	Comment
93		9α,15(S)-Dihydroxy-11-keto-2,3-dinor-thromba-5-cis,13-trans-dienoic acid	$C_{16}H_{20}O_6$	340	TXB₂ (i.v.)	Human urine[109]		(t,i)	
94		9α,15(S)-Dihydroxy-11-keto-2,3,4,5-tetranor-thromb-13-trans-enoic acid	$C_{16}H_{26}O_6$	314	TXB₂ (i.v.)	Human urine[109] Rabbit urine and plasma[242]		(t,i,l,j)	
95		9α,15(S)-Dihydroxy-11-keto-2,3-dinor-thromb-13-trans-enoic acid	$C_{18}H_{30}O_6$	342	TXB₂ (i.v.)	Human urine[109]		(t,i,l)	
96		9α,15(S)-Dihydroxy-11-keto-thromb-5-cis,13-trans-diene-1,20-dioic acid	$C_{20}H_{30}O_8$	398	TXB₂ (i.v.)	Human urine[109]		(t,k,m)	
97		9α,15(S)-Dihydroxy-11-keto-2,3-dinor-thromb-5-cis,13-trans-diene-1,20-dioic acid	$C_{18}H_{26}O_8$	370	TXB₂ (i.v.)	Human urine[109]		(t,i,m)	
98		9α-Hydroxy-11,15-diketo-thromb-5-cis-enoic acid	$C_{20}H_{32}O_6$	368	TXB₂ (i.v.)	Human urine[109, 241] Human plasma[241]		(t,f)	
99		9α,15(S)-Dihydroxy-11-keto-thromb-5-cis-enoic acid	$C_{20}H_{34}O_6$	370	TXB₂ (i.v.)	Human urine[109]		(t,f,g,h)	

No.	Structure	Name	Formula	Mass	Compound	Source	References
100		9α-Hydroxy-11,14-diketo-2,3-dinor-thromb-5-*cis*-enoic acid	$C_{18}H_{28}O_6$	340	TXB$_2$ (i.v.)	Human urine[109] Monkey urine[110]	(t,i,f,g)
101		9α-Hydroxy-11,15-diketothromb-5-*cis*-ene-1,20-dioic acid	$C_{20}H_{30}O_8$	398	TXB$_2$ (i.v.)	Human urine[109]	(t,f,g,k,m)
102		9α-Hydroxy-11,15-diketo-2,3-dinor-thromb-5-*cis*-ene-1,20-dioic acid	$C_{18}H_{26}O_8$	370	TXB$_2$ (i.v.)	Human urine[109]	(t,i,f,g,m)
103		9α-Hydroxy-11,15-diketo-2,3-dinor-thrombanoic acid	$C_{18}H_{31}O_6$	343	TXB$_2$ (i.v.)	Human urine[109]	(t,i,i,f,g)
104		9α-Hydroxy-11,15-diketo-2,3,4,5-tetranor-thrombanoic acid	$C_{16}H_{24}O_6$	314	TXB$_2$ (i.v.)	Human urine[109]	(t,i,i,i,f,g)
105		9α-Hydroxy-11,15-diketo-2,3,4,5-tetranor-thromb-1,20-dioic acid	$C_{16}H_{24}O_8$	344	TXB$_2$ (i.v.)	Human urine[109] Monkey urine[110]	(t,i,i,i, f,g,k,m)
106		9α-Hydroxy-11,15-diketo-2,3,18,19-tetranor-thromb-5-*cis*-ene-1,20-dioic acid	$C_{16}H_{22}O_8$	342	TXB$_2$ (i.v.)	Human urine[109]	(t,i,f,g, k,m,n)

Table 2 (continued)
PROSTAGLANDIN METABOLITES

	Structure	Chemical name	Molecular formula	Mol wt	Substrate	Source	Biological action	Presumed metabolic route from primary prostaglandin[a]	Comment
107		6-(1,3-dihydroxypropyl)-7-hydroxy-10-keto-3-cis-pentadecanoic acid	$C_{18}H_{32}O_6$	344	TXB₂ (i.v.)	Human urine[109] Monkey urine[110]		(i,f,g)*	*C-11, C-12-Dihydroxylation mechanism unknown
108		9α,15(S)-dihydroxy-11,20-dioxo-2,3-dinor-5-cis,13-trans-dienoic acid	$C_{18}H_{26}O_7$	354	TXB₂ (i.v.)	Human urine[109]		(t,i)	Not precisely identified
109		9α,11-Dihydroxy-15-keto-thromb-5-cis-enoic acid (15-keto-13,14,Dihydro-TXB₂)	$C_{20}H_{34}O_6$	370	TXB₂	Guinea pig lung[112] Human plasma[241]		(f,g)	
110		8-(1,3-dihydroxypropyl)-9-hydroxy-12-keto-5-cis-heptadecaenoic acid	$C_{20}H_{36}O_6$	372	TXB₂	Guinea pig liver[110] (in vitro only)		(f,g)*	*C-11, C-12 Dihydroxylation mechanism unknown Not found in guinea pig urine[110]
111		9,11,15-Trihydroxy-6-ketoprost-13-enoic acid (6-keto-PGF₁α)	$C_{20}H_{34}O_6$	370	PGI₂ (i.v.) PGI₂	Monkey urine[116] Human urine[121,252] Human plasma b,245 Rabbit liver[119] Rat bile[125] Rabbit plasma[248] Rabbit lung[244] Cat plasma[248]			Formed by hydrolysis of PGI₂
112		9,11,15-Trihydroxy-6-keto-1,2-dinorprosta-13-enoic acid (2,3-dinor-6-keto-PGF₁α)	$C_{18}H_{30}O_6$	342	PGI₂ (i.v.) and 6-keto-PGF₁α (i.v.) PGI₂	Monkey urine[116] Human urine[118,121] Rat urine[117,120] Rabbit liver[119] Rat bile[125] Rabbit kidney b,181,243		(i)	Major metabolite (human urine)[252] Levels increased in tumor-bearing rats[246]

No.	Name	Formula	MW	Precursor	Source	Ref.	Notes
113	9,11-Dihydroxy-6,15-diketo-prostanoic acid (15-keto-13,14-dihydro-6-keto-PGF$_{1\alpha}$)	C$_{20}$H$_{34}$O$_6$	370	PGI$_2$ (i.v.) PGI$_2$	Rabbit plasma[248] Monkey urine[116] Human urine[232] Cat plasma[248] Rat bile[125] Human plasma[245] Human urine[116]	(f,g)	
114	9,11-Dihydroxy-6,15-diketo-1,2-dinorprostan-3-oic acid	C$_{18}$H$_{30}$O$_6$	342	PGI$_2$ (i.v.) and 6-keto-PGF$_{1\alpha}$ (i.v.)	Human urine[116] Rat urine[117,247] Human urine[121,232]	(i,f,g)	Increased in tumor-bearing rats[246]
115	9,11,19-Trihydroxy-6,15-diketo-1,2-dinorprostan-3-oic acid	C$_{18}$H$_{30}$O$_7$	358	PGI$_2$ PGI$_2$ (i.v.)	Rat bile[125] (free, or as glucuronide) Rabbit kidney[b,186,243] Rabbit lung[b,244] Monkey urine[116] Rat urine[117]	(i,f,g,j)	Major metabolite (as glucuronide) Not found in rabbit liver[119]
116	9,11-Dihydroxy-6,15-diketo-1,2-dinorprosta-3,20-dioic acid	C$_{18}$H$_{28}$O$_8$	372	PGI$_2$ (i.v.) and 6-keto-PGF$_{2\alpha}$ (i.v.) PGI$_2$	Monkey urine[116] Rat urine[117,247] Human urine[121,232] Rat bile[125]	(i,f,g,k,m)	
117	9,11,15,19-tetrahydroxy-6-keto-1,2-dinorprosta-13-en-3-oic acid (Dinor-ω-hydroxy-6-keto PGF$_{1\alpha}$)	C$_{18}$H$_{28}$O$_7$	356	PGI$_2$ (i.v.) and 6-keto-PGF$_{1\alpha}$ (i.v.)	Rat urine[117,120,247]	(i,j)	

Table 2 (continued)
PROSTAGLANDIN METABOLITES

	Structure	Chemical name	Molecular formula	Mol wt	Substrate	Source	Biological action	Presumed metabolic route from primary prostaglandin[a]	Comment
118		9,11,20-trihydroxy-6,15-diketo-1,2-dinorprostan-3-oic acid	$C_{18}H_{30}O_7$	358	PGI_2(i.v.) and 6-keto-$PGF_{1\alpha}$(i.v.)	Monkey urine[116] Rat urine[117, 247]		(i,f,g,k)	
119		9,11,-Dihydroxy-6,15-diketo-1,2,19,20-tetranorprosta-3,18-dioic acid	$C_{16}H_{24}O_8$	344	PGI_2(i.v.) and 6-keto-$PGF_{1\alpha}$(i.v.)	Rat urine[117, 247] Human urine[242]		(i,f,g,n, k,m)	
120		11,15-Dihydroxy-2,3,4,5,6-pentanor-prosta-13-enoic acid γ-lactone (Pentanor-$PGF_{1\alpha}$-γ-lactone)	$C_{15}H_{24}O_4$	268	PGI_2 and 6-keto-$PGF_{1\alpha}$ PGI_2(i.v.)	Rabbit liver[119] and kidney[134] Rat bile[125] Rat urine[125] Rabbit kidney[b, 188, 243]		(i,i,*)	*Oxidative decarboxylation of α-keto acid
121		11,15-Dihydroxy-6,9-diketoprost-13-enoic acid (6-keto-PGE_1)	$C_{20}H_{32}O_6$	368	PGI_2 and 6-keto-$PGF_{1\alpha}$	Rabbit liver[119] Human platelets[122] Human kidney[b, 140]	Inhibits platelet aggregation (potency similar to PGI_2)[123]	(b)	Not appreciably formed in human plasma from PGI_2(i.v.)[251]
122		9,11,15,19-Tetrahydroxy-6-ketoprost-13-enoic acid (19-hydroxy-6-keto-$PGF_{1\alpha}$)	$C_{20}H_{34}O_7$	386	PGI_2 and 6-keto-$PGF_{1\alpha}$	Rabbit liver[119]		(j)	

No.	Name	Formula	MW	Derived from	Source	Biological activity	Ref.	Comments
123	15-Hydroxy-11-ketoprosta-5,9,13-trienoic acid (PGJ₂)	$C_{20}H_{30}O_4$	334	PGD₂	Human serum albumin[128] Aqueous solution[127,129]	Inhibits cell proliferation[127-131] Inhibits platelet aggregation[127,129] Inhibits smooth muscle contraction[129]		Not known if enzyme-catalyzed
124	15-Hydroxy-11-ketoprosta-5,9,12-trienoic acid (Δ¹²-13,14-dihydro-PGJ₂)	$C_{20}H_{31}O_4$	335	PGD₂ and PGJ₂	Human plasma[128,132] Cat plasma[127]	Inhibits cell proliferation[127,132]		
125	7,9,13-Trihydroxydinorprost-3,11-dienoic acid (Dinor-PGF₁α)	$C_{18}H_{30}O_5$	326	PGD₂ (i.v.)	Monkey urine[133] Human urine[134,135]		(i,u)	Major F-ring metabolite
126	7,13-Dihydroxy-9-keto-dinorprost-3,11-dienoic acid (Dinor-PGD₂)	$C_{18}H_{28}O_5$	324	PGD₂ (i.v.)	Monkey urine[133] Human urine[134,135]		(i)	d
127	7,13,18-Trihydroxy-9-ketodinorprost-3,11-dienoic acid	$C_{20}H_{32}O_6$	368	PGD₂ (i.v.)	Monkey urine[133]		(i,k)	d
128	7-Hydroxy-9,13-diketodinorprost-3-enoic acid	$C_{18}H_{28}O_5$	324	PGD₂ (i.v.)	Monkey urine[133] Human urine (mastocytosis)[136]		(f,g,i)	d
129	7,18-Dihydroxy-9,13-diketodinorprost-3-enoic acid	$C_{18}H_{28}O_6$	340	PGD₂ (i.v.)	Monkey urine[133] Human urine (mastocytosis)[134]		(f,g,i,k)	Major D-ring metabolite d

Table 2 (continued)
PROSTAGLANDIN METABOLITES

	Structure	Chemical name	Molecular formula	Mol wt	Substrate	Source	Biological action	Presumed metabolic route from primary prostaglandin[a]	Comment
130		7-Hydroxy-9,13-diketodinorprost-3-ene-1, 18-dioic acid	$C_{18}H_{26}O_7$	354	PGD_2 (i.v.)	Monkey urine[133] Human urine (mastocytosis)[136]		(f,g,i, k,m)	d
131		5,11-Dihydroxy-7-ketotetranorprost-9-enoic acid	$C_{16}H_{26}O_5$	298	PGD_2 (i.v.)	Monkey urine[133]		(i,l,j)	
132		5,16-Dihydroxy-7,11-diketotetranorprostanoic acid	$C_{16}H_{26}O_6$	314	PGD_2 (i.v.)	Monkey urine[133]		(f,g,i, l,j,k)	
133		5-Hydroxy-7,11-diketo-tetranorprosta-1,16-dioic acid	$C_{16}H_{24}O_7$	328	PGD_2 (i.v.)	Monkey urine[133]		(f,g,i,l, i,k,m)	
134		7-Hydroxy-9,13-diketo (dinor, ω-dinor) prost-3-ene-1,16,dioic acid	$C_{16}H_{22}O_7$	326	PGD_2 (i.v.)	Monkey urine[133] Human urine (mastocytosis)[136]		(f,g,i,k, m,n)	d
135		7-Hydroxy-9-keto (dinor, ω-tetranor) prost-3-ene-1,14-dioic acid	$C_{14}H_{20}O_6$	284	PGD_2 (i.v.)	Monkey urine[133] Human urine (mastocytosis)[136]		(f,g,h,i,k, m,s,n)	d
136		9α,11,15(S)-Trihydroxy-2,3-dinorthromb-13-*trans*-enoic acid. (2,3-Dinor-TXB$_1$)	$C_{18}H_{32}O_4$	344	TXB_2 (i.v.)	Rabbit urine[242] Rabbit plasma[242]		(i,l)	Minor product in plasma

No.	Structure	Name	Molecular formula	MW	Precursor	Source	Mechanism	Comments
137		9α,11,15(S)-Trihydroxy-2,3,4,5-tetranorthromb-13-*trans*-enoic acid. (2,3,4,5-Tetranor-TXB$_1$.)	$C_{16}H_{28}O_6$	316	TXB$_2$ (i.v.)	Rabbit urine[242] Rabbit plasma[242]	(i,j)	
138		5,9α,11α,15(S)-Tetrahydroxy-6-ketoprost-13-enoic acid. (5-Hydroxy-6-keto-PGF$_{1α}$)	$C_{20}H_{34}O_7$	386	PGI$_2$	Rabbit kidney[b,243]	(Epoxygenase)	Probably generated from 5(6)-oxido-PGI$_1$, via P-450 dependent mixed function oxidase system[243] Inhibited by SKF 525A but not by metyrapone.
139		9,11,19-Trihydroxy-6,15-diketo-1,2-dinorprostanoic acid.	$C_{18}H_{30}O_7$	358	6-keto-PGF$_{1α}$ (i.v.)	Rat urine[247]	(i,f,g,j)	Minor product
140		9,11,15-Trihydroxy-6,19-diketo-1,2-dinorprost-13-enoic acid.	$C_{18}H_{28}O_7$	356	6-keto-PGF$_{1α}$ (i.v.)	Rat urine[247]	(j) followed by oxidation at C$_{19}$	
141		9,11-Dihydroxy-6,15-diketoprost-13-enoic acid. (6,15-diketo-PGF$_{1α}$)	$C_{20}H_{32}O_6$	368	PGI$_2$, PGG$_2$, Arachidonic acid	Bovine mesenteric arteries and veins[b,249] Monkey lung[b,250] Porcine aorta[b,253] Fetal calf aorta and ductus arteriosus[b,254]	Probably (f) or PGI$_2$ isomerase	Poor metabolism by 6-keto-PGF$_{1α}$; Bicyclic 15-keto-PGI$_2$ may be intermediate. 15-Hp-PGI$_2$ is also a possible intermediate from PGG$_2$.[253,254]
142		9,11,15-Trihydroxy-6-keto-1,2-dinorprosta-3,20-dioic acid.	$C_{18}H_{30}O_8$	374	PGI$_2$ (i.v.)	Human urine[252]	(i,g,k,m)	
143		9,11,15-Trihydroxy-6-keto-1,2,19,20-tetranorprost-13-en-3,18-dioic acid.	$C_{16}H_{24}O_8$	344	PGI$_2$ (i.v.)	Human urine[252]	(i,n,k,m)	

Table 2 (continued)
PROSTAGLANDIN METABOLITES

	Structure	Chemical name	Molecular formula	Mol wt	Substrate	Source	Biological action	Presumed route (Fig. 1)	Comments
144		9,11,15,20-Tetrahydroxy-6-keto-1,2-dinorprostanoic acid.	$C_{18}H_{30}O_8$	374	PGI_2 (i.v.)	Human urine[252]		(i,g,k)	
145		9,11,15-Trihydroxy-6-keto-1,2,19,20-tetranorprost-3,18-dioic acid	$C_{16}H_{26}O_8$	346	PGI_2 (i.v.)	Human urine[252]		(i,g,n,k,m)	
146		9,11,15-Trihydroxy-6-keto-1,2,19,20-tetranorprost-3,18-dioic acid-γ-lactone.	$C_{16}H_{24}O_7$	328	PGI_2 (i.v.)	Human urine[252]		(i,g,n,k,m,p)	
147		9,11,15-Trihydroxy-6-keto-1,2-dinorprosta-13,17-dienoic acid. (Δ^{17}-1,2-dinor-6-keto-$PGF_{1\alpha}$)	$C_{18}H_{28}O_6$	340	PGI_3	Human urine[141,142]		(i)	Major metabolite
148		7,11-Dihydroxy-5-keto-tetranorprosta-9,13-dienoic acid.	$C_{16}H_{24}O_5$	296	PGE_3	Rat urine[143]		(i,j,j)	Major metabolite

No.	Structure	Name	Formula	MW	Parent	Source	Biological activity	Specificity (u) with β-specificity	Comments
149		9α,11β,15(S)-Trihydroxy-prosta-5,13-dienoic acid. (9α,11β-PGF₂)	$C_{20}H_{34}O_5$	354	PGD₂ (i.v.), PGD₂	Human urine[134,135] Human plasma[135] Rabbit liver[137] Bovine lung[b,149]	Systemic blood pressure elevation in rat:[135] Platelet aggregation inhibition;[137] Constriction of canine coronary artery strips[137,147] and human bronchial smooth muscle[148]	(u,f,g,i,k,m,n)	Major metabolite
150		7α,9β-Dihydroxy-13-keto (dinor,ω-dinor) prost-3-ene-1,16-dioic acid.	$C_{16}H_{24}O_7$	328	PGD₂ (i.v.)	Human urine[134]		(u,i)	
151		7α,9β,13(S)-Trihydroxy-dinorprost-3,11-dienoic acid. (7α,9β-Dinor-PGF₂)	$C_{18}H_{30}O_5$	326	PGD₂ (i.v.)	Human urine[134]		(u,i)	
152		7α,9β,13(S)-Trihydroxy-dinorprost-3-enoic acid. (7α,9β-dihydrodinor-PGF₂)	$C_{18}H_{32}O_5$	328	PGD₂ (i.v.)	Human urine[134]		(u,g,i)	
153		7α,9β-Dihydroxy-13 keto-dinorprost-3-enoic acid.	$C_{18}H_{30}O_5$	326	PGD₂ (i.v.)	Human urine[134]		(u,f,g,i)	
154		5α,7-Dihydroxy-11-keto (tetranor,ω-dinor) prosta-1,14-dioic acid.	$C_{14}H_{22}O_7$	302	PGD₂ (i.v.)	Human urine[136]		(u,f,g,i,l,i,n,k,m)	
155		5α,7β-Dihydroxy-11-keto-tetranor-prosta-1,16-dioic acid.	$C_{16}H_{26}O_7$	330	PGD₂ (i.v.)	Human urine[134]		(u,f,g,i,l,i,k,m)	
156		7α,9β-Dihydroxy-13-keto (dinor,ω-dinor)prost-3-ene-1,16-dioic acid.	$C_{16}H_{27}O_7$	328	PGD₂ (i.v.)	Human urine[134]		(u,f,g,i,n,k,m)	
157		5-Hydroxy-7,11-diketo-tetranorprostanoic acid.	$C_{16}H_{26}O_5$	298	PGD₂ (i.v.)	Human urine[136] (mastocytosis)		(f,g,i,l,i)	
158		7,13-Dihydroxy-9-keto (dinor,ω-dinor)prost-3-ene-1,16-dioic acid.	$C_{14}H_{24}O_7$	328	PGD₂ (i.v.)	Human urine[134] (mastocytosis)		(g,i,n,k,m)	Major PGD-ring metabolite (mastocytosis)

Table 2 (continued)
PROSTAGLANDIN METABOLITES

	Structure	Chemical name	Molecular formula	Mol wt	Substrate	Source	Biological action	Presumed route (Fig. 1)	Comments
159		7,13-Dihydroxy-9-keto (dinor,ω-dinor)prost-3-ene-1,16-dioic acid-γ-lactone.	$C_{18}H_{22}O_6$	298	PGD_2 (i.v.)	Human urine[136] (mastocytosis)		(g,i,n,k,m,p)	
160		11,15-Dihydroxy-9-ketoprosta-5,13-dienoic acid (PGE_2)	$C_{20}H_{32}O_5$	352	$PGF_{2\alpha}$	Human kidney[b,140]		(b)	
161		7,9,13-Trihydroxy-dinorprost-3-ene-1,18-dioic acid.	$C_{18}H_{30}O_7$	358	$PGF_{2\alpha}$ (i.v.)	Monkey urine[145]		(g,i,k,m)	Unusual metabolite
162		9,11,15,20-Tetrahydroxyprost-5,13-dienoic acid. (20-Hydroxy-$PGF_{2\alpha}$)	$C_{20}H_{34}O_6$	370	$PGF_{2\alpha}$	Pregnant rabbit lung[b,152]		(k)	
163		9,11,15-Trihydroxyprost-5,13-dien-1,20-dioic acid. (20-Carboxy-$PGF_{2\alpha}$)	$C_{20}H_{32}O_7$	384	$PGF_{2\alpha}$	Pregnant rabbit lung[b,152]		(k,m)	
164		9,11,20-Trihydroxy 15-ketoprost-5-enoic acid. (Dihydroketo-20-hydroxy-$PGF_{2\alpha}$)	$C_{20}H_{34}O_6$	370	$PGF_{2\alpha}$	Pregnant rabbit lung[b,152]		(f,g,k)	
165		9,11-Dihydroxy-15-keto-prost-5-en-1,20-dioic acid. (Dihydroketo-20-carboxy-$PGF_{2\alpha}$)	$C_{20}H_{32}O_7$	384	$PGF_{2\alpha}$	Pregnant rabbit lung[b,152]		(f,g,k,m)	

No.	Name	Formula	MW	Related	Source	Footnote	Comments
166	11,15,20-Trihydroxy-9-ketoprost-5,13-dienoic acid. (20-Hydroxy-PGE$_2$)	C$_{20}$H$_{32}$O$_6$	368	PGE$_2$	Sheep vesicular glands[b,151] Pregnant rabbit lung[b,152] Rabbit liver[b,163] Ram seminal fluid[b,167]	(k)	May be formed by P-450 system (in rabbit liver)[163]
167	11,15-Dihydroxy-9-ketoprost-5,13-dien-1,20-dioic acid. (20-Carboxy-PGE$_2$)	C$_{20}$H$_{30}$O$_7$	382	PGE$_2$	Pregnant rabbit lung[b,152]	(k,m)	
168	11,20-Dihydroxy-9,15-diketoprost-5-enoic acid. (Dihydroketo-20-hydroxy-PGE$_2$)	C$_{20}$H$_{34}$O$_6$	370	PGE$_2$	Pregnant rabbit lung[b,152]	(f,g,k)	
169	11,15,20-Trihydroxy-9-ketoprost-13-enoic acid. (20-Hydroxy-PGE$_1$)	C$_{20}$H$_{34}$O$_6$	370	PGE$_1$	Pregnant rabbit lung[b,163,165] Rabbit liver[b,163] Ram seminal fluid[b,167]	(k)	May be formed by P-450 system (in rabbit liver)[163]
170	11,15-Dihydroxy-9-ketoprost-13-en-1,20-dioic acid. (20-Carboxy-PGE$_1$)	C$_{20}$H$_{32}$O$_7$	384	PGE$_1$	Pregnant rabbit lung[b,152]	(k,m)	
171	11,20-Dihydroxy-9,15-diketoprostanoic acid. (Dihydroketo-20-hydroxy-PGE$_1$)	C$_{20}$H$_{34}$O$_6$	370	PGE$_1$	Pregnant rabbit lung[b,152]	(f,g,k)	
172	11-Deoxy-13,14-dihydro-15-keto-11β,16ξ-cyclo-PGE$_2$ (BC-PGE$_2$;bicyclo-PGE$_2$)[a]	C$_{20}$H$_{30}$O$_4$	334	PGE$_2$ (i.v.) and PGE$_2$	Human plasma[157,161,169] Dog plasma[155] Sheep plasma[155,157] Guinea pig lung[156]	(f,g) followed by non-enzymatic cyclization	Stable metabolite of PGE$_2$, ketodihydro-PGE$_2$, and ketodihydro-PGA$_2$. Used for radioimmunoassay of PGE$_2$ metabolites

Table 2 (continued)
PROSTAGLANDIN METABOLITES

	Structure	Chemical name	Molecular formula	Mol wt	Substrate	Source	Biological action	Presumed route (Fig. 1)	Comments
173		11,15,18-Trihydroxy-9-ketoprost-13-enoic acid. (18-Hydroxy-PGE$_1$)	C$_{20}$H$_{34}$O$_6$	370	PGE$_1$	Rabbit liver[b,165] Rat liver[b,166]		(j)	
174		11,15,18-Trihydroxy-9-ketoprost-5,13-dienoic acid (18-Hydroxy-PGE$_2$)	C$_{20}$H$_{32}$O$_6$	368	PGE$_2$	Rabbit liver[b,166] Rat liver[b,166]		(j)	
175		5,9,11,15-Tetrahydroxy-6-ketoprost-13-enoic acid. (5-Hydroxy-6-keto-PGF$_{1\alpha}$)	C$_{20}$H$_{34}$O$_7$	386	PGI$_2$	Rabbit kidney[b,185]		(Epoxygenase)	Unusual metabolite

[a]See Table 1 for details of metabolic reactions.
[b]In vitro production of metabolite.
[c]Presumed substrate.
[d]Stereochemistry at C$_{10}$ not established in human.

Table 3
LIPOXYGENASE METABOLITES

Structure		Chemical name	Molecular formula	Mol wt	Substrate	Source	Biological action	Presumed catalytic route	Comments
	176	5(S)-Hydroxy-6-*trans*,8-*cis*,11-*cis*,14-*cis*-eicosatetraenoic acid (5-HETE)	$C_{20}H_{32}O_3$	320	5-HpETE	Human polymorphonuclear leukocytes[b,255]		Peroxidase	
	177	5(S),6(S)-5(6)-Epoxy-7-*trans*,9-*trans*,11-*cis*,14-*cis*-eicosatetraenoic acid. (LTA₄)	$C_{20}H_{30}O_3$	318	5-HpETE	Porcine leukocytes[b,198]		AA 5-lipoxygenase and dehydrase	Mechanism probably involves stereo-specific elimination of the C-10-hydrogen of 5-HpETE.[233]
	178	5(S),12(R)-Dihydroxy-6-*cis*,8-*trans*,10-*trans*,14-*cis*-eicosatetraenoic acid. (LTB₄)	$C_{20}H_{32}O_4$	336	LTA₄, 5-HpETE	Rat kidney[b,176,257] Human lung[a,180] Human neutrophils[b,213] Rat neutrophils[b,213] Human, guinea pig, bovine, sheep, rabbit, rat and dog plasma[b,224] Guinea pig liver[b,226] Guinea pig lung[a,226] Guinea pig kidney[b,226] Human polymorphonuclear leukocytes[b,255] Human liver[b,226]	Potent neutrophil chemotaxis and chemokinesis[214] Increases vascular permeability[215] Stimulates leukocyte aggregation and adhesion to vascular endothelium;[216] Modulates pain responses;[217] Stimulates generation of suproxides in neutrophils;[213] 100-times less potent bronchoconstrictor than LTC₄ (through release of TXA₂)[222] Chemotactic toward polymorphonuclear cells, eosinophils, and monocytes;[222] Possible ionophoric activity.[222]	LTA₄ hydrolase	Major metabolite (rat kidney and isolated glomeruli[176,257] Possible leukotactic mechanism in plasma[224] Inhibited by LTA₃, (inhibits covalent coupling of LTA₄ to hydrolase.[213] U-60257[222] and eicosapentaenoic acid (probable site: LTA₄ hydrolase)[228]
	179	5(S),12(R)-and 5(S),12(S)-Dihydroxy-6-*trans*,8-*trans*,10-*trans*,14-*cis*-eicosatetraenoic acid. (6-*trans*-LTB₄ and 12-epi-6-*trans*-LTB₄)	$C_{20}H_{32}O_4$	336	LTA₄; LTC₄; LTD₄; LTE₄; 5-HpETE	Human lung[b,180] Human leukocytes[b,193,200,255] Porcine leukocytes[b,198]		From LTA₄: nonenzymic hydrolysis[223] From other LTs: myeloperoxidase mediated oxidation (S-chlorosulfonium ion is probable intermediate) followed by carbo-cation formation[193,200]	No spasmogenic or immunoreactive activity.[193,200] Inhibitors: catalase, azide[193,200]

Table 3 (continued)
LIPOXYGENASE METABOLITES

Structure		Chemical name	Molecular formula	Mol wt	Substrate	Source	Biological action	Presumed catalytic route	Comments
	180	5(S),12(R)-Dihydroxy-19,20-dinor-6-cis,8-trans,10-trans,14-cis-eicosatetraen-1,18-dioic acid. (18-Carboxy-19,20-dinor-LTB₄)	$C_{18}H_{26}O_6$	338	LTB₄[b,171]	Rat liver[b,171]			
	181	5(S),12(R),20-Trihydroxy-6-cis,8-trans,10-trans,14-cis-eicosatetraenoic acid. (20-Hydroxy-LTB₄)	$C_{20}H_{32}O_5$	352	LTB₄; LTB₄(i.v.); 8,15-DHETE (i.v.)	Rat liver[b,171,186] Human leukocytes[b,173,194-196] Rat pleural exudates[181] Monkey plasma[190] Monkey urine[190] Rabbit urine[190] Rabbit plasma[190]	Similar bronchoconstrictor activity to LTB₄ (guinea pig lung strips)[195] Approximately 50 times less potent than LTB₄ in promoting leukocyte adhesion to endothelial cells (hamster cheek pouch);[195] Chemotactic agent for human polymorphonuclear leukocytes[194]	ω-Hydroxylase	May be mediated by P-450 (rat liver microsomes) Leukotactic and chemotactic actions Inhibited by ETYA (eicosatetraynoic acid) and NDGA (nordihydroguaiaretic acid).[194]
	182	5(S),12(R)-Dihydroxy-6-cis,8-trans,10-trans,14-cis-eicosatetraen-1,20-dioic acid. (20-Carboxy-LTB₄)	$C_{20}H_{30}O_6$	368	LTB₄	Rat liver[b,171] Human neutrophils[b,173] Human polymorphonuclear leukocytes[b,194]	Similar bronchoconstrictor activity to LTB₄ (Guinea pig lung strips);[195] Approximately 50 times less potent than LTB₄ in promoting leukocytes adhesion to endothelial cells (hamster cheek pouch)[195] Chemotactic agent to human polymorphonuclear leukocytes[194]	ω-Hydroxylase	Leukotactic and Chemotactic activity inhibited by ETYA and NDGA.[194]
	183	5(S),12(R),19-Trihydroxy-6-cis,8-trans,10-trans,14-cis-eicosatetraenoic acid. (19-Hydroxy-LTB₄)	$C_{20}H_{32}O_5$	352	LTB₄	Rat liver[b,186]			May be mediated by P-450 (rat liver microsomes)[186]
	184	5(S),12(R),20-and 5(S),12(S),20-Trihydroxy-6-trans,8-trans,10-trans,14-cis-eicosatetraenoic acid.	$C_{20}H_{34}O_5$	354	6-trans-LTB₄ and 12-epi-6-trans-LTB₄	Human leukocytes[b,221]		ω-Hydroxylase	

No.	Compound	Formula	MW	Related / precursor	Enzyme	Source	Biological activity / Remarks
185	(20-Hydroxy-6-*trans*-LTB₄ and 20-hydroxy-12-epi-6-*trans*-LTB₄)			6-*trans*-LTB₄ and 12-epi-6-*trans*-LTB₄	ω-Hydroxylase	Human leukocytes[b,221]	
186	5(S),12(R) and 5(S),12(S)-Dihydroxy-6-*trans*,8-*trans*,10-*trans*,14-*cis*-eicosatetraen-1,20-dioic acid. (20-Carboxy-6-*trans*-LTB₄ and 20-Carboxy-12-epi-6-*trans*-LTB₄)	$C_{20}H_{30}O_6$	368				
	5(S),12(S)-Dihydroxy-6-*trans*,8-*cis*,10-*trans*,14-*cis*-eicosatetraenoic acid. (5(S),12(S)-DHETE; 5(S),12(S)-6-*trans*-8-*cis*-LTB₄)	$C_{20}H_{32}O_4$	336	5-HETE	12-Lipoxygenase	Human platelet sonicates[229]	Minimal polymorphonuclear leukocyte aggregating activity[229]
187	5,12,20-Trihydroxy-6,8,14-eicosatrienoic acid or 5,12,20-Trihydroxy-8,10,14-eicosatrienoic acid. (5,12,20-THETrE)	$C_{20}H_{34}O_5$	354	6-*trans*-LTB₄	Δ⁶-or Δ¹⁰-reductase	Human leukocytes[220]	Major metabolite
188	5(S)-Hydroxy-6(R)-S-glutathionyl-7-*trans*,9-*trans*,11-*cis*,14-*cis*-eicosatetraenoic acid. (LTC₄)	$C_{30}H_{47}O_9N_3S$	625	LTA₄, LTD₄	Glutathione-S-transferase (LTD₄ to LTC₄ is catalyzed by same enzyme in presence of low concentration of glutathione)[235]	Rat kidney[b,176,235,257] Human lung[b,180] Guinea pig lung[192]	Potent vasoconstrictor and bronchoconstrictor;[222] Negative inotropic effect in cardiac contractions;[222] Selective peripheral airway agonist (more potent than histamine);[237] Increases vascular permeability and stimulates mucus secretion;[223,237] Direct hypotensive effect on systemic arteries;[231] Potently induces guinea pig ileum contractions (SRS-like)[236] Stimulates luteinizing hormone release from rat anterior pituitary cells[222] — Component of SRS-A[222] FPL-55712 does not inhibit LTC₄, induced contraction of guinea pig parenchymal strips;[237] Indomethacin blocks bronchoconstrictor effects of i.v.-administered LTC₄ (so effect is probably due to TXA₂ release) but enhances bronchoconstrictor effect of aerosol-administered LTC₄;[222] LTC₄ and LTD₄ probably bind to separate receptors.[256]

Table 3 (continued)
LIPOXYGENASE METABOLITES

	Structure	Chemical name	Molecular formula	Mol wt	Substrate	Source	Biological action	Presumed catalytic route	Comments
189		5(S)-Hydroxy-6(R)-S-glutathionyl-7-trans, 9-trans-11-trans,14-cis-eicosatetraenoic acid. (11-trans-LTC₄)	$C_{30}H_{47}O_9N_3S$	625	LTC₄ (i.v.)	Rabbit brain[172] Rat urine[177]		Δ^{11} isomerization (after addition of thiyl radicals).[264]	
190		5(S)-Hydroxy-6(R)-S-glutathionyl-7-trans, 9-trans,11-cis,14-cis-eicosatetraenoic acid sulfoxides. (S-Diastereomeric-LTC₄ sulfoxides)	$C_{30}H_{47}O_{10}N_3S$	641	LTC₄	Human polymorphonuclear leukocytes[b,193] Rat lung[b,187]	Little or no spasmogenic activity[193] (1% of activity of LTC₄ to guinea pig ileum)[200] Immunoreactive to LTC₄.[200]	H₂O₂/myeloperoxidase followed by hydrolysis (S-chlorosulfonium ion is probable intermediate)	Biosynthesis inhibited by catalase, azide.
191		5(S),15(S)-Dihydroxy-6(R), S-glutathionyl-7-trans, 9-trans,13-trans,11-cis-eicosatetraenoic acid. (15-Hydroxy-Δ^{13}-trans-LTC₃)	$C_{30}H_{47}O_{10}N_3S$	641	LTC₄	Mouse liver[b,197]	Contraction of guinea pig ileum (less potent than LTC₄).[197]	Lipoxygenase and peroxidase	
192		5(S)-Hydroxy-6(R)-S-cysteinylglycyl-7-trans, 9-trans,-11-cis,14-cis-eicosatetraenoic acid. (LTD₄)	$C_{25}H_{40}O_6N_2S$	496	LTC₄; LTC₄ (i.v.); LTA₄	Rat kidney[b,168,176,235,257] Rat liver[b,174] Rat peritoneal cells[b,188] Rat urine[177] Guinea pig trachea[b,189] Human lung[b,180,182] Ferret lung[b,181] Guinea pig lung[b,183,192] Rat lung[b,187]	Potent vasoconstrictor and bronchoconstrictor;[222,223] Induces guinea pig ileum contractions (SRS-like);[212,236] Selective peripheral airway agonist (more potent than LTC₄ or histamine);[237] Direct hypotensive effects on systemic arteries;[237] Increases vascular permeability;[223,237] and stimulates mucus secretion;[223] Negative inotropic effect on cardiac contraction.[222]	γ-glutamyl transpeptidase	Component of SRS-A; (the major SRS-A of human lung);[236] Conversion occurs even in liver cells deficient in uptake of cysteinyl leukotrienes;[174] Reaction reversible and accompanied by γ-glutamyl amino acid formation[235] FPL-55712 inhibits contractile response (by low concentrations of LTD₄) of lung parenchymal strips[256]

No.	Structure	Name	Formula	MW		Source	Mechanism	Notes
193		5(S)-Hydroxy-6(R)-S-cysteinylglycyl-7-*trans*,9-*trans*,11-*trans*,14-*cis*-eicosatetraenoic acid. (11-*trans*-LTD₄)	$C_{25}H_{40}O_6N_2S$	496	LTC₄ (i.v.)	Rat urine[77]	Δ¹¹-isomerization followed by γ-glutamyl transpeptidase	
194		5(S)-Hydroxy-6(R)-S-cysteinylglycyl-7-*trans*,9-*trans*-11-*cis*,14-*cis*-eicosatetraenoic acid sulfoxides. (S-Diastereomeric-LTD₄ sulfoxides)	$C_{25}H_{40}O_7N_2S$	512	LTD₄	Human polymorphonuclear leukocytes[b,193]	H₂O₂/myeloperoxidase (S-chlorosulfonium ion is probable intermediate)[193]	No spasmogenic activity[193]
195		5(S)-Hydroxy-6(R)-S-cysteinylglycyl-7-*trans*,9-*trans*,11-*trans*,14-*cis*-eicosatetraenoic acid sulfoxides. (S-Diastereomeric 11-*trans*-LTD₄ sulfoxides)	$C_{25}H_{40}O_7N_2S$	512	LTD₄	Human polymorphonuclear leukocytes[b,193]	H₂O₂/myeloperoxidase (S-chlorosulfonium ion is probable intermediate)[193,200]	No spasmogenic activity[193]. Immunoreactive to LTD₄[193]
196		5(S)-Hydroxy-6(R)-S-cysteinyl-7-*trans*,9-*trans*,11-*cis*,14-*cis*-eicosatetraenoic acid (LTE₄)	$C_{23}H_{37}O_5N\,S$	439	LTD₄;LTC₄; LTA₄; LTC₅(i.v.)	Rat kidney[b,161,176,235,257]; Rat liver[b,174]; Human urine[77,184]; Human lung[b,180,182]; Rat urine[77]; Ferret lung[b,182]; Rat lung[b,182]; Guinea pig lung[b,182]; Porcine kidney[b,212]; Guinea pig alveolar macrophage[b,183]; Guinea pig trachea[b,212]	(from LTD₄): LTD₄-dipeptidase	Potent bronchoconstrictor and vasoconstriction;[222] Contraction of guinea pig tracheal smooth muscle more potent than LTC₄ or LTD₄;[256] Increases vascular permeability and stimulates mucus secretion[222,223] Induces guinea pig ileum contractions (SRS-like), less potent than LTC₄[212,238] Negative inotropic effects on cardiac contractions[222] — Component of SRS-A[212,238] Major metabolite of LTC₄ in human urine; Conversion occurs even in liver cells deficient in uptake of cysteinyl leukotrienes[174] LTD₄ and LTE₄ probably bind to separate receptors
197		5(S)-Hydroxy-6(R)-S-cysteinylglycyl-7-*trans*,9-*trans*,11-*trans*,14-*cis*-eicosatetraenoic acid. (11-*trans*-LTE₄)	$C_{23}H_{37}O_5N\,S$	439	11-*trans*-LTD₄; LTC₄(i.v.)	Porcine kidney[b,212]; Rat urine[77]	Dipeptidase	Induced contraction of guinea pig ileum (SRS-like) and guinea pig parenchymal strips (less potent than LTE₄[238] and much less potent than LTC₄);[212] Increased permeability of cutaneous vasculature (less potent than LTE₄[238]

Table 3 (continued)
LIPOXYGENASE METABOLITES

	Structure	Chemical name	Molecular formula	Mol wt	Substrate	Source	Biological action	Presumed catalytic route	Comments
198		5(S)-Hydroxy-6(R)-S-cysteinyl-7-trans,9-trans,11-cis,14-cis-eicosatetraenoic acid sulfoxides. (S-diastereomeric LTE$_4$ sulfoxides)	C$_{23}$H$_{37}$O$_6$N S	455	LTE$_4$	Human polymorphonuclear leukocytes[b,193]	(No spasmogenic activity)	H$_2$O$_2$/myeloperoxidase (S-chlorosulfonium ion is probable intermediate[193])	
199		5-Hydroxy-6(S)-(2-acetamido-3-thiopropionyl)-7-trans,9-trans,11-cis,14-cis-eicosatetraenoic acid. (N-Acetyl-LTE$_4$)	C$_{25}$H$_{39}$O$_6$N S	481	LTE$_4$ (i.v.) LTE$_4$	Rat liver[170] Rat liver[b,175] Rat kidney[b,175] Rat spleen[b,175] Rat skin[b,175] Rat lung[b,175] Rat urine[177] Rat feces[177]			Major metabolite (rat liver)
200		5-Hydroxy-6-S-(2-acetamido-3-thiopropionyl)-7-trans,9-trans,11-trans,14-cis-eicosatetraenoic acid. (N-Acetyl-11-trans-LTE$_4$)	C$_{25}$H$_{39}$O$_6$N S	481	LTC$_4$ (i.v.)	Rat urine[177] Rat feces[179]			
201		5(S)-Hydroxy-6(R)-S-cysteinyl-γ-glutamyl-7-trans,9-trans,11-cis,14-cis-eicosatetraenoic acid. (LTF$_4$)	C$_{28}$H$_{44}$O$_8$N$_2$ S	568	LTE$_4$	Porcine kidney[b,233] Rat kidney[b,235]	Induced contraction of isolated guinea pig ileum (less potent than LTE$_4$)[233]	γ-glutamyl-transpeptidase	
202		14(15)-Epoxy-5-cis,8-cis,10-trans,12-trans-eicosatetraenoic acid. (14(15)-LTA$_4$)	C$_{20}$H$_{30}$O$_3$	318	15-HpETE	Rabbit reticulocytes[b,199] Human leukocytes[b,231,232] Porcine leukocytes[b,263]		12-Lipoxygenase (as 14(15)-LTA synthase)	
203		14(R),15(S)-and 14(S),15(S)-Dihydroxy-5-cis,8-cis,10-trans,12-trans-eicosatetraenoic acid. (14,15-LTB$_4$)	C$_{20}$H$_{32}$O$_4$	336	15-HpETE	Human leukocytes[b,231]	Inhibits NK cell cytotoxicity and superoxide generation[223]	Nonenzymic hydrolysis of 14(15)-LTA$_4$	

No.	Name	Formula	MW	Precursor	Source	Biological activity	Enzyme
204	15-Hydroxy-14-S-glutathionyl-5-cis,8-cis,10-trans,12-trans-eicosatetraenoic acid. (14,15-LTC₄)	$C_{30}H_{47}O_9N_3S$	625	15-HpETE	Human leukocytes[b,232]	Contracts guinea pig ileum[232]	Glutathione-S-transferase (from 14(15)-LTA₄)
205	15-Hydroxy-14-S-cysteinylglycyl-5-cis,8-cis,10-trans,12-trans-eicosatetraenoic acid. (14,15-LTD₄)	$C_{25}H_{40}O_6N_2S$	496	15-HpETE	Human leukocytes[b,232]	Contracts guinea pig ileum[232]	γ-glutamyl transpeptidase (from 14,15-LTC₄)
206	8(R),15(S)-and 8(S),15(S)-Dihydroxy-5-cis,9-trans,11-trans,13-trans-eicosatetraenoic acid. (8,15-LTB₄)	$C_{20}H_{32}O_4$	336	15-HpETE	Human leukocytes[b,231] Porcine leukocytes[b,263]		12-Lipoxygenase (via 14(15)-LTA₄)
207	5,6-Dihydroxy-7-trans,9-trans,11-cis,14-cis-eicosatetraenoic acid. (5,6-DHETE)	$C_{20}H_{32}O_4$	336	LTA₄, 5-HpETE	Human lung[b,180] Porcine leukocytes[b,198] Mouse liver[b,225]		Nonenzymic hydrolysis; A separate 5,6-DHETE is formed by mouse liver cytosolic epoxide hydrolase[225]
208	5(S),12(S)-Dihydroperoxy-6-trans,8-cis,10-trans,14-cis-eicosatetraenoic acid. (5(S),12(S)-DHpETE)	$C_{20}H_{32}O_6$	368	12-HpETE	Porcine leukocytes[b,98]		Arachidonate 5-lipoxygenase
209	5(S),15(S)-Dihydroperoxy-6-trans,8-cis,11-cis,13-trans-eicosatetraenoic acid. (5,15-DHpETE)	$C_{20}H_{32}O_6$	368	15-HpETE	Porcine leukocytes[b,198]		Arachidonate 5-lipoxygenase
210	12(S),20-Dihydroxy-5-cis,8-cis,10-trans,14-cis-eicosatetraenoic acid. (12,20-DHETE)	$C_{20}H_{32}O_4$	336	12(S)-HETE	Human neutrophils[230]		ω-hydroxylase
211	5(S),12(S),20-Trihydroxy-6-trans,8-cis,10-trans,14-cis-eicosatetraenoic acid. (5(S),12(S),20-THETE)	$C_{20}H_{32}O_5$	352	5(S),12(S)-DHETE	Human leukocytes[231]		ω-hydroxylase

Table 3 (continued)
LIPOXYGENASE METABOLITES

	Structure	Chemical name	Molecular formula	Mol wt	Substrate	Source	Biological action	Presumed catalytic route	Comments
212		3-Hydroxy-11(12)-epoxy-5-cis,9-trans,14-cis-eicosatetraenoic acid. (Hepoxilin A$_3$)	$C_{20}H_{32}O_4$	336	12-HpETE	Rat lung[b, 202, 206] Rat pancreatic islets[b, 205]	Insulin secretagogue[b, 205]	Catalyzed by hemin and hemoglobin (pancreas)	
213		10-Hydroxy-11(12)-epoxy-5-cis,8-cis,14-cis-eicosatetraenoic acid. (Hepoxilin B$_3$)	$C_{20}H_{32}O_4$	336	12-HpETE			Catalyzed by hemin and hemoglobin (pancreas)	
214		8,11,12-Trihydroxy-5-cis,9-trans,14-cis-eicosatetraenoic acid. (Trioxilin A$_3$)	$C_{20}H_{34}O_5$	354	12-HpETE	Rat lung[202, 206]		Epoxide hydratase	
215		10,11,12-Trihydroxy-5-cis,8,cis,14-cis-eicosatrienoic acid. (Trioxilin B$_3$)	$C_{20}H_{34}O_5$	354	12-HpETE	Rat lung[b, 202, 206]		Epoxide hydratase	
216		5,6,15-Trihydroxy-7-trans,9-trans,11-cis,13-trans-eicosatetraenoic acid. (Lipoxin A)	$C_{20}H_{32}O_5$	352	15-HpETE; 5,15-DHETE	Human leukocytes[b, 178, 210] Human reticulocytes[b, 207]	In human neutrophils, stimulates superoxide anion generation and degranulation with no aggregation response;[178] Stimulates contraction of pulmonary tissue and extravasation of plasma from postcapillary venules[178] Contracts lung strips (slow onset and long lasting),[178] Inhibits NK cell activity against K562 target cells (no effect on cyclic AMP levels or target cell binding)[178]	5-Lipoxygenase; epoxidase	
217		5(S),14(R),15(S)-Trihydroxy-6-trans,8-cis,10-trans,12-trans-eicosatetraenoic acid. (Lipoxin B)	$C_{20}H_{32}O_5$	352	15-HpETE	Human leukocytes[b, 178, 210] Human reticulocytes[b, 207]	Inhibits NK cell activity against K562 target cells (no effect on cyclic AMP levels or target cell binding)[178]	5-Lipoxygenase; epoxidase	Inhibited by ETYA, NDGA, anaerobiosis

No.	Structure	Name	Formula	MW	Derivative name	Source	Enzyme	Comments
218		5(S),14(R),15(S)- and 5(S),14(S),15(S)-Trihydroxy-6-*trans*,8-*trans*,10-*trans*,12-*trans*-eicosatetraenoic acid. (8-*trans*-Lipoxin B and 14(S)-8-*trans*-lipoxin B)	$C_{20}H_{32}O_5$	352	15-HpETE	Human leukocytes[b,210]	Epoxidase	
219		5(R),6(R)- and 5(S),6(S)-6(9)-Oxy-5,11,15-trihydroxyprosta-13-*trans*-enoic acid. (5-Hydroxy-PGI₁,β and 5-Hydroxy PGI₁α)	$C_{20}H_{34}O_6$	370	5(6)-EpETrE	Ram seminal vesicles[191]	Cyclooxygenase and epoxidase	
220		10-Hydroxy-11(12)-epoxy-8-*trans*-heptadecenoic acid. (10,11(12)-HEpHE)	$C_{17}H_{30}O_4$	298	12-HpHDE (12-Hydroperoxy-heptadecadienoic acid)	Fetal calf aorta[b,211]	DGLA cyclo-oxygenase	
221		10,11,12-Trihydroxy-8-*trans*-heptadecenoic acid. (10,11,12-THHE)	$C_{17}H_{32}O_5$	316	12-HpDHE; 10,11(12)-HEpHE	Fetal calf aorta[b,211]	Epoxide hydrolase	
222		5(S),6(R)-Dihydroxy-7-*trans*,9-*trans*,11-*cis*-eicosatrienoic acid. (LTB₃)	$C_{20}H_{34}O_4$	338	LTA₃	Human neutrophils[b,203]	LTA-hydrolase	Poor conversion

Table 3 (continued)
LIPOXYGENASE METABOLITES

	Structure	Chemical name	Molecular formula	Mol wt	Substrate	Source	Biological action	Presumed catalytic route	Comments
223	H OH COOH / CH CONHCH₂ COOH / NH CO(CH₂)₂ CH COOH / NH₂ / S–CH₂	5(S)-Hydroxy-6(R)-S-glutathionyl-7-*trans*, 9-*trans*,11-*cis*-eicosatrienoic acid. (LTC₃)	$C_{30}H_{46}O_9N_3S$	627	LTD₃	Guinea pig lung[b,203] Guinea pig liver[b,203]	Contracts smooth muscle and guinea pig ileum (approximately equipotent with LTC₄)[218,219]	γ-glutamyl transpeptidase reversal by high tissue concentration of glutathione	Reaction does not occur in guinea pig liver and kidney *in vitro*[203] FPL-55712 inhibits LTD₃-induced SRS activity
224	H OH COOH / CHCONHCH₂COOH / NH₂ / S–CH₂	5(S)-Hydroxy-6(R)-cysteinylglycyl-7-*trans*, 9-*trans*,11-*cis*-eicosatrienoic acid. (LTD₃)	$C_{25}H_{42}O_6N_2S$	498	LTC₃ (i.v.) LTC₃	Guinea pig lung[b,203] Mouse liver[209] Mouse bile[209] Mouse small intestine[209] Mouse lung[209] Rat kidney[b,234]	Contracts smooth muscle[209]	γ-glutamyl transpeptidase	Not formed in guinea pig liver and kidney (inhibition of γ-glutamyl transpeptidase by endogenous[203] glutathione) Inhibited by anthglutin[234]
225	H OH COOH / CHCOOH / NH₂ / S–CH₂								

No.	Name	Formula	MW	Precursor/Derivative	Source	Biological activity	Enzyme/Catalysis	
226	5(S)-Hydroxy-6(R)-S-cysteinyl-7-*trans*,9-*trans*,11-*cis*-eicosatrienoic acid. (LTE₃)	$C_{23}H_{38}O_5NS$	440	LTC₃ (i.v.) LTD₃	Monkey blood[208] Mouse liver[209] Mouse bile[209] Mouse small intestine[209] Mouse kidney[209] Mouse lung[209] Porcine kidney[b,211] Rat kidney[b,234] Guinea pig kidney[b,203] Guinea pig liver[b,203]		LTD₃-dipeptidase	Inhibited by anthglutin[234]
227	5(S),12(S)-Dihydroxy-6-*trans*,8-*cis*,10-*trans*,14-*cis*,17-*cis*-eicosapentaenoic acid. (5,12-DHEPE; 5(S),12(S)-6-*trans*-8-*cis*-LTB₅)	$C_{20}H_{30}O_4$	334	5-HEPE	Human platelet sonicates[229]	Minimal polymorphonuclear leukocyte aggregating activity[229]	12-Lipoxygenase	
228	5(S)-Hydroxy-6(R)-S-cysteinyl-7-*trans*,9-*trans*,11-*cis*,14-*cis*,17-*cis*-eicosapentaenoic acid. (LTE₅)	$C_{23}H_{35}O_5$	391	LTD₅	Porcine kidney[b,212]		γ-glutamyl-transpeptidase	
229	5(S),15(S)-Dihydroperoxy-6-*trans*,8-*cis*,11-*cis*,13-*trans*,17-*cis*-eicosapentaenoic acid. (5,15-DHpEPE)	$C_{20}H_{30}O_6$	368	15-HpEPE	Porcine leukocytes[b,227]		5-lipoxygenase	
230	8-Hydroxy-11(12)-epoxy-5-*cis*,9-*trans*,14-*cis*,17-*cis*-eicosatetraenoic acid. (Hepoxilin A₄)	$C_{20}H_{30}O_4$	334	12-HpEPE	Rat platelets[b,201]	Insulin secretagogue[201]	Catalyzed by hemoglobin and hematin	
231	10-Hydroxy-11(12)-epoxy-5-*cis*,8-*cis*,14-*cis*,17-*cis*-eicosatetraenoic acid. (Hepoxilin B₄)	$C_{20}H_{30}O_4$	334	12-HpEPE	Rat platelets[b,201]	Insulin secretagogue (more stable than hepoxilin A₄)[201]	Catalyzed by hemoglobin and hematin	
	8,11,12-Trihydroxy-5-*cis*,9-*trans*,14-*cis*,17-*cis*-eicosatetraenoic acid. (Trioxilin A₄)	$C_{20}H_{32}O_5$	352	12-HpEPE	Rat platelets[b,201] Rat lung[b,202,205]		Epoxide hydratase	

Table 3 (continued)
LIPOXYGENASE METABOLITES

	Structure	Chemical name	Molecular formula	Mol wt	Substrate	Source	Biological action	Presumed catalytic route	Comments
232		10,11,12-Trihydroxy-5-*cis*, 8-*cis*,14-*cis*,17-*cis*-eicosatetraenoic acid. (Trioxilin B₄)	$C_{20}H_{32}O_5$	352	12-HpEPE	Rat platelet[b, 201] Rat lung[b, 202, 206]		Epoxide hydratase	
233		5,6,15-Trihydroxy-7-*trans*, 9-*trans*,11-*cis*,13-*trans*, 17-*cis*-eicosapentaenoic acid. (Lipoxene A)	$C_{20}H_{30}O_5$	350	15-HpEPE	Porcine leuko-cytes[b, 227]		5-lipoxygenase; Epoxidase; Epoxide hydrolase	
234		5(S),14(S),15(S)-Trihydroxy-6-*trans*,8-*cis*, 10-*trans*,12-*trans*,17-*cis*-eicosapentaenoic acid. (Lipoxene B)	$C_{20}H_{30}O_5$	350	15-HpEPE	Porcine leuko-cytes[b, 227]		5-Lipoxygenase; epoxidase	

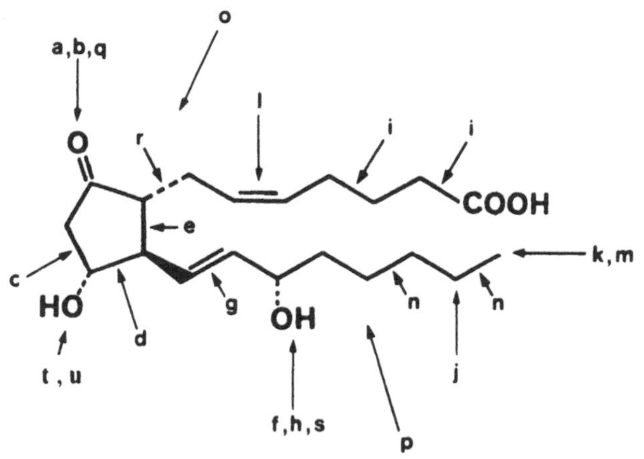

FIGURE 1.   Target sites of enzymes which metabolize prostaglandins.
For details of PG metabolism, see Tables 1 and 2.

REFERENCES

1. **Lee, S. C. and Levine, L.**, Prostaglandin metabolism. I. Cytoplasmic reduced nicotinamide adenine dinucleotide phosphate-dependent and microsomal reduced nicotinamide adenine. Dinucleotide-dependent prostaglandin E 9-ketoreductase activities in monkey and pigeon tissues, *J. Biol. Chem.*, 249, 1369—1375, 1974.

2. **Lee, S. C. and Levine, L.**, Prostaglandin metabolism. II. Identification of two 15-hydroxyprostaglandin dehydrogenase types, *J. Biol. Chem.*, 250, 548—552, 1975.

3. **Hamberg, M. and Samuelsson, B.**, Metabolism of prostaglandin E_2 in guinea pig liver, *J. Biol. Chem.*, 246, 1073—1077, 1971.

4. **Hensby, C. N.**, Reduction of prostaglandin E_2 to prostaglandin $F_{2\alpha}$ by an enzyme in sheep blood, *Biochim. Biophys. Acta*, 348, 145—154, 1974.

5. **Leslie, C. A. and Levine, L.**, Evidence for the presence of a prostaglandin E_2-9-keto reductase in rat organs, *Biochem. Biophys. Res. Commun.*, 52, 717—724, 1973.

6. **Lee, S. C. and Levine, L.**, Purification and properties of chicken heart prostaglandin delta 13-reductase, *Biochem. Biophys. Res. Commun.*, 61, 14—21, 1974.

7. **Lee, S. C. and Levine, L.**, Purification and regulatory properties of chicken heart prostaglandin 9-ketoreductase, *J. Biol. Chem.*, 250, 4549—4555, 1975.

8. **Lee, S. C., Pong, S. S., Katzen, D., Wu, K. Y., and Levine, L.**, Distribution of prostaglandin E 9-ketoreductase and types I and II 15-hydroxyprostaglandin dehydrogenase in swine kidney medulla and cortex, *Biochemistry*, 14, 142—145, 1975.

9. **Kaplan, L., Lee, S. C., and Levine, L.**, Partial purification and some properties of human erythrocyte prostaglandin 9-ketoreductase and 15-hydroxyprostaglandin dehydrogenase, *Arch. Biochem. Biophys.*, 167, 287—293, 1975.

10. **Stone, K. J. and Hart, M.**, Inhibition of renal PGE_2-9-ketoreductase by diuretics, *Prostaglandins*, 12(2), 197—207, 1976.

11. **Stone, K. J. and Hart, M.**, Prostaglandin-E_2-9-ketoreductase in rabbit kidney, *Prostaglandins*, 10, 273—288, 1975.

12. **Coceani, F. and Bishai, I.**, Demonstration of prostaglandin 15-hydroxy dehydrogenase (15-PGDH), delta 13-reductase (13-PGR), and 9-keto-reductase (9K-PGR) in the frog spinal cord, *Adv. Prostaglandin Thromboxane Res.*, 2, 836—837, 1976.

13. **Pace-Asciak, C. and Miller, D.**, Prostaglandins during development. II. Identification of prostaglandin 9-hydroxy dehydrogenase activity in adult rat kidney homogenates, *Experientia*, 30, 590—592, 1974.

14. **Pace-Asciak, C. and Cole, S.**, Inhibitors of prostaglandin catabolism. I. Differential sensitivity of 9-PGDH, 13-PGR and 15-PGDH to low concentrations of indomethacin, *Experientia*, 31, 143—145, 1975.

15. **Pace-Asciak, C.,** Activity profiles of prostaglandin 15- and 9-hydroxydehydrogenase and 13-reductase in the developing rat kidney, *J. Biol. Chem.,* 250, 2795—2800, 1975.

16. **Pace-Asciak, C. and Domazet, Z.,** 9-Hydroxyprostaglandin dehydrogenase activity in the adult rat kidney, regional distribution and sub-fractionation, *Biochim. Biophys. Acta,* 380, 338—343, 1975.

17. **Polet, H. and Levine, L.,** Metabolism of prostaglandins, E, A, and C, in serum, *J. Biol. Chem.,* 250, 351—357, 1975.

18. **Polet, H. and Levine, L.,** Serum prostaglandin A1 isomerase, *Biochem. Biophys. Res. Commun.,* 45(5), 1169—1176, 1971.

19. **Polet, H. and Levine, L.,** Partial purification and characterization of prostaglandin A isomerase from rabbit serum, *Arch. Biochem. Biophys.,* 168, 96—103, 1975.

20. **Jones, R. I., Cammock, S., and Horton, E. W.,** Partial purification and properties of cat plasma prostaglandin A isomerase, *Biochim. Biophys. Acta,* 208, 588—601, 1972.

21. **Anggard, E., Larsson, C., and Samuelsson, B.,** The distribution of 15-hydroxy prostaglandin dehydrogenase and prostaglandin-13-reductase in tissues of the swine, *Acta Physiol. Scand.,* 81, 396—404, 1971.

22. **Nakano, J., Montague, B., and Darrow, B.,** Metabolism of prostaglandin E₁ in human plasma, uterus, and placenta, in swine ovary and in rat testicle, *Biochem. Pharmacol.,* 20, 2512—2514, 1971.

23. **Granstrom, E.,** Metabolism of prostaglandin $F_{2\alpha}$ in guinea pig lung, *Eur. J. Biochem.,* 20, 451—458, 1971.

24. **Granstrom, E.,** Metabolism of prostaglandin $F_{2\alpha}$ in swine kidney, *Biochim. Biophys. Acta,* 239, 120—125, 1971.

25. **Saeed, S. A. and Roy, A. C.,** Purification of 15-hydroxy prostaglandin dehydrogenase from bovine lung, *Biochem. Biophys. Res. Commun.,* 47, 96—102, 1972.

26. **Braithwaite, S. S. and Jarabak, J.,** Studies on a 15-hydroxyprostaglandin dehydrogenase from human placenta, purification and partial characterization, *J. Biol. Chem.,* 250, 2315—2318, 1975.

27. **Thaler-Dao, H., Saintot, M., Baudin, G., Descomps, B., and Crastes de Paulet, A.,** Purification of the human placental 15-hydroxy prostaglandin dehydrogenase: properties of the purified enzyme, *FEBS. Lett.,* 48(2), 204—208, 1974.

28. **Schlegel, W., Demers, L. M., Hildebrandt, S. T., and Behrman, H. R.,** Partial purification of human placental 15-hydroxy-prostaglandin dehydrogenase: kinetic properties, *Prostaglandins,* 5, 417—433, 1974.

29. **Paulsrud, J. R. and Miller, O. N.,** Inhibition of 15-OH prostaglandin dehydrogenase by several diuretic drugs, *Fed. Proc.,* 33, 590, 1974.

30. **Limas, C. J. and Cohn, J. N.,** Regulation of myocardial prostaglandin dehydrogenase activity. The role of cyclic 3′,5′-AMP and calcium ions, *Proc. Soc. Exp. Biol. Med.,* 142, 1230—1234, 1973.

31. **Vonkeman, H., Nugteren, D. H., and Van Dorp, D. A.,** The action of prostaglandin 15-hydroxyde-hydrogenase on various prostaglandins, *Biochim. Biophys. Acta,* 187, 581—583, 1969.

32. **Anggard, E. and Samuelsson, B.,** Purification and properties of a 15-hydroxyprostaglandin dehydrogenase from swine lung, *Ark. Kem.,* 25, 293—300, 1966.

33. **Nakano, J., Anggard, E., and Samuelsson, B.,** 15-Hydroxy-prostanoate dehydrogenase. Prostaglandins as substrates and inhibitors, *Eur. J. Biochem.,* 11, 386—389, 1969.

34. **Marrazzi, and Matschinsky,** *Pharmacologist,* 13, 292, 1971.

35. **Crutchley, D. J. and Piper, P. J.,** Prostaglandin inactivation in guinea pig lung and its inhibition, *Br. J. Pharmacol.,* 52, 197—203, 1974.

36. **Hansen, H. S.,** Inhibition by indomethacin and aspirin of 15-hydroxyprostaglandin dehydrogenase in vitro, *Prostaglandins,* 8, 95—105, 1974.

37. **Sun, F. F., Armour, S. B., Bockstanz, V. R., and McGuire, J. C.,** Studies on 15-heteroxyprostaglandin dehydrogenase from monkey lung, *Adv. Prostaglandin Thromboxane Res.,* 1, 163—169, 1976.

38. **Tai, H. H., Tai, C. L., and Hollander, C. S.,** Regulation of prostaglandin metabolism: inhibition of 15-hydroxyprostaglandin dehydrogenase by thyroid hormones, *Biochem. Biophys. Res. Commun.,* 57, 457—468, 1974.

39. **Larsson, C. and Anggard, E.,** Formation and metabolism of prostaglandins in the rabbit kidney. Regional differences of the enzymes involved, *Acta Pharmacol. Toxicol.,* (KBH) 31, 107, 1972.

40. **Johnson, M., Davison, P., and Ramwell, P. W.,** Carnitine-dependent beta-oxidation of prostaglandins, *J. Biol. Chem.,* 247, 5656—5658, 1972.

41. **Nakano, J. and Morsy, N. H.,** Beta-Oxidation of prostaglandins E₁ and E₂ in rat lung and kidney homogenate, *Clin. Res.,* 19, 142, 1971.

42. **Israelsson, U., Hamberg, M., and Samuelsson, B.,** Biosynthesis of 19-hydroxy-prostaglandin A₁, *Eur. J. Biochem.,* 11, 390—394, 1969.

43. **Hamberg, M. and Israelsson, U.,** Metabolism of prostaglandin E₂ in guinea pig liver. I., *J. Biol. Chem.,* 245, 5107—5114, 1970.

44. **Anggard, E. and Larsson, C.,** The sequence of the early steps in the metabolism of prostaglandin E₁, *Eur. J. Pharmacol.,* 14, 66—70, 1971.

45. **Anggard, E. and Samuelsson, B.,** Metabolism of prostaglandin E₁ in guinea pig lung: the structure of two metabolites, *J. Biol. Chem.,* 239, 4097—4102, 1964.

46. **Samuelsson, B.,** Prostaglandins and related factors. XXVII. Synthesis of tritium-labeled prostaglandin E₁ and studies on its distribution and excretion in the rat, *J. Biol. Chem.,* 239, 4091—4096, 1964.

47. **Anggard, E., Green, K., and Samuelsson, B.,** Synthesis of tritium-labeled prostaglandin E₂ and studies on its metabolism in guinea pig lung, *J. Biol. Chem.,* 240, 1932—1940, 1965.

48. **Hamberg, M. and Samuelsson, B.,** On the metabolism of prostaglandins E₁ and E₂ in man, *J. Biol. Chem.,* 246(22), 6713—6721, 1971.

49. **Anggard, E. and Samuelsson, B.,** The metabolism of prostaglandin E₃ in guinea pig lung, *Biochemistry,* 4, 1864—1871, 1965.

50. **Taylor, P. L. and Kelly, R. W.,** 19-Hydroxylated E prostaglandins as the major prostaglandins of human semen, *Nature (London),* 250, 665—667, 1974.

51. **Hamberg, M.,** Metabolism of prostaglandins in rat liver mitochondria, *Eur. J. Biochem.,* 6, 135—146, 1968.

52. **Hamberg, M. and Wilson, M.,** Structures of new metabolites of prostaglandin E₂ in man, *Adv. Biosci.,* 9, 39—48, 1973.

53. **Green, K.,** Metabolism of prostaglandin E₂ in the rat, *Biochemistry,* 10, 1072—1086, 1971.

54. **Hamberg, M. and Samuelsson, B.,** The structure of the major urinary metabolite of prostaglandin E₂ in man, *J. Am. Chem. Soc.,* 91, 2177—2178, 1969.

55. **Pace-Asciak, C., Morawska, K., and Wolfe, L. S.,** Metabolism of prostaglandin F₁α by the rat stomach, *Biochim. Biophys. Acta,* 218, 288—295, 1970.

56. **Granstrom, E. and Samuelsson, B.,** The structure of the main urinary metabolite of prostaglandin F₂α in the guinea pig, *Eur. J. Biochem.,* 10, 411—418, 1969.

57. **Granstrom, E.,** On the metabolism of prostaglandin F₂α in female subjects: structures of two metabolites in blood, *Eur. J. Biochem.,* 27, 462—469, 1972.

58. **Granstrom, E., Inger, U., and Samuelsson, B.,** The structure of a urinary metabolite of prostaglandin F₁α in the rat. XXIX. Prostaglandins and related factors, *J. Biol. Chem.,* 240, 457—461, 1965.

59. **Green, K.,** Structures of urinary metabolites of prostaglandin F₂α in the rat, *Acta Chem. Scand.,* 23, 1453—1455, 1969.

60. **Green, K.,** Metabolism of prostaglandin F₂α in the rat, *Biochim. Biophys. Acta,* 231, 419—444, 1971.

61. **Sun, F. F. and Stafford, J. E.,** Metabolism of prostaglandin F₂α in rhesus monkeys, *Biochim. Biophys. Acta,* 369(1), 95—110, 1974.

62. **Dimov, V. and Green, K.,** The metabolism of prostaglandin F₃α in the rat, *Biochem. Biophys. Acta,* 306, 257—269, 1973.

63. **Granstrom, E. and Samuelsson, B.,** On the metabolism of prostaglandin F₂α in female subjects. II. Structures of six metabolites, *J. Biol. Chem.,* 246, 7470—7485, 1971.

64. **Granstrom, E.,** Structures of C14 metabolites of prostaglandin F₂α, *Adv. Biosci.,* 9, 49—60, 1973.

65. **Granstrom, E. and Samuelsson, B.,** Structure of a deoxyprostaglandin in man, *J. Am. Chem. Soc.,* 94, 4380—4381, 1972.

66. **Svanborg, K. and Bygdeman, M.,** Metabolism of prostaglandin F₂α in the rabbit, *Eur. J. Biochem.,* 28, 127—135, 1972.

67. **Sun, F. F.,** Metabolism of prostaglandin in F₂α in the rat, *Biochim. Biophys. Acta,* 348, 249—262, 1974.

68. **Hamberg, M. and Samuelsson, B.,** On the metabolism of prostaglandins E₁ and E₂ in the guinea pig, *J. Biol. Chem.,* 247, 3495—3502, 1972.

69. **Hamberg, M. and Samuelsson, B.,** The structure of a urinary metabolite of prostaglandin E₂ in the guinea pig, *Biochem. Biophys. Res. Commun.,* 34, 22—27, 1969.

70. **Samuelsson, B., Granstrom, E., Green, K., and Hamberg, M.,** Metabolism of prostaglandins, *Ann. N.Y. Acad. Sci.,* 180, 138—163, 1971.

71. **Granstrom, E. and Samuelsson, B.,** The structure of a urinary metabolite of prostaglandin F₂α in man, *J. Am. Chem. Soc.,* 91, 3398—3400, 1969.

72. **Granstrom, E. and Samuelsson, B.,** On the metabolism of prostaglandin F₂α in female subjects, *J. Biol. Chem.,* 246, 5254—5263, 1971.

73. **Granstrom, E.,** On the metabolism of prostaglandin F₂α in female subjects. Structure of two C14 metabolites, *Eur. J. Biochem.,* 25, 581—589, 1972.

74. **Attallah, A., Duchesne, M. J., Osawa, Y., and Lee, J. B.,** Metabolism of PGA₁: isolation and characterization of the major renal metabolites, *Adv. Prostaglandin Thromboxane Res.,* 2, 860, 1976.

75. **Hamberg, M. and Samuelsson, B.,** Prostaglandins in human seminal plasma, *J. Biol. Chem.,* 241, 257—263, 1966.

76. **Jones, R. L. and Cammock, S.,** Purification, properties, and biological significance of prostaglandin A isomerase, *Adv. Biosci.,* 9, 61—70, 1973.

77. **Jones, R. L.,** 15-Hydroxy-9-oxoprosta-11,13-dienoic acid as the product of a prostaglandin isomerase, *J. Lipids Res.,* 13, 511—518, 1972.

78. **Jones, R. L.,** Properties of a new prostaglandin, *Br. J. Pharmacol.,* 45, 144P-145P, 1972.

79. **Jones, R. L.,** A prostaglandin isomerase in cat plasma, *Biochem. J.,* 119, 64P—65P, 1970.

80. **Anggard, E.,** The biological activities of three metabolites of prostaglandin E_1, *Acta Physiol. Scand.,* 66, 509—510, 1966.

81. **Nakano, J.,** Relationship between the chemical structure of prostaglandins and their vasoactivities in dogs, *Br. J. Pharmacol.,* 44, 63—70, 1972.

82. **Wassermann, M. A.,** Bronchopulmonary effects of prostaglandin $F_{2\alpha}$ and three of its metabolites in the dog, *Prostaglandins,* 9(6), 959—973, 1975.

83. **Pike, J. E., Kupiecki, F. P., and Weeks, J. R.,** Biological activity of the prostaglandins and related analogs, in *Prostaglandins: Proceedings of the Second Nobel Symposium, Stockholm,* Bergstron, S. and Samuelsson, B., eds., Wiley Interscience, New York, 1967, 161—171.

84. **Dawson, W., Lewis, R. L., McMahon, R. E., and Sweatman, W. J. F.,** Potent bronchoconstrictor activity of 15-keto prostaglandin $F_{2\alpha}$, *Nature (London),* 250, 331—332, 1974.

85. **Hensby, C. N.,** The enzymatic conversion of prostaglandin D_2 to prostaglandin $F_{2\alpha}$, *Prostaglandins,* 8, 369—375, 1974.

86. **Ziboh, V. A., Lord, J., and Penneys, N. S.,** Metabolism of PGE_2 by human and rat skin, *Adv. Prostaglandin Thromboxane Res.,* 2, 881—882, 1976.

87. **Boot, J. R., Dawson, W., and Harvey, J.,** Comparative biological activity of prostaglandin E_2 and its C20 metabolites on smooth muscle preparations, *Adv. Prostaglandin Thromboxane Res.,* 2, 958, 1976.

88. **Crutchley, D. J. and Piper, P. J.,** Comparative bioassay of prostaglandin E_2 and its three pulmonary metabolites, *Br. J. Pharmacol.,* 54, 397—399, 1975.

89. **McGiff, J. C., Terragno, N. A., Strand, J. C., Lee, J. B., Lonigro, A. J., and Ng, K. K. F.,** Selective passage of prostaglandins across the lung, *Nature (London),* 223, 742—745, 1969.

90. **Ferreira, S. H. and Vane, J. R.,** Prostaglandins: their disappearance from and release into the circulation, *Nature (London),* 216, 868—873, 1967.

91. **Vane, J. R.,** The release and fate of vasoactive hormones in the circulation, *Br. J. Pharmacol.,* 35, 209—242, 1969.

92. **Willis, A. L. and Cornelsen, M.,** Repeated injection of prostaglandin E_2 in rat paws induced chronic swelling and a marked decrease in pain threshold, *Prostaglandins,* 3, 353—357, 1973.

93. **Ferreira, S. H.,** Prostaglandins, aspirin-like drugs and analgesia, *Nature (London) New Biol.,* 240, 200—203, 1972.

94. **Kuhn, D. C. and Willis, A. L.,** Prostaglandin E_2, inflammation, and pain threshold in rat paws, *Br. J. Pharmacol.,* 49, 183P—184P, 1973.

95. **Williamson, H. E., Bourland, W. A., and Marchand, G. R.,** Inhibition of ethacrynic acid-induced increase in renal blood flow by indomethacin, *Prostaglandins,* 8, 297—301, 1974.

96. **Bailie, M. D., Barbour, J. A., and Hook, J. B.,** Effects of indomethacin on furosemide-induced changes in renal blood flow, *Proc. Soc. Exp. Biol. Med.,* 148(4), 1173—1176, 1975.

97. **Williamson, H. E., Bourland, W. A., and Marchand, G. R.,** Inhibition of furosemide-induced in renal blood flow by indomethacin, *Proc. Soc. Exp. Biol. Med.,* 148, 164—167, 1975.

98. **Alam, N. A., Clary, P., and Russell, P. T.,** Depressed placental prostaglandin E_1 metabolism in toxemia of pregnancy, *Prostaglandins,* 4, 363—370, 1973.

99. **Nakano, J. and Proncan, A. V.,** Metabolic degradation of prostaglandin E_1 in the lung and kidney of rats in endotoxin shock, *Proc. Soc. Exp. Biol. Med.,* 144, 506—508, 1973.

100. **Armstrong, J. M., Blackwell, G. J., Flower, R. J., McGiff, J. C., and Mullane, K.,** Possible contribution of prostaglandins to genetic hypertension in rats: identification of a biochemical lesion, *Br. J. Pharmacol.,* 55, 244P, 1975.

101. **Westbrook, C. and Jarabak, J.,** Purification and partial characterization of an NADH-linked delta 13-15-ketoprostaglandin reductase from human placenta, *Biochem. Biophys Res. Commun.,* 66, 541—546, 1975.

102. **Stone, K. J. and Hart, M.,** Inhibition of renal PGE_2-9-ketoreductase by diuretics, *Prostaglandins,* 12(2), 197—207, 1976.

103. **Levine, L., Wu, K.-Y., and Herrmann, H.,** Decreased levels of an inhibitor of prostaglandin E 9-keto reductase activity in chick dystrophic breast muscle, *Nature (London),* 260, 791—793, 1976.

104. **Oien, H. G., Ham, E. A., Zanetti, M. E., Olm, E. H., and Kuehl, F. A., Jr.,** A 15-hydroxyprostaglandin dehydrogenase specific for prostaglandin A in rabbit kidney, *Proc. Natl. Acad. Sci. U.S.A.,* 73, 1107—1111, 1976.

105. **Kupfer, D. and Navarro, J.,** Metabolism of prostaglandin A_1 by hepatic microsomal monooxygenase P-450 system in guinea pig and rat, *Life Sci.,* 18, 507—513, 1976.

106. **Tai, H. H. and Hollander, C. S.,** Regulation of prostaglandin metabolism: activation of 15-hydroxyprostaglandin dehydrogenase by chlorpromazine and imipramine related drugs, *Biochem. Biophys. Res. Commun.,* 68, 814—820, 1976.

107. **Hamberg, M.,** On the absolute configuration of 19-hydroxy-prostaglandin B₁, *Eur. J. Biochem.,* 6, 147—150, 1968.

108. **Westwick, J.,** The effect of pulmonary metabolites of prostaglandins E₁, E₂, and F₂α on ADP-induced aggregation of human and rabbit platelets (proceedings), *Br. J. Pharmacol.,* 58(2), 297P—298P, 1976.

109. **Roberts, L. J., II, Sweetman, B. J., and Oates, J. A.,** Metabolism of thromboxane B₂ in man, *J. Biol. Chem.,* 256, 8384—8393, 1981.

110. **Roberts, L. J., II, Sweetman, B. J., and Oates, J. A.,** Metabolism of thromboxane B₂ in the monkey, *J. Biol. Chem.,* 253, 5305—5318, 1978.

111. **Pace-Asciak, C. R. and Edwards, N. S.,** Tetranor thromboxane B₂ is the principal urinary catabolite formed after i.v. infusion of thromboxane in the rat, *Biochem. Biophys. Res. Commun.,* 97, 81—86, 1980.

112. **Anhut, H., Peskar, B. A., and Bernauer, W.,** Release of 15-keto-13,14-dihydro-thromboxane B₂ and prostaglandin D₂ during anaphylaxis as measured by radioimmunoassay, *Naunyn-Schmiedeberg's Arch. Pharmacol.,* 305, 247—252, 1978.

113. **Roberts, L. J., II, Sweetman, B. J., Payne, N. A., and Oates, J. A.,** Metabolism of thromboxane B₂ in man. Identification of the major urinary metabolite, *J. Biol. Chem.,* 252, 7415—7417, 1977.

114. **Roberts, L. J., II, Sweetman, B. J., Morgan, J. L., Payne, N. A., and Oates, J. A.,** Identification of the major urinary metabolite of thromboxane B₂ in the monkey, *Prostaglandins,* 13, 631—647, 1977.

115. **Kindahl, H.,** Metabolism of thromboxane B₂ in the cynomologus monkey, *Prostaglandins,* 13, 619—629, 1977.

116. **Roberts, L. J., II, Brash, A. R., and Oates, J. A.,** Metabolic fate of thromboxane A₂ and prostacyclin: prostaglandins and the cardiovascular system, in *Advances in Prostaglandin, Thromboxane, and Leukotriene Research,* Vol. 10, Oates, J., Ed., Raven Press, New York, 1982, 211—225.

117. **Sun, F. F. and Taylor, B. M.,** Metabolism of prostacyclin in rat, *Biochemistry,* 17, 4096—4101, 1978.

118. **Rosenkranz, B., Fischer, C., Reiman, I., Weimer, K. E., Beck, G., and Frölich, J. C.,** Identification of the major metabolite of prostacyclin and 6-ketoprostaglandin F₁α in man, *Biochim. Biophys. Acta,* 619, 207—213, 1980.

119. **Wong, P. Y.-K., Malik, K. U., Desiderio, D. M., McGiff, J. C., and Sun, F. F.,** Hepatic metabolism of prostacyclin (PGI₂) in the rabbit; formation of a potent novel inhibitor of platelet aggregation, *Biochem. Biophys. Res. Commun.,* 93, 486—494, 1980.

120. **Pace-Asciak, C. R., Carrara, M. C., and Domazet, Z.,** Identification of the main urinary metabolites of 6-ketoprostaglandin F₁α (6K-PGF₁α) in the rat, *Biochem. Biophys. Res. Commun.,* 78, 115—121, 1977.

121. **Rosenkranz, B., Fischer, C., Weimer, K. E., and Frolich, J. C.,** Metabolism of prostacyclin and 6-ketoprostaglandin F₁α in man, *J. Biol. Chem.,* 255, 10194—10198, 1980.

122. **Wong, P. Y.-K., Lee, W. H., Chao, P. H.-W., Reiss, R. F., and McGiff, J. C.,** Metabolism of prostacyclin by 9-hydroxyprostaglandin dehydrogenase in human platelets. Formation of a potent inhibitor of platelet aggregation and enzyme purification, *J. Biol. Chem.,* 255, 9021—9024, 1980.

123. **Wong, P. Y.-K., McGiff, J. C., Sun, F. F., and Lee, W. H.,** 6-Ketoprostaglandin E₁ inhibits the aggregation of human platelets, *Eur. J. Pharmacol.,* 60, 245—248, 1979.

124. **Wong, P. Y.-K., McGiff, J. C., Dagen, L., Malik, K. V., and Sun, F. F.,** Metabolism of prostacyclin in the rabbit kidney, *J. Biol. Chem.,* 254, 12—14, 1979.

125. **Taylor, B. M. and Sun, F. F.,** Tissue distribution and biliary excretion of prostacyclin metabolites in the rat, *J. Pharmacol. Exp. Therap.,* 214, 24—30, 1980.

126. **Roberts, L. J., II,** Comparative metabolism and fate of eicosanoids, in *CRC Handbook of the Eicosanoids: Prostaglandins and Related Lipids,* Vol. 1, CRC Press, Boca Raton, Fla., in press, 1987.

127. **Mahmud, I., Smith, D. L., Willis, A. L., Whyte, M. A., Nelson, J. T., Cho, D., Tokes, L. G., and Alvarez, R.,** On the identification and biological properties of prostaglandin J₂, *Prostaglandins Leukotrienes Med.,* 16, 131—146, 1984.

128. **Fitzpatrick, F. A. and Wynalda, M. A.,** Albumin catalyzed metabolism of prostaglandin D₂, *J. Biol. Chem.,* 258, 11713—11718, 1983.

129. **Fukushima, M., Kato, T., Ota, K., Arai, Y., and Narumiya, S.,** 9-Deoxy-Δ⁹-prostaglandin D₂: a prostaglandin D₂ derivative with potent antineoplastic and weak smooth muscle contracting activities, *Biochem. Biophys. Res. Commun.,* 109, 626—633, 1982.

130. **Smith, D. L., Willis, A. L., and Mahmud, I.,** Eicosanoid effects on cell proliferation in vitro: relevance to atherosclerosis, *Prostaglandins Leukotrienes Med.,* 16, 1—10, 1984.

131. **Mahmud, I., Alvarez, R., Miller, F., Nelson, J. T., Cho, D., Tokes, L. G., Smith, D. L., Whyte, M. A., and Willis, A. L.,** Properties of Prostaglandin J₂, *Fed. Proc.,* 43, 980, 1984.

132. **Kikawa, Y., Narumiya, S., Fukushima, M., Wakatiuka, M., and Hayaishi, O.,** 9-Deoxy-Δ⁹,Δ¹²-13,14-dihydroprostaglandin D₂, a metabolite of PGD₂ formed in human plasma, *Proc. Natl. Acad. Sci. U.S.A.,* 81, 1317—1321, 1984.

133. **Ellis, C. K., Smigel, M. D., Oates, J. A., Oelz, O., and Sweetman, B. J.,** Metabolism of prostaglandin D₂ in the monkey, *J. Biol. Chem.,* 254, 4152—4163, 1979.

134. **Liston, T. E. and Roberts, L. J., II,** Metabolic fate of radiolabelled prostaglandin D_2 in a normal human male volunteer, *J. Biol. Chem.,* 260, 13172—13180, 1985.

135. **Liston, T. E. and Roberts, L. J., II,** Transformation of prostaglandin D_2 to 9α, 11β-(15S)-trihydroxy-prosta-(5Z,13E)-dien-1-oic acid ($9\alpha,11\beta$-prostaglandin F_2): a unique biologically active prostaglandin produced enzymatically *in vivo* in humans, *Proc. Natl. Acad. Sci. U.S.A.,* 82, 6030—6034, 1985.

136. **Roberts, L. J., II and Sweetman, B. J.,** Metabolic fate of endogenously synthesized prostaglandin D_2 in a human female with mastocytosis, *Prostaglandins,* 30, 383—400, 1985.

137. **Pugliese, G., Spokas, E. G., Marcinkiewicz, E., and Wong, P. Y.-K.,** Hepatic transformation of prostaglandin D_2 to a new prostanoid, $9\alpha,11\beta$-prostaglandin F_2, that inhibits platelet aggregation and constricts blood vessels, *J. Biol. Chem.,* 260, 14621—14625, 1985.

138. **Grandström, E., Kindahl, M., and Swahn, M.-L.,** Profiles of prostaglandin metabolites in the human circulation: Identification of late-appearing, long-lived products, *Biochim. Biophys. Acta,* 713, 46—60, 1982.

139. **Morrison, A. R., McLaughlin, L., Bloch, M., and Needleman, P.,** A novel cyclooxygenase metabolite of arachidonic acid, *J. Biol. Chem.,* 259, 13579—13583, 1984.

140. **Hassid, A., Sebrosky, A., and Dunn, M. J.,** Metabolism of prostaglandins by human renal enzymes: presence of 9-hydroxyprostaglandin dehydrogenase activity in human kidney, in *Advance in Prostaglandin, Thromboxane, and Leukotriene Research,* Vol. 11, Samuelsson, B., Paoletti, R., and Ramwell, P., Eds., Raven Press, New York, 1983, 499—504, 1983.

141. **Hamazaki, T., Fischer, S., Urakaze, M., Sawazake, S., Weber, P. C., and Yano, S.,** Comparison of the urinary metabolite of PGI_3 (PGI_3-M) between a Japanese fishing village and a farming village, in *Proceedings, 6th Annual Conference on Prostaglandins and Related Compounds,* Florence, Italy, 1986, p. 38.

142. **Fischer, S. and Weber, P. C.,** Prostaglandin I_3 is formed in vivo in man after dietary eicosapentaenoic acid, *Nature,* 307, 165—168, 1984.

143. **Diczfalusy, U. and Hamberg, M.,** Identification of the major urinary metabolite of prostaglandin E_3 in the rat, *Biochim. Biophys. Acta,* 878, 387—393, 1986.

144. **Sun, F. F.,** Metabolism of prostaglandin $F_{2\alpha}$ in the rat, *Biochim. Biophys. Acta,* 348, 249—262, 1974.

145. **Sun, F. F. and Stafford, J. E.,** Metabolism of prostaglandin $F_{2\alpha}$ in rhesus monkeys, *Biochim. Biophys. Acta,* 369, 95—110, 1974.

146. **Granström, E. and Samuelsson, B.,** On the metabolism of prostaglandin $F_{2\alpha}$ in female subjects, *J. Biol. Chem.,* 246, 5254—5263, 1971.

147. **Roberts, L. J., II, Seibert, K., Liston, T. E., Tantengco, M. V., and Robertson, R. M.,** PGD_2 is transformed by human coronary arteries to $9\alpha,11\beta$-PGF_2 which contracts human coronary artery rings, in *Advances in Prostaglandin, Thromboxane, and Leukotriene Research,* in press.

148. **Seibert, K., Sheller, J. R., and Roberts, L. J., II,** $9\alpha,11\beta$-Prostaglandin F_2: formation and metabolism by human lung and contractile effects on human bronchial smooth muscle, *Proc. Natl. Acad. Sci. U.S.A.,* in press.

149. **Watanabe, K., Iguchi, Y., Iguchi, S., Arai, Y., Hayaishi, O., and Roberts, L. J., II,** Stereospecific conversion of prostaglandin D_2 to (5Z,13E)-15(S)-9,11,15-trihydroxyprosta-5,13-dienoic acid (9,11-prostaglandin F_2) and of prostaglandin H_2 to prostaglandin F_2 by bovine lung prostaglandin F synthase, *Proc. Natl. Acad. Sci. U.S.A.,* 83, 1583—1587, 1986.

150. **Fitzgerald, G. A.,** Thromboxane biosynthesis and antagonism in man, in *Proceedings, Sixth International Conference on Prostaglandins and Related Compounds,* Florence, Italy 1986, p. 12.

151. **Oliw, E. H. and Hamberg, M.,** Characterization of prostaglandin E_2 20-hydroxylase of sheep vesicular glands, *Biochim. Biophys. Acta,* 879, 113—119, 1986.

152. **Powell, W. S. and Solomon, S.,** Formation of 20-hydroxyprostaglandins by lungs of pregnant rabbits, *J. Biol. Chem.,* 253, 4609—4616, 1978.

153. **Okimura, T., Nakayama, R., Sago, T., and Saito, K.,** Identification of prostaglandin E metabolites from primary cultures of rat hepatocytes, *Biochim. Biophys. Acta,* 837, 197—207, 1985.

154. **Rosenkranz, B., Fischer, C., Boeynaems, J.-M., and Frölich, J. C.,** Metabolic disposition of prostaglandin E_1 in man, *Biochim. Biophys. Acta,* 750, 231—236, 1983.

155. **Bothwell, W., Verburg, M., Wynalda, M., Daniels, E. G., and Fitzpatrick, F. A.,** A radioimmunoassay for the unstable pulmonary metabolites of prostaglandin E_1 and E_2: an indirect index of the in vivo disposition and pharmacokinetics, *J. Pharm. Exp. Therap.,* 220, 229—235, 1982.

156. **Granström, E. and Kindahl, H.,** Radioimmunologic determination of 15-keto-13,14-dihydro-PGE_2: a method for its stable degradation product, 11-deoxy-15-keto-13,14-dihydro-11β,16-cyclo PGE_2, in *Advances in Prostaglandin and Thromboxane Research,* Vol. 6, Samuelsson, B., Ramwell, P. W., and Paoletti, R. W., Eds., 1980, 181—182.

157. **Demers, L. M., Brennecke, S. P., Mountford, L. A., Brunt, J. D., and Turnbull, A. C.,** Development and validation of a radioimmunoassay for prostaglandin E_2 metabolite levels in plasma, *J. Clin. Endocrin. Metab.,* 57, 101—106, 1983.

158. **Simberg, N.**, The metabolism of prostaglandin E₂ in perinatal rabbit lungs, *Prostaglandins*, 26, 275—285, 1983.

159. **Glance, D. G., Elder, M. G., and Myatt, L.**, Uptake, transfer and metabolism of prostaglandin E₂ in the isolated perfused human placental cotyledon, *Prostaglandins, Leukotrienes and Medicine*, 21, 1—14, 1986.

160. **Ekblad, U., Erkkola, R., and Uotila, P.**, Prostaglandin E₂ is only slightly metabolized in the fetal circulation of perfused human placenta, *Prostaglandins, Leukotrienes and Medicine*, 8, 481—488, 1982.

161. **Brennecke, S. P., Castle, B. M., Demers, L. M., and Turnbull, A. C.**, Maternal plasma prostaglandin E₂ metabolite levels during human pregnancy and parturition, *Br. J. Obstet. Gynaecol.*, 92, 345—349, 1985.

162. **Miller, M. J. S., Spokas, E. G., and McGiff, J. C.**, Metabolism of prostaglandin E₂ in the isolated perfused kidney of the rabbit, *Biochem. Pharmacol.*, 31, 2955—2960, 1982.

163. **Kostas, K. P., Theoharides, A. D., Kupfer, D., and Coon, M. J.**, Hydroxylation of prostaglandins by inducible isozymes of rabbit liver microsomal cytochrome P-450, *J. Biol. Chem.*, 257, 11221—11229, 1982.

164. **Morrison, A. R. and Crowley, J. R.**, Identification of an unusual cyclooxygenase metabolite of arachidonic acid in rabbit renal medulla, *Arch. Biochem. Biophys.*, 234, 413—417, 1984.

165. **Holm, K. A., Koop, D. R., Coon, M. J., Theoharides, A. D., and Kupfer, D.**, ω-1 and ω-2-Hydroxylation of prostaglandins by rabbit hepatic microsomal cytochrome P-450 isozyme 6, *Arch. Biochem. Biophys.*, 243, 135—143, 1985.

166. **Holm, K. A., Engell, R. J., and Kupfer, D.**, Regioselectivity of hydroxylation of prostaglandins by liver microsomes supported by NADPH versus H₂O₂ in methylcholanthrene-treated and control rats; formation of novel prostaglandin metabolites, *Arch. Biochem. Biophys.*, 237, 477—489, 1985.

167. **Oliw, E. H., Fahlstadius, P., and Hamberg, M.**, Isolation and biosynthesis of 20-hydroxyprostaglandins E₁ and E₂ in ram seminal fluid, *J. Biol. Chem.*, 261, 9216—9221, 1986.

168. **Sraer, J., Bens, M., Oudinet, J. P., and Ardaillou, R.**, Bioconversion of leukotriene C₄ by rat glomeruli and papilla, *Prostaglandins*, 31, 909—921, 1986.

169. **Punzengraber, C., Stanek, B., Sinzinger, H., and Silberbauer, K.**, Bicycloprostaglandin E₂ metabolite in congestive heart failure and relation to vasoconstrictor neurohumoral principles, *Am. J. Cardiol.*, 57, 619—623, 1986.

170. **Hagmann, W., Denzlinger, C., Rapp, S., Weckbecker, G., and Keppler, D.**, Identification of the major endogenous leukotriene metabolite in the bile of rats as N-acetyl leukotriene E₄, *Prostaglandins*, 36, 239—251, 1986.

171. **Harper, T. W., Garrity, M. J., and Murphy, R. C.**, Metabolism of leukotriene B₄ in isolated rat hepatocytes. Identification of a novel 18-carboxy-19,20-dinor leukotriene B₄ metabolite, *J. Biol. Chem.*, 261, 5414—5418, 1986.

172. **Spector, R. and Goetzl, E. J.**, Leukotriene C₄ transport and metabolism in the central nervous system, *J. Neurochem.*, 46, 1308—1312, 1986.

173. **O'Flaherty, J., Kosfeld, S., and Nishihara, J.**, Binding and metabolism of leukotriene B₄ by neutrophils and their subcellular organelles, *J. Cell Physiol.*, 126, 359—370, 1986.

174. **Weckbecker, G. and Keppler, D. C.**, Leukotriene C₄ metabolism by hepatoma cells deficient in the uptake of cysteinyl leukotrienes, *Eur. J. Biochem.*, 154, 559—562, 1986.

175. **Bernström, K. and Hammarström, S.**, Metabolism of leukotriene E₄ by rat tissues: formation of N-acetyl leukotriene E₄, *Arch. Biochem. Biophys.*, 244, 486—491, 1986.

176. **Wong, P. Y.-K., Chao, P. H.-W., and Spokas, E. G.**, Metabolism of leukotriene A₄ in the rat kidney and isolated glomeruli, in *Advances in Prostaglandin, Thromboxane, and Leukotriene Research*, Vol. 15, Hayaishi, O. and Yamamoto, S., Eds., Raven Press, New York, 1985, 423—426.

177. **Hammarström, S., Örning, L., Bernström, K., Gustaffson, B., Norin, E., and Kaijser, L.**, Metabolism of leukotriene C₄ in rats and humans, in *Advances in Prostaglandin, Thromboxane, and Leukotriene Research*, Vol. 15, Hayaishi, O., and Yamamoto, S., Eds., Raven Press, New York, 1985, 185—188.

178. **Serhan, C. N., Fahlstadius, P., Dahlen, S.-E., Hamberg, M., and Samuelsson, B.**, Biosynthesis and biological activities of lipoxins, in *Advances in Prostaglandin, Thromboxane, and Leukotriene Research*, Vol. 15, Hayaishi, O., and Yamamoto, S., Eds., Raven Press, New York, 1985, 163—166.

179. **Örning, L., Norin, E., Gustaffson, B., and Hammarström, S.**, In vivo metabolism of leukotriene C₄ in germ free and conventional rats. Fecal excretion of N-acetylleukotriene E₄, *J. Biol. Chem.*, 261, 766—771, 1986.

180. **Sirois, P., Brousseau, Y., Chagnon, M., Gentile, J., Gladu, M., Salari, H., and Borgeat, P.**, Metabolism of leukotrienes by adult and fetal human lungs, *Exp. Lung. Res.*, 9, 17—30, 1985.

181. **Taylor, B. M. and Sun, F. F.**, Disappearance and metabolism of leukotriene B₄ during carrageenan-induced pleurisy, *Biochem. Pharmacol.*, 34, 3495—3498, 1985.

182. **Aharony, D., Dobson, P. T., and Krell, R. D.,** In vitro metabolism of [³H]-peptide leukotrienes in human and ferret lung: a comparison with the guinea pig, *Biochem. Biophys. Res., Commun.,* 131, 892—898, 1985.

183. **Paterson, N. A., McIver, D. J., and Schurch, S.,** Zymosan enhances leukotriene D₄ metabolism by porcine alveolar macrophages, *Immunology,* 56, 153—159, 1985.

184. **Örning, L., Kaijser, L., and Hammarström, S.,** In vivo metabolism of leukotriene C₄ in man: urinary excretion of leukotriene E₄, *Biochem. Biophys. Res. Commun.,* 130, 214—220, 1985.

185. **Wong, P. Y.-K., Malik, K. U., Taylor, B. M., Schneider, W. P., McGiff, J. C., and Sun, F. F.,** Epoxidation of prostacyclin in rabbit kidney, *J. Biol. Chem.,* 260, 9150—9153, 1985.

186. **Newton, J. F., Eckardt, R., Bender, P. E., Leonard, T., and Straub, K.,** Metabolism of leukotriene B₄ in hepatic microsomes, *Biochem. Biophys. Res. Commun.,* 128, 733—738, 1985.

187. **Harper, T. W., Westcott, J. Y., Voelkel, N., and Murphy, R. C.,** Metabolism of leukotrienes B₄ and C₄ in the isolated perfused rat lung, *J. Biol. Chem.,* 259, 14437—14440, 1984.

188. **Aharony, D. and Dobson, P.,** Discriminative effect of gamma-glutamyl transpeptidase inhibitors of metabolism of leukotriene C₄ in peritoneal cells, *Life Sci.,* 35, 2135—2142, 1984.

189. **Snyder, D. W., Aharony, D., Dobson, P., Tsai, B. S., and Krell, R. D.,** Pharmacological and biochemical evidence for metabolism of peptide leukotrienes by guinea-pig airway smooth muscle in vitro, *J. Pharm. Exp. Ther.,* 231, 224—229, 1984.

190. **Serafin, W. E., Oates, J. A., and Hubbard, W. C.,** Metabolism of leukotriene B₄ in the monkey. Identification of the principal nonvolatile metabolite in the urine, *Prostaglandins,* 27, 899—911, 1984.

191. **Oliw, E. H.,** Metabolism of 5(6) Oxido-eicosatrienoic acid by ram seminal vesicles. Formation of two stereoisomers of 5-hydroxyprostaglandin I₁, *J. Biol. Chem.,* 259, 2716—2721, 1984.

192. **Sirois, P. and Brousseau, Y.,** Leukotriene transformation of guinea pig lungs, *Prostaglandins, Leukotrienes, and Medicine,* 10, 133—143, 1983.

193. **Lee, C. W., Lewis, R. A., Tauber, A. I., Mehrotra, M., Corey, E. J., and Austen, K. F.,** The myeloperoxidase-dependent metabolism of leukotrienes C₄, D₄, and E₄ to 6-trans-leukotriene B₄ diastereomers and the subclass specific S-diastereomeric sulfoxides, *J. Biol. Chem.,* 258, 15004—15010, 1983.

194. **Jubiz, W., Radmark, O., Malmsten, C., Hansson, G., Lindgren, J. A., Palmblad, J., Uden, A. M., and Samuelsson, B.,** A novel leukotriene produced by stimulation of leukocytes with formylmethionylleucylphenylalanine, *J. Biol. Chem.,* 257, 6106—6110, 1982.

195. **Hansson, G., Lindgren, J. A., Dahlen, S. E., Hedqvist, P., and Samuelsson, B.,** Identification and biological activity of novel ω-oxidized metabolites of leukotriene B₄ from human leukocytes, *FEBS Lett.,* 130, 107—112, 1981.

196. **Salmon, J. A., Simmons, P. M., and Palmer, R. M. J.,** Synthesis and metabolism of leukotriene B₄ in human neutrophils measured by specific radioimmunoassay, *FEBS Lett.,* 146, 18—22, 1982.

197. **Örning, L. and Hammarström, S.,** Isolation and characterization of 15-hydroxylated metabolites of leukotriene C₄, *FEBS Lett.,* 153, 253—256, 1983.

198. **Ueda, N., Kaneko, S., Yoshimoto, T., and Yamamoto, S.,** Purification of arachidonate 5-lipoxygenase from porcine leukocytes and its reactivity with hydroperoxyeicosatetraenoic acids, *J. Biol. Chem.,* 261, 7982—7988, 1986.

199. **Bryant, R. W., Schewe, T., Rapoport, S. M., and Bailey, J. M.,** Leukotriene formation by a purified reticulocyte lipoxygenase enzyme. Conversion of arachidonic acid and 15-hydroperoxyeicosatetraenoic acid to 14,15-leukotriene A₄, *J. Biol. Chem.,* 260, 3548—3555, 1985.

200. **Lee, C. W., Lewis, R. A., Corey, E. J., Barton, A., Oh, H., Tauber, A. I., and Austen, K. F.,** Oxidative inactivation of leukotriene C₄ by stimulated human polymorphonuclear leukocytes, *Proc. Natl. Acad. Sci. U.S.A.,* 79, 4166—4170, 1982.

201. **Pace-Asciak, C. R.,** Formation of hepoxilin A₄, B₄ and the corresponding trioxilins from 12(S)-hydroperoxy-5,8,10,14,17-icosapentaenoic acid, *Prostaglandins, Leukotrienes, and Medicine,* 22, 1—9, 1986.

202. **Pace-Asciak, C. R.,** Hemoglobin and hemin-assisted transformation of hydroperoxy-5,8,10,14-eicosatetraenoic acid, *Biochim. Biophys. Acta,* 793, 485—488, 1984.

203. **Hammarström, S.,** Metabolism of leukotriene C₃ in the guinea pig. Identification of metabolites formed by lung, liver and kidney, *J. Biol. Chem.,* 256, 9573—9578, 1981.

204. **Parker, C. W., Delkoch, M. M., Huber, M. M., and Falkenhein, S. F.,** Formation of the cysteinyl form of slow reacting substance (leukotriene E₄) in human plasma, *Biochem. Biophys. Res. Commun.,* 97, 1038—1046, 1980.

205. **Pace-Asciak, C. R. and Martin, J. M.,** Hepoxilin, a new family of insulin secretagogues formed by intact rat pancreatic islets, *Prostaglandins, Leukotrienes, and Medicine,* 16, 173—180, 1984.

206. **Pace-Asciak, C. R., Granström, E., and Samuelsson, B.,** Arachidonic acid epoxides. Isolation and structure of two hydroxy epoxide intermediates in the formation of 8,11,12- and 10,11,12-trihydroxyeicosatrienoic acids, *J. Biol. Chem.,* 258, 6835—6840, 1983.

207. **Kühn, H., Wiesner, R., and Stender, H.,** The formation of products containing a conjugated tetraenoic system by pure reticulocyte lipoxygenase, *FEBS Lett.,* 177, 255—259, 1984.

208. **Hammarström, S., Bernström, K., Orning, L., Dahlen, S. E., Hedqvist, P., Smedegard, G., and Revenäs, B.,** Rapid in vivo metabolism of leukotriene C_3 in the monkey, *Macaca iris, Biochem. Biophys. Res. Commun.,* 101, 1109—1115, 1981.

209. **Appelgren, L. E. and Hammarström, S.,** Distribution and metabolism of ^{3}H-labelled leukotriene C_3 in the mouse, *J. Biol. Chem.,* 257, 531—535, 1982.

210. **Serhan, C. N., Hamberg, M., Samuelsson, B., Morris, J., and Wishka, D. G.,** On the stereochemistry and biosynthesis of lipoxin B, *Proc. Natl. Acad. Sci. U.S.A.,* 83, 1983—1987, 1986.

211. **Funk, C. D. and Powell, W. S.,** Conversion of 8,11,14-eicosatrienoic acid to 11,12-epoxy-10-hydroxy-8-heptadecenoic acid by aorta, *Prostaglandins,* 25, 299—309, 1983.

212. **Bernström, K. and Hammarström, S.,** Metabolism of leukotriene D by porcine kidney, *J. Biol. Chem.,* 256, 9579—9582, 1981.

213. **Evans, J. F., Nathaniel, D. J., Zamboni, R. J., and Ford-Hutchinson, A. W.,** Leukotriene A_3. A poor substrate but a potent inhibitor of rat and human neutrophil leukotriene A_4 hydrolase, *J. Biol. Chem.,* 260, 10966—10970, 1985.

214. **Ford-Hutchinson, A. W., Bray, M. A., Doig, M. V., Shipley, M. E., and Smith, M. J. H.,** Leukotriene B, a potent chemokinetic and aggregating substance released from polymorphonuclear leukocytes, *Nature,* 286, 264—265, 1980.

215. **Bray, M. A., Cunningham, F. M., Ford-Hutchinson, A. W., and Smith, M. J. H.,** Leukotriene B_4. A mediator of vascular permeability, *Br. J. Pharmacol.,* 72, 483—486, 1981.

216. **Bray, M. A., Ford-Hutchinson, A. W., and Smith, M. J. H.,** Leukotriene B_4. An inflammatory mediator in vivo, *Prostaglandins,* 22, 213—222, 1981.

217. **Rackham, A. and Ford-Hutchinson, A. W.,** Inflammation and pain sensitivity: effects of leukotrienes D_4, B_4 and prostaglandin E_1 in the rat paw, *Prostaglandins,* 25, 193—203, 1983.

218. **Jakschik, B. A., Sams, A. R., Sprecher, H., and Needleman, P.,** Fatty acid structural requirements for leukotriene biosynthesis, *Prostaglandins,* 20, 401—410, 1980.

219. **Hammarstrom, S.,** Conversion of 5,8,11-eicosatrienoic acid to leukotrienes C_1 and D_1, *J. Biol. Chem.,* 256, 2275—2279, 1980.

220. **Powell, W. S.,** Novel pathway for the metabolism of 6-trans-leukotriene B_4 by human polymorphonuclear leukocytes, *Biochem. Biophys. Res. Commun.,* 136, 707—712, 1986.

221. **Powell, W. S.,** Properties of leukotriene B_4 20-hydroxylase from polymorphonuclear leukocytes, *J. Biol. Chem.,* 259, 3082—3089, 1984.

222. **Samuelsson, B.,** Leukotrienes: mediators of immediate hypersensitivity reactions and inflammation, *Science,* 220, 568—575, 1983.

223. **Samuelsson, B.,** Leukotrienes and related compounds, in *Advances in Prostaglandin, Thromboxane, and Leukotriene Research,* Vol. 15, Hayaishi, O. and Yamamoto, S., Eds., Raven Press, New York, 1985, 1—9.

224. **Fitzpatrick, F., Haeggström, J., Granström, E., and Samuelsson, B.,** Metabolism of leukotriene A_4 by an enzyme of blood plasma: a possible leukotactic mechanism, *Proc. Natl. Acad. Sci. U.S.A.,* 80, 5425—5429, 1983.

225. **Haeggström, J., Meijer, J., and Rädmark, O.,** Leukotriene A_4. Enzymatic conversion into 5,6-dihydroxy-7,9,11,14-eicosatetraenoic acid by mouse liver cytosolic epoxide hydrolase, *J. Biol. Chem.,* 261, 6332—6337, 1986.

226. **Haeggström, J., Rädmark, O., and Fitzpatrick, F. A.,** Leukotriene A_4-hydrolase activity in guinea pig and human liver, *Biochim. Biophys. Acta,* 835, 378—384, 1985.

227. **Wong, P. Y.-K., Hughes, R., and Lam, B.,** Lipoxene: a new group of trihydroxypentaenes of eicosapentaenoic acid derived from porcine leukocytes, *Biochem. Biophys. Res. Commun.,* 126, 763—772, 1985.

228. **Prescott, S. M.,** The effect of eicosapentaenoic acid on leukotriene B production by human neutrophils, *J. Biol. Chem.,* 259, 7615—7621, 1984.

229. **Lee, T. H., Mencia-Huerta, J.-M., Shih, C., Corey, E. J., Lewis, R. A., and Austen, K. F.,** Characterization and biologic properties of 5,12-dihydroxy derivatives of eicosapentaenoic acid, including leukotriene B_5 and the double lipoxygenase product, *J. Biol. Chem.,* 259, 2383—2389, 1984.

230. **Marcus, A. J., Safier, L. B., Ullman, H. L., Broekman, M. J., Islam, N., Oglesby, T. D., and Gorman, R. R.,** 12S,20-Dihydroxyicosatetraenoic acid. A new icosanoid synthetized by neutrophils from 12S-hydroxyicosatetraenoic acid produced by thrombin or collagen-stimulated platelets, *Proc. Natl. Acad. Sci. U.S.A.,* 81, 903—907, 1984.

231. **Lundberg, U., Rädmark, O., Malmsten, C., and Samuelsson, B.,** Transformation of 15-hydroperoxy-5,9,11,13-eicosatetraenoic acid into novel leukotrienes, *FEBS Lett.,* 126, 127—132, 1981.

232. **Sok, D. E., Han, C.-O., Shieh, W.-R., Zhou, B.-N., and Sih, C. J.,** Enzymatic formation of 14,15-leukotriene A and C(14)-sulfur-linked peptides, *Biochem. Biophys. Res. Commun.,* 104, 1363—1370, 1982.

233. **Bernström, K. and Hammarström, S.,** A novel leukotriene formed by transpeptidation of leukotriene E, *Biochem. Biophys. Res. Commun.,* 109, 800—804, 1982.

234. **Ormstad, K., Uehara, N., Orrenius, S., Örning, L., and Hammarström, S.,** Uptake and metabolism of leukotriene C_3 by isolated rat organs and cells, *Biochem. Biophys. Res. Commun.*, 104, 1434—1440, 1982.

235. **Anderson, M. E., Allison, R. D., and Meister, A.,** Interconversion of leukotrienes catalyzed by purified gamma-glutamyl transpeptidase: concomitant formation of leukotriene D_4 and gamma-glutamyl acids, *Proc. Natl. Acad. Sci. U.S.A.*, 79, 1088—1099, 1982.

236. **Lewis, R. A., Austen, K. F., Drazen, J. M., Clark, D. A., Marfat, A., and Corey, E. J.,** Slow reacting substance of anaphylaxis: identification of leukotrienes C-1 and D from human and rat sources, *Proc. Natl. Acad. Sci. U.S.A.*, 77, 3710—3714, 1980.

237. **Drazen, J. M., Austen, K. F., Lewis, R. A., Clark, D. A., Goto, G., Marfat, A., and Corey, E. J.,** Comparative airway and vascular activities of leukotrienes C-1 and D *in vivo* and *in vitro*, *Proc. Natl. Acad. Sci. U.S.A.*, 77, 4354—4358, 1980.

238. **Lewis, R. A., Drazen, J. M., Austen, K. F., Clark, D. A., and Corey, E. J.,** Identification of the C(6)-S-conjugate of leukotriene A with cysteine as a naturally occurring slow reacting substance of ana-phylaxis (SRS-A). Importance of the 11-*cis*- geometry for biological activity, *Biochem. Biophys. Res. Commun.*, 96, 271—277, 1980.

239. **Robinson, C. and Hoult, J. R. S.,** Thromboxane analogues inhibit the metabolism of thromboxane B_2 in perfused guinea-pig lung, *Biochim. Biophys. Acta*, 754, 190—200, 1983.

240. **Moore, P. K. and Hoult, J. R. S.,** Selective action of aspirin- and sulphasalazine-like drugs against prostaglandin synthesis and breakdown, *Biochem. Pharmacol.*, 31, 969—971, 1982.

241. **Westlund, P., Granström, E., Kumlin, M., and Nordenström, A.,** Identification of 11-dehydro-TXB2 as a suitable parameter for monitoring thromboxane production in the human, *Prostaglandins*, 31, 929—960, 1986.

242. **Westlund, P., Kumlin, M., Nordenström, A., and Granström, E.,** Circulating and urinary thromboxane B_2 metabolites in the rabbit: 11-dehydro-thromboxane B_2 as parameter of thromboxane production, *Prostaglandins*, 31, 413—443, 1986.

243. **Wong, P. Y.-K., Malik, K. U., Taylor, B. M., Schneider, W. P., McGiff, J. C., and Sun, F. F.,** Epoxidation of prostacyclin in the rabbit kidney, *J. Biol. Chem.*, 260, 9150—9153, 1985.

244. **Wong, P. Y.-K., McGiff, J. C., Sun, F. F., and Malik, K. U.,** Pulmonary metabolism of prostacyclin (PGI$_2$) in the rabbit, *Biochem. Biophys. Res. Commun.*, 83, 731—738, 1978.

245. **Rosenkranz, B., Fischer, C., and Frölich, J. C.,** Prostacyclin metabolites in human plasma, *Clin. Pharmacol. Ther.*, 29, 420—424, 1981.

246. **Huang, Y. S., Martineau, A., Falardeau, P., and Davignon, J.,** Fatty acid composition of tissue phospholipids and prostaglandin excretion in hyperlipidemia induced in rats by implantation of the mam-motropic pituitary tumor MET-F4, *Lipids*, 18, 412—422, 1983.

247. **Sun, F. F., Taylor, B. M., Suter, D. M., and Weeks, J. R.,** Metabolism of prostacyclin. Urinary metabolite profile of 6-keto-PGF$_{1\alpha}$ in rat, *Prostaglandins*, 17, 753—759, 1979.

248. **Forstermann, U., Neufang, B., and Hertting, G.,** Metabolism of 6-keto-prostaglandin F$_{1\alpha}$ and prostacyclin to 6,15-diketo-13,14-dihydroprostaglandin F$_{1\alpha}$-like material in cats and rabbits, *Biochim. Biophys. Acta*, 712, 684—691, 1982.

249. **Wong, P. Y.-K., Sun, F. F., and McGiff, J. C.,** Metabolism of prostacyclin in blood vessels, *J. Biol. Chem.*, 253, 5555—5557, 1978.

250. **McGuire, J. F. and Sun, F. F.,** Metabolism of prostacyclin. Oxidation by rhesus monkey lung 15-hydroxyl prostaglandin dehydrogenase, *Arch. Biochem. Biophys.*, 189, 92—96, 1978.

251. **Jackson, E. K., Goodman, R. P., Fitzgerald, G. A., Oates, J. A., and Branch, R. A.,** Assessment of the extent to which exogenous prostaglandin I_2 is converted to 6-keto-prostaglandin E_1 in human subjects, *J. Pharmacol. Exp. Ther.*, 221, 183—187, 1982.

252. **Brash, A. R., Jackson, E. K., Saggese, C. A., Lawson, J. H., Oates, J. A., and Fitzgerald, G. A.,** Metabolic disposition of prostacyclin in humans, *J. Pharmacol. Exp. Ther.*, 226, 78—87, 1983.

253. **Salmon, J. A., Smith, D. R., and Cottee, F.,** The major product of PGG$_2$ metabolism by aortic microsomes is 6,15-dioxo PGF$_{1\alpha}$, *Prostaglandins*, 17, 747—752, 1979.

254. **Powell, W. S.,** Formation of 6 oxoprostaglandin F$_{1\alpha}$ and monohydroxyicosatetraenoic acids from arachi-donic acid by fetal calf aorta and ductus arteriosis, *J. Biol. Chem.*, 257, 9457—9464, 1982.

255. **Sun, F. F., McGuire, J. C., Kane, K., and Feinmark, S. J.,** Enzymatic reactions of leukotriene biosynthesis in human polymorphonuclear leukocytes, in *Advances in Prostaglandin, Thromboxane and Leukotriene Research*, Vol. 15, Hayaishi, O. and Yamamoto, S., Eds., Raven Press, New York, 1985, 181—183.

256. **Austen, K. F. and Lewis, R. A.,** Subclass specific receptors for sulfidopeptide leukotrienes, in *Advances in Prostaglandin, Thromboxane and Leukotriene Research*, Vol. 15, Hayaishi, O. and Yamamoto, S., Eds., Raven Press, New York, 1985, 329—332.

257. **Wong, P. Y.-K., Chao, P. H.-W., and Spokas, E. G.**, Metabolism of leukotriene A_4 in the rat kidney and isolated glomeruli, in *Advances in Prostaglandin, Thromboxane and Leukotriene Research*, Vol. 15, Hayaishi, O., and Yamamoto, S., Eds., Raven Press, New York, 1985, 423—426.

258. **Edwards, N. S. and Pace-Asciak, C. R.**, Identification of a 'urinary-type' metabolite of prostaglandin $F_{2\alpha}$ in the rat circulation, *Biochim. Biophys. Acta*, 711, 369—371, 1982.

259. **Parker, C. W.**, Lipoxygenases and leukotrienes, *J. Allergy Clin. Immunol.*, 74, 343—348, 1984.

260. **Jakschik, B. A., Kuo, C. G., and Wei, Y. F.**, Enzymatic formation of leukotrienes, in *Biochemistry of Arachidonic Acid Metabolism*, Lands, W. E. M., Ed., Martinus Nijhoff Publishing, Boston, 1985, 51—75.

261. **Papatheofanis, F. J. and Lands, W. E. M.**, Lipoxygenase mechanisms, in *Biochemistry of Arachidonic Acid Metabolism*, Lands, W. E. M., Ed., Martinus Nijhoff Publishing, Boston, 1985, 9—39.

262. **Massiot, J. G., Soberman, R. J., Ackerman, N. R., Heavey, D., Roberts, L. J., II, and Austen, K. F.**, Potential therapeutic uses of inhibitors of leukotriene generation and function, *Prostaglandins*, 32, 481—494, 1986.

263. **Yamamoto, S., Ueda, N., Yokoyama, C., Kaneko, S., Shinto, F., Yoshimoto, T., Oates, J. A., and Brash, A. R.**, Dioxygenase and leukotriene A synthase activities of arachidonate 5- and 12-lipoxygenases purified from porcine leukocytes, *Proceedings, Sixth International Conference on Prostaglandins and Related Compounds*, Abstracts Book, p. 6, 1986.

264. **Atrache, V., Sok, S. E., Pai, J. K., and Sih, C. J.**, Formation of 11-*trans* slow reacting substances, *Proc. Natl. Acad. Sci. U.S.A.*, 78, 1523—1526, 1981.

265. **Willis, A. L. and Stone, K. J.**, Properties of prostaglandins, thromboxanes, their precursors, intermediates, metabolites, and analogs — a compendium, in *Handbook of Biochemistry and Molecular Biology*, Vol. II, Fasman, G. D., Ed., CRC Press, Boca Raton, 1976, 312—423.

Index

INDEX

A

AA, see Arachidonic acid
Acetaminophen (4'-hydroxyacetanilide), 27
ACTH, see Adrenocorticotrophic hormone
Acylglucosylceramides, 94
Adenosine diphosphate (ADP), 148
Adenosine monophosphate (AMP), 31
Adenosine triphosphate (ATP), 126—127
Adjuvant arthritis, 27
Adrenic acid, 47, 54—55, 69—71
Adrenocorticotrophic hormone (ACTH), 10
Airway smooth muscle, 138
Albumin, 122, 133, 135, 137
  TXA_2 binding to, 239
Amantadine, 149
Amikacin, 153
Aminoglycosides, 152—153, see also specific types
p-Aminohippuric acid, 226
Amphipathic proteins, 104
Amphiphilic cationic drugs, 149—151, see also specific types
Anaphylaxis, 138
Anesthetics, see also specific types
  enzyme interactions penetrated by, 148
  inhibition of phospholipase A_2 by, 147—149
Angiotensin II, 135
Anionic phospholipids, 145, see also specific types
Antagonists, see Inhibition
Anterior uvea, 226
  prostaglandin transport and, 221
Antibiotics, see also specific types
  inhibition of phospholipases by, 151—153
Anti-inflammatory agents, see also specific types
  nonsteroidal, see Nonsteroidal anti-inflammatory agents
  steroid, 155—156, see also specific types
Antimalarial drugs, 149—151, see also specific types
Antipodes, 17
Antirheumatic gold salt, 27
Apoenzyme, 163
Arachidic acid, 119
Arachidonate, 135
  conversion of linoleate to, 100
  esterification of, 123
  liberation of, from phospholipids, 126
  metabolites of, in platelets, 28
  in platelet phospholipids, 11, 120—127
Arachidonic acid, 10, 133, 193, see also Arachidonate
  conversion of 8,11,14-eicosatrienoic acid to, 103
  conversion of linoleic acid to, 100, 106
  conversion to prostaglandin products, 155
  Δ^5-desaturase and, 91
  dietary intake of, 90
  8,15-dihydroperoxy derivatives of, 24
  EFA deficiency and, 101

inhibition of, 149, 156
  linoleic acid and, 100
  metabolism of, in lung, 135
  mobilization of, 153
  oxygenation products, 47, 50—51, 60—61
  production of, 87
Arterial blood, 136
Arthritis
  adjuvant, 27
  rheumatoid, 155, 156
Aspirin, 26—27
  inhibition of prostaglandin synthesis by, 26, 153
  PLA_2 and, 155
Assays, see also specific types
  direct, 135—136
  indirect, 136—138
Atherosclerosis obliterans, 33
Autooxidation of unsaturated fatty acids, 22

B

Benoxaprofen, 29
Benzocaine, 148
Biliary excretion, of prostaglandins, 221
Biosynthesis, see Prostaglandins, biosynthesis; Synthesis
1,7-Bis p-aminophenoxy heptane, 149—151
4,4'-Bis (diethylaminoethoxy) α,β-diethyldiphenylethane, 149, 150
Bleomycin chloride, 152
Bleomycin hydrochloride, 151—152
Blood-brain barrier, 227
Bradykinin, 135, 138
Brain, 224, 227
  basal turnover of prostaglandins in, 10
  monoamine oxidase inhibitor in, 28
  prostaglandin accumulation in, 216
  prostaglandin transport and, 218
Brain cortex $PGF_{2\alpha}$, 29
Bromocresol green, 29, 218, 226
Bronchoconstriction, 138
BW755C, 28
By-products, see also specific types
  of prostaglandin synthesis, 22—23
  of thromboxane synthesis, 22—23

C

Carageenin, 27
Carbenicillin, 151, 152
Carbon-133, 14-arachidonic acid, 135
Carbon atoms, 17
Cardiolipin, 145
Carriers, 179
Cascade of desaturation/elongation, 10
Catabolism of lysosomal lipids, 151

Q

R

S